Geology and Mineralogy Research Developments

An Introduction to Aluminosilicates

GEOLOGY AND MINERALOGY RESEARCH DEVELOPMENTS

Additional books and e-books in this series can be found on Nova's website under the Series tab.

GEOLOGY AND MINERALOGY RESEARCH DEVELOPMENTS

AN INTRODUCTION TO ALUMINOSILICATES

NERO REGINA BLEVINS

EDITOR

nova science publishers
New York

Library of Congress Cataloging-in-Publication Data

ISBN: 978-1-53617-250-8

Published by Nova Science Publishers, Inc. † New York

CONTENTS

PREFACE

An Introduction to Aluminosilicates first reviews the main advances in methods of obtaining synthetic montmorillonite for environmental and biomedical areas. The perspectives and main challenges in the application of these materials is critically analyzed.

The authors next study Polyhydroxy metal complexes of Al, Fe, Zr, Cr and Ti (inorganic pillarizing agents) for the synthesis of pillared clays.

Information is provided on the possibility of using amorphous aluminosilicates as a modifying additive in the formulation of cement-based dry building mixtures. The physicochemical properties of the additive are considered, and the results of XRD and DTA analyzes are presented.

The authors investigate the molecular orientation and surface morphology of organized molecular films with regard to solid-state structures for organo-modified aluminosilicates by a surface pressure-area isotherm, in-plane and out-of plane X-ray diffraction, and atomic force microscopy.

A new method for the synthesis of amorphous mesoporous aluminosilicates with acidic properties and a narrow pore size distribution in the range of 2-7 nm is also explored.

The urease immobilization of composite adsorbents polyacrylamide–bentonite, polyacrylamide–chitosan, and polyacrylamide–chitosan–bentonite is assessed, produced from polyacrylamide as a hydrogel, bentonite as an aluminosilicate mineral, and chitosan as a polysaccharide.

Following this, the authors provide a brief overview of literature data where crystalline aluminosilicates have been utilized for the capture and immobilization of radionuclides.

The physiochemical properties of as-synthesized layered aluminosilicate are characterized by small angle PXRD, FTIR, porosity studies, thermal studies, 27Al MAS NMR, and morphological studies.

Subsequently, this compilation addresses the way in which magnetic nanoparticles have proven to be frequently occurring building blocks for hybrid structure construction and, moreover, have shown innovative prospects as multifunctional adsorbents.

Modern methods for the synthesis of highly dispersed micro- and micromesoporous mordenite with a high degree of crystallinity, as well as high crystallinity mordenite with a hierarchical porous structure, methods for modifying the mentioned crystalline aluminosilicates as a result of postsynthetic treatments, and modern adsorption and catalytic systems based on them are described.

Next, the high activity and selectivity of high crystallinity zeolite Y with a hierarchical porous structure in the synthesis of practically important oligomers of various unsaturated compounds is established.

The authors highlight that the aluminosilicate present in kaolin may also be used in the production of zeolitic materials, and some details of this application are explained.

The adsorptive features of PAAm-Ch-Z for Th4+ are investigated in view of its dependency on pH, concentration, time, temperature and ionic strength. The parameters derived from the compatibility of experimentally obtained data to Langmuir, Freundlich and Dubinin-Radushkevich, van t'Hoff, pseudo frist-/second-order and Weber Morris models are utilized in the evaluation of adsorption and its thermodynamics and kinetics.

Lastly, the authors consider a possible mechanism for the generation of an acetyl-zeolite intermediate via transfer from different acetyl donors through theoretical studies using a cluster model of H-ZSM-5 zeolite designed by three TO4 tetrahedral units.

Chapter 1 - Montmorillonite (MMT) is a phyllosilicate belonging to the 2:1 smectite groups. This material has a layer-charge density in a range of 0.2 to 0.6 per unit cell. Its negative charge is a consequence of the isomorphic substitution of Al^{3+} for Mg^{2+}. Furthermore, MMT has a high cation exchange capacity, due to the presence of cations in the interlamellar space. This material may be considered multifunctional, because of its characteristics and great availability in nature. Therefore, it has applications in several fields from the environmental to the biomedical one. Nevertheless, montmorillonite in a bentonite form (the most predominant natural form), from different deposits, differs on chemical and mineralogical composition, affecting its properties such as surface charge; cationic-exchange and structural capacity; and microstructural and textual parameters. Thus, synthetic montmorillonites with controlled and reproducible characteristics are more promising phyllosilicates than the natural ones. In addition, the high degree of purity of synthetic montmorillonites makes them more safely applied in areas that require precise control of some characteristics, for example, in the catalysis and biomedical technology. In this sense, the present chapter intends to review the main advances in methods of obtaining synthetic montmorillonite for environmental and biomedical areas. Therefore, the perspectives and main challenges in the application of these materials will be critically analyzed.

Chapter 2 - Clay minerals are the main constituents of clays, being formed by aluminosilicates in fibrous structures and crystalline lamellae with equivalent spherical diameter less than or equal to 2 μm. Clay minerals from the smectites group have layers organized as tetrahedral sheets of silica and octahedral alumina, condensed and polymerized. The cations adsorbed on the surface of the clay and between the lamellae can be exchanged without compromising the crystalline structure and contribute to modify some physicochemical properties. In this way, clays are very versatile materials and have varied applications due to their swelling, adsorption and ion exchange properties. The applications of clay minerals in heterogeneous catalysis have been extensively studied in the last 80 years. Through some modifications, it is possible to improve catalytic performance by increasing access to their active sites. In recent decades, pillared clays have been widely used in many applications, in particular in the cracking of petroleum and the pyrolysis of waste. They are a class of modified clays, synthesized from the exchange of the cations of the original clay by inorganic molecules, usually complexes of metal cations introduced into the interlamellar space of the clay minerals. After heating, this system undergoes dehydration and dehydroxylation processes, forming metal oxide pillars strongly bonded to layers of the minerals, producing high resistance and thermal stability as well as increasing porosity, specific surface area and basal spacing (d001). The polyhydroxy metal complexes of Al, Fe, Zr, Cr and Ti are the inorganic pillarizing agents which have been studied for the synthesis of pillared clays. The natural clays have Brønsted and Lewis acidity, but in pillared clays the, Lewis acidity is predominant.

Chapter 3 - Information is giving on the possibility of using amorphous aluminosilicates as a modifying additive in the formulation of cement-based dry building mixtures (DBM) with the cement. The physicochemical properties of the additive are considered, the results of XRD and DTA analyzes are presented.

The regularities of the structure formation of the cement composite with the use of additives based on amorphous aluminosilicates in the formulation are established. He is additionally consist in the formation of calcium-sodium hydrosilicates and minerals of the zeolite group in reducing the amount of free lime and increasing the amount of chemically bound water reducing the total and capillary porosity and increasing the contraction and gel porosity.

It has been established that the introduction of an amorphous aluminosilicate-based additive into the formulation of adhesive DBM helps to accelerate the increase in strength of tile adhesive. A model of the strength of the cement composite has been compiled, allowing you to choose the optimal content of components in the formulation of the dry building mixture. It was revealed that tile glue made on the basis of the developed formulation is crack resistant, resistant to slipping, and has high adhesive strength. The regularities of changes in the stress state of the adhesive layer from the action of temperatures for various climatic conditions of operation are established. The scope of the developed adhesive dry building mix is determined depending on the climatic conditions of use.

The stress distribution in the adhesive layer based on the developed composition of tile glue from the action of temperature in various climatic regions. It were found that the maximum values of tensile stresses σx occurring in the adhesive layer at the border with the substrate and at the border with the ceramic tile are observed in March.

It were revealed that in the center of the adhesive joint the maximum values of tangential stresses are observed in March (Penza) and in February (Yakutsk), and in the marginal zone in September (Penza) and in October (Yakutsk). For the conditions of Penza, the maximum values of shear stresses in the adhesive layer at the border with the substrate are observed in February, and at the border with ceramic tile in September. For the conditions of Yakutsk, the maximum values of tangential stresses in the adhesive layer are observed at the border with the substrate and at the border with ceramic tile in October. It was revealed that tile adhesive based on the developed adhesive DBM is crack resistant for Russian cities, located in a dry moisture zone and climatic subareas IIB.

The formulation of a cement-based adhesive dry building mix designed for gluing tiles to surfaces and containing Portland cement, quartz sand of the Ukhta deposit with the optimal ratio of fractions, the additive based on amorphous aluminosilicates, the plasticizing additive Kratasol PFM and the redispersible powder Neolith P 4400. Tile adhesive made on the basis of the developed DBM is characterized by the following indicators: adhesion strength upon tearing after 28 days of air-dry hardening is more than R_{adg}> 1,4 MPa, frost resistance grade F50, frost resistance grade of the contact zone $F_{kz}50$, water absorption with capillary suction W = 1.43 kg/m^2 H 0.5, tile sliding is less than 0.3 mm, shrink strain 0.029% - 0.034%.

Chapter 4 - In the first half of this chapter, it is investigated the molecular orientation and surface morphology of organized molecular films with regard to solid-state structures for organo-modified aluminosilicates (montmorillonites) by surface pressure-area (π–A) isotherm, in-plane and out-of plane X-ray diffraction (XRD), and atomic force microscopy (AFM). From the results of out-of plane XRD, formation of highly ordered layer structure was confirmed in these clay Langmuir–Blodgett (LB) film. Further, two-dimensional lattice of long alkyl chain of organo-modified parts packed hexagonally or orthorhombically in the films. Surface morphology of Z-type monolayers on solid of organo-modified aluminosilicates indicates heterogeneous modification ratio in montmorillonite surface by AFM observation. In addition, monolayer behavior on the water surface and mesoscopic morphological formation on solid of mixed films of organo-modified aluminosilicate and poly-(L-lactide) (PLLA) were investigated by π-A isotherm and AFM. Collapsed surface pressures are independent on the mixed ratio, and indicate almost constant value in their isotherms. This tendency is a peculiarity of immiscible system. From the result of AFM observation, phase separated structure was confirmed in mesoscopic scales. This phase separated morphology remarkably varied with mixed ratio.

In the second half of this chapter, ultrathin films of regularly adsorbed biological molecules have been fabricated by means of the modified LB method using an organo-modified aluminosilicate. In this combined method, clay LB films play a template role in the formation of the lysozyme thin layer. Chemisorption of biological molecules to the anionic montmorillonite surface was confirmed by a comparison between infrared spectra of multilayers of organo-clay and of chemisorbed lysozyme. The surface morphology of these monolayers was observed by atomic force microscopy. In the case of organo-modified aluminosilicate, it is also indicated that chemisorbed enzyme activity is possible to maintain until 160°C at attempting to evaluation of the heat-resistance. The maintaining of enzyme activity at high temperature region is possible together with the maintaining of three-dimensional steric structure of enzyme. This maintaining of steric structure of enzyme is based on the formation of aggregate of adsorbed enzyme induced by epitaxial-growth from aluminosilicate surface and the construction of sandwiched structure of enzyme between organo-aluminosilicates.

Chapter 5 - A new method *without using templates* synthesis of amorphous mesoporous aluminosilicates with acidic properties and a narrow pore size distribution in the range of 2-7 nm has been developed. The physicochemical properties of samples of amorphous mesoporous ASM aluminosilicate with a different Si/Al molar ratio (40, 80, 120) were studied.

The catalytic properties of samples of amorphous mesoporous aluminosilicate ASM with different Si/Al molar ratios (40, 80, 120) were studied in the synthesis of practically important pyridines (by the interaction of C_2–C_5 alcohols with formaldehyde and ammonia, cyclocondensation of acetaldehyde and propionic aldehyde with ammonia), dialkylquinolines and alkyltetrahydroquinolines (by reaction of aniline with C_3, C_4 aldehydes) and alkyldihydroquinolines (by interaction of aniline with ketones, acetone and acetophenone). It is found that mesoporous aluminosilicate ASM sample with a molar ratio of Si/Al = 40, which has the highest acidity among the studied samples, exhibits the highest activity and selectivity in these reactions.

Chapter 6 - The objectives of this work were the preparation, characterization, and investigation in urease immobilization of composite adsorbents polyacrylamide–bentonite (PAAmB), polyacrylamide–chitosan (PAAmCh), and polyacrylamide–chitosan–bentonite (PAAmChB) produced from polyacrylamide as a hydrogel, bentonite as an aluminosilicate mineral, and chitosan as a polysaccharide. The formation of the composites was proved by FT-IR, XRD, DSC, and SEM analysis.

The effect of concentration, time, and temperature on urease immobilization on the composites was investigated by conductometry. The monolayer adsorption capacities obtained from the Langmuir model were 93 mg g^{-1} for PAAmB, 174 mg g^{-1} for PAAmCh and 95 mg g^{-1} for PAAmChB. The immobilization was time-dependent with pseudo-second-order kinetics, but it was not temperature dependent.

The activity dependence of urease immobilized on the composites to the substrate (urea) concentration, pH and temperature was evaluated with reference to the activity measured for free urease. The V_{max} and K_m values derived from the Michaelis–Menten graphs were 11 units mg^{-1} and 128 mM for the free enzyme, 11 units mg^{-1} and 243 mM for PAAmB, 20 units mg^{-1} and 104 mM for PAAmCh, and 19 units mg^{-1} and 102 mM for

PAAmChB. The optimum pH and temperature were determined for the composites and free enzymes.

Chapter 7 - Crystalline aluminosilicates have a wide range of applications such as media for the capture and/or immobilization of radionuclides. Several different types of metal-exchanged zeolites have been successfully demonstrated as effective sorbents for gaseous, aqueous, or solid-state capture and immobilization of radionuclides including iodine [e.g., $I_{2(g)}$, $CH_3I_{(g)}$], xenon, krypton, radium (e.g., ^{226}Ra and ^{228}Ra), selenium (e.g., $^{79}Se^{6+}$), technetium (e.g., $^{99}TcO_4^-$), strontium (e.g., $^{90}Sr^{2+}$), cesium (e.g., $^{135}Cs^+$, $^{137}Cs^+$), mixed alkali halide streams (i.e., iodide and chloride salts), and actinides (e.g., U^{6+}, Th^{4+}). These zeolites include *sodalite* (SOD), *mordenite* (MOR or Z-type), *faujasite* (FAU), *Linde type-A* (LTA), *cancrinite* (CAN), *gismondine* (GIS), and *heulandite* (or clinoptilolite) (HEU). The structural building blocks of these seven materials vary from cages (e.g., α-cages, β-cages, ε-cages) to channels linking various cage types and these are mostly comprised of aluminosilicates. This chapter will delve into the crystallographic differences between these framework materials and provide a brief overview of literature data where these seven crystalline aluminosilicates have been utilized for the capture and/or immobilization of radionuclides. Metal-exchanged zeolites (e.g., MOR, FAU, LTA) have often been the source of study for gaseous capture of iodine where the more promising compositions include Ag-exchanged sorbents; these types of materials can be made in a variety of robust forms including berl saddles and extrudates. Sodalite can be used as an immobilization strategy for halide waste streams (e.g., Cl, I), including those from electrochemical reprocessing of used nuclear fuel, as well as TcO_4^-. Cancrinite has been demonstrated to enhance sorption capacity of sediments for several problematic ionic radionuclides including Cs^+, SeO_4^{2-}, and TcO_4^-. Actinides such as U^{6+} and Th^{4+} have been successfully captured from solutions using MOR, FAU, LTA, and HEU-type aluminosilicates. Finally, mixtures of aluminosilicates have been included in a set of zeolite materials that show promise as sorbents for ^{226}Ra and ^{228}Ra removal from mine waters.

Chapter 8 - Layered aluminosilicates are well known in the name of zeolite, which are generated by the basic crystallographic building unit of TO_4 (T = Si, Al) tetrahedral, shared corners of their unit among itself with two-dimensional manner. The dimensional array of layered aluminosilicate was defined channels and porous structure. Si^{4+} created a neutral framework, whereas Al^{3+} created a negative charge in the framework and neutralized by alkali metal cations.

Layered aluminosilicates synthesized in the source of the tetrahedral atom of 'Si' and 'Al'/coating seeds, surfactant as the template or structure-directing agents (SDAs), mineralizer as maintaining pH, the solvent, etc. The mixture makes the gel and is heated at a temperature in the range of 140-150°C for a time period. The factors influencing the formation of layered aluminosilicates were the time period of synthesis, the pH, the total amount of solvent, SDA, gel composition, and temperature. The formation mechanism followed the liquid-crystal templating/ADOR (assembly-disassembly-organization-reassembly) mechanism. However, the partial inadequate experimental details of formation mechanism were a blockade of custom design and complete exploration of aluminosilicate in an experimental approach. To increases the pressure on a scientist to synthesize the layered aluminosilicate in the green and rational pathway to decrease hazardous waste.

The Physiochemical properties of as-synthesized layered aluminosilicate have been Characterized by Small angle PXRD, FTIR, Porosity studies, Thermal studies, ^{27}Al MAS NMR, and Morphological studies.

Layered aluminosilicate acting as Bronsted and Lewis acid, well-defined porous and its channels, undergo adsorption and desorption phenomenon. Most of its parameters, like porosity, size, and morphology, were tunable.

It has used as a Heterogeneous catalyst, adsorption-desorption, separation, membrane, petrochemical industry, drug delivery, tissue engineering, and sensor. In forthcoming, it explored in the field of advanced modern construction material, electronic, and agricultural engineering.

Chapter 9 - Thanks to recent biosynthesis achievements, the development of versatile protocols in architectural design and control has advanced the research in adsorbent chemistry beyond the conventional one, which has not processed multifunctional heterostructures before. It should also be pointed out that usually novel and state-of-the art biosynthesis protocols allow for the simplification of experimental design and apply renewable and ubiquitous biological materials, thereby minimizing energy requirements and waste generation. In recent decades, magnetic nanoparticles have proven to be frequently occuring building blocks for hybrid structure construction and moreover, have shown innovative prospects as multifunctional adsorbents. Such magnetic heterostructures are commonly prepared by using aluminosilicates (zeolites), silicates, carbon or biopolymer shells as linkers for further functionalization or protective coatings. Recently, some studies have illustrated that nanoparticles have the ability to enter organisms during ingestion or inhalation and can translocate within the body to various organs and tissues to exert toxicity effect. In order to depress the toxic effects of nanoparticles, various surface coatings have been employed. This can greatly reduce the contact of nanoparticles with living bodies and thus their toxicity impact. Therefore, immobilized metal oxides nanoparticles have been videly used in environmental applications and have shown promising performance in pollutant removal and toxicity mitigation. Among the most widely-used nanoparticles, magnetic nanoparticles of iron and manganese oxides have sparked an immense interest in research of engineering applications for the treatment of contaminated water or subsurface environments. These nanoparticles not only have a large removal capacity, fast kinetics and high reactivity for specific pollutants removal due to their extremely small particle size and thus high surface area to volume ratio, but also one more important aspect: they are economically feasible. Their cleanup process of pollutants removal is mostly controlled by complexation. Zeolites occuring in nature are highly porous aluminosilicates of volcanic origin. They have been applied for the immobilization of Fe – and Mn-oxide nanoparticles in current research. Zeolite rock is essentially composed of an active component, i.e., industrially most frequently used clinoptilolite as well as some other geological minerals, potentially nanodispersed in rock matrix, such as volcanic glass, feldspar, quartz and clays, which may reach up to several tens % of the rock. Pharmaceuticals, which are known today as emerging contaminants (ECs), can enter the environment through a number of pathways and can be further distributed to various environmental media. One prominent pathway could be the use of wastewater sludge or wastewater for field fertilization and irrigation. In aquatic environments, a large variety of these compounds and their metabolites have been detected. For this reason, the objective of the work presented was to examine several natural, commercial or even waste products in selected pharmaceutical (diclofenac DCF) removal, by using synthetic water solutions in laboratory measurements. Several surface treated zeolite samples, such as FeO(OH)-zeolite, MnOx - zeolite, carbonized and surfactant coated zeolite, were also evaluated and compared for DCF removal efficiency. In contrast to a voluminous organic and dissociated DCF molecule, another model pollutant Zn(II) cation was chosen for the evaluation of those zeolite samples, which did not indicate any or only low DCF uptake ability. It was chosen particularly to demonstrate the samples removal efficiency to a different type of pollutant. Subsequently, some complementary analytical methods (XPS, XRD, SEM, SEM-EDS, BET, MAS NMR) were used for the characterization of the material studied.

Chapter 10 - This review discusses modern methods for the synthesis of highly dispersed micro- and micromesoporous mordenite with a high degree of crystallinity, as well as high crystallinity mordenite with a hierarchical porous structure, the granules of which represent single crystal intergrowths of various sizes, methods for modifying the mentioned crystalline aluminosilicates as a result of postsynthetic treatments (ion exchange, dealumination), and modern adsorption and catalytic systems.

Chapter 11 - The preparation of Chitosan (Ch)-Zeolite (Z) composite inter linked by ephycholorhydrine entrapped in polyacrylamide (PAAm) so called PAAmChZ in a ternary combination were achieved. The mineral (Z) was clinoptilolite with the SiO_2/Al_2O_3 ratio value of 4.7. For the structural characterization of the composite FT-IR, XRD, SEM, TGA, BET-Porosity, cation exchange capacity (CEC) and point of zero charge (PZC) analysis were used. Swelling characteristics in view of capacity and kinetics were tested in water and in solution of KNO_3 0.25 M.

FT-IR, XRD, SEM and PZC analysis confirmed the formation of composite in ternary combination. The results of swelling study showed that the adsorption kinetics obeyed Schott's second-order model the overall swelling mechanism was Fickian, i.e., the process was diffusion controlled.

Adsorptive features of PAAm-Ch-Z for Th^{4+} was investigated in view of its dependency on pH, concentration, time, temperature and ionic strength. The parameters derived from the compatibility of experimentally obtained data to Langmuir, Freundlich and Dubinin-Radushkevich (DR), van t'Hoff, pseudo frist-/second-order and Weber

Morris (intra particle diffusion) models were utilized in the evaluation of adsorption and its thermodynamics and kinetics.

The results pointed out that the adsorption increased with increasing solution pH and that Th^{4+} had high affinity to the composite. The kinetics studies proved that the adsorption could be explained by the pseudo second-order model and the adsorption mechanism was via ion exchange and/or complex formation. The changes in values of adsorption enthalpy, entropy and Gibbs free energy were $\Delta H > 0$, $\Delta S > 0$ and $\Delta G < 0$ implying the overall adsorption process was endothermic, increasing entropy and spontaneous.

Chapter 12 - Kaolin is a source of aluminosilicate with a great technological potential of applications in several industrial sectors. This mineral is found in large quantities in the earth's crust. Its use has enabled the production of materials that involve different particle sizes, including nanotechnology. The aluminosilicate in the form of crystal present in kaolin can be transformed into amorphous materials through thermal treatments. Amorphous aluminosilicates, also called metakaolin, may react with other compounds and form new structures by means of the application of different physical and chemical processes, according to industrial interests. Thus, based on the importance and wide use of industrial kaolin, this chapter gives priority emphasis on two materials that may be produced from metakaolin: geopolymers and synthetic zeolites. The application of kaolin to the production of geopolymers has been studied in order to obtain geopolymer matrices with different application potentials. The geopolymer technology may be applied in the production of cementitious matrices of high compressive strength and in the production of porous matrices for the application in the adsorption of substances. Zeolites are another promising technology that has been highlighted in the last decades, mainly for the application as molecular sieves. It is important to highlight that the aluminosilicate present in kaolin may also be used in the production of zeolitic materials. Some details of this application will also be explained in the course of this chapter.

Chapter 13 - The high activity and selectivity of high crystallinity zeolite Y with a hierarchical porous structure in the synthesis of practically important oligomers of various unsaturated compounds (n- and isopentenes, oct-1-ene, cyclohexene, cyclooctene, norbornene, indene, and α-methylstyrene) was established. The following advantages of the mentioned zeolite in comparison with the microporous zeolite H-Y are shown: high activity in the conversions of unsaturated compounds; oligomerization products obtained in his presence have a wider molecular weight distribution; provides highly selective dimers and trimers of cyclohexene and cyclooctene, in which the cycle of the initial monomer remains unchanged.

Chapter 14 - Zeolites are well known microporous silicoaluminates with extensive uses as heterogeneous catalysts. Particularly, the acid form of ZSM-5 zeolite has left evidence of its high activity due to its participation in several chemical processes. Acyl-transfer reactions are among the most fundamental processes in biochemistry and organic chemistry in which zeolites play an efficient role due to its unique porous structure and acidity. It has been reported that an acylated surface has been generated as an active intermediate in several acylation reactions, but to the best of our knowledge, no rational mechanism has been suggested for its generation. Thus, in this chapter, the authors will consider a possible mechanism for the generation of an acetyl-zeolite intermediate due to transfer from different acetyl donors by theoretical studies using a cluster model of H-ZSM-5 zeolite designed by three TO_4 tetrahedral units (3T-cluster). Based on Intrinsic Reaction Coordinate (IRC) method, the authors discuss the atomic reorganization that evolves from acetyl donors adsorbed over 3T-cluster to the acetylated zeolite intermediate. Moreover, the authors confirm this arrangement and support bounded species and broken bonds based on the concept of Global Electron Density Transfer (GEDT) and Density of States (DOS), Projected Density of States (PDOS) and Overlay Population Density of States (OPDOS) diagrams Thus, for the first time, a concerted mechanism has been proposed for the generation of the acetylated-zeolite intermediate, based on first-principle methods.

In: An Introduction to Aluminosilicates
Editor: Nero Regina Blevins
ISBN: 978-1-53617-250-8

Chapter 1

MONTMORILLONITE-BASED ALUMINOSILICATE MATERIALS: PROPERTIES AND APPLICATIONS

Domingos Lusitâneo Pier Macuvele[1,2,*], PhD,
Ana Paula Fagundes[1],
Rozineide Aparecida Antunes Boca Santa[1], PhD,
Jonas Valente Matsinhe[3], PhD, Natan Padoin[1], PhD,
Cíntia Soares[1], PhD and Humberto Gracher Riella[1], PhD
[1]Graduate Program in Chemical Engineering,
Federal University of Santa Catarina, Florianópolis, SC, Brazil
[2]Department of Natural Sciences and Mathematics,
University of Rovuma-Extension of Niassa, Lichinga, Mozambique
[3]Department of Chemical Engineering, Eduardo Mondlane University,
Maputo, Mozambique

ABSTRACT

Montmorillonite (MMT) is a phyllosilicate belonging to the 2:1 smectite groups. This material has a layer-charge density in a range of 0.2 to 0.6 per unit cell. Its negative charge is a consequence of the isomorphic substitution of Al^{3+} for Mg^{2+}. Furthermore, MMT has a high cation exchange capacity, due to the presence of cations in the interlamellar space. This material may be considered multifunctional, because of its characteristics and great availability in nature. Therefore, it has applications in several fields from the environmental to the biomedical one. Nevertheless, montmorillonite in a bentonite form (the most predominant natural form), from different deposits, differs on chemical and mineralogical

[*] Corresponding Author's E-mail: lusitaneom24@gmail.com.

composition, affecting its properties such as surface charge; cationic-exchange and structural capacity; and microstructural and textual parameters. Thus, synthetic montmorillonites with controlled and reproducible characteristics are more promising phyllosilicates than the natural ones. In addition, the high degree of purity of synthetic montmorillonites makes them more safely applied in areas that require precise control of some characteristics, for example, in the catalysis and biomedical technology. In this sense, the present chapter intends to review the main advances in methods of obtaining synthetic montmorillonite for environmental and biomedical areas. Therefore, the perspectives and main challenges in the application of these materials will be critically analyzed.

Keywords: synthetic montmorillonite, nanocomposites, zeolites, adsorption, bone tissue

1. Introduction

Clay minerals are hydrous aluminosilicates in which predominate clay-sized soils. The clay minerals are generally classified into three types, based upon the arrangement of tetrahedral and octahedral sheets in their basic structure (Barton, Christopher Karathanasis 2002).

Montmorillonite is a phyllosilicate particle that belongs to the 2:1 types of smectites. Naturally, it occurs in the form of bentonite. This clay mineral possesses special properties such as cation exchange capacity, total negative charge, porous structure, and hydrophilic nature. For these reasons, MMT has been received special attention concerning applications in modern science and technology. However, the use of clay minerals is not recent. Some reports mentioned the applications of clay-based minerals by ancient Egyptians in pigments, traditional medicine, and cosmetics.

In the last years, these applications have risen exponentially due to the advance of instrumental techniques to assess the structure of materials, such as MMT. MMT is also applied in environmental fields to remove contaminants specifically, heavy metals, organic emerging contaminants, dyes, and others (Fagundes et al. 2019; Chen et al. 2017; Parolo et al. 2008). Furthermore, it is used as a flame retardant in conventional synergistic flame-retardants, which are mostly toxic. Moreover, MMT has been reported as a promisor material for biomedical applications, for example, drug delivery, tissue engineering, and others.

MMT, in form of bentonite from different deposits, does not present a constant physicochemical characteristic, surface charge, and textural properties. Therefore, the use of natural MMT, for some applications, such as catalysis and biomedical technology, is limited by a required material with controlled composition and textural properties. In addition, to understand MMT interactions at a molecular level it is important to use higher purity materials in order to avoid interferences. For these reasons, the synthesis of MMT from analytical grade reagents has been reported. The hydrothermal synthesis of MMT is the simplest manner to obtain it. Furthermorein the last years, the solid-state mechanosynthesis was introduced.

This chapter will review the main methods applied to obtain the synthetic montmorillonite as well as the special attention given to the applications of montmorillonite in environmental and biomedical fields. Furthermore, some challenges and unsolved questions were critically analyzed.

2. MONTMORILLONITE

2.1. Structure

The chemical representation of montmorillonite is $(Na,Ca)_{0,3}(Al,Mg)_2Si_4O_{10}(OH)_2 \cdot nH_2O$. A 2:1 aluminosilicate (Figure 1) is characterized by its crystal structure encompassing one octahedral aluminum-centered layer between two tetrahedral silicon-centered sheets. Its formation occurs by weathering that causes chemical changes in the volcanic ash. Montmorillonite, the main component of bentonite, belongs to the general family of clay minerals. Its structure and special characteristics result in the large specific surface area, adsorption property, rheological properties, and swelling properties in polar solvents.

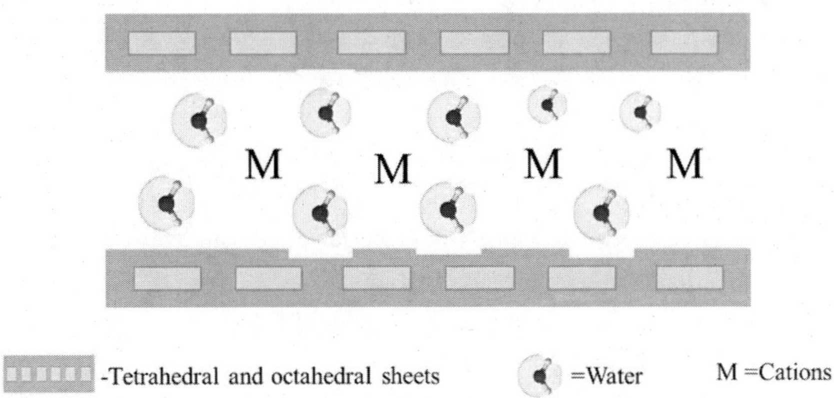

Figure 1. The structure of montmorillonite clay.

Prepared montmorillonites may be characterized by different physicochemical methods. Fourier transform infrared spectroscopy is a method used often to investigate the structure, bonds, and chemical properties of clay minerals. Besides this method, there are some others such as the techniques of diffuse reflectance (DRIFT) attenuated total reflectance (ATR), specular reflectance, powder X-ray diffraction analysis (PXRD), particle size distribution (PSD), thermogravimetric analysis (TGA), scanning electron microscope (SEM), and transmission electron microscopy (TEM). Figure 2 represents a SEM image of natural bentonite showing irregular shape of MMT particles.

TEM is a powerful technique to analyze materials from nanoparticles to microparticles. In the case of MMT-based materials, TEM is used to analyze the morphology, shape of particles, and other structural information.

Figure 2. SEM micrographs of bentonite from Mozambique

XRD is a unique analytical technique that provides structural information of materials such as chemical composition, crystal structure, crystallite, and layer size. In the case of MMT-based materials, the most important peak is located between 2Θ (5 to 7) that corresponds to 001 reflection (Nones et al. 2016a).

Analytical tools, specially Infrared spectroscopy (IR), Magnetic Resonance nuclear (MRN), and Raman. IR, provide structural information regarding functional groups in order to understand other structural aspects of MMT, MRN allows knowing the number of hydrogens and carbons as well as how they are bonded.

2.2. Properties

An important property of montmorillonite is the interleaving ability of other components between its layers. The imperfection of the crystal lattice and isomorphic substitution induce a negative net charge that leads to the adsorption of inorganic cations such as Na^+, K^+, Ca^{2+}, Mg^{2+} in the interlayer space. In addition, it is non-toxic and it has a high surface area, high porosity, high temperature, and solvent resistance. Moreover, montmorillonite displays excellent swelling properties and cation exchange capacities. Among the plethora of nanomaterials, clay minerals are an important, widely abundant, low-cost, and non-toxic class of materials with unique properties such as intercalation, swelling, ion exchange, and adsorption properties.

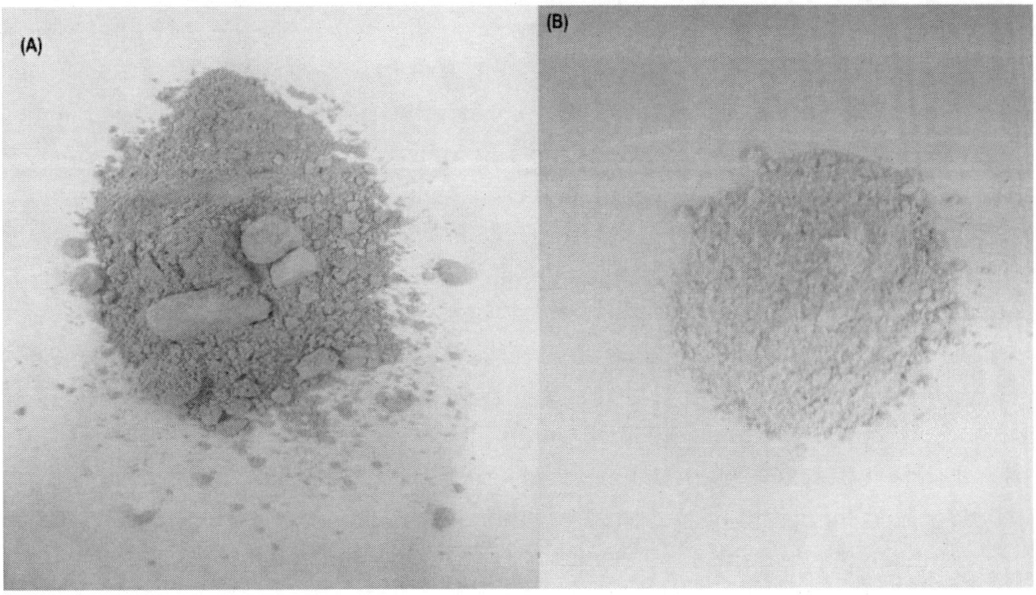

Figure 3. Bentonite – (A) Brazilian bentonite and (B) Mozambican bentonite.

Some types of bentonites (60 - 95% of MMT) present different colors depending on the mineral compositions. For example, the Brazilian bentonite, extracted in Santa Catarina, is brown because of the high composition of iron oxides, whereas, the Mozambican bentonite is greenish-white, due to the high quantity of sodium and low quantity of iron oxides (Figure 3).

2.3. Synthetic Montmorillonite

The montmorillonite is belonging to the smectite group. Naturally, it occurs mainly in the bentonite, varying from 60% to 99% of montmorillonite. Montmorillonite is applied in several fields from environmental to biomedical ones. However, natural montmorillonite is not homogenous in their characteristics and most of them have some impurities that can interfere directly on the understanding of some phenomena at the molecular level. In addition, some applications of montmorillonite, such as the biomedical one, require materials of a certain degree of purity.

In this sense, the synthesis of the montmorillonite by means of the analytical grade of starting materials is very important to applications and to understand some insights at the molecular level. The first attempt on the synthesis of montmorillonite was achieved in 1954 by a Japanese scientist (Otsubo and Kato 1954). Otsubo and Kato (1954) synthesized a pure phase of montmorillonite with Zn or Al partially substituted on the ZnO_2-Al_2O_3-SiO_2 system. Later, another hydrothermal synthesis of montmorillonite was performed but the focus was to optimize the reactional parameters, such as time, pH, pressure, mineralizing agent (OH^- or F^-), and Mg contents, in order to obtain a montmorillonite with a higher purity (O. Yu. Golubeva et al. 2013; Zhang et al. 2010; Le Forestier et al. 2010; Harder 1972; Olga Yu Golubeva, Brazovskaya, and Shamova 2018). High temperatures, ranging from 300 to 400, and pressure, between 2 to 120 MPa, were needed during a reasonable time to achieve montmorillonite with a higher crystallinity. The montmorillonite with a higher degree of crystallinity was produced by hydrothermal synthesis under high pressure (120 MPa) during 28 days. The synthesized product presented similar features with natural montmorillonite.

However, some properties, such as macroscopic swelling, differed with natural montmorillonite (Le Forestier et al. 2010). Another approach to enhance the hydrothermal synthesis is the change of mineralizing agent, from OH^- to F^- (Marc Reinholdt et al. 2001; M. Reinholdt et al. 2005; Marc Reinholdt et al. 2013). The presence of F^- as a mineralizing agent favors the formation of a crystallized montmorillonite phase. Interestingly, the F^- affects positively the hydrothermal synthesis even in low quantities (M. Reinholdt et al. 2005).

The mechanosynthesis route was evaluated in the synthesis of montmorillonite. The solid-state of mechanosynthesis does not require the use of solvents, making this route green and promisor. Fahami (2017) reported for the first time the application of mechanosynthesis in the production of montmorillonite. The author observed that synthesized montmorillonite presented a higher crystallinity and twisted microlayers in a rose-shaped platy structure, suggesting that this material is promisor for the most diverse applications.

Furthermore, the natural montmorillonite due to its structure may be modified resulting in new materials with enhanced properties (Macuvele, Nones, Matsinhe, Bezerra, et al. 2017). Organic cations due to the exchangeable cations may intercalate into the interlayer space causing an expansion of the interlayer space. These cations do not only interleave in the interlamellar space, but reinforce the surface of the material. Typically, montmorillonite may be prepared by an aqueous solution or solid-state reaction. However, the aqueous solution is the most common method for preparing it in the laboratory or even on an industrial scale.

Thus, the synthesized clay is easily obtained by contact with surfactants. In general, some common modifiers that synthesize organo-montmorillonite are cationic surfactants, such as octadecyltrimethylammonium chloride (OTAC), cetyltrimethylammonium chloride (CTAC), cetyltrimethylammo-nium bromide (CTAB), and hexadecyl dimethyl benzyl ammonium chloride (HDBAC) (Nones et al. 2016b; Cipolletti et al. 2014).

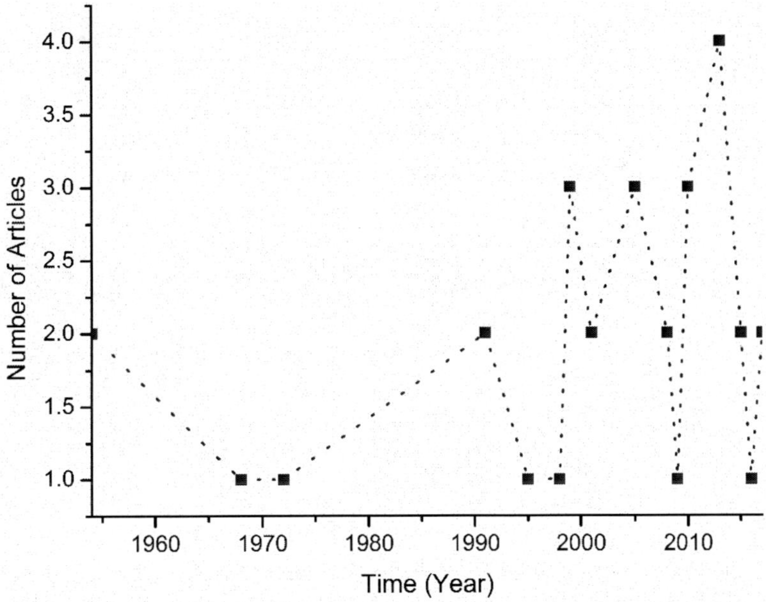

Figure 4. Number of main articles that synthesized montmorillonite from analytical grades reactants from 1954 to 2017.

In general, there are two types of montmorillonite synthesis, hydrothermal and mechanical synthesis. However, the hydrothermal synthesis, due to its simplicity, remains the most used method in the hydrothermal synthesis of montmorillonite (Figure 4). In addition, the natural montmorillonite may be modified together with several organic and inorganic compounds in order to enhance its properties.

3. Applications

The montmorillonite, due to its properties, may be applied in various fields such as environmental remediation and biomedical ones. In the environmental area, MMT is applied as an adsorbent for removal of many types of contaminants specially dyes, heavy metals, emerging contaminants, and among others. Furthermore, MMT has been reported as a promising material for bone tissue engineering and drug delivery system. Figure 5 shows a resume of some main applications of MMT-based materials.

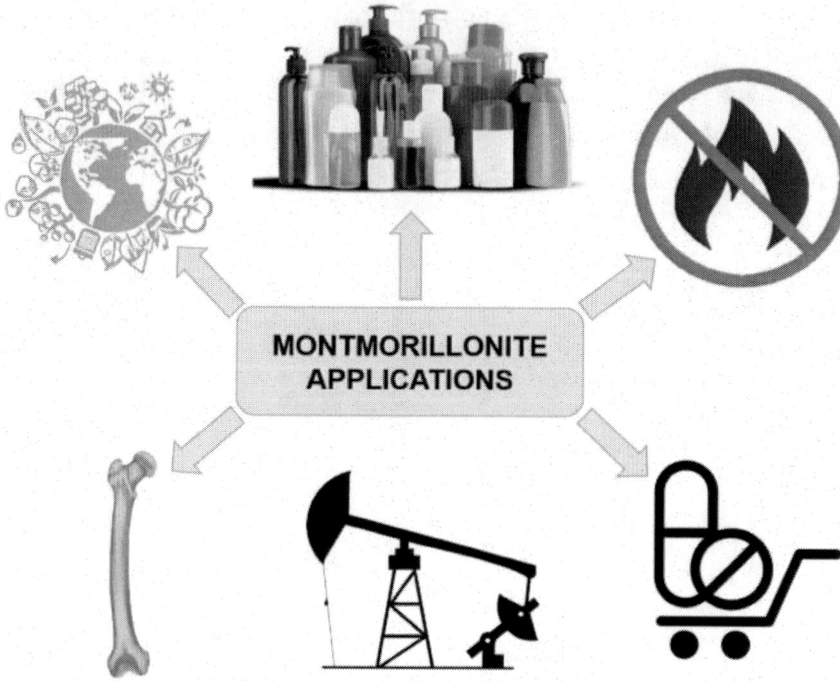

Figure 5. Applications of MMT-based materials in fields from the environmental to biomedical ones.

3.1. Environmental Applications

3.1.1. Treatment of Wastewater

Nowadays, modern societies are challenged by the pollution and contamination of water, air, and soil. In order to overcome these problems, several attempts have been done, mainly to purify the contaminated water. The use of materials (nanoparticles and microparticles) has been discussed due to its efficiency as purifiers (da Silva et al. 2019; Z. Li et al. 2010). However, some natural materials, such as clays, are very interesting to be applied as adsorbents because of its properties and high availability in nature. Clays, in its natural and modified forms, have been extensively employed for the removal of contaminants from different wastewaters. Following this line of reasoning, it is possible to highlight the bentonite as a remarkable absorbent because of its surface charge and surface area.

However, bentonites may have different characteristics and colors depending on their origin and location. For example, Saudi Arabia's activated bentonite is a calcium montmorillonite equivalent to a "Texas bentonite" in the USA or Fuller's earth in the United Kingdom. Saudi activated bentonite has already been used for the removal of lead and cobalt from aqueous solutions. The results showed that sorption of these ions in the Saudi clay was affected by stirring time, initial contaminant concentration, and pH of the mixture (S. Al-Shahrani 2010; S. S. Al-Shahrani 2014).

Many studies have been reported in literature on the removal of various metal ions using raw and modified montmorillonite, such as Pb (II), Cu (II), Cd (II), Cr (III), Cr (VI), Ni (II), Mn (II), Zn (II), Hg (II), Ba (II), Co (II), As, Sr, and Va. The adsorption of montmorillonites may occur from van der Waals or electrostatic interactions between the clay surface and the contaminants. Furthermore, the type of exchangeable cations present in the interlamellar region also influences the adsorption capacity. Some operational factors may affect the adsorption capacity of these clays They are the structural characteristics of the pollutants, the pH of the solution, the interaction time, ionic strength, and the swelling characteristics of montmorillonites.

Thus, the montmorillonite has an excellent adsorption capacity, which is controlled by the composite's nature and pore structure. However, in its fresh state, it has limited applications due to its high hydrophilicity, making it difficult to interact with a series of highly nonpolar materials.

Therefore, structural modifications have been the object of research for the improvement of these properties. One of them is the intercalation of a species in the interlayer region of a clay mineral with preservation of the stratified structure. Another technique used is an acid activation. This process increases the surface area and acidity of clay minerals. It also eliminates many impurities.

These modifications consist of interleaving organic molecules between the structural or adsorbed layers on the surface of the particles. Because of it, the bentonites expand, changing it from hydrophilic to hydrophobic and thus enabling a wide range of applications.

The focus of the surface modification of clays is notable for science because organoclays are essential to developing polymer nanocomposites and new applications.

Currently, the most highlighted application of the organophilic clays is in the area of polymeric nanocomposites. Its hydrophobic nature reduces the surface energy and makes the clay increase its compatibility with the polymer matrix. In addition, the efficiency in the adsorption of organic molecules by organophilic clays may be increased by the use of long-chain alkylammonium radicals. These functions, as dispersants and stabilizers for the suspension of clay particles in a solution, prevent them from settling, increasing the area available for adsorption of the organic molecule.

In addition to the application of natural or modified montmorillonite as an adsorbent, in recent years, clay-based nanostructured materials are being implemented for water treatment.

Nanocomposite materials comprise a few numbers of phases in which at least one of the phases has a size in the nanoscale range. In addition to have an even larger surface area, these materials display distinctive features, such as surface functional group, pore volume, and pore size (Yadav, Gadi, and Kalra 2019).

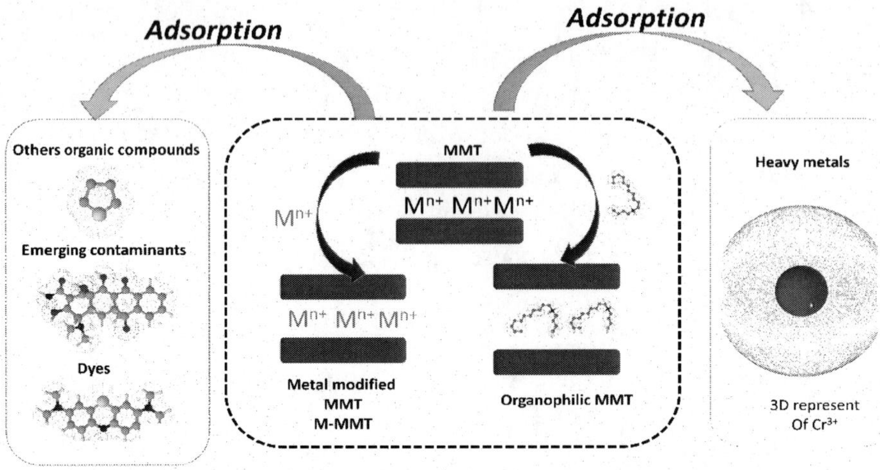

Figure 6. Schematic illustration of MMT and its modified derivatives with respective environmental applications.

Bentonite and its organo-modified derivatives have relevant application in the removal of pollutants of several natures. The adsorptive property of these materials is attributed in part to their lamellar structure and to their easily modifiable surface chemistry to suit the bonding with various organic compounds. Additionally, in recent years, the application of these materials as adsorbents of emerging contaminants has emerged in the literature. Figure 6 highlights MMT modifications and its applications on the removal of contaminants with different natures.

3.1.2. Adsorbents for Mycotoxins

Contamination of grains by fungi is also a worldwide issue, mainly because of the cereals that are used for animal feed. These fungi produce toxic chemicals called mycotoxins. These substances may be formed during the growing, harvesting, drying or storage of grains, becoming a threat to animal health due to their mutagenicity and carcinogenicity. Thus, mycotoxins are the most common pollutants in animal feed, causing reduced immunity and complications in the kidneys, liver, nervous system, and gastrointestinal system. Mycotoxins may also be transferred to the human body through the consumption of animal derivatives (milk, eggs, etc.), which could also harm human health.

Recently, the adsorption process is being considered a more suitable option for the treatment of mycotoxins, since they are sustainable, low cost and highly efficient (Y. Li et al. 2018). Thus, several types of adsorbents have been studied for the removal of these mycotoxins. In the same manner, the bentonite has attracted attention due to its adsorption performance, high chemical stability, advantages of biocompatibility, natural abundance, and low cost.

Some studies have shown that when montmorillonite was added to the dairy cow diet, it could adsorb mycotoxin aflatoxin B1 (AFB_1) (Nones et al. 2017). In addition, no changes were identified in the milk properties such as milk quality, fat, protein, and composition (Maki et al. 2016). In another study, mycotoxin aflatoxin B_1 (AFB_1) and mycotoxin zearalenone (ZER) were simultaneously adsorbed by organo-montmorillonites modified with mono and di alkyl cationic surfactants (G. Wang et al. 2018). Moreover, the organophilic treatments of bentonite may protect stem cells against damage caused by AFB_1 (Nones et al. 2016b).

Other approaches include the modification of MMT by transition metal cations to enhance the adsorption of mycotoxins. For example, MMT was modified by Zn leading to the improvement of adsorption of AFB_1. MMT-Zn was also able to protect the cells against the AFB1 (Nones et al. 2017).

MMT is a promising material to be applied as protective agents against mycotoxins. The treatment of MMT, such as organophilic, thermal and metal ions modifications, increased the ability to adsorb and its subsequent protective effects. Many mechanisms have been proposed to understand the interactions with MMT-based materials and mycotoxins. Some of these mechanisms include cation exchange, intercalation, hydrophobic effects, and electrostatic interactions.

3.3. Flame Retardants Additives

Flame retardants additives are substances (natural or synthetic) that are added in materials in order to retard or eliminate the flame propagation. These materials are commonly incorporated in a polymer matrix. Due to the organic nature, polymers have higher flammability than most of the polymer-based materials. Many materials, inorganic or organic ones, may be used as flame-retardants. The inorganic materials has some advantages such as thermal stability, low cost, and non–toxicity. The later is very compatible with the matrix polymer. Although all the advantages of inorganic flame-retardants, it has some negative outcomes such as poor interaction with polymer matrix, that may result in aggregations or agglomerations, and restrictions for higher loads. On the other side, many organic flame-retardants have in their structure halogens and phosphorus, leading them to release harmful and toxic compounds during their combustion.

The flame inhibition or retardation of these materials are directly related to the polymer combustion process. There are four keys to be considered during the combustion of polymeric-based materials: (1) heat, (2) oxygen, (3) fuel, and (4) radical reaction. The success of materials such as flame retardant must inhibit one or more keys mentioned above. Furthermore, besides of their flame inhibition, the ideal flame-retardant materials do not affect negatively other properties of polymers such as mechanical, chemical, and tribological properties.

Montmorillonite is a promising material to be considered as a flame-retardant. Besides of its efficiency as a flame retardant, this material is cheap and occurs naturally as a bentonite (60 - 95%). It may be reinforced by several articles published from 2000 until nowadays (Figure 7). According to the literature, montmorillonite in polymer matrix enhances the flame resistance by the formation of carbonaceous silicate char on the surface of polymer/montmorillonite nanocomposite during burning. This creates barriers against heat and mass transfer.

Furthermore, montmorillonite alone in polymer matrix delays mass loss. Montmorillonite is supposed to induce a chemical crosslink of the decomposing polymer due to long residence time for polymers radicals. Clays have also some catalytic properties to aromatize hydrocarbons. Those polymer/montmorillonite nanocomposites cannot be considered commercial flame-retardant products, because its slow burning is not possible to extinguish. In addition, the polymer/montmorillonite cannot meet the regulatory test on their own (Morgan 2006). However, montmorillonite in a polymer matrix is a type of flame-retardant according to the definition, but to meet commercial standards must pass some regulatory tests. The montmorillonite to meet most of regulatory tests must be mixed or modified by/with other substances. However, the topic about the regulations of flame retardants is out of the scope of this chapter, for more details, the Morgan (2006)'s review may be read.

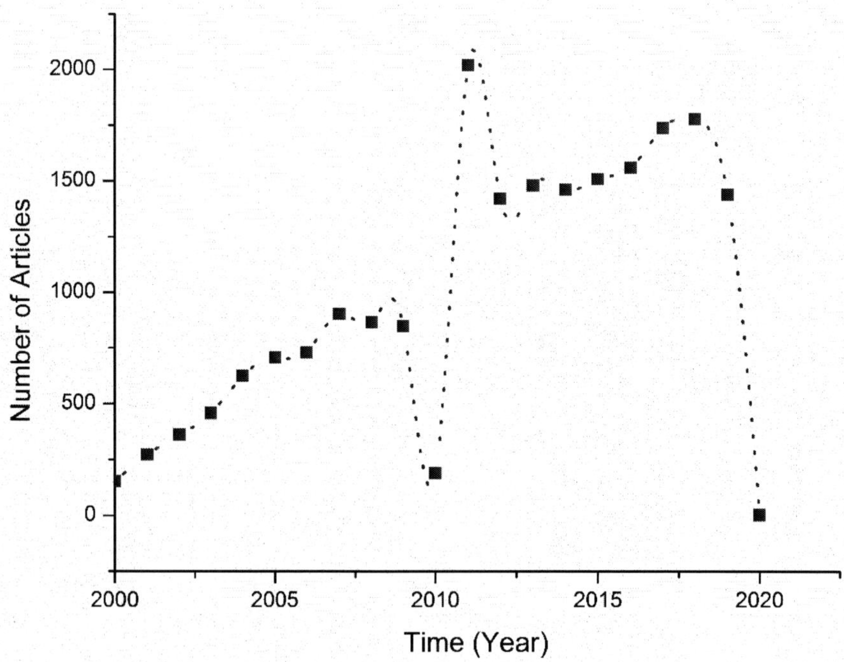

Figure 7. Number of main articles with applications of MMT and their derivatives as flame-retardant additive from 2000 to 2020.

In order to enhance the flame retardancy properties of montmorillonites, many attempts have been reported. This includes modification by chitosan, phosphor-compounds, metal oxide, hydroxides nanoparticles, and others. The improvement of flame-retardant properties was observed when nano-montmorillonite was functionalized by intumescent flame-retardant, 2-(2-(5,5-dimethyl-1,3,2-dioxaphosphinyl-2-ylamino)ethy-amino)-N,N, N-triethyl-2-oxoethanaminium chloride (c-MMT) was added in polyurethane matrix (PU). The efficiency of c-MMT compared with a non-modified MMT was attributed to a higher exfoliation and a better dispersion on polymer matrix (Huang et al. 2010). In addition, c-MMT induced a better dispersion of char on PU/c-MMT nanocomposite after combustion (Huang et al. 2010). Another study achieved the enhancement of flame-retardants properties by modification of MMT using phosphorylated chitosan. The fire resistance was attributed to the formation of compact char protection in condensed phase during burning (Liu et al. 2019). Yan, Xu, and Wang (2018) reported that the addition of organophilic MMT (OMMT) with polyphosphate (OPEA) flame retardant enhanced the flame resistance.

The synergistic effect of OMMT was correlated with the formation in the condensed phase of cross-linked and aromatic structures in the char layers. This correlation is important for the formation of a stable barrier against heat and mass transfer (Yan, Xu, and Wang 2018). Recently, OMMT has also been combined with graphene oxide. Besides the improvement of flame-retardants properties, the OMMT/graphene oxide improved the mechanical properties. Similarly as mentioned before, the flame resistance was attributed to the formation of a complete and compact protective char layer (X. Wang et al. 2018). Wang, Ge, and Jiang (2019), for the first time, reported Ternary system (carboxymethycellulose-Na, MMT, and graphene oxide) self-extinguish-ment containing MMT.

MMT based-materials are very promisor to be applied as flame-retardants in polymers. However, as mentioned above, only MMT as flame retardants for commercial purpose does not meet all regulators tests. Thus, the flame properties were improved significantly by the modification of MMT with metal Ions; phosphor-compounds; metal oxide and hydroxide nanoparticles; graphene and their derivatives; biopolymers; and others. This made this material prospective for commercial proposes.

3.2. Drug Delivery Systems

Nowadays, modern society is facing many environmental problems. For this reason, many fields (including pharmaceutical technology) are researching innovative technologies that may reduce the environmental impacts. Nevertheless, the advanced sustainable technologies are directly connected with the development of materials. In addition, scientists are focusing on renewable and non-toxic materials to challenge the accumulation of hazardous waste.

Drug delivery systems (DDSs) are devices or pharmaceutical formulations that enable the introduction of a therapeutic substance in the body. They also enhance their efficiency and safety by controlling the rate, time, and place of release of these drugs in the body. After the administration, DDSs start to liberate the active ingredients. Thus, the drug is transported across biological barriers to reach the site of action.

Many materials, such as clay minerals (natural or synthetics), metal oxides, nanomaterials, polymers, and nanocomposites, may be applied as DDSs. Following an environmental perspective, the natural materials, mainly clays, are extensively demonstrated in the literature due to its application in DDS (Joshi et al. 2009).

Clay minerals are used in the medical field since a long time ago, due to their availability in nature. Among the plethora of clays, montmorillonite is an important, widely abundant, low-cost, and non-toxic class of clays with unique properties as mentioned previously.

The interactions of drugs and montmorillonite-based DDS may occur in different manners depending on chemical nature of the drug molecule. Montmorillonite is cationic but their mean charge is negative due to the contribution of surface or external charge. Those may interact with certain drugs by ion exchange and electrostatic force. The possibility of intercalation of an organic molecule in the interlayer space is an important property because it opens the possibility of development of inorganic-organic hybrid materials that possess inorganic and organic properties in a single material.

The nonselective β-adrenergic, timol maleate (Figure 8) was successfully intercalated into the interlayer space of montmorillonite at acid pH. The in vitro release in gastric fluid (pH = 1.2) showed that about 43% was released (Joshi et al. 2009). This result suggests that montmorillonite may be used as DDSs of timol maleated in oral administration. Another study evaluated the montmorillonite based DDSs and it revealed that an improvement in the release of ibuprofen was achieved (Dziadkowiec et al. 2017). In this same study, the drug molecule was intercalated in the interlayer space of montmorillonite. More recently, the montmorillonite was modified by CTAB and this modification allowed the intercalation of curcumin and methotrexate (Kar et al. 2019). Furthermore, organoclay particles acted as delivery of curcumin and methotrexate what enhanced the performance of them to kill cancer cells. The strategy to induce folate receptor expression, using a non-toxic dose of curcumin, facilitate folate receptor over-expressed cancer cells to internalize and accumulate folate-antagonist methotrexate in high concentration within cells. This phenomenon allows the use of low drug concentration, which is not harmful to surrounding normal cells, whereas this same concentration is high enough to damage and kill cancer cells (Kar et al. 2019).

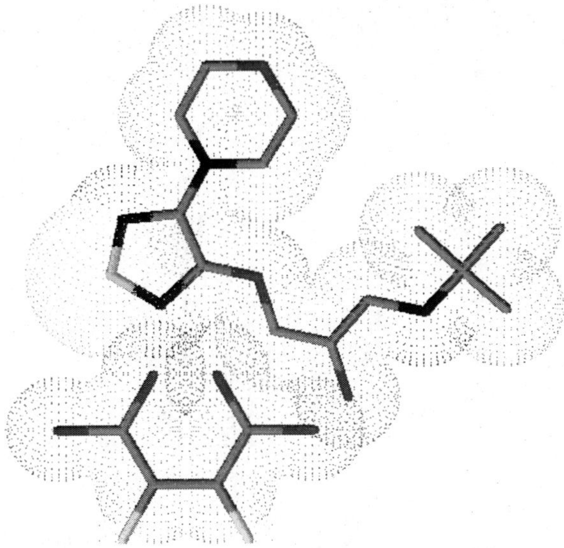

Figure 8. 3D structure of timol maleate.

Synthetic montmorillonites have been reported as a precursor for DDS materials (Olga Yu. Golubeva, Pavlova, and Yakovlev 2015). Unlike natural montmorillonites, synthetic montmorillonites properties, such as morphology, pore size, and the total charge magnitude, may be controlled by varying synthesis parameters to improve their performance depending on applications. In vitro release of vitamin B1 was performed in hydrothermally synthesized montmorillonite and natural one K10. The results showed that synthetic montmorillonite released 54 Wt% and 19 Wt% of Vitamin B1 in simulated gastric (SGF) and intestinal (SIF) fluids (Figure 10). However, in natural one K10, these values reached 37 wt.% and 11 wt.% respectively. This difference may be correlated with the chemical composition of both types of montmorillonites. Furthermore, in both types of montmorillonites, VB1 was adsorbed in the interlayer space by intercalation mechanism, the main mechanism of organic molecules with a certain charge when they interact with smectite-based materials (Olga Yu. Golubeva, Pavlova, and Yakovlev 2015).

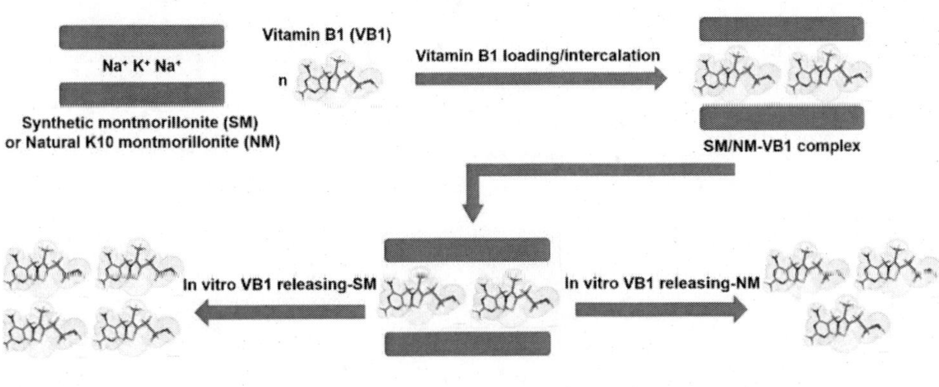

Figure 10. Synthetic and K10 montmorillonite as DDS for Vitamin B1, adapted from Olga Yu, Golubeva, Pavlova and Yakovlev (2015).

In summary, the montmorillonite-based materials are promising DDS because of their remarkable properties. Natural and synthetic materials were applied as DDS, leading to improve the safety, toxicity, and efficiency of drugs.

3.4. Bone Tissue Engineering

Tissue engineering is a multidisciplinary area that combines engineering, material methods, biochemical and physicochemical aspects to regenerate or replace an organ or biological tissue. Nowadays, tissue engineering has become the main area that deals with replacement and construction of new artificial organs. Materials used in tissue engineering must combine some keys characteristics such as chemical, biological, and mechanical properties.

Bone tissue engineering has received attention recently due to the structural role of bones as the body support. In addition, bone tissue engineering has attracted more interest due to the current limitations in surgery and failure of artificial hips and joints. Besides that, some biomaterials do not meet all the requirements for materials to be used in bone tissue engineering, as the ones mentioned above mechanical, biological, and chemical properties.

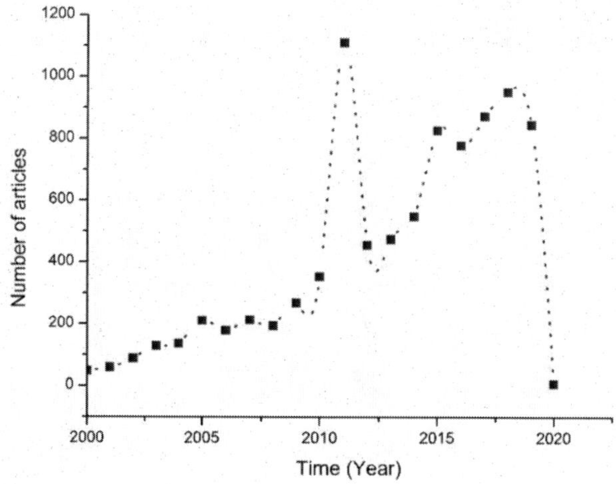

Figure 11. Number of main articles that used MMT-based materials for prospective applications in bone tissue engineering from 2000 to 2020.

Currently, many materials have been developed for applications in bone tissue engineering. Those materials include polymers (synthetic and natural), ceramics, polymer nanocomposites, alloys, and others (Macuvele, Nones, Matsinhe, Lima, et al. 2017). Polymeric nanocomposites are the most interesting materials to be applied in bone tissue engineering, due to their combination of unique characteristics of each component of the nanocomposite. For example, polymers provide some support properties and some nanofillers that, besides the enhancement of tribo-mechanical properties, improve the cell-biomaterial interactions. Nowadays, in bone tissue engineering, there are many types of nanofillers, including bioceramics (mainly the hydroxyapatite), metal oxides, clays, and clay minerals. Furthermore, the application of MMT-based materials in bone tissue engineering has risen significantly in the last decades. These trends corroborate considerably the number of articles published in the last twenty years (Figure 11).

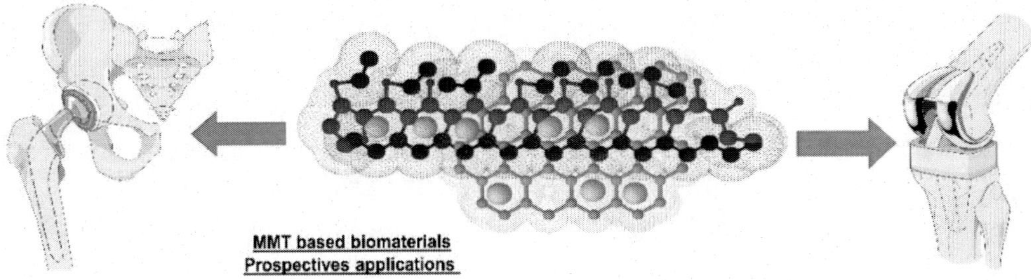

Figure 12. Schematic illustration of MMT-based composite for prospective applications in total or partial joint replacement. Adapted from Macuvele et al. (2019).

Clay and clay minerals have been applied in the biomedical field since a long time in ancient Egypt. Nowadays, montmorillonite minerals have been incorporated in the polymer matrix to fabricate polymer nanocomposite with enhanced mechanical and biological properties for prospective applications in total (Figure 12) or partial joint replacement (Macuvele et al. 2019). Montmorillonite, besides the increment of mechanical properties, enhances bioactivity, biodegradability, and textural properties of polymer nanocomposites-based scaffolds (Jamshidi, Akbari, and Nourmohammadi 2019).

It was developed a methacrylated glycol chitosan-montmorillonite nanocomposite hydrogel for bone tissue engineering. The addition of montmorillonite resulted in the formation of an interconnected microstructured porous. This may promote native cell infiltration, proliferation, and in situ differentiation in the absence of any growth factors, small molecular drugs, or genes (Cui et al. 2019). This behavior was correlated with the intercalation chemistry that is the main characteristics of montmorillonite-based polymer nanocomposites. Similar trends were observed when the MMT was incorporated in the poly-l-lactic acid (PLLA). Besides the enhancement of biomaterial-cell interaction, the MMT induced a formation of a showed three-dimensional interconnected porous structure that is important for the tissue growth within scaffolds (Shuai et al. 2019). Our previous work demonstrated that organo-modified MMT compatibilized the ultra-high molecular weight of the polyethylene-based (golden standard of total hip replacement) composite, leading in the enhancement of tribological, mechanical, and hemacompatibility properties (Macuvele et al. 2019).

The MMT together with hydroxyapatite also induced the growth of apatite in the simulated blood fluid (SBF). This observation is an indication of bioactivity, which may make this material promising in bone tissue engineering applications. Similar research reported that when MMT was incorporated in the biopolymer blend, Xylan/Chitosan was significantly attached with the expression of filopodia what resulted in formations on the surface of the needle such as apatite and osteoblasts (Ali et al. 2019). This together with other characteristics make the Xylan/Chitosna/MMT an ideal scaffold for bone tissue engineering applications (Ali et al. 2019). Recently, MMT was modified by Sr^{2+} and incorporated in chitosan matrix to porous three-dimensional scaffolds via freeze-drying method. Sr^{2+}MMT/Chitosan and MMT/chitosan presented a considerable index of osteoblasts proliferation and viability, without statistical difference between both scaffolds. However, cells on Sr^{2+}-modified scaffolds induced production of significant DNA concentrations rather than non-modified scaffolds. This may suggest that the Sr^{2+} has a role in cell parameters. Furthermore, the Sr^{2+} has a great role in bone formation when this is in an ideal concentration (8.7 and 87.6 ppm) (Pavlidou and Papaspyrides 2008). It corroborates with the results because the Sr^{2+} was found in this range in the supernatant (Koç Demir, Elçin, and Elçin 2018a). This study not only showed that MMT enhances cell proliferation and viability, but also showed that it may release metal cations with great physiological importance (Koç Demir, Elçin, and Elçin 2018b).

In summary, MMT-based biomaterials have attracted attention in the academia and industry due to their remarkable properties. As stated previously, MMT may be incorporated in matrix macromolecules to produce biomaterials with prospective application in bone regeneration, augmentation, bone repair, and in total or partial bone replacement. Figure 13 illustrates the resume of main applications in bone tissue engineering.

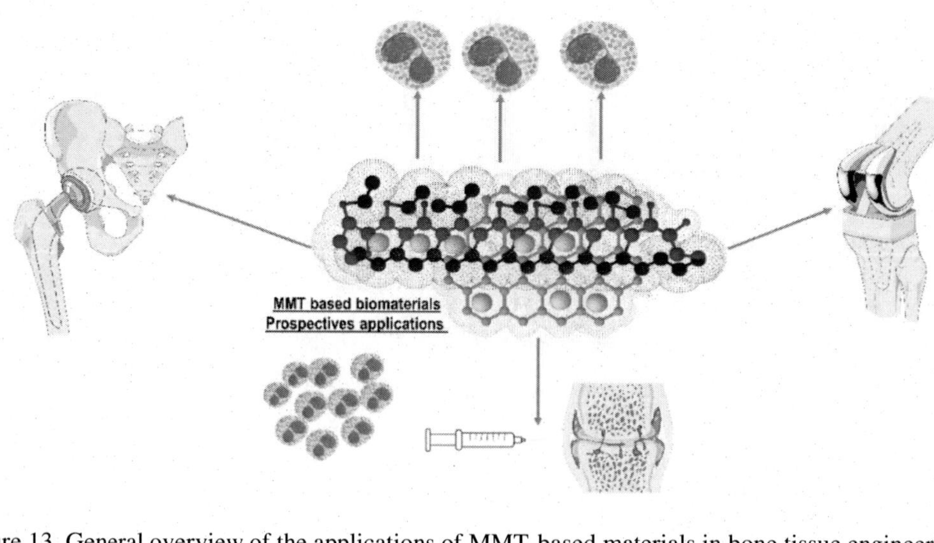

Figure 13. General overview of the applications of MMT-based materials in bone tissue engineering.

3.5. Other Applications

Montmorillonite also has been used in other applications, being a versatile material with an adequate profile to obtain products or even high added value inputs. Currently, a remarkable area of application of organophilic clays is polymeric nanocomposites one. However, there are several industrial-scale applications. These applications include many industrial uses such as: rheological additives in drilling fluids, foundry bond clays for pelletizing iron ores; sealants in irrigation ditches; a suspending agent for fertilizers, insecticides, and farm ponds; in the thickening of paints, moldings, refractory varnishes, adhesives, earthen dams, landfill liners, paints, and grease; clarifying in filters, cosmetics, and personal care products.

3.5.1. Drilling Fluids

Drilling fluids play an important role in oil and gas well drilling operations. Montmorillonite is used worldwide as a water-based drilling fluid additive to control rheology and filtration loss during drilling. For this process, high viscosity is required and the mud must be thixotropic. In addition, fluid viscosity must be high to transport particulate matter from the bottom of the hole to the surface. A drilling fluid must also exhibit all of these properties over a wide temperature range. Thus, this clay is easy to swell and disperse in water, facilitating rheology control and fluid loss during drilling (Zhuang, Zhang, and Jaber 2019).

3.5.2. Dermocosmetic Products

The use of clays in dermo-cosmetic began to be applied even in prehistory, being used for wound healing and skin cleansing. In addition, earlier reports show that clays were used as a therapeutic agent for wound care and bleeding suppression, as well as for aesthetic purposes in the mummification of corpses and as an anti-inflammatory.

Nowadays, due to the versatility of these materials, the presence of clay minerals may be identified in a wide variety of dermo-cosmetic products, especially facial creams, sunscreens, make-up items, skin cleansers, and shampoos. The properties of montmorillonite guarantee chemical stability, controlled release, and dermo-cosmetic or pharma-ceutical efficiency. Moreover, they are widely used in the cosmetic industry for the stabilization of emulsions (Moraes et al. 2017).

3.5.3. Zeolite Synthesis

Zeolites are alluminosilicates that have unique structural features of pores and channels, resulting in lower densities when compared to other materials with similar elemental compositions. This material occurs in nature and may be synthesized in the laboratory. The applications of natural zeolites are not enough explored because of the impurities and structural defects on them. For this reason, the synthesis of zeolites has paramount importance to prepare them for the most varied range of applications. Zeolites are applied in the catalysis to obtain value-added, adsorbent, and ion exchangers products.

The hydrothermal synthesis of zeolites is the most explored method due to its simplicity. Analytical grade reactants are generally used to obtain zeolites (Moreira et al. 2019). However, in the last decades, by means of environmental movements' pressure, some other sources of Al/Si emerged in the literature (Ojumu, Du Plessis, and Petrik 2016; Querol et al. 2002; H. Wang et al. 2016). Most of them include the use of coal bottom ash, coal fly ash, and types of waste that contain Al/Si in their composition.

Montmorillonite in its composition possesses Al and Si, which makes these materials a candidate to be a precursor of zeolitic materials. In addition, clays and zeolites have some similarities such as large cation exchange capacities; open, low-density structures; and reversible hydration and dehydration.

The orthorhombic zeolite N was synthesized from montmorillonites at a lower temperature. The obtained zeolite N presented the same properties of others obtained using conventional Al/Si source. In addition, the synthesis was undertaken in the pilot and industrial scale, showing that the MMT is an interesting source for Al/Si (Mackinnon, Millar, and Stolz 2010). Another study provided a deep review in obtaining zeolites (sodalite, faujasite and A-type zeolite) from clays by means of montmorillonite, hydrothermal, and pre-fusion hydrothermal conversion processes. The clays used as a source of Al/Si were successfully converted into zeolites. In additionof an extra amount of Al source played an important role on zeolitization process (Belviso et al. 2017).

4. Remarks and Perspectives

Montmorillonite-based materials are promising for biomedical and environmental applications. This material has some special properties such as low price and toxicity, availability in nature as bentonite, high cation capacity exchange, and others. In some applications, mainly in catalysis and biomedical technology, montmorillonite must have higher purity. Therefore, in the last decades, more attention has been given to optimizing the routes to synthesize montmorillonite from analytical grade reactants. The results have shown that the synthetic montmorillonite has controllable particle sizes, morphology, and physic-chemical and textural properties. Thus, it is possible to have a deeper understanding of the interactions at a molecular level without interference of impurities. In addition, montmorillonite has been reported as a flame-retardant, drug release system, adsorbent, and filler for bone tissue engineering. Future research could be directed to: (1) understand montmorillonite interactions between biological macromolecules applying molecular dynamics simulations; (2) modify montmorillonite using plant extracts and studying its prospective application as a flame-retardant, a filler for polymer nanocomposite as well as applications in drug delivery systems, and bone tissue engineering.

Acknowledgments

This work was supported by the National Council for Scientific and Technological Development (CNPq, Brazil) and the Coordination for the Improvement of Higher Education Personnel (CAPES, Brazil). We also thank LCME/UFSC for SEM analysis.

References

Achaby, M. El, H. Ennajih, F. Z. Arrakhiz, A. El Kadib, R. Bouhfid, E. Essassi, and A. Qaiss. 2013. "Modification of Montmorillonite by Novel Geminal Benzimidazolium Surfactant and Its Use for the Preparation of Polymer Organoclay Nanocomposites." *Composites Part B: Engineering* 51: 310–17. https://doi.org/10.1016/j.compositesb.2013.03.009.

Al-Shahrani, S. 2010. "Removal of Lead from Aqueous Solutions Using Saudi Activated Bentonite." *WIT Transactions on Ecology and the Environment* 135: 277–88. https://doi.org/10.2495/WP100241.

Al-Shahrani, S.S. 2014. "Treatment of Wastewater Contaminated with Cobalt Using Saudi Activated Bentonite." *Alexandria Engineering Journal* 53 (1): 205–11. https://doi.org/10.1016/J.AEJ.2013.10.006.

Ali, Asif, Saleheen Bano, Satish S. Poojary, Dhruv Kumar, and Yuvraj Singh Negi. 2019. "Effect of Incorporation of Montmorillonite on Xylan/Chitosan Conjugate Scaffold." *Colloids and Surfaces B: Biointerfaces* 180 (August): 75–82. https://doi.org/10.1016/J.COLSURFB.2019.04.032.

Barton, Christopher Karathanasis, Tasos. 2002. "A Novel Method for Measurement and Characterization of Soil Macroporosity." *Commun. Soil Sci. Plant Anal* 33: 1305–22.

Belviso, Claudia, Francesco Cavalcante, Giancarlo Niceforo, and Antonio Lettino. 2017. "Sodalite, Faujasite and A-Type Zeolite from 2:1dioctahedral and 2:1:1 Trioctahedral Clay Minerals. A Singular Review of Synthesis Methods through Laboratory Trials at a Low Incubation Temperature." *Powder Technology* 320 (October): 483–97. https://doi.org/10.1016/J.POWTEC.2017.07.039.

Chen, Yu, Jingdong Peng, Huan Xiao, Huanjun Peng, Lingli Bu, Ziyu Pan, Yan He, Fang Chen, Xiang Wang, and Shiyu Li. 2017. "Adsorption Behavior of Hydrotalcite-like Modified Bentonite for Pb2+, Cu2+and Methyl Orange Removal from Water." *Applied Surface Science* 420: 773–81. https://doi.org/10.1016/j.apsusc.2017.05.138.

Cipolletti, Valeria, Maurizio Galimberti, Marco Mauro, and Gaetano Guerra. 2014. "Organoclays with Hexagonal Rotator Order for the Paraffinic Chains of the Compensating Cation. Implications on the Structure of Clay Polymer

Nanocomposites." *Applied Clay Science* 87: 179–88. https://doi.org/10.1016/j.clay.2013.11.001.

Cui, Zhong-Kai, Soyon Kim, Jessalyn J. Baljon, Benjamin M. Wu, Tara Aghaloo, and Min Lee. 2019. "Microporous Methacrylated Glycol Chitosan-Montmorillonite Nanocomposite Hydrogel for Bone Tissue Engineering." *Nature Communications* 10 (1): 1–10. https://doi.org/10.1038/s41467-019-11511-3.

Dziadkowiec, Joanna, Rola Mansa, Ana Quintela, Fernando Rocha, and Christian Detellier. 2017. "Preparation, Characterization and Application in Controlled Release of Ibuprofen-Loaded Guar Gum/Montmorillonite Bionanocomposites." *Applied Clay Science* 135 (January): 52–63. https://doi.org/10.1016/J.CLAY.2016.09.003.

Fagundes, Ana Paula, DLP Macuvele, Natan Padoin, Cíntia Soares, and Humberto Gracher Riella. 2019. "A Novel Ultrahigh-Molecular-Weight Polyethylene-Based Nanocomposite for Contaminants Adsorption in Aqueous Systems." *Materials Letters* 240: 197–200. https://doi.org/10.1016/j.matlet.2018.12.102.

Fahami, Abbas. 2017. "*Mechanochemical activation of synthetic nanosized 2-D layered Materials: preparation and characterization.*" Texas State University. https://doi.org/10.1017/CBO9781107415324.004.

Forestier, Lydie Le, Fabrice Muller, Frédéric Villieras, and Manuel Pelletier. 2010. "Textural and Hydration Properties of a Synthetic Montmorillonite Compared with a Natural Na-Exchanged Clay Analogue." *Applied Clay Science* 48 (1–2): 18–25. https://doi.org/10.1016/J.CLAY.2009.11.038.

Golubeva, O. Yu., N. Yu. Ul'yanova, T. G. Kostyreva, I. A. Drozdova, and M. V. Mokeev. 2013. "Synthetic Nanoclays with the Structure of Montmorillonite: Preparation, Structure, and Physico-Chemical Properties." *Glass Physics and Chemistry* 39 (5): 533–39. https://doi.org/10.1134/s1087659613050088.

Golubeva, Olga Yu., Svetlana V. Pavlova, and Alexander V. Yakovlev. 2015. "Adsorption and in Vitro Release of Vitamin B1 by Synthetic Nanoclays with Montmorillonite Structure." *Applied Clay Science* 112–113 (August): 10–16. https://doi.org/10.1016/J.CLAY.2015.04.013.

Golubeva, Olga Yu, Elena Yu Brazovskaya, and Olga V. Shamova. 2018. "Biological Activity and Sorption Ability of Synthetic Montmorillonite Modified by Silver/Lysozyme Nanoparticles." *Applied Clay Science* 163 (October): 56–62. https://doi.org/10.1016/J.CLAY.2018.07.015.

Harder, Hermann. 1972. "The Role of Magnesium in the Formation of Smectite Minerals." *Chemical Geology* 10 (1): 31–39. https://doi.org/10.1016/0009-2541(72)90075-7.

Huang, Guobo, Jianrong Gao, Yujing Li, Liang Han, and Xu Wang. 2010. "Functionalizing Nano-Montmorillonites by Modified with Intumescent Flame Retardant: Preparation and Application in Polyurethane." *Polymer Degradation and Stability* 95 (2): 245–53. https://doi.org/10.1016/J.POLYMDEGRADSTAB.2009.08.013.

Ismadji, Suryadi, Felycia Edi Soetaredjo, and Aning Ayucitra. 2015. "Natural Clay Minerals as Environmental Cleaning Agents." In *Clay Materials for Environmental Remediation*, 5–37. https://doi.org/10.1007/978-3-319-16712-1.

Jamshidi, Seyedeh M, B Akbari, and J Nourmohammadi. 2019. "Nanoclay Reinforced Starch-Polycaprolactone Scaffolds for Bone Tissue Engineering." *Journal of Tissues and Materials* 2: 55–63. https://doi.org/10.22034/jtm.2019.168671.1008.

Joshi, Ghanshyam V., Bhavesh D. Kevadiya, Hasmukh A. Patel, Hari C. Bajaj, and Raksh V. Jasra. 2009. "Montmorillonite as a Drug Delivery System: Intercalation and in Vitro Release of Timolol Maleate." *International Journal of Pharmaceutics* 374 (1–2): 53–57. https://doi.org/10.1016/J.IJPHARM.2009.03.004.

Kar, Subrata, Banani Kundu, R.L. Reis, Ruma Sarkar, Papiya Nandy, Ruma Basu, and Sukhen Das. 2019. "Curcumin Ameliorates the Targeted Delivery of Methotrexate Intercalated Montmorillonite Clay to Cancer Cells." *European Journal of Pharmaceutical Sciences* 135 (July): 91–102. https://doi.org/10.1016/J.EJPS.2019.05.006.

Koç Demir, Aysel, Ayşe Eser Elçin, and Yaşar Murat Elçin. 2018a. "Strontium-Modified Chitosan/Montmorillonite Composites as Bone Tissue Engineering Scaffold." *Materials Science and Engineering: C* 89 (August): 8–14. https://doi.org/10.1016/J.MSEC.2018.03.021.

——. 2018b. "Strontium-Modified Chitosan/Montmorillonite Composites as Bone Tissue Engineering Scaffold." *Materials Science and Engineering: C* 89 (August): 8–14. https://doi.org/10.1016/J.MSEC.2018.03.021.

Li, Yan, Guangyan Tian, Gongyue Dong, Shansong Bai, Xiaoyu Han, Jinsheng Liang, Junping Meng, and Hong Zhang. 2018. "Research Progress on the Raw and Modified Montmorillonites as Adsorbents for Mycotoxins: A Review." *Applied Clay Science* 163 (October): 299–311. https://doi.org/10.1016/J.CLAY.2018.07.032.

Li, Zhaohui, Laura Schulz, Caren Ackley, and Nancy Fenske. 2010. "Adsorption of Tetracycline on Kaolinite with PH-Dependent Surface Charges." *Journal of Colloid and Interface Science* 351 (1): 254–60. https://doi.org/10.1016/j.jcis.2010.07.034.

Liu, Xiaodong, Jia Guo, Wufei Tang, Hongfei Li, Xiaoyu Gu, Jun Sun, and Sheng Zhang. 2019. "Enhancing the Flame Retardancy of Thermoplastic Polyurethane by Introducing Montmorillonite Nanosheets Modified with Phosphorylated Chitosan." *Composites Part A: Applied Science and Manufacturing* 119 (April): 291–98. https://doi.org/10.1016/J.COMPOSITESA.2019.02.009.

Mackinnon, Ian D.R., Graeme J. Millar, and Wanda Stolz. 2010. "Low Temperature Synthesis of Zeolite N from Kaolinites and Montmorillonites." *Applied Clay Science* 48 (4): 622–30. https://doi.org/10.1016/J.CLAY.2010.03.016.

Macuvele, D.L.P., J. Nones, J.V. Matsinhe, A.T. Bezerra, M.M. Lima, E.S.W. Santos, M.A. Fiori, and H.G. Riella. 2017. "Simultaneous Effect of Reaction Temperature and

Concentration of Surfactant in Organo-Bentonite Synthesis." *Materials Science Forum* 899: 42–47. https://doi.org/10.4028/www.scientific.net/MSF.899.42.

Macuvele, D L P, Guilherme Colla, Karina Cesca, Luiz F.B. Ribeiro, César E. da Costa, Janaína Nones, Everton R. Breitenbach, et al. 2019. "UHMWPE/HA Biocomposite Compatibilized by Organophilic Montmorillonite: An Evaluation of the Mechanical-Tribological Properties and Its Hemocompatibility and Performance in Simulated Blood Fluid." *Materials Science and Engineering: C* 100 (July): 411–23. https://doi.org/10.1016/J.MSEC.2019.02.102.

Macuvele, D L P, Janaína Nones, Jonas V Matsinhe, Marla M Lima, Cíntia Soares, Márcio A Fiori, and Humberto G Riella. 2017. "Advances in Ultra High Molecular Weight Polyethylene/Hydroxyapatite Composites for Biomedical Applications: A Brief Review." *Materials Science and Engineering: C* 76 (July): 1248–62. https://doi.org/10.1016/J.MSEC.2017.02.070.

Maki, C.R., A.P.A. Monteiro, S.E. Elmore, S. Tao, J.K. Bernard, R.B. Harvey, A.A. Romoser, and T.D. Phillips. 2016. "Calcium Montmorillonite Clay in Dairy Feed Reduces Aflatoxin Concentrations in Milk without Interfering with Milk Quality, Composition or Yield." *Animal Feed Science and Technology* 214 (April): 130–35. https://doi.org/10.1016/J.ANIFEEDSCI.2016.02.014.

Moraes, Jemima Daniela Dias, Silvana Raquel Alina Bertolino, Silvia Lucia Cuffini, Diego Fernando Ducart, Pedro Eriberto Bretzke, and Gislaine Ricci Leonardi. 2017. "Clay Minerals: Properties and Applications to Dermocosmetic Products and Perspectives of Natural Raw Materials for Therapeutic Purposes—A Review." *International Journal of Pharmaceutics* 534 (1–2): 213–19. https://doi.org/10.1016/J.IJPHARM.2017.10.031.

Moreira, Jarina Costa, Rozineide. A. Antunes Boca Santa, Giorgio L. Miraglia, Cíntia Soares, and Humberto Gracher Riella. 2019. "Evaluation of Different Reaction Systems to Obtain Zeolite 4A via Reverse Microemulsion." *Microporous and Mesoporous Materials* 279 (May): 262–70. https://doi.org/10.1016/J.MICROMESO.2018.12.042.

Morgan, Alexander B. 2006. "Flame Retarded Polymer Layered Silicate Nanocomposites: A Review of Commercial and Open Literature Systems." *Polymers for Advanced Technologies* 17 (4): 206–17. https://doi.org/10.1002/pat.685.

Nones, Janaína, Jader Nones, Anicleto Poli, Andrea Gonçalves Trentin, Humberto Gracher Riella, and Nivaldo Cabral Kuhnen. 2016a. "Organophilic Treatments of Bentonite Increase the Adsorption of Aflatoxin B1 and Protects Stem Cells Against Cellular Damage." *Colloids and Surfaces B: Biointerfaces*. https://doi.org/10.1016/j.colsurfb.2016.05.061.

———. 2016b. "Organophilic Treatments of Bentonite Increase the Adsorption of Aflatoxin B1and Protect Stem Cells against Cellular Damage." *Colloids and Surfaces B: Biointerfaces* 145: 555–61. https://doi.org/10.1016/j.colsurfb.2016.05.061.

Nones, Janaína, Anita Solhaug, Gunnar Sundstøl Eriksen, Domingos Lusitâneo Pier Macuvele, Anicleto Poli, Cíntia Soares, Andrea Gonçalves Trentin, Humberto Gracher Riella, and Jader Nones. 2017. "Bentonite Modified with Zinc Enhances Aflatoxin B 1 Adsorption and Increase Survival of Fibroblasts (3T3) and Epithelial Colorectal Adenocarcinoma Cells (Caco-2)." *Journal of Hazardous Materials* 337: 80–89. https://doi.org/10.1016/j.jhazmat.2017.04.068.

Ojumu, Tunde V., Pieter W. Du Plessis, and Leslie F. Petrik. 2016. "Synthesis of Zeolite A from Coal Fly Ash Using Ultrasonic Treatment - A Replacement for Fusion Step." *Ultrasonics Sonochemistry* 31: 342–49. https://doi.org/10.1016/j.ultsonch.2016.01.016.

Otsubo, Y, and C Kato. 1954. "Hydrothermal Synthesis of Montmori-llonite Type Silicates. III. Synthesis of Zinc-Montmorillonite." *J. Chem. Soc. Jpn* 75: 456–59. https://doi.org/https://doi.org/10.1246/nikkashi1948.75.456.

Parolo, M. E., M. C. Savini, J. M. Vallés, M. T. Baschini, and M. J. Avena. 2008. "Tetracycline Adsorption on Montmorillonite: PH and Ionic Strength Effects." *Applied Clay Science* 40 (1–4): 179–86. https://doi.org/10.1016/j.clay.2007.08.003.

Pavlidou, S., and C.D. Papaspyrides. 2008. "A Review on Polymer–layered Silicate Nanocomposites." *Progress in Polymer Science* 33 (12): 1119–98. https://doi.org/10.1016/J.PROGPOLYMSCI.2008.07.008.

Querol, X., N. Moreno, J. C. Umaa, A. Alastuey, E. Hernández, A. López-Soler, and F. Plana. 2002. "Synthesis of Zeolites from Coal Fly Ash: An Overview." *International Journal of Coal Geology* 50 (1–4): 413–23. https://doi.org/10.1016/S0166-5162(02)00124-6.

Reinholdt, M., J. Miehé-Brendlé, L. Delmotte, R. Le Dred, and M.-H. Tuilier. 2005. "Synthesis and Characterization of Montmorillonite-Type Phyllosilicates in a Fluoride Medium." *Clay Minerals* 40 (2): 177–90. https://doi.org/10.1180/0009855054020164.

Reinholdt, Marc, Jocelyne Brendlé, Marie-Hélène Tuilier, Serge Kaliaguine, and Emmanuelle Ambroise. 2013. "Hydrothermal Synthesis and Characterization of Ni-Al Montmorillonite-Like Phyllosilicates." *Nanomaterials* 3 (1): 48–69. https://doi.org/10.3390/nano3010048.

Reinholdt, Marc, Jocelyne Miehé-Brendlé, Luc Delmotte, Marie Hélène Tuilier, Ronan Le Dred, Robert Cortès, and Anne Marie Flank. 2001. "Fluorine Route Synthesis of Montmorillonites Containing Mg or Zn and Characterization by XRD, Thermal Analysis, MAS NMR, and EXAFS Spectroscopy." *European Journal of Inorganic Chemistry*, no. 11: 2831–41. https://doi.org/10.1002/1099-0682(200111)2001:11<2831::AID-EJIC2831>3.0.CO;2-6.

Shuai, Cijun, Yang Li, Pei Feng, Wenjing Yang, Zhenyu Zhao, and Wei Liu. 2019. "Montmorillonite Reduces Crystallinity of Poly-l-Lactic Acid Scaffolds to Accelerate

Degradation." *Polymers for Advanced Technologies*, no. April: 2425–35. https://doi.org/10.1002/pat.4690.

Silva, Anderson Felipe Viana da, Ana Paula Fagundes, Domingos Lusitâneo Pier Macuvele, Elita Fontanele Urano de Carvalho, Michelangelo Durazzo, Natan Padoin, Cíntia Soares, and Humberto Gracher Riella. 2019. "Green Synthesis of Zirconia Nanoparticles Based on Euclea Natalensis Plant Extract: Optimization of Reaction Conditions and Evaluation of Adsorptive Properties." *Colloids and Surfaces A: Physicochemical and Engineering Aspects*, September, 123915. https://doi.org/10.1016/j.colsurfa.2019.123915.

Sun, Zhiming, Yuri Park, Shuilin Zheng, Godwin A. Ayoko, and Ray L. Frost. 2013. "XRD, TEM, and Thermal Analysis of Arizona Ca-Montmorillonites Modified with Didodecyldimethylammonium Bromide." *Journal of Colloid and Interface Science* 408 (October): 75–81. https://doi.org/10.1016/J.JCIS.2013.07.007.

Wang, Cheng, Xuesong Ge, and Yijun Jiang. 2019. "Synergistic Effect of Graphene Oxide/Montmorillonite-Sodium Carboxymethycellulose Ternary Mimic-Nacre Nanocomposites Prepared via a Facile Evaporation and Hot- Pressing Technique." *Carbohydrate Polymers* 222 (October): 115026. https://doi.org/10.1016/J.CARBPOL.2019.115026.

Wang, Gaofeng, Yushan Miao, Zhiming Sun, and Shuilin Zheng. 2018. "Simultaneous Adsorption of Aflatoxin B1 and Zearalenone by Mono- and Di-Alkyl Cationic Surfactants Modified Montmorillonites." *Journal of Colloid and Interface Science* 511 (February): 67–76. https://doi.org/10.1016/J.JCIS.2017.09.074.

Wang, Haixia, Xiuru Wang, Zhenghe Xu, and Mingliang Zhang. 2016. "Synthetic Zeolite from Coal Bottom Ash and Its Application in Cadmium and Nickel Removal from Acidic Wastewater." *Desalina-tion and Water Treatment* 57 (54): 26089–100. https://doi.org/10.1080/19443994.2016.1160438.

Wang, Xuebao, Bo Bi, Jiangwei Liu, Shousheng Yang, Liang Zhou, Lingang Lu, Yong Wang, Fan Xu, and Run Huang. 2018. "Halogen-Free Intumescent Flame-Retardant Ethylene-Vinyl Acetate Copolymer System Based on Organic Montmorillonite and Graphene Nanosheets." *Journal of Applied Polymer Science* 135 (23): 1–10. https://doi.org/10.1002/app.46361.

Yadav, Vijay Bahadur, Ranu Gadi, and Sippy Kalra. 2019. "Clay Based Nanocomposites for Removal of Heavy Metals from Water: A Review." *Journal of Environmental Management* 232 (February): 803–17. https://doi.org/10.1016/J.JENVMAN.2018.11.120.

Yan, Long, Zhisheng Xu, and Xinghua Wang. 2018. "Synergistic Effects of Organically Modified Montmorillonite on the Flame-Retardant and Smoke Suppression Properties of Transparent Intumescent Fire-Retardant Coatings." *Progress in Organic Coatings* 122 (September): 107–18. https://doi.org/10.1016/J.PORGCOAT.2018.05.016.

Zhang, Di, Chun-Hui Zhou, Chun-Xiang Lin, Dong-Shen Tong, and Wei-Hua Yu. 2010. "Synthesis of Clay Minerals." *Applied Clay Science* 50 (1): 1–11. https://doi.org/10.1016/J.CLAY.2010.06.019.

Zhuang, Guanzheng, Zepeng Zhang, and Maguy Jaber. 2019. "Organoclays Used as Colloidal and Rheological Additives in Oil-Based Drilling Fluids: An Overview." *Applied Clay Science* 177 (September): 63–81. https://doi.org/10.1016/J.CLAY.2019.05.006.

In: An Introduction to Aluminosilicates
Editor: Nero Regina Blevins
ISBN: 978-1-53617-250-8

Chapter 2

PILLARED CLAYS, A CLASS OF HETEROGENEOUS CATALYSTS DERIVED FROM MODIFIED CLAY MINERALS: PROPERTIES AND USES IN WASTE PYROLYSIS

Ana C. S. Serra[1], Monica R. C. Marques[1,*] and Jacyra G. Faillace[2]
[1]Chemistry Institute, State University of Rio de Janeiro, Rio de Janeiro,RJ, BR
[2]Chemistry Coordination, Federal Institute of Science and Technology of Rio de Janeiro, Rio de Janeiro, RJ, BR

ABSTRACT

Clay minerals are the main constituents of clays, being formed by aluminosilicates in fibrous structures and crystalline lamellae with equivalent spherical diameter less than or equal to 2 μm. Clay minerals from the smectites group have layers organized as tetrahedral sheets of silica and octahedral alumina, condensed and polymerized. The cations adsorbed on the surface of the clay and between the lamellae can be exchanged without compromising the crystalline structure and contribute to modify some physicochemical properties. In this way, clays are very versatile materials and have varied applications due to their swelling, adsorption and ion exchange properties. The applications of clay minerals in heterogeneous catalysis have been extensively studied in the last 80 years. Through some modifications, it is possible to improve catalytic performance by increasing access to their active sites. In recent decades, pillared clays have been widely used in many applications, in particular in the cracking of petroleum and the pyrolysis of waste. They are a class of modified clays, synthesized from the exchange of the cations of the original clay by inorganic molecules, usually complexes of metal cations introduced into the interlamellar space of the clay minerals. After heating, this system undergoes dehydration and

[*]Corresponding Author's Email:monicamarquesuerj@gmail.com.

dehydroxylation processes, forming metal oxide pillars strongly bonded to layers of the minerals, producing high resistance and thermal stability as well as increasing porosity, specific surface area and basal spacing (d001). The polyhydroxy metal complexes of Al, Fe, Zr, Cr and Ti are the inorganic pillarizing agents which have been studied for the synthesis of pillared clays. The natural clays have Brønsted and Lewis acidity, but in pillared clays the, Lewis acidity is predominant.

Keywords: clay minerals, pillared clays, catalyst, pyrolysis, cracking, plastic waste, petroleum, oil sludge, heavy gas oil

INTRODUCTION

Clays are materials that have been used since the dawn of humanity. The earliest and most documented uses of clays for abstract cave art to make pigments and for medicinal purposes. The use in making ceramic articles for various purposes has also been observed (Bergaya et al. 2013, 823).

Research suggests that clays participated in the process of formation of small molecules on earth millions of years ago, promoting catalytic reactions with amino acids linked to the origin of life. Such rudimentary precursor molecules gradually evolved into more complex molecules and progressed to generate more elaborate life forms such as plants and animals (Saladino et al. 2004; Stern and Jedrzejas, 2008). Clays are also described as natural catalysts in the formation and thermal evolution of oil (John and Shimoyana 1972; Johns 1979). In addition, clays are components of various types of soils and sediments, playing an important catalytic function in chemical soil transformation processes (Fripiat and Cruz-Cumplido 1974).

The use of clays in different applications has increased considerably since the start of the 20th century. With the increasing search for technological solutions that are environmentally friendly and the development of green chemistry, these materials have gained prominence. The abundance of world clay reserves, allied to their low price and ease of disposal without harming the environment partly explains the increased demand for these materials (Bergaya and Lagaly 2006, 1). With the advancement of nanotechnology, nanoclays have been developed to improve the characteristics and broaden utilization of clays in various applications (Zeng et al. 2005). Figure 1 shows a diagram of the different applications of clays in different areas of knowledge.

The term clay has several meanings and applications, depending on the area of knowledge. Clay corresponds to minerals that are less than 2 μm in size. This particle size definition derives from petrographic studies carried out by optical microscopy in the late 19th century. In these studies, when crystals smaller than 2 μm in size could not be distinguished, they were classified by the generic name “clay” (Velde and Meunier 2008, 1). The term clay can also be used to define materials that are formed from the physical (degradation) and chemical (decomposition) modifications undergone by certain rock types (Santos 1989). Other definitions include clays as natural, earthy, fine-grained materials that, when moistened, exhibit plasticity and which harden when dried or heated. In general, the term clay refers to soil particles that have a diameter of less than 2 μm and which may include different types of minerals, such as clay minerals, which are the main constituents, in addition to quartz, feldspar, carbonates, oxides, metals and even organic matter (Bergaya and Lagaly 2006, 3-9).The presence of water allows clays to exhibit plasticity. Other properties of clay minerals include the anisotropy of the layers or particles, the ease with which the outer surface, and often the inner surface, can be modified (by adsorption and ion exchange) and hardening by drying or heating (Bergaya, Theng and Lagaly 2006,8).

The chemical structure of clay minerals was better determined after the development of optical microscopy, X-ray diffraction and chemical methods from the first half of the 20th century (Ross and Shannon 1925; 1926; Hendricks 1929; Hendricks and Fry 1930; Ross and Kerr 1930, 1931). Linus Pauling was the first to determine the structure of micas and related materials, providing the basis for elaboration of the structure of phyllosilicates (Pauling 1930, 1930a). This model proposes that the structural units of the phyllosilicates are formed by extremely stable SiO_4 tetrahedra, which are polymerized forming two-dimensional sheets. The lamellar nature of clays inspired its generic name: phyllosilicates. In Greek, phyllon means 'leaf', or lamella (Luna and Schuchardt 1999).

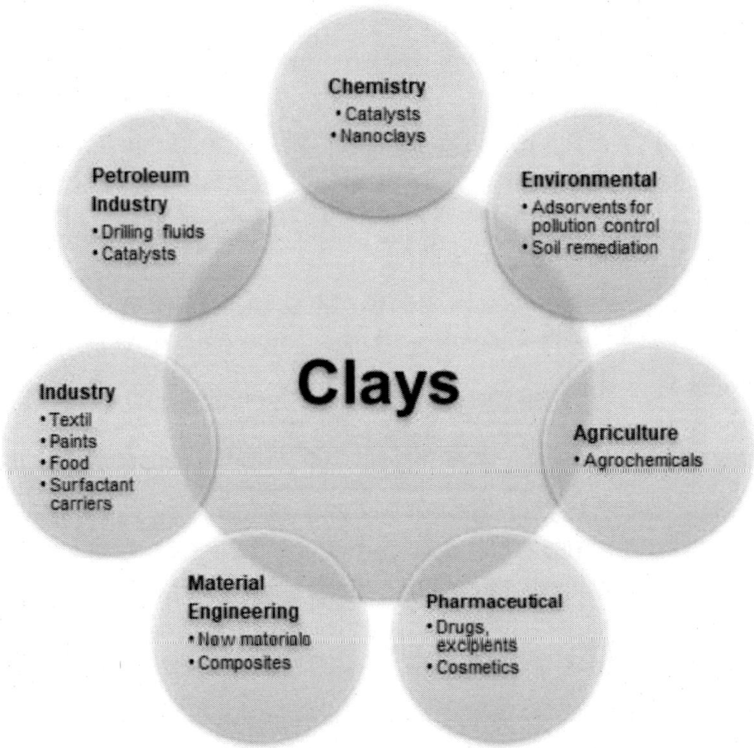

Figure 1. Applications of clays in different areas of knowledge.

Tetrahedral sheets are formed by Si^{4+} tetrahedra coordinated to four O^{2-} atoms. Each tetrahedron shares three of its oxygen atoms with three neighboring tetrahedra (basal oxygen). The tetrahedral structures are linked together to form continuous hexagonal sheets. Common tetrahedral cations are Si^{4+}, Al^{3+}, and Fe^{3+}.The octahedral sheets are composed of Al^{3+}, Fe^{2+} or Mg^{2+} octahedra coordinated to six O^{2-} or OH^{-} anions. The octahedral structures are also connected to form sheets of hexagonal or pseudohexagonal symmetry. The tetrahedral and octahedral sheets are condensed with each other by sharing an oxygen atom (apical oxygen) (Brigatti, Galan and Theng 2013, 21-22). In a unit cell composed of 20 oxygen atoms and 4 hydroxyls, there are eight tetrahedral and six octahedral sites. When all these sites are occupied by cations, the phyllosilicate is called trioctahedral. When only two-thirds are occupied by cations, it is a dioctahedral phyllosilicate (Luna and Schuchardt 1999). Non-crystalline or X-ray amorphous materials including organic matter are referred as 'associated phases' whether or not they impart plasticity to clay. Accordingly, allophane would qualify as an associated phase since this group of hydrated aluminosilicates lacks long-range order; but allophane be regarded as a clay mineral because of past usage. The same applies to imogolite, an aluminosilicate with long-range order in one direction (Bergaya and Lagaly 2006, 9-10). Another type includes fibrous clay minerals such as sepiolite and palygorskite (Galán 1996).

It was Grim in 1962 who first proposed the classification of clay minerals. In this classification, clay minerals are divided according to the way the tetrahedral and octahedral sheets are arranged forming layers, into: type 1:1 layers (kaolinite group and halloysite group); and type 2:1 layers (illite group, smectite group, and vermiculite), mixed layers (chlorite group) and chain structures (sepiolite and palygorskite). Type 1:1 clay minerals have layers composed of a condensed tetrahedral sheet (T) and an octahedral sheet (O). This is known as the TO layer. Type 2:1 clay minerals consist of an octahedral sheet inserted between two tetrahedral sheets (TOT). This group includes smectite, pyrophyllite, vermiculite, mica and illite (Uddin 2017).

In smectite group, the bonding between two silica sheets is very weak, which permits the water and exchangeable ions to enter. This leads to the development of swelling capacity. The name Smectite was suggested as a group name by the Clay Minerals Group of the Mineralogical Society of Great Britain for all clay minerals with an expanding lattice, e.g., montmorillonite, beidellite, nontronite, saponite, and bentonite.The cation exchange capacity, specific surface area, and adsorption capacity of smectites are considerably larger than those of other clay families (Uddin 2017).

Clays and clay minerals have a small particle size and complex porous structure with high specific surface area, which allows strong physical, chemical interactions with dissolved species. These interactions are due to the electrostatic repulsion, cristallinity, and adsorption or specific cation exchange reactions. The highly porous surface area that possesses attractive force suggests that the bonding power will also be high (Uddin 2017).

In this section, we focus on the main properties exhibited by the smectite group because natural swelling clay minerals are the most used for pillaring process (Vicente, Gil and Bergaya 2013, 526-527).

Cations from octahedral and tetrahedral sheets may be replaced by other cations of the same or different ionic charges without altering the crystal structure. This process is called isomorphic substitution. However, if the cation has a different charge than the substituted cation, a charge is generated on the surface of the layer. According to Vaccari (1999), natural clay minerals can be cationic or anionic. While cationic clays are much more abundant, anionic ones are easier to synthesize. Anionic clay minerals are positively charged on the surface of the sheets and present anions between the layers to balance these charges. Cationic clay minerals have a negatively charged surface and have cations between the layers to balance these charges. Some type 2:1 clays, such as montmorillonite, have a positive charge deficiency in their crystal structure, caused by isomorphic substitutions, resulting in an excess of negative charges distributed across the surface of the sheets. These substitutions may be Si^{4+} by Al^{3+} at the tetrahedral sites, Al^{3+} by Mg^{2+} or Mg^{2+} by Li^{+} (or a vacancy) at the octahedral sites. The excess of resulting negative charge is counterbalanced by hydrated alkaline cations like Na^{+} and K^{+}, or earth alkaline cations like Ca^{2+} and Mg^{2+}, which can be reversibly exchanged for other cations (Teixeira-Neto and Teixeira-Neto 2009). The chemical composition and formula of the unit cell of the "theoretical" or extreme montmorillonite of the series is $(Al_{3,33}Mg_{0,67})Si_8O_{20}(OH)_4.M^{+1}{}_{0,67}$, where M^{+1} is a monovalent cation. This formula shows that the unit cell has a negative electric charge due to isomorphic replacement of the Al^{3+} by Mg^{2+} (Coelho, De Souza Santos and De souza Santos 2007). The M^{+} cation that balances the negative charge is called an exchangeable cation. According to Varma (2002), the layer charge varies between $0.5 < x < 1.2$.

The cation exchange capacity (CEC) of an aluminosilicate is the quantity of ions, particularly cations, that it can adsorb and exchange. This capacity results from the imbalance of electrical charges in the crystalline structure due to isomorphic substitutions, broken chemical bonds at the edges of particles, and the interaction of H_3O^{+} ions with the charges at these broken bonds. To neutralize these charges, there are exchangeable cations, which are electrostatically fixed along the faces and between the structural layers (Santos 1989). The exchange power of a cation will be greater the higher its valence and the lower its hydration are. Higher valence ions are more strongly attracted to the material because it has a higher ionic charge. The exchangeable cation content, expressed in milliequivalents of cation per 100 g of clay, defines the CEC. Clays of the smectite group, including montmorillonite, exhibit a high cation exchange capacity of 80 to 150 meq/100 g clay (Santos 1989; Gomes 1988).Cations within the crystalline lamellae and especially the cations between the structural layers can be exchanged for other cations present in an aqueous solution without modifying the crystal structure of the clays. When the clay mineral is dry, the cations located between the lamellae are in the hexagonal cavities of the tetrahedral layers, so the tetrahedral-octahedral-tetrahedral (TOT) clays have a higher CEC than tetrahedral-octahedral (TOT) clays such as kaolinites. Due to the presence of water, these cations move into the interlamellar space and thus can be easily exchanged. Group 2:1 clay minerals differ mainly regarding the intercalated cations. For example, in micas the cation is potassium, while in smectites there are different hydrated cations. Organic matter and other components of the clay fraction have high CEC (300 meq/100 g). Thus, it is important to treat the clay fraction for the removal of interferents in order to study the catalytic behavior of the clay mineral (Santos 1989). The cation exchange capacity is an important property of clays, because cation exchange allows chemical modifications. These modifications directly influence the clay's physicochemical properties and possible technological applications, such as heterogeneous catalysis (Coelho, De Souza Santos and De souza Santos, 2007, Teixeira-Neto and Teixeira-Neto 2009).

Clays disperse evenly and form a paste upon contact with water, because water molecules are strongly attracted to the surfaces of clay minerals. The interaction of water with clays occurs differently between the different types of clay minerals. In the 2:1 clay minerals of the smectite group, the adsorption mechanism occurs by the interaction of water with the cations present in the interlamellar region. Depending on the valence of these cations, more or fewer layers of water are formed around them (Varma 2002). Higher valence cations adsorb less water, and lower valence cations tend to adsorb more water molecules, forming layers that move the lamellae away from the clay mineral, causing greater spacing. For cations of the same valence, the greater the volume of the ion, the weaker its electric field will be in solution, and consequently the lower the degree of hydration (Cases et al. 1997). Thus, for the same series of ions, the hydrated radius is generally inversely proportional to the ionic radius of the crystal. In the case of monovalent cations, selectivity follows the following sequence: H^+>Cs^+>Rb^+>NH^{4+}>K^+>Na^+>Li^+. For bivalent cations the sequence is as follows: Ba^{2+}>Sr^{2+}>Ca^{2+}>Mg^{2+}. In type 2:1 clay minerals such as montmorillonite, when large monovalent and divalent exchangeable cations occupy the space between layers, spacing can reach a value of about 22,5 Å (Coelho, De Souza Santos and De souza Santos 2007). The swelling capacity of a clay mineral is as important as its CEC, since this influences the intercalation capacity and consequently the chemical modification of the clay mineral.

Clay minerals play an important role in preserving environmental quality by acting as adsorbent materials for the removal of toxic metal ions in aqueous solution (Srinivasan 2011; De Aguiar, Novaes and Guarino 2002). Their use as adsorbents has advantages over many other commercially available adsorbents in terms of low cost, plentiful availability, high specific surface area, excellent adsorption, non-toxic properties and high ion exchange potential (Uddin, 2017). Most clay minerals are negatively charged and are very effective and widely used to adsorb metal cations in solutions. Heavy metal uptake involves a number of complex adsorption mechanisms, such as direct bonding between metal cations and the surface of clay minerals, surface complexation, and ion exchange etc. (Churchman et al. 2006). In many situations, pretreatment is required to increase the adsorption capacity of clays, i.e., to modify them to increase their absorption capacity. This pretreatment improves surface area, pore volume and number of acidic sites on the surface. Through this pretreatment or modification, clay minerals become hydrophobic and organophilic, thus increasing the adsorption of nonionic organic compounds (Ismadji, Soetaredjo and Ayucitra 2015).

An important property of clays is their acidity. Vaccari (1999) reported that due to the small particle size, clay minerals can have both Lewis acid and Brønsted acid sites. Brønsted acidity is given by hydroxyl groups in their structure and by the dissociation of water in the solvation of interlamellar cations. Thus, this acid character can be changed depending on the type of interlamellar cation, i.e., the more polarizing the cation, the greater the dissociation capacity of water and the higher the acidity will be (Shriver and Atkins 2008, 140-141). However, Lewis acidity is imparted by the Al^{3+} and Fe^{3+} cations at the crystal edges (Santos 1989; Gomes 1988). In addition, Fe^{3+} is a site that has oxidizing character (Vaccari 1999). Information on the nature of acid sites in solids is essential for assessing their catalytic potential.

By determining the textural properties of a material, information is obtained about its porous structure. In a clay mineral, a set of layers is referred to as a 'particle', and a set of particles as an 'aggregate'. The distribution of pore sizes in clay minerals can be divided into three ranges, as interlayer pores, inter-particle pores and inter-aggregate pores (Bergaya and Lagaly 2006, 11). Microporous clays can be studied by observing the interlayer region, with pore size below 2 nm (<20 A°). Mesoporous clays can be the consequence of the spatial organization of the layers within the particles, with pore size between 2 and 50 nm (20 to 500 A°) and macroporous clays have arrangement with aggregate sets and pore size greater than 50 nm (>500 A°). Heterogeneous solid-fluid interaction processes depend directly on the specific surface area of the solid reagent. This area is defined as the total surface area of the particles per unit of mass. The surface area of clays is normally measured by N_2 adsorption-desorption isotherms at 77K. The porosity of the solid may cause hysteresis cycles to arise because the mechanism in desorption differs from what occurs in adsorption. The form presented by these isotherms can provide information regarding the porous structure of the material (Gregg and Sing 1982). Different mathematical methods can be applied to determine the value of the specific area. The most common are the Brunauer, Emmett and Teller model (BET), developed for multilayers; and the Langmuir model, developed for monolayers or for chemical adsorption (Gil, Korili and Vicente 2008). In addition to these models, the Barrett, Joyner and Halenda model (BJH) is also used. This model is commonly employed to ascertain the average size and pore size distribution along with the specific area (Gregg and Sing 1982). The specific area of a montmorillonite has values ranging from 300 to 800 m^2/g (Santos 1989).

Modified Clay Minerals: Pillared Clays

Historical Perspective

Applications of clay minerals as heterogeneous catalysts have long been known (Houdry et al. 1938). The research on the synthesis of pillared clays was based on the pioneering work of Barrer and MacLeod (1955), who obtained microporous materials by replacing the exchangeable cations between smectite layers with tetraalkylammonium ions. However, it was later found that alkylammonium decomposes below 250°C, causing the pillared clay structure to collapse (Loeppert Jr. and Mortland 1979). Syntheses using other organic pillars were also performed to obtain permanent porosity and utility as molecular sieves and catalysts despite thermal instability at moderate temperatures (>250°C) (Mortland and Berkheiser 1976).

The use of clays intercalated with aqueous solutions of metal cations is not recent. In 1940, an American bentonite was intercalated with $FeCl_3$ and $NiCl_2$ solutions to obtain a heterogeneous catalyst (Broughton 1940). Subsequently, polymeric (hydr)oxycations were used as intercalating agents for clay minerals by Sawhney (1968). However, the article of the Brindley and Sempels (1977) is considered as the first report of the preparation of an inorganic pillared clay. The procedure consisted of exchanging the Na^+ ions in the smectite for oligomeric (hydr)oxy aluminum cations, which were converted to aluminum oxides by heating. The aluminum oxide species acted as pillars in the space between the smectite layers, keeping them permanently apart. The final pillar structure presented basal spacing between 17 to 18 Å and thermal stability up to 500◦C. Using similar methods, Lahav et al. (1978) independently prepared analogous pillared clays containing aluminum hydroxy ions.

In the 1960s, interest in pillared clays shifted to synthetic zeolites (Breck 1980, 391-422), because they have better thermal, hydrothermal activity, selectivity, and stability for oil cracking reactions. However, with the global crisis caused by the rise in oil prices that began in 1973, the interest in potential cracking materials with pores larger than those of zeolites (6 to 9 Å) used as catalysts in fluid cracking catalysts (FCC) intensified. Pillared clays have a similar structure to zeolites, but with larger pores, and act as selective catalysts in the conversion of large molecules existing in heavy oil fractions (Pergher and Sprung, 2005). The reason is that zeolitic catalysts have a limited micropore size and are quickly poisoned by metals in the heavy oil fractions.

In addition, by varying the size of the pillar or the spacing between the pillars, or both, pore size can be adjusted to suit the specific application (Gil, Korili, and Vicente 2008). Thus, pillared clays offer new possibilities for catalysis of larger molecules, such as those found in crude oil residues.

Definition of Pillared Clays and Pillaring Concept

Pillared clays are a class of modified clays, synthesized from the exchange of the interlayer cations of the original clay with oligomers from hydrated metal cations, organic or inorganic complexes.. After heating, this system goes through dehydration/ dehydroxylation processes, resulting in stable pillars composed of rigid structural metal oxides that keep the clay layers apart (Ding et al. 2001).

Clay pillaring is a process that converts lamellar clay minerals into highly porous structures. According to the definition of the International Union of Pure and Applied Chemistry (IUPAC), in a technical report presented by Schoonheydt et al. (1999), pillaring is defined as the process in which a lamellar compound is transformed into a thermally stable micro and/or mesoporous material, without destruction of the lamellar structure during the process. Therefore, the term pillaring is associated with the formation and preparation of catalytically active micro and or mesoporous materials called pillared interlayered clays (PILC) (Vicente, Gil, and Bergaya 2013, 524). The pillaring procedure begins with the addition of water to the smectite, where the interlayer cations are hydrated, causing swelling between the layers. This is followed by the interchange of the interlayered cations by voluminous cationic species such as inorganic metal cation complexes, polyoxo or polyhydroxy cations, organic cations or other species (Vicente, Gil, and Bergaya 2013, 533-534; Mishra 2010, 100-102). After intercalation, drying is performed, followed by calcination of the expanded clay, where dehydration and dehydroxylation of the intercalating species occurs, forming a rigid oxide structure called a pillar. The connection between pillars and lamellae is stable even when the material is heated to high temperatures. The acid character of the pillars may change depending on the functional groups formed on the outside of the pillars and the pillar-lamella bond, which also produces acidic sites (Guerra et al. 2006; Carvalho 2002). The pillar structures distributed between the smectite layers form a microporous system of two-dimensional channels comparable to those of zeolites (Kloprogge 1998). These zeolite-like materials with high thermal stability and high surface areas are called pillared interlayered clays (PILC). The micropore structure is tailored by the nature of the host material and of the pillaring species and may have pore sizes larger than those of zeolites (Mishra 2010, 99-100).

According to Pergher Fornes and Corna (1999), for pillaring to occur, three criteria must be met: 1) the intercalating species need to be large enough to promote spacing between the layers (basal spacing increase) and prevent collapse between them during calcination; 2) the distance between the pillars must be sufficient to allow the molecules to enter, both for their catalytic action and for adsorption to occur, so that the entry of molecules depends on both interlamellar distance (pillar size) and interpillar distance (pillar density); and 3) the layers must be sufficiently rigid that the desired distance between the pillars is maintained, since otherwise the flexibility of the layers can close the spaces between the pillars.

The pillaring step can be done by internal synthesis or external synthesis. In internal synthesis, the pillaring agent is prepared between the layers, simulating a natural process. In external synthesis, the pillaring agent is prepared separately, followed by the cation exchange process with the interlayer cations in the clay. The method employing external synthesis is the most appropriate because it is more versatile and allows the preparation of the pillaring agent in a pure system under more controlled conditions (Pergher, Fornes and Corna 1999). The last step of pillaring is calcination. During calcination, three different situations can occur: (i) pillaring; (ii) polyhydroxycation causing degradation to the hydroxide layer (14.0 Å, pseudo chlorite); or (iii) polyhydroxycation that is temporary (no pillaring occurs). During pillaring, dehydration occurs at 300°C, dehydroxylation at 300-400°C and clay degradation at temperatures higher than 800-900°C. The stability of the pillared clay as a function of temperature will depend on the nature of the clay and the composition of the pillar. For example, there is increased stability when using a mixture of polyhydroxycation or doping small amounts of some other element (Vaccari 1999).

The properties of pillared clay can be changed according to the choice of different parameters, such as nature of pillaring agent (the metal used for pillaring), type of clay and particle size, pillaring procedure, heat treatment etc. (Vaccari 1999). This type of material is attractive because it can have high surface area, a broad spectrum of properties (structural, chemical, catalytic, etc.) and controlled internal structure, with reactive sites and/or species chosen that have desired applications or structures produced for specific physical or chemical processes.

Properties of Pillared Clays

The properties of pillared clays depend on the preparation conditions of the intercalating compounds, the intercalation method, washing, drying and the clay used.

Acidity

Pillared clays have both Lewis (electron-pair receptor) and Brønsted (proton donor) acidity on their surface. Brønsted acidity is generally related to OH^- sylanol groups originating from the layers of the clay structure. Lewis acidity is attributed to metal oxide pillars (Ming-Yuan, Zhonghui and Enze, 1988). The quantity and strength of acid sites are related to the types of clays and metal oxide pillars (Ding et al. 2001).

The sheets of non-pillared clays have only a weak acid character. Acid sites on the sheets are found at exchangeable cation sites and on the ends of the sheets or on defects in crystal structure. In the latter case, structural hydroxyls are exposed to the surface, producing Brønsted acidity (Luna and Schuchardt 1999). The process of pillaring leads to a significant increase in clay acidity. This increase in the acid character of clays is attributed to two factors: (a) growth in the specific area relative to non-pillared clay; and (b) the emergence of new acid sites in the introduced molecular pillars and the binding sites between the pillars and the sheets (Bagshaw, Cooney 1993). The type of hydroxycation used in cation exchange also influences with this acidity, in the order: Ti> Zr> Al> Fe and Ni (Ming-Yuan, Zhonghui and Enze 1988). In addition, the greater the number of pillars, the greater the number of acid sites (Pergher, Fornes and Corna 1999). The temperature of heat treatment also has a strong influence on the acidity of pillared clays. With increasing temperature (300-500°C), Brønsted acidity decreases due to dehydroxylation, thus increasing Lewis acidity. Therefore, when calcination occurs at high temperatures (>300°C), the predominant acidity is that of Lewis (Pergher, Fornes and Corna 1999). Acid activation of clays prior to pillaring is another viable way to improve acidity, particularly Brønsted acidity (Mokaya and Jones 1995). In addition, increased acidity is accompanied by increased pore volume and mean pore diameter (Mokaya et al. 1993; Falaras et al. 2000).

To investigate surface acidity in pillared clay samples, the most commonly used methods are the study of adsorbed ammonia or pyridine by infrared spectroscopy (IR) and programmed temperature desorption (TPD) of adsorbed ammonia (Zubkov et al. 1994). By choosing adsorbed substrates with different basicities, it is possible to determine the number and strength of acid sites. The type, amount and strength of the surface acidity of pillared clays are critical because they are directly related to catalytic activities and determine which type of reaction can be catalyzed.

Porous Structure

In a pillared clay, a micropore is defined by the distance between the layers of the clay mineral, and the lateral distance, which is the space between the metal oxide pillars. These distances can be controlled in order to vary pore size by adjusting various parameters involved in the synthesis of pillared clays (Gil, Korili, and Vicente 2008). Pore size depends on synthesis conditions, such as the type of clay used (Occelli, Senders, and Lynch 1997), the cation exchange capacity (CEC) of clays (Cheng and Yang 1997), the type of metal oxide pillars (Gil, Massinon and Grange 1995) and the heat treatment temperature (Gil and Montes 1994).

Pillared clays have large surface area, high pore volume and pore size ranging from micro (<20 Å) to mesoporous (20 to 500 Å). The pore size of pillared clays is larger than that of zeolites commonly used for oil cracking (6 to 9 Å). Because of this, their catalytic action is facilitated for cracking larger molecules such as heavy hydrocarbons, as they cannot enter the microporous system of zeolites. The selective effect caused by pore size facilitates the diffusion of reagents and other products in the pillared clays. Some authors report that compared to single metal oxide pillars, intercalating clay layers with mixed oxide pillars is a more efficient method for increasing basal spacing (Ding et al. 2001). The introduction of surfactants during intercalation and the use of certain clay drying methods such as lyophilization and supercritical drying may lead to an increase in pore size (Ding et al. 2001). The distribution of the pillars in the pillared clays generates a porous structure similar to that of zeolites, but with larger pores, increasing the access of larger reactive molecules to the interlamellar catalytic regions (Ding et al. 2001). Interpillar distance has more influence on pore size distribution than interlamellar distance. This distance depends on the number of pillars introduced into the clay mineral, which mainly depends on the cation exchange capacity, in addition to the ionic charge of intercalating species and their size (Gil, Korili, and Vicente 2008).

Thermal Stability

The thermal stability of the pillared clays depends on the type of clay, the type of metal oxide pillars, the stability of the pillar-sheet connection and the density of the pillars in the interlayer region (Pergher, Fornes and Corna 1999; Ding et al. 2001). Thermal and hydrothermal stability are important mainly under high temperature conditions found in oil cracking reactions and for maintaining high catalytic activity.

The natural stability of montmorillonite, which is part of the smectite group, ranges from 500°C to 700°C. In general, the porous structure of pillared clays can be obtained up to 400-500°C (Gil and Montes 1994), but some are stable at higher temperatures. An additional increase in heating temperature can cause the collapse of the clay layers due to sintering of the pillars and dehydroxylation of clay mineral sheets (Zhu, Ma and Lu 1999). The stability of the pillar-sheet connection depends on the preparation characteristics of the pillars and the characteristics of the clay itself. The intercalation of clay layers with mixed oxide pillars is also reported as an efficient way to improve the thermal stability of pillar clays, mainly using aluminum pillars mixed with other metals (Vicente, Gil, and Bergaya 2013, 532-533). The increase in the density of the pillars in the interlayer region is associated with increased thermal stability of pillared clays. The higher the pillar density, the better will be the pillar distribution in the interlayer space, which increases the thermal stability. This occurs through techniques that improve the diffusion of macrocations in the interlamellar space (Tang et al. 1995; Pergher, Fornes and Corna 1999; Ding et al. 2001; Gil et al. 2008).

Because of this series of properties, pillared clays have attracted interest for both catalysis (Emam 2013; Kurian and Kavitha 2016; Baloyi, Ntho and Moma 2018) and adsorption applications (Uddin 2017).

Pillared Clays with Metals and Transition Metals

A huge variety of polynuclear cations can be used in the preparation of pillared clays. In particular, transition metal pillared clays are interesting because the pillar itself is catalytically active due to the presence of redox and acid-base characteristics (Mishra 2010, 100). Most of the works found in the literature deal with the introduction of polyoxocations, which have good thermal stability, as well as higher specific area in relation to natural clay.

The most used metals are aluminum, zirconium, iron, chromium and titanium. New types of pillars can also be designed according to the desired application for the material.

Intense research on pillared interlayered clays in recent decades includes several review articles and books reporting synthesis procedures and the characterization and applications of these solids (Gil, Gandía and Vicente, 2000; A Gil, Korili and Vicente 2008; Gil et al. 2010; Ding et al. 2001; Serwicka and Bahranowski 2004; Centi and Perathoner 2008; Kloprogge, Duong and Frost 2005; Bergaya, Aouad and Mandalia 2006; Vicente, Gil, and Bergaya 2013, 523-557; Kausar et al. 2018; Adeyemo, Adeoye, Bello, 2017; Herney-Ramirez, Vicente, Madeira, 2010).

This section will cover the pillared clays of aluminum, iron and chrome.

Aluminum Pillared Clays (Al-PILC)

The hydrolysis behavior of metal cations, especially aluminum, is well known (Baes and Mesmer 1976, 1-489). Because of this, aluminum polyhydroxycations are the most discussed in the literature. Generally, pillaring agents are prepared from their chlorides, and then hydrolyzed by the addition of sodium hydroxide at an [OH]/[Al] molar ratio between 1.0 and 2.5 (Kloprogge 1998; Pergher, Fornes and Corna 1999).

The composition of aluminum pillar solutions strongly depends on the preparation conditions. The degree of hydrolysis or basicity (OH/Al ratio) is an important factor controlling the pH of the solution and therefore the nature of the aluminum species present. Species in solution can be formed by different monomers, but it is assumed that the species with the highest Al is the Keggin ion $[Al_{13}O_4(OH)_{24}(H_2O)_{12}]^{7+}$. This ion is formed by an AlO_4 tetrahedron surrounded by 12 aluminum hydroxide octahedra and has five isomers (Casey 2006). The size of this chemical species ranges from 0.8 to 1 nm (Pergher, Fornes and Corna 1999). Its geometric structure was deduced from X-ray diffraction patterns by Keggin (1934), whose name was given to this structure. Recently, Zu et al. (2017) developed a convenient method for preparing pillared clays containing Keggin-Al_{30} ions from montmorillonite (Mt) and a hydrolyzed Al (III) chloride solution. In a later work, Zu et al. (2018) reported that montmorillonite pillared with Keggin-Al_{30} ions is more thermally stable than that pillared with Keggin-Al_{13} ions. The meso and microporous structure of the Keggin-Al_{30} ion pillared clay remains stable up to 800°C.

The relationship between the basicity and the aluminum species formed in the pillar solution showed that at low OH/Al molar ratios, monomer formation occurs. With increasing basicity, there is a decrease in monomeric species (Al_{mono}) and an increase in Al_{13} species, which are the metallic oligomers of interest for the pillaring process. At molar ratios greater than 2.5, there is a decrease in Al_{13} species and the appearance of high molecular weight polymeric species (Al_{poli}) (Furrer, Ludwig and Schindler 1999; Sterte 1988).

Guerra et al. (2006) observed that the variation of aluminum concentration in the preparation of the pillaring agent produces pillared clays with the following properties:

- different basal spacing, which increases with increasing aluminum concentration (natural clay is the smallest),
- higher percentage of aluminum and lower percentage of Ca, Na and K in the clay composition;
- different number of pillars;
- higher specific area, total pore volume and mean pore diameter with increasing of aluminum;
- higher thermal stability with increasing aluminum concentration. The presence of Keggin ions (Al_{13}) was detected both in prepared pillar clays and in pillaring agents with different aluminum concentrations.

The temperature variation in the preparation of the pillaring agent influences the properties of the pillared clay. Guerra et al. (2005) showed that with increasing temperature in the preparation of Keggin ions for pillaring of a montronite (a dioctahedral smectite, with CEC of 58 meq/100 g of clay), there was a reduction in the interlamellar distance, number of pillars, specific area and total pore volume. Thus, the concentration of Al (III) in Al_{13} predominates over monomeric and polymeric species at temperatures lower than 50°C (Pergher, Fornes and Corna 1999).

Aging time of the pillaring agent and the concentration of the clay suspension also influence the properties of pillared clay. Pergher and Sprung (2005) performed pillaring of a montmorillonite with aluminum polyhydroxycations aged for 1 h, 6 and 15 days. In addition, they added the pillaring agent to the dry clay, suspended in 1 g clay/15 mL water and suspended in 1 g clay/100 mL water. The incorporation of aluminum was much higher in the clay in which the pillaring agent aged for 1 h and the clay suspension with a concentration of 1 g/100 mL of water were used. However, the basal spacing was smaller, indicating that the incorporated oligomers were smaller. The pillared clay with pillaring agent aged for 6 days presented the largest basal spacing. Moreover, the dilution of the clay suspension generated clays with greater basal spacing and higher incorporation of aluminum.

Microwave irradiation during the preparation of the pillaring solution and in the intercalation step significantly decreases the formation time of polyoxocations, as well as the amount of water required. The energy needed for microwave processing is also lower, Irradiation provides the necessary dispersion of cations in a more concentrated solution. During intercalation, it favors the rapid spread of species for exchange (Fetter and Bosch 2010, 2; González et al. 2017).

The intercalation/pillaring of clay minerals with solutions containing two cations has been investigated since the 1980s. One of the pillar cations is generally aluminum. The second inorganic cation is then added with varying molar fractions to improve the thermal, adsorbent and catalytic properties of the pillared products. Incorporation of elements such as Ga, as well as transition metals such as Cr, Fe, Cu, Mo, Zr, Ru and lanthanides in pillar solutions with aluminum has been reported (Vicente, Gil, and Bergaya 2013,532-533; Ding, Zuo and Qi 2015).

Aluminum pillared clays have applications as catalysts in various types of reactions, such as CO oxidation (Tomul and Balci 2009; Li, Li, and Suzuki 2013), dehydration (Fatimah and Rapei 2015; Marosz, Kowalczyk and Chmielarz, 2019), isomerization (Peng Quan et al. 2013), and alkylation (Faghihian and Mohammadi 2014), in addition to having adsorption properties for heavy metal removal from aqueous solutions (Ma et al. 2015; Mnasri-Ghnimi and Frini-Srasra 2019).

Iron Pillared Clays (Fe-PILC)

The hydrolysis behavior of Fe^{3+} ions is similar to that of Al^{3+} and Cr^{3+}, forming polymeric cations (Pergher, Fornes and Corna 1999). Studies suggest that a cationic species structurally analogous to the Keggin Al_{13} ion, called Fe_{13}= $[Fe_{13}O_4(OH)_{24}(HO)_{12}]^{7+}$, is formed in the basic hydrolysis of aqueous iron (III) solutions (Bradley and Kydd 1993). Fe_{13} ions are extremely labile and decompose rapidly, but the addition of iron (II) stabilizes Fe_{13} ions. The spacing between layers of smectite interspersed with Fe_{13} ion is 9.7 Å. Iron pillared clays are inexpensive to prepare, have acidic properties and pillars that can be catalytically active (Kloprogge, 1998).

Tzou (1983) described the preparation of Fe-PILC from solutions of $FeCl_3$, $Fe(NO_3)_3$, $Fe(ClO_4)_3$ and $Fe_2(SO_4)_3$ with different OH/Fe ratios, aging times and temperatures. In the case of Cl^- and NO_3^- salts, increasing the [OH]/[Fe] ratio from 0.0 to 1.0 resulted in an increase in layer spacing from 3 to 14 Å. Another slow increase was observed when the [OH]/[Fe] ratio was increased to 2.5. Specific surface area was found to increase from 244 to 343 $m^{2/}g$ after calcination at 550°C.

Yurdakoç et al. (2008) synthesized clays with iron and chromium using $CrCl_3$, $FeCl_3$, $Cr(acetylacetonate)_3$ and $Fe(acetylacetonate)_3$ as pillar precursors. The use of metal halide as a precursor of the pillaring and [OH]/[Fe] 2.0 ratio has been found to be more effective compared to acetylacetonate complexes. Thus, pillared clays with larger specific surface area, pore volume and mesopore size around 4 nm were obtained. They also observed that in pillared clays with calcined iron and chromium at a temperature of 200 ºC, Lewis acidity was much higher than Brønsted acidity.

A new meso–microporous delaminated structure containing pillared fragments in an iron pillared clay structure was obtained. The specific surface area and the highest porosity of iron pirated delaminated clays were 215.7 m^2/g and 0.29 cm^3/g, respectively. Mesopores in the delaminated structure were the major contributor to total surface area and porosity, and most of them were preserved after calcination at 500°C (Yuan et al. 2006). Microwave treatment has been widely used in material science in recent years, making the preparation of pillared clays faster than traditional methods (Barakan and Aghazadeh 2019).

Iron pillared clays have applications as catalysts in various types of reactions, such as selective NO_x catalytic reduction (Dong et al. 2018). heavy gas oil cracking (Faillace et al.

2017), heavy oil catalytic aquathermolysis processes (Starshinova et al. 2019), Fenton and photo-Fenton processes (Herney-Ramirez et al. 2010; Catrinescu et al. 2012), Friedel-Crafts alkylations (Manju 2005), catalytic wet air oxidation (CWAO) (Sassi et al. 2018), catalytic wet peroxide oxidation (CWPO) (Bankovic et al. 2009) and as adsorbents (Pires and Pinto 2010; Pandey and Saini 2018).

Chromium Pillared Clays (Cr-PILC)

Chromium is an interesting transition metal for catalysis applications, mainly due to the variety of its oxidation states. The basis for chromium activity in a broad spectrum of reactions is the variability of oxidation states, coordination environments and degree of polymerization of chromium species (Weckhuysen, Wachs and Schoonheydt 1996). Basic hydrolysis of chromium salts leads to the formation of different types of oligomeric polyoxocations, such as $[Cr_2(OH)_2(H_2O)_8]^{4+}$ (dimers), $[Cr_3(OH)_4(H_2O)_9]^{5+}$ (trimers), $[Cr_4(OH)_6(H_2O)_{11}]^{6+}$ (open tetramers) $[Cr_4(OH)_5O(H_2O)_{10}]^{5+}$ (closed tetramers), or even hexamers (Carr 1985; Toranzo et al. 1997; Mata et al. 2007). The presence of these compounds is a prerequisite for the preparation of wider spaced chromium pillared clays (Mata et al. 2007).

The physicochemical characteristics of chromium pillared clays are affected by several factors, such as the preparation method, type of clay used, strength of the base used, stoichiometric ratio of the base and chromium ions [OH]/[Cr], the chromium ion stoichiometric coefficient for clay quantity in grams (mmol Cr/clay in g), temperature and aging time (Pinnavaia et al. 1985; Tzou and Pinnavaia 1988; Volzone et al. 1993, Kirksi et al. 1997; Akçay 2004).

Chromium pillared clays are mainly prepared using oligomeric species derived from chromium hydrolysis as pillaring agents due to their stability (Tzou and Pinnavaia 1988). However, other pillaring agents such as Cr (III) acetate complexes (Mishra and Parida 1998a and 1998b) and Cr (III) oxyformat (Yun, Hwang and Choe 2000) have also been used.

The size of the oligomers and the extent of formation in the pillar solution depend on aging time and temperature. Hydrolysis of chromium ions is a slow process at room temperature. However, at high temperatures (60-90 ^oC), the hydrolysis process is completed within one or two days (Tzou and Pinnavaia 1988; Drljaca et al. 1997; Sychev, De Beer and Van Santen 1997; Arafaui, Frini-Srasra and Srasra 2009). Tzou and Pinnavaia (1988) observed that in the intercalation with hydrolyzed oligomers, an increase in temperature of the pillaring agent, increase of the [OH]/[Cr] ratio and longer hydrolysis time (aging) favored the formation of pillared clays with greater basal spacing. The values obtained ranged from 18 to 27 Å. Chromium hydroxide precipitation in the pillar solution can be prevented by using a weaker base such as Na_2CO_3 or by decreasing the NaOH concentration (Tzou and Pinnavaia, 1988). The [OH]/[Cr] ratio regulates the hydrolysis process and generation of polycationic species (Brindley and Yamanaka 1979; Carr 1985).

High layer spaced pillared clays (23-27.6 Å) are obtained when used in the hydrolysis process with [OH]/[Cr] ratio in the range 1.5-2.5 (Tzou and Pinnavaia, 1988; Arafoui, Frini- Srasra and Srasra 2009).

The effect of the stoichiometric coefficient (mmol Cr/clay in g) on the textural properties of chromium pillared clay has also been studied (Volzone 1995; Gyftopoulou et al. 2005). Chromium pillared smectites were synthesized using Cr/clay ratios in the range of 0.5 to 20 mmol/g. As the Cr/clay ratio increased, the thermal stability and basal spacing of the chromium pillared clays increased. A value of 20.7 Å was obtained for the molar ratio 20 mmolCr/g clay (Volzone 1995). Tzou and Pinnavia (1988) synthesized a Cr-PILC with basal spacing of 27.6 Å using the Cr/clay ratio of 50 mmoL/g from a thermally aged pillar solution at 95 °C.

Clays with different composition, structure and charge density were used for the preparation of chromium-pillared clays (Volzone 2001; Gyftopoulou et al. 2005; Yurdakoc et al. 2008; Tomul and Balci, 2009). The incorporation of chromium depends significantly on the charge density of the layer, which in turn affects the thermal stability and textural properties of the chromium pillared clay (Tomul and Balci 2009).

Heating conditions (temperature and atmosphere) strongly influence the structural properties of these solids. The instability of the chromium pillars, with the consequent decrease in the basal spacing, was observed by some authors above 200°C (Drljaca et al. 1992) and above 420°C (Volzone, 1995) when calcined in air. Tzou and Pinnavaia (1988) also observed that the calcination of polyhydroxy cations for the formation of chromium pillared clay should be done in the absence of air (vacuum or inert gas) to avoid oxidation of chromium as volatile oxides. Chromium (III) was later found to oxidize chromium (VI) in the presence of air (Roulia 2005). The structure of the solid is more stable when calcination is performed in an inert atmosphere instead of air (Volzone and Cesio, 2003). Calcination under ammonia atmosphere also formed stable pillared clays above 500°C (Yun, Hwang and Choe 2000). Mishra and Parida (1998a), using chromium (III) acetate, synthesized thermally stable chromium pillared clay with pillar height > 17.5 Å calcined at 500°C Roulia (2005).

Chromium pillared clays were synthesized by adding NaOH to the solution and Cr (III) salt until the pH approached 13, then acidifying the solution with $HClO_4$ to pH 2.2, followed by reflux for 1.5 h. Addition of acid and base with further reflux during preparation reduced the aging period of the oligomer. Pillared clays produced after calcination at 475°C under nitrogen atmosphere showed higher basal spacing and increased surface area compared to the non-reflux method. Bodman et al. (2003) prepared chromium pillared clays from $Cr(NO_3)_3$ and Na_2CO_3, calcined at 500 °C in a nitrogen atmosphere and with a basal spacing of 24.2 Å, for the hydrocracking of heavy petroleum distillation residues.

The results confirmed the high efficiency and stability of the catalyst at 450 °C in cracking hydrocarbons to form compounds with lower molecular masses, especially at longer reaction times. Chromium pillared clays were also used in coal hydrocracking, presenting high conversion percentages in liquid products (Bodman et al. 2002; Gyftopoulou et al. 2005).

The thermal and catalytic properties of pillared clays can be enhanced by the use of mixed metal pillaring agents. While one of the pillar component metals provides greater thermal stability, catalytic activity sites are attributed to the second component. In certain cases, it has been observed that materials with mixed metal oxide exhibit improvements in surface area and acidity compared to single-pillar materials. Using iron (III) and chromium (III) acetate complexes, pillared clays with both chromium and iron with thermal stability up to 500°C were synthesized. Fe-Cr pillared clays presented basal spacing of 16.8 Å after calcination at 500°C. Increasing the selectivity of the cumene cracking reaction to generate benzene was dependent on decreasing Fe/Cr ratio and lower reaction temperature. This was due to the higher number of Brønsted acid sites at lower reaction temperatures (Mishra and Parida 1998b).

Studies have shown that the thermal stability of chromium-containing pillared clays is enhanced by the addition of aluminum to the pillars (Gil, Gandía and Vicente 2000; Ding et al. 2001). Toranzo et al. (1997) prepared mixed Al-Cr pillars by co- basic hydrolysis of their salts. Tomul and Balci (2009) prepared Cr-Al pillared clays by mixing different amounts of solutions containing each of the hydroxylated species. Mata et al. (2007) reported efficiency of Al-Cr pillared clay catalyst in propene oxidation using oligomers obtained by co-hydrolysis of the corresponding metal chlorides. De Léon et al. (2014) reported high propene yields in the propane oxidative dehydrogenation reaction at 450 ° C using Cr-Al pillared clays.

Tomul (2016) recently investigated the use of ultrasound treatment during the preparation of Fe-Cr pillared solutions. Fe-Cr pillared solutions were aged in less than 20 minutes compared to the hours or days required with the conventional method. The characteristics and catalytic behavior of solid pillars resulted in the catalytic degradation of phenol with hydrogen peroxide, confirming the efficiency of ultrasound treatment. Baloyi, Ntho and Moma (2018) used ultrasound treatment both in the aging stage and in the intercalation of pillaring agents to synthesize pillars with chromium and aluminum pillars. Results showed high efficiency in catalytic oxidation of phenol, superior in relation to the use of separate chromium and aluminum pillared clays.

Chromium pillared clays have applications as catalysts in various types of reactions, such as dehydrogenation (Pinnavaia Tzou and Landau 1985; De León 2014), cracking (Mishra and Parida 1998a), isomerization (Chareonsiriwat and Chavasiri 2017), oxidation reactions (Mata et al. 2007; Tomul and Balci 2009; Baloyi, Ntho and Moma 2018), and combustion (LI et al. 2018) in addition to their adsorption properties (Georgescu et al. 2018).

Applications of Clays and Pillared Clays in Waste Pyrolysis

Petroleum Wastes

Petroleum has been the subject an intense search since the end of the nineteenth century (Thomas, 2004, 1-2), since it is a fundamental raw material for the social and economic development of nations. At the same time, oil exploration, production, refining and distribution activities generate huge amounts of waste daily, with different levels of hazard, thus having a relevant impact on the environment (Cordes et al. 2016; Brown et al. 2017). The most common wastes in the oil industry generated in large quantities are mud produced at different stages of the extraction process, sediment from the bottom of oil storage tanks and their derivatives, oil sludges and oil emulsified solids (Xu et al. 2014; Jafarinejad 2017, 19-78).

Oily material is the most abundant waste, and when accompanied by sand or rock dust it is classified as oily solid material, traditionally called 'oil sludge.' In addition to oil components, oil sludge includes a variety of different chemicals, especially heavy organic compounds, water and solid particles of various sizes (Paněk et al. 2014). Oil sludge is a complex mixture typically consisting of oil (30-50%), water (30-50%) and solids (10-12%) with varying composition (Zheng et al., 2012), making it difficult to reuse. In oil sludge, the oily phase contains 40 to 60% saturated hydrocarbons, 25 to 40% aromatics, and 10 to 15% asphaltenes. In aromatics, benzene is often present, partially responsible for the flammability of oil sludge (Valle, Guimarães and Sampaio 2004). Oil sludge can cause major damage to the environment and threats to human health if not properly treated (Elektorowicz and Muslat, 2008).

Heavy gas oil (HGO) is a high viscosity residue from the petroleum refining process, containing paraffins and aromatics. During the refining process, initially by distillation at atmospheric pressure, the high boiling heavy fraction rich in high molecular weight compounds is collected at the base of the fractionation tower. This waste, subjected to distillation at reduced pressure, is then separated into three types of gas oil: light gas oil (atmospheric), heavy gas oil (vacuum gas oil) and residual gas oil (vacuum waste) (Hunt, Philp and Kvenvolden 2002).

Plastic Wastes

Plastics are in high demand due to their widespread use. Plastics are employed in a wide range of applications, including the automotive, packaging, agriculture, food, leisure and construction industries, due to their lightness, high chemical stability and low degradability (Harshal and Lawankar 2013). Plastic consumption is increasing worldwide at a rate of 4% per year due to its durability, lightness and low cost (Miandad et al. 2016). According to the Plastics Europe Market Research Group (PEMRG), worldwide plastics production reached nearly 350 million tonnes in 2017. China is the largest producer of plastics, followed by Europe and the United States, Canada and Mexico, which are NAFTA member countries - North American Free Trade Agreement (Plastic Europe 2018). Up to 2015, global production of plastics (including polymeric resins and fibers) had reached about 7.8 billion tonnes, half of which had been produced in the last 13 years (Geyer, Jambeck and Law 2017). Jambeck et al. (2015) estimated that 250 million tonnes of poorly managed plastic waste could reach the oceans by 2025, with unpredictable effects on the health of marine life, and consequently on human health (De Sá et al. 2018).

The term plastic comes from the Greek verb 'plassein', which means to mold, because plastics are very versatile and easy to mold and shape. Plastics are long repeated chains of linked molecules. Such repeated chains are commonly referred to as polymers. Polymers are defined as large organic molecules consisting of many small molecules (repeated chain units), known as monomers. Therefore, PP-propylene is a polymer with repeated propylene (propene) units (C_3H_6=C_3H_6). A typical polymer may include tens of thousands (or hundreds of thousands) of monomers. When such a chemical formulation is manufactured, certain additives are mixed to give it specific shape and properties. Therefore, a polymer mixed with an additive is a plastic material (Al-Salem 2019a, 7). Additives blended with plastics during their manufacture to improve their properties and performance include fillers, dyes, thermal stabilizers and plasticizers (Andrady and Neal, 2009).

Plastics have often replaced more traditional materials such as wood, metal, glass, leather, paper and rubber because they can be lighter, stronger, corrosion resistant, acid and base resistant and durable, besides their insulating properties. Plastic materials can be divided into two main groups according to their thermal behavior: thermosets and thermoplastics (Noreña et al. 2012, 151). Thermoplastics soften when exposed to heat and can be molded and shaped; This heating process can be repeated several times. These plastics are used to make containers, packaging, trash bags and other non-durable goods (Al-Salem et al. 2009). Some examples are high and low density polyethylene (HDPE, LDPE), polystyrene (PS), polypropylene (PP), polyvinyl chloride (PVC), polyethylene terephthalate (PET) and poly methyl methacrylate (PMMA). In contrast, thermosets solidify irreversibly when heated, as an irreversible network of crosslinked covalent bonds is formed, providing hard, durable, strong and heat resistant products. This is the case of unsaturated polyurethane (UP), unsaturated polyesters and alkyd, phenolic and epoxy resins. For this reason, they are mainly used in automobiles, construction adhesives, furniture, kitchen appliances, paints and coatings. A third group of plastics, similar to rubber, are called elastomers. They are formed by lightly crosslinked polymer chains in a lesser extent than thermosets, giving these materials elastic properties and relatively good strength (Noreña et al. 2012, 151).

The largest groups in the total production of non-fibrous plastics are PE (36%), PP (21%) and PVC (12%), followed by PET, PU and PS (<10% each). Polyester, mostly PET, is responsible for 70% of all PP&A (polyester, polyamide and acrylic) fiber production. Together, these seven groups represent 92% of all plastics ever made. Approximately 42% of all non-fibrous plastics are used for packaging, composed predominantly of PE, PP and PET (Geyer, Jambeck, and Law, 2017). About 6% of world oil production is allocated to the plastics processing industry (Al-Salem 2019a, 3).

HDPE is a thermoplastic made of petroleum with chains having low branch content. It is formed by carbon and hydrogen atoms joined together to form a high molecular weight compound. Its melting temperature is approximately 126°C and its density is between 0.941 and 0.965 g/cm^3. The polymer chain may be 500,000-1,000,000 carbon units long. The linearity of the chains explains their high molecular weight and high crystallinity, implying high density. In addition, such features make chain orientation, alignment and packing more efficient. (Kumar, Panda and Singh 2011). The difference between HDPE and LDPE is the degree of branching. The mechanical properties differ dramatically between them. Molecular weight, molecular weight distribution and number of branches determine many of the mechanical and chemical properties of the final product. High density polyethylene is more crystalline than low density polyethylene because it contains little or no branching. Thus, it has stronger intermolecular forces and higher tensile strength than lower density polyethylene. HDPE is also harder and more opaque and can withstand slightly higher temperatures. HDPE is used in a variety of applications ranging from oil storage containers, detergent bottles, milk bottles, to chemical resistant piping systems, because it is more durable and safer compared to steel or PVC (Kumar, Panda and Singh 2011).

LDPE is a branched chain polyethylene with density between 0.910 and 0.925 g/cm^3. It is a non-completely crystalline solid with crystallinity degree in the range of 50 to 60%. This feature generates various properties, such as opacity, tensile strength, tear strength, stiffness, chemical resistance and low temperature flexibility. The most common types of LDPE are low density linear polyethylene (LLDPE) and low density branched polyethylene (BLDPE). They differ in their density, degree of branching and availability of functional groups on the surface. (Kumar and Sangeeta, 2015). LDPE is used for packaging, trays and plastic bags for food and non-food items. It is also used as a protective coating on paper, textiles and other plastics. However, LDPE is slowly degraded after disposal, which pollutes the environment and disturbs the ecosystem (Sen and Raut 2015).

With regard to plastic waste, polyolefins have now become an environmental problem due to their intense use, generating a huge amount of degradation resistant post-consumer waste. These materials, which are mostly non-biodegradable, have long life in the environment, which may exceed 100 years. Incorrect disposal in inappropriate places such as dumps, rivers, slopes, among others, increases the environmental impact of these wastes (Paraschiv et al. 2015). In particular, polyethylene plastic wastes from containers used to store fuel and toxic wastes in the petroleum industry are important sources of contamination and require differentiated treatment and disposal (Joppert Jr., Da Silva and Marques 2015).

Waste Treatment ???

Several technologies have been investigated and applied for the treatment of these plastic wastes. The processes used to treat solid plastic waste (PSW) result in the production of chemicals, fuels and energy. There are four main methods for treating solid plastic waste (Noreña et al. 2012, 153; Al-Salem, 2019b, 31):

- Primary (extrusion) - Involves the reintroduction of clean plastic solid waste into a new extrusion cycle. In this process, the plastic is melted and processed based on its properties in order to produce products of similar material. Reusing plastic is preferable to recycling because it requires less energy and resources.
- Secondary (mechanical recycling) - Involves plastic reprocessing by melting, grinding or granulating. Separation, washing and waste preparation are essential. It is an economical and viable route for the recovery of plastic waste for transformation into new products such as carpet, clothing and bottle fibers.
- Tertiary (chemical recycling) - Refers to techniques used to break down plastic polymers into their basic monomers for the production of petrochemicals and polymers. These chemical recycling products may also be useful as fuels. It is the most sustainable method.
- Quaternary (incineration and energy recovery) - Consists of burning waste to produce energy in the form of heat, steam and electricity. It results in a volume reduction of 90 to 99%, which reduces landfilling. However it produces large quantities of polluting gases.

Among the tertiary recycling methods are pyrolysis (thermal decomposition of polymer chains), gasification (decomposition under oxygen or stream atmosphere to produce synthesis gas) and hydrogenation (hydrogen chain breakage). Recycling and energy recovery rates are still very low. In 2017, only 7.2 and 9.6% by weight of plastic waste generated worldwide was recycled and incinerated, respectively. In addition, about 60% of all plastics ever produced (4.9 billion tonnes) has been discarded and is accumulating in landfills or in the environment. This accumulation can cause irreversible damage to natural ecosystems (Geyer, Jambeck and Law 2017). Incinerating plastics releases hazardous substances such as heavy metals, dioxins and furans, which can cause adverse health effects (Zhou et al. 2018). Among the routes of chemical recycling, pyrolysis is that have presented greatest technological advance (Lopez et al. 2017).

Pyrolysis allows direct fuel production or monomer recovery using simple technologies with less environmental impact (Lopez et al. 2017; Al-Salem, Evangelisti and Lettieri 2014; Al-Salem et al. 2017).

Pyrolysis is an attractive alternative since it can be employed to obtain a wide range of products, including coal, oil/wax and gaseous products (López et al. 2011; Chen et al. 2014; Lopez et al. 2017). This is also an interesting option because it allows the processing of waste containing mixed plastics or even contaminated plastic material, for which mechanical recycling is not feasible (Alston et al. 2011; Adrados et al. 2012; Paraschiv et al. 2015).

The generation of oil sludge is an inevitable problem during operations in the oil industry. Due to its toxicity and adverse environmental effect, oil sludge needs effective treatment. A variety of technologies are being developed for oil recovery and sludge disposal, and some of them have been applied on an industrial scale (Hu, Li e Zeng 2013; Tyagi and Lo 2013). Pyrolysis has also proved to be a promising alternative for converting oil sludge into value-added products (Yang et al. 2018; Gong et al. 2018).

Pyrolysis

Thermal Pyrolysis

The decomposition of a compound by the sole action of heat is called pyrolysis. It occurs at high temperatures where macromolecules break down into molecules with low molecular weight. It is usually performed in an inert atmosphere, generating three fractions: a gas fraction composed of volatile hydrocarbons; a liquid fraction composed of condensable hydrocarbons; and a solid fraction composed of coke products and unprocessed inorganic compounds (Undri et al. 2013).

Thermal pyrolysis is an endothermic process where temperatures ranging from 300 to 900 °C are applied. Waste pyrolysis is affected by a number of parameters such as temperature, which influences the conversion rate of the pyrolysis reaction. In addition, it also depends on the type of plastic waste used, chemical composition of the petroleum waste and the catalyst used (Miandad et al. 2016, Al-Salem et al. 2017).

In recent decades, pyrolysis has stood out for treating waste to obtain valuable products with low environmental impact (Hu et al. 2013; Gong et al. 2018). In the pyrolysis of polyolefins and elastomers, the degradation of polymeric chains occurs, converting them into petrochemical or fuel feedstocks when subjected to high temperatures (Sharuddin et al. 2016; Sogancioglu et al. 2017). According to Bhaskar et al. (2003), pyrolysis of high hydrocarbon plastic waste is greatly favored due to the high conversion rates to liquid products (pyrolytic oil). The compounds present in pyrolytic oil can be used as fuel or as an input in refineries. Pyrolysis is also reportedly an efficient technique for the treatment of petroleum refining waste (Demirbas et al. 2016).

Thermal Catalytic Pyrolysis

Catalytic pyrolysis is conducted under heating in the presence of specific catalysts. Compared to thermal pyrolysis, this process can both reduce the reaction temperature and control the formation of products. Thus, using catalysts it is possible to consume less energy, have a shorter reaction time and obtain good selectivity of products (Miandad et al. 2016).

The most common type of catalyst used is heterogeneous, since the fluid product mixture can easily be separated from the solid catalyst. Thus, the heterogeneous catalyst is economically preferable because in many cases it can also be reused (Sharuddin et al. 2016). Catalyst characteristics, such as BET specific surface area, pore size, pore volume and acidity, are the main factors affecting the activity of any catalyst in the pyrolysis process. Some of the catalysts used are zeolites (Miandad et al. 2017; Milato, França and Marques 2019), FCC (Zang et al. 2015; Ma et al. 2018) and non-zeolitic catalysts such as alumina, silica-alumina (Luo et al. 2000), MCM-41 (Miskolczi, Wu, and Williams 2017) and pillared clays (De Stefanis et al. 2013; Jia et al. 2017; Li et al. 2017).

This process can produce light olefins from a wide variety of wastes, such as post-consumer plastic, formed by polymers like polyethylene (Santos et al. 2019). On the other hand, heavy oils from petroleum refining can produce pyrolytic oil containing high-value light hydrocarbons such as gasoline and diesel fractions (Demirbas et al. 2017). Heavy oil studies have been conducted since the 1960s, but with the oil crisis in the late 1970s, studies on catalytic pyrolysis of heavy oils were particularly conducted in search of alternative fuels (Meng et al. 2005; Demirbas et al. al. 2017).

Co-pyrolysis

When pyrolysis is performed with two or more materials, the process is called co-pyrolysis. The materials to be pyrolyzed together alter the degradation mechanisms and influence the chemical composition of the products.

Co-pyrolysis of plastic waste has been studied with various types of materials, such as coal (Sharma and Ghoshal 2010; Melendi-Espina et al. 2019), biomass (Zhang et al. 2014; Uzoejinwa et al. 2018), waste heavy oil (Siddiqui and Redhwi 2009; Assumpção, Carbonell and Marques 2011; Tan et al. 2014), heavy gas oil (Joppert Jr., Da Silva and Marques 2015; Faillace et al. 2017), waste engine oil (Uçar, Ahmet and Özkan 2016) and shale oil (Bozkurt et al. 2017).

Mixtures of these residues influence the degradation mechanism and chemical composition of the pyrolytic oil. Mixing plastic and fossil waste results in a higher production of pyrolytic oil compared to separately pyrolyzed waste. In the case of heavy oil and polyolefins, the synergistic effect favors increased pyrolytic oil production (Önal et al. 2014). This can happen due to the hydrogen released from the polymeric material, stabilizing the free radicals generated in the cracking of the bonds. In addition, the presence of oily material with a certain fluidity favors solvency and consequently the separation of polymer chains (Bozkurt et al. 2017).

The co-pyrolysis technique has been extensively studied because it is an efficient way to convert waste plastics and other hydrocarbon sources, especially petroleum waste and heavy oils, into fuel with energy potential (Miskolczi and Ateş, 2016; Lin et al. 2016; Schubert et al. 2019). Other benefits of this method include reduced waste volume, chemical recovery and fossil fuel substitution (Onenç et al. 2012; Chen et al. 2014; Jamradloedluk and Lertsatitthanakorn 2014).

Mechanisms of Cracking Reactions

Thermal Cracking

Attempts to elucidate the thermal cracking mechanism of polyethylene date to the late 1940s (Oakes and Richarhds 1949). Thermal cracking reactions follow a mechanism via free radicals, breaking covalent bonds by the action of heat, producing free radicals (Noreña 2012, 162).

Cracking reactions involve breaking only the C-C bond of the hydrocarbon molecules. These reactions are endothermic, that is, they require heat to occur. Cracking occurs through the mechanism of free radicals. They are formed by the separation of a shared electron pair from the structure. Separation of the electron pair can be done by breaking the C-C or C-H bond. The mechanisms proposed for the reactions involving pyrolysis processes are classified as follows (Inguanzo et al. 2002):

Unimolecular reactions: R-R→ R. + R’
Hydrogen abstraction: R. + R’-H → R-H + R’
Termination reactions: R’ + R’. → R-R

These reactions can result in the formation of various products, such as saturated, unsaturated and aromatic hydrocarbons, depending on the raw material applied and the temperature. Authors suggest that pyrolysis reactions of organic materials follow the radical mechanism. However, series reactions occur, and possibly in large numbers, in parallel and sequentially. These reactions occur not only in the gas phase but also in the liquid and solid phases (Poutsma 2000). In this broad spectrum of possible reactions that occur in pyrolysis systems, the following can be highlighted:

1. radical formation reactions (homolysis);
2. radical interconversion reactions, in which one radical is consumed and another is formed (molecular rearrangement, β bond breakdown and hydrogen subtraction); and

3. radical consumption reactions (rapid gas phase radical combination that only depends on the reaction chemistry, where liquid phase radical combination is slower since it depends on the temperature and viscosity of the reaction medium).

Aguado and Serrano (1999, 73-124) mentioned that thermal decomposition of polymers occurs through a radical mechanism, associated with three processes:

- random polymer breakdown: occurs at any point in the molecular structure, leading to the formation of primary products, which may also undergo cracking reactions, also random;
- molecular chain breakdown: occurs at the ends and may generate simpler chain primary products or new polymer chains, provided that depolymerization occurs when the molecule formed is an initiating monomer; and
- separation of molecular structures: when these structures tend to form lower molecular weight molecules.

These authors also mention that the thermal degradation of plastics in inert atmosphere occurs preferentially through radical mechanism. Depolymerization of these residues results from the cleavage at the end of the molecule or through its random breakdown. In most situations, there is random splitting in the molecular structure leading to the formation of complex mixtures. Thermal degradation of polymers tends to initially occur at the weakest points of the molecular structure (small double branches, oxygenated groups), promoting the formation of primary radicals, which can trigger polymerization reactions. Some unwanted elements or components present in the waste, such as chlorine, sulfur and trace elements, may pose a problem in polymer processing. Hydrocracking, hydrogenation refining treatment and high pressure hydrogenation are modern technologies used to improve this process in order to respond to market changes and reduce the attainment of heavier fuel fractions (Rana et al. (2007).

Catalytic Cracking

The reaction mechanisms of catalytic cracking of polymer chains follow similar paths as catalytic cracking of hydrocarbons in oil refineries. These mechanisms have been studied for several years. The catalytic cracking process occurs at temperatures high enough for parallel thermal cracking reactions to occur. Catalytic cracking reactions occur either by protonation of carbon atoms in the polymer chain (protons from Brönsted acid sites), or by abstraction of hydride ions from the polymer chain by Lewis acid sites. The resulting ions can be stabilized by reactions of β fission, isomerization or hydrogen transfer (Noreña et al. 2012, 162). The reactions are more or less favored depending on the temperature (Kumar, Panda and Singh 2011

The surface area and porous structure of the catalyst also play an important role. The breaking of the polymer chains begins on the outer surface of the catalyst. Small enough fragments can enter the pores, where additional cracking reactions occur, resulting in small gas molecules (Noreña 2012, 162). Unlike thermal cracking, a particular catalyst can act selectively to obtain specific products.

In the presence of catalysts, heterogeneous catalytic cracking occurs on the surface of the molten polymer and solid catalysts. The main reaction steps are catalyst surface diffusion, catalyst adsorption, chemical reaction, catalyst desorption, and liquid phase diffusion. The dominant rate is controlled by elementary reactions of polymer bonds to the catalyst active site. But catalyst selectivity regarding the raw material is important. Selectivity is affected by the molecular size and shape of the raw materials, intermediates and products formed (Takuma et al. 2000).

ABOUT OUR RESEARCH

Pyrolysis Using Different Wastes and Catalysts

The research group of Monica R.C. Marques (UERJ/CI/LABTAM) has been carrying out consolidated work in the area of co-pyrolysis of waste using heavy gas oil (Carneiro 2011; Assumpção et al. 2011; Joppert Jr., Da Silva and Marques 2015; Faillace et al. 2017) and oil sludge (Silva et al. 2017; Milato, França and Marques 2019). In the co-pyrolysis of offshore heavy oil sludge and heavy gas oil with polyolefins, reported in the above works, we obtained pyrolytic liquid rich in light hydrocarbon fractions, besides presenting medium calorific power, suggesting similarity with fuel oil. Thus, with a single process it is possible to treat two wastes of difficult degradation.

Carneiro (2011) studied the influence of polyethylene on co-pyrolysis with heavy gas oil from the Campos Basin, Brazil. Heavy gas oil was mixed with post-consumption LDPE waste and heated to 550°C. The results showed that with the increase in the amount of post-consumer polyethylene added to heavy gas oil, there was an increase in the hydrocarbon fraction of the diesel oil generated in the pyrolytic liquid by co-pyrolysis. Qualitatively, the fraction of diesel oil was directly proportional to the amount of LDPE added to heavy gas oil. The percentage of diesel oil in the pyrolytic liquid was highest when 50% plastic waste was added. However, above this value, a decrease in the diesel fraction percentage was observed, with wax formation with a characteristic odor. Wax formation was also reported by Kiran, Ekinci and Snape (2000), when polyethylene was pyrolysated. In addition, Shinb and Seo et al. (2002) observed that depending on the waste to be pyrolyzed and the pyrolysis temperature, the pyrolytic liquid may be composed of low molecular weight or long chain hydrocarbons such as waxes. This study also showed that plastic waste can be used from mechanical recycling or with energy treatment for the production of fuel oils.

Assumpção (2011) performed study co-pyrolysis of post-consumption polypropylene with heavy gas oil from the Campos Basin. The results obtained in the temperature range of 400 to 500 °C showed that higher temperature favors the increase of pyrolytic liquid yield in the diesel distillation range. On the other hand, a large increase in the amount of PP caused a decrease in the total yield of the product. The results showed the potential of co-pyrolysis for chemical recycling of post-consumption polypropylene articles.

The production of pyrolytic liquid depends on the relationship between the parameters involved in the pyrolysis process (Miandad et al. 2016). The factorial design method (FDM) was used to obtain a higher yield of the diesel fuel fraction (C_9 to C_{23}) through co-pyrolysis of high density polyethylene (HDPE) and heavy gas oil (HGO) residues. Reaction parameters such as temperature, catalyst amounts and HDPE were optimized using FDM. Keeping the amount of HGO constant, a higher fuel diesel fraction yield of 87% was obtained under two different conditions: (a) temperature = 550 °C without catalyst; and (b) temperature = 500°C with 25% catalyst (0.2 g HDPE). Importantly, no residue was produced under both conditions and the FCC catalyst generated an equal fraction yield using a lower temperature. Thus, catalytic co-pyrolysis proved to be an environmentally friendly approach for recycling waste oil-contaminated plastic packaging waste (HDPE waste) from gas stations (Joppert Jr., Da Silva and Marques 2015).

As mentioned earlier, the literature demonstrates that the use of catalysts can lower the reaction temperature and provide greater selectivity in the distribution of pyrolysis products.

Silva et al. (2017) studied the effect of temperature and the use of an acid catalyst (K-10® clay) on polyethylene (PE) co-pyrolysis with oil sludge on pyrolytic oil production. The addition of K-10 clay at 450°C increased gas production from oil sludge pyrolysis. As an acid catalyst, K-10 clay favors the cracking of oil sludge in low molecular weight compounds such as C_1 to C_5 hydrocarbons. Pyrolytic oil production from pyrolysis of oily sludge and HDPE was higher than production from pyrolysis using oily sludge and LDPE. The solvency of oil sludge causes the decompression of HDPE chains, which favors the breakdown of complex molecules into low molecular weight compounds. The pyrolytic oil from catalytic co-pyrolysis of oil sludge had high content of light paraffinic hydrocarbons, with low olefin and aromatic content. These results demonstrated that catalytic co-pyrolysis can potentially be used to convert the mixture of waste PE and oil sludge into a high diesel fraction oil.

Mesoporous catalysts are currently used in the petroleum and petrochemical industry to convert heavy to light hydrocarbons. These materials, with low acidity and narrow pore size distribution, generate much higher yields of liquid products than microporous zeolites (López et al. 2011a). In this context, pillared clays can also be a good alternative, since they have high thermal stability, high surface area, mesoporosity, acid sites and lower cost than traditional mesoporous catalysts..

Millato et al. (2019) evaluated pyrolysis of oil sludge from the petroleum industry in contact with three mesoporous zeolite catalysts (CBV 720, 760 and 780 at 450°C to obtain oil rich in paraffins. The zeolite with the lowest acidity (CBV-780) produced 56% oil fraction with the highest yield of light hydrocarbons (96%). Compared with the thermal pyrolysis of this waste, the use of mesoporous zeolitic catalysts increased the production of light hydrocarbons and reduced the production of aromatic compounds in the pyrolysis from oil sludge. The products obtained in the presence of mesoporous zeolites showed selectivity for conversion of light hydrocarbons with decreased content of aromatic compounds.

Faillace et al. (2017) evaluated the effect of iron and aluminum pillared clays on the quality of the diesel fraction generated in the pyrolysis of HGO and HDPE wastes. The influence of temperature and selectivity on the formation of C_{10}-C_{23} linear aliphatic hydrocarbons was also evaluated. The results of co-pyrolysis catalyzed by pillared clays were compared with the results using a zeolite as cracking substance. The catalytic pyrolysis of HGO and the HGO/HDPE mixture produced a higher pyrolytic oil yield than that of thermal pyrolysis at 450 ° C. Co-pyrolysis of HGO and HDPE at 450°C using Fe-PILC-Fe-300 as catalyst (fractionated natural clay containing fractional iron calcined at 300 °C) produced the largest amount of light linear hydrocarbons (nC_{10} to nC_{23}). Thus, this catalyst was the most selective for the production of diesel fraction at lower temperature than using commercial zeolite as catalyst. In addition, pillared clay catalysts have the advantage of being less expensive than zeolitic catalysts.

Li et al. (2017) conducted a study of pyrolysis of a polyolefin mixture analyzing the influence on the process of iron, titanium and zirconium and aluminum pillars. Their results showed a high yield of pyrolytic oil with diesel fraction and large amount of H_2 in the gaseous product when using iron pillared clay. In addition, as also noted by Faillace et al. (2017), they observed that pyrolysis oil yields were higher with the use of iron pillared clay, suggesting that secondary over-cracking of primary pyrolytic intermediate species was reduced in the presence of the catalyst.

CONCLUSION

Recycling post-consumer plastic waste for energy recovery through an economic and eco-friendly process is a challenge. Many studies have shown that the pyrolysis process has excellent potential to convert plastic waste into products that can be used as fuels and feedstocks for particular petrochemical processes. The flexibility that pyrolysis provides in terms of product selectivity can be achieved by adjusting parameters appropriately and using catalysts. The use of catalysts is in accordance with the principles of green chemistry, since it allows the use of lower operating temperatures while increasing the selectivity of product distribution, making the process more sustainable. In addition, pillared clay catalysts may be a good alternative for use in these processes since they have good physicochemical characteristics for catalysis and are inexpensive. Chemical recycling helps to minimize the environmental impact caused by improper disposal of post-consumer plastic waste. In addition, it also allows the reuse of a non-renewable natural resource such as petroleum through the use of fuel fractions obtained in this process.

REFERENCES

Adeyemo, Aderonke Ajibola, Idowu Olatunbosun Adeoye, and Olugbenga Solomon Bello. 2017 "Adsorption of dyes using different types of clay: a review."*Applied Water Science* 7(2):543-568. Accessed August 4, 2019. doi:10.1007/s13201-015-0322-y.

Adrados, A., I. De Marco, B. M. Caballero, A. López, M. F. Laresgoiti, and A. Torres. 2012. "Pyrolysis of plastic packaging waste: A comparison of plastic residuals from material recovery facilities with simulated plastic waste." *Waste Management* 32(5):826-832. Accessed September 16, 2019. doi:10.1016/j.wasman.2011.06.016.

Aguado, Jose, and David P. Serrano. 1999. "Thermal processes." In *Feedstock Recycling of Plastic Wastes*." Cambridge, UK: Royal Society of Chemistry.

Akçay, Mehmet. 2004."The catalytic acylation of alcohols with acetic acid by using Lewis acid character pillared clays." *Applied Catalysis A: General* 269(1-2):157-160. doi:10.1016/j.apcata.2004.04.010.

Al-Salem, S. M. 2019a. "Introduction." In *Plastics to Energy: Fuel, Chemicals, and Sustainability Implications,* Edited by S.M. Al-Salem, 3–20. William Andrew Publishing. doi:10.1016/B978-0-12-813140-4.00001-7.

Al-Salem, S. M. 2019b."Major Technologies Implemented for Chemicals and Fuel Recovery." In *Plastics to Energy: Fuel, Chemicals, and Sustainability Implications,* Edited by S.M. Al-Salem, 21-44. William Andrew Publishing. doi:10.1016/B978-0-12-813140-4.00002-9.

Al-Salem, S. M., A. Antelava, A. Constantinou, G. Manos, and A. Dutta. 2017. "A review on thermal and catalytic pyrolysis of plastic solid waste (PSW)."*Journal of Environmental Management* 197:177-198. Accessed September 16, 2019. doi:10.1016/j.jenvman.2017.03.084.

Al-Salem, S. M., P. Lettieri, and J. Baeyens. 2009. "Recycling and recovery routes of plastic solid waste (PSW):A review." *Waste management* 29(10):2625-2643. Accessed September 13, 2019. doi:10.1016/j.wasman.2009.06.004.

Al-Salem, S. M., S. Evangelisti, and P. Lettieri. 2014."Life cycle assessment of alternative technologies for municipal solid waste and plastic solid waste management in the

Greater London area." *Chemical Engineering Journal* 244:391-402. Accessed September 16, 2019. doi:10.1016/j.cej.2014.01.066.

Alston, Sue M., Allan D. Clark, J. Cris Arnold, and Bridget K. Stein. 2011. "Environmental impact of pyrolysis of mixed WEEE plastics part 1: experimental pyrolysis data." *Environmental science & technology* 45(21): 9380-9385. Accessed September 16, 2019. doi:10.1021/es201664h.

Andrady, Anthony L., and Mike A. Neal. 2009. "Applications and societal benefits of plastics."*Philosophical Transactions of the Royal Society B*: Biological Sciences 364(1526):1977-1984. Accessed September 13, 2019. doi: 10.1098/rstb.2008.0304.

Arfaoui, Sonia, Najoua Frini-Srasra, and Ezzeddine Srasra. 2009. "Synthesis and characterization of hydroxy-chromium pillared bentonite."*Surface Engineering and Applied Electrochemistry* 45(3):239-245. Accessed August 30, 2019. doi:10.3103/S1068375509030119.

Assumpção, Luiz C. F. N., Montserrat M. Carbonell, and Mônica R. C. Marques. 2011. "Co-pyrolysis of polypropylene waste with Brazilian heavy oil." *Journal of Environmental Science and Health, Part A* 46(5):461-464. Accessed October 10, 2018. doi: 10.1080/10934529.2011.551724.

Assumpção, Luiz CFN, Montserrat M. Carbonell, and Mônica RC Marques. 2011. "Co-pyrolysis of polypropylene waste with Brazilian heavy oil." *Journal of Environmental Science and Health, Part A* 46(5): 461-464. Accessed October 10, 2019. doi: 10.1080/10934529.2011.551724.

Atkins, P. W., and D. F. Shriver. Trans. 2008. *Química Inorgânica*. 4 ed. Porto Alegre: Artmed Editora S/A.

Baes Jr, C. F., and R. E. Mesmer. 1976. *The Hydrolysis of Cations*, 1-489. New York: John Wiley & Sons.

Bagshaw, Stephen A., and Ralph P. Cooney. 1993. "FTIR surface site analysis of pillared clays using pyridine probe species." *Chemistry of Materials* 5(8):1101-1109. Accessed August 18, 2019. doi:10.1021/cm00032a013.

Baloyi, Jeffrey, Thabang Ntho, and John Moma. 2018. "A Novel Synthesis Method of Al/Cr Pillared Clay and its Application in the Catalytic Wet Air Oxidation of Phenol."*Catalysis Letters* 148(12):3655-3668. Accessed August 12, 2019. doi:10.1007/s10562-018-2579-x.

Banković, Predrag, Aleksandra Milutinović Nikolić, Nataša Jović-Jovičić, Jasmina Dostanić, Željko Čupić, Davor Lončarević, and Dušan M. Jovanović. 2009. "Synthesis, characterization and application of Al, Fe-pillared clays."*Acta Physica Polonica A* 115(4): 811-815. Accessed August 29, 2019. doi:10.12693/APhys PolA.115.811.

Barakan, Shima, and Valeh Aghazadeh. 2019. "Synthesis and characterization of hierarchical porous clay heterostructure from Al, Fe-pillared nano-bentonite using microwave and ultrasonic techniques." *Microporous and Mesoporous Materials* 278:138-148. Accessed August 27, 2019. doi:10.1016/j.micromeso.2018.11.031.

Barrer, RM and, and D. M. MacLeod. 1955. "Activation of montmorillonite by ion exchange and sorption complexes of tetra-alkyl ammonium montmorillonites." *Transactions of the Faraday Society* 51:1290-1300. Accessed August 18, 2019 doi:10.1039/TF9555101290.

Bergaya, F., A. Aouad, and T. Mandalia. 2006. "Pillared clays and clay minerals.' In *Handbook of Clay Science* edited by Faïza Bergaya, Benny K.G. Theng, Gerhard Lagaly, 393-421. Elsevier. doi:10.1016/S1572-4352(05)01012-3.

Bergaya, F., K. Beneke, R. W. Berry, G. Lagaly, and K. B. Tankersley. 2013. "Clay Science: A Young Discipline and a Great Perspective." In *Developments in Clay Science,* vol.5, edited by Faïza Bergaya, Gerhard Lagaly, 819-855. Elsevier. doi:10.1016/B978-0-08-098258-8.00028-6.

Bergaya, Faiza, and Gerhard Lagaly. 2006."General introduction: clays, clay minerals, and clay science." *Developments in clay science* edited by Faïza Bergaya, Benny K.G. Theng, Gerhard Lagaly: 1-18. Elsevier. doi:10.1016/S1572-4352(05)01001-9.

Bhaskar, Thallada, Md Azhar Uddin, Kazuya Murai, Jun Kaneko, Kenji Hamano, Toshiaki Kusaba, Akinori Muto, and Yusaku Sakata. 2003. "Comparison of thermal degradation products from real municipal waste plastic and model mixed plastics." *Journal of Analytical and Applied Pyrolysis* 70(2): 579-587. Accessed September 22, 2019. doi:10.1016/S0165-2370(03)00027-5.

Bodman, S. D., W. R. McWhinnie, V. Begon, I. Suelves, M-J. Lazaro, T. J. Morgan, A. A. Herod, and R. Kandiyoti. 2002. "Metal-ion pillared clays as hydrocracking catalysts (I): Catalystic preparation and assessment of performance at short contact times." *Fuel* 81:449–459. Accessed August 12, 2019. doi:10.1016/S0016-2361(03)00126-1.

Bodman, S. D., W. R. McWhinnie, V. Begon, M. Millan, I. Suelves, M-J. Lazaro, A. A. Herod, and R. Kandiyoti. 2003. "Metal-ion pillared clays as hydrocracking catalysts (II): effect of contact time on products from coal extracts and petroleum distillation residues." *Fuel* 82(18): 2309-2321. Accessed August 12, 2019. doi:10.1016/S0016-2361(03)00126-1.

Bozkurt, P. A.; Tosun, O.; Canel, M. 2017. "The synergistic effect of co-pyrolysis of oil shale and low density polyethylene mixtures and characterization of pyrolysis liquid." *Journal of the Energy Institute* 90(3): 355-362. Accessed October 10, 2019. doi:10.1016/j.joei.2016.04.007.

Bradley, Susan M., and Ronald A. Kydd. 1993. "Comparison of the species formed upon base hydrolyses of gallium (III) and iron (III) aqueous solutions: the possibility of existence of an $[FeO_4Fe_{12}(OH)_{24}(H_2O)_{12}]^{7+}$ Polyoxocation."*Journal of the Chemical Society, Dalton Transactions* 15: 2407-2413. Accessed August 4, 2019. doi:10.1039/DT9930002407.

Breck, D. W. 1980. “Potential uses of natural and synthetic zeolites in industry.” In: *The properties and application of zeolites,* edited by Towsend, R.P., 391-422. London: The Chemical Society.

Brigatti, Maria Franca, Emilio Galan, and B. K. G. Theng. 2013. “Structure and mineralogy of clay minerals.” In *Developments in clay science,* vol. 5, edited by Faïza Bergaya, Gerhard Lagaly, 21-81. Elsevier. doi:10.1016/B978-0-08-098258-8.00002-X.

Brindley, G. W., and R. E. Sempels. 1977. “Preparation and properties of some hydroxy-aluminium beidellites.”*Clay Minerals* 12(3): 229-237. Accessed August 20, 2019. http://citeseerx.ist.psu.edu/viewdoc/download?doi=10.1.1.615.9954&rep=rep1&type=pdf.

Brindley, G. W., and S. Yamanaka. 1979 “A study of hydroxy-chromium montmorillonites and the form of the hydroxy-chromium polymers.” *American Mineralogist* 64(7-8):830-835.

Broughton, Geoffrey. 1940. “Catalysis by Metallized Bentonites.” *The Journal of Physical Chemistry* 44(2): 180-184. Accessed August 22, 2019. doi:10.1021/j150398a005.

Brown, David M., Matthijs Bonte, Richard Gill, James Dawick, and Peter J. Boogaard. 2017. “Heavy hydrocarbon fate and transport in the environment.” *Quarterly Journal of Engineering Geology and Hydrogeology* 50(3): 333-346. Accessed September 13, 2019. doi:10.1144/qjegh2016-142.

Carneiro, Débora da S., and Mônica R. da C. Marques. 2011 “Có-pirólise de resíduos de polietileno com gasóleo pesado da bacia de campos.”*Polímeros: Ciência e Tecnologia* 21(5): 347-352. Accessed October 10, 2018. doi: 10.1590/S0104-14282011005000074 [Co-pirolysis of Polyethylene Waste with the Campos Bacia Heavy Oil *Polímeros: Ciência e Tecnologia* 21(5):347-352.]

Carr, R. Melville. 1985. “Hydration states of interlamellar chromium ions in montmorillonite.”*Clays and Clay Minerals* 33(4):357-361. Accessed August 29, 2019. doi:10.1346/CCMN.1985.03304.

Carvalho, Wagner Alves. 2002. “Oxidação de ciclohexano em fase gasosa catalisada por argilas pilarizadas com ferro e cromo.”*Eclética Química* 27(spe):353-365. Accessed August 13, 2019. doi:10.1590/S0100-46702002000200029 [Ciclohexane oxidation in gas phase using iron and chromium pillared clays as catalysts *Eclética Química* 27(spe):353-365.]

Cases, J. M,I. Bérend, M. Francois, J. P. Uriot, L.J. Michot, and F. Thomas. 1997. “Mechanism of Adsorption and Desorption of Water Vapor by Homoionic Montmorillonite: 3. The Mg^{2+}, Ca^{2+}, Sr^{2+} and Ba^{2+} Exchanged Forms.” *Clays Clay Miner* 45(1): 8-22. Accessed June, 24, 2019. doi:10.1346/CCMN.1997.0450102.

Casey, William H. 2006 “Large aqueous aluminum hydroxide molecules.” *Chemical reviews* 106(1):1-16. Accessed August 4, 2019. doi:10.1021/cr040095d.

Catrinescu, Cezar, Daniela Arsene, Petru Apopei, and Carmen Teodosiu. 2012. “Degradation of 4-chlorophenol from wastewater through heterogeneous Fenton and photo-Fenton process, catalyzed by Al–Fe PILC." *Applied Clay Science* 58:96-101. Accessed August 27, 2019. doi:10.1016/j.clay.2012.01.019.

Centi, Gabriele, and Siglinda Perathoner. 2008. "Catalysis by layered materials: A review." *Microporous and mesoporous materials* 107(1-2):3-15. Accessed August 6, 2019. doi:10.1016/j.micromeso.2007.03.011.

Chareonsiriwat, Laddawan, and Warinthorn Chavasiri. 2017. "Convenient method for the transformation of epoxide to aldehyde and acetonide mediated by Cr-PLM." *Synthetic Communications* 47(4): 257-267. Accessed August 24, 2019. doi:10.1080/00397911.2016.1254799.

Chen, Dezhen, Lijie Yin, Huan Wang, and Pinjing He. 2014. "Pyrolysis technologies for municipal solid waste: a review." *Waste management* 34(12): 2466-2486.

Chen, Guanyi, Xiaoxiong Zhang, Wenchao Ma, Beibei Yan, and Yanbin Li. 2014: "Co-pyrolysis of corn-cob and waste cooking-oil in a fixed bed reactor with HY upgrading process." *Energy Procedia* 61: 2363-2366. Accessed October 2, 2018. doi: 10.1016/j.egypro.2014.12.00.

Cheng, Linda S., and Ralph T. Yang. 1997. "Tailoring micropore dimensions in pillared clays for enhanced gas adsorption." *Microporous materials* 8(3-4):177-186. Accessed August 11, 2019. doi:10.1016/S0927-6513(96)00067-3.

Coelho, Antonio Carlos Vieira, Pérsio de Souza Santos, and Helena de Souza Santos. 2007. "Argilas Especiais: Argilas Quimicamente Modificadas–Uma Revisão."*Química Nova* 30(5):1282-1294. Accessed June, 24, 2019. http://quimicanova.sbq.org.br/imagebank/pdf/Vol30No5_1282_41-RV06145.pdf [Special Clays: Chemically Modified Clays–A Review *Química Nova* 30(5):1282-1294.].

Cordes, Erik E., Daniel OB Jones, Thomas A. Schlacher, Diva J. Amon, Angelo F. Bernardino, Sandra Brooke, Robert Carney, Danielle M. DeLeo, Katherine M. Dunlop, Elva G Escobar-Briones and Andrew R. Gates, 2016. "Environmental impacts of the deep-water oil and gas industry: a review to guide management strategies."*Frontiers in Environmental Science* 4:1-26. Accessed September 13, 2019. doi:10.3389/fenvs.2016.00058.

De Aguiar, M. R. M. P., Amanda Cardoso Novaes, and Alcides Wagner Serpa Guarino. 2002. "Remoção de metais pesados de efluentes industriais por alumino-ssilicatos."*Química Nova* 25(6/B): 1145-1154. Accessed June, 26, 2019. doi:10.1590/S0100-40422002000700015 [Removal Of Heavy Metals From Wastewaters By Aluminosilicate *Química Nova* 25(6/B): 1145-1154.]

De León, María Andrea, Carolina De Los Santos, Luis Latrónica, Ana María Cesio, Cristina Volzone, Jorge Castiglioni, and Marta Sergio. 2014. "High catalytic activity at low temperature in oxidative dehydrogenation of propane with Cr–Al pillared clay." *Chemical Engineering Journal* 241:336-343. Accessed August 12, 2019. doi:10.1016/j.cej.2013.10.086.

De Sá, Luís Carlos, Miguel Oliveira, Francisca Ribeiro, Thiago Lopes Rocha, and Martyn Norman Futter. "Studies of the effects of microplastics on aquatic organisms: what do we know and where should we focus our efforts in the future?." 2018: *Science of the total environment* 645:1029-1039. Accessed September 13, 2019. doi:10.1016/j.scitotenv.2018.07.207.

De Stefanis, Adriana, Patrizia Cafarelli, Fortunata Gallese, Elisabetta Borsella, Agostino Nana, and Giorgio Perez. 2013. "Catalytic pyrolysis of polyethylene: A comparison between pillared and restructured clays." *Journal of Analytical and Applied Pyrolysis* 104: 479-484. Accessed September 25, 2019. doi:10.1016/j.jaap.2013.05.023.

Demirbas, Ayhan, Abdullah Bafail, and Abdul-Sattar Nizami. 2016. "Heavy oil upgrading: Unlocking the future fuel supply." *Petroleum Science and Technology* 34(4):303-308. Accessed September 22, 2019. doi:10.1080/10916466.2015.1136949.

Demirbas, Ayhan, Khalid Al-Ghamdi, Nejdet Sen, Avni Aslan, and Walid M. Alalayah. 2017. "Gasoline-and diesel-like products from heavy oils via catalytic pyrolysis." *Petroleum Science and Technology* 35(15):1607-1613. Accessed September 25, 2019. doi:10.1080/10916466.2017.1336768.

Ding, Menglin, Shufeng Zuo, and Chenze Qi. 2015 "Preparation and characterization of novel composite AlCr-pillared clays and preliminary investigation for benzene adsorption." *Applied Clay Science* 115:9-16. Accessed August 4, 2019. doi:10.1016/j.clay.2015.07.020.

Ding, Zhe, J. Theo Kloprogge, Ray L. Frost, G. Q. Lu, and H. Y. Zhu. 2001. "Porous clays and pillared clays-based catalysts. Part 2: a review of the catalytic and molecular sieve applications." *Journal of Porous Materials* 8(4): 273-293. Accessed August 6, 2019. doi:10.1023/A:101311303091.

Ding, Zhe, J. Theo Kloprogge, Ray L. Frost, G. Q. Lu, and H. Y. Zhu. 2001. "Porous clays and pillared clays-based catalysts. Part 2: a review of the catalytic and molecular sieve applications." *Journal of Porous Materials* 8(4):273-293. Accessed August 15, 2019. doi: 10.1023/A:101311303091.

Dong, Shi-Lin, Ya-Xin Su, L. I. U. Xin, Qian-Cheng Li, Min-Hao Yuan, Z. H. O. U. Hao, And Wen-Yi Deng. 2018 "Experimental study on selective catalytic reduction of NO by C3H6 over Fe/Ti-PILC catalysts." *Journal of Fuel Chemistry and Technology* 46(10):1231-1239. Accessed August 27, 2019. doi:10.1016/S1872-5813(18)30051-3.

Drljaca, A., J. R. Anderson, L. Spiccia, and T. W. Turney. 1992 "Intercalation of montmorillonite with individual chromium (III) hydrolytic oligomers." *Inorganic Chemistry* 31(23):4894-4897. Accessed August 30, 2019. doi:10.1021/ic00049a033.

Drljaca, A., J. R. Anderson, Leone Spiccia, and Terry W. Turney. 1997. "A new method for generating chromium (III) intercalated clays." *Inorganica chimica acta* 256(1):151-154. Accessed August 30, 2019. doi:10.1016/S0020-1693(96)05421-7.

Elektorowicz, M., and Z. Muslat, Z. 2008. "Removal of heavy metals from oil sludge using ion exchange textiles." *Environmental technology* 29(4):393-399. Accessed September 13, 2019. doi:10.1080/09593330801984290.

Emam, Eman A. 2013. "Clays as catalysts in petroleum refining industry." ARPN *Journal of Science and Technology* 3(4): 356-375. Accessed August 6, 2019. https://pdfs.semanticscholar.org/2672/bf56a1ccafe3b73bab4e10ae48cfa15146d8.pdf.

Faghihian, Hossein, and Mohammad Hadi Mohammadi. 2014. "Acid activation effect on the catalytic performance of Al-pillared bentonite in alkylation of benzene with olefins." *Applied Clay Science* 93:1-7. Accessed August 4, 2019. doi:10.1016/j.clay.2014.02.026.

Faillace, Jacyra Guimarães, Claudinei Fernandes de Melo, Soraia Peres Lima de Souza, and Mônica Regina da Costa Marques. "Production of light hydrocarbons from pyrolysis of heavy gas oil and high density polyethylene using pillared clays as catalysts. 2017."*Journal of Analytical and Applied Pyrolysis* 126:70-76. Accessed October 10, 2018. doi:10.1016/j.jaap.2017.06.023.

Falaras, Polycarpos, Fani Lezou, Georgios Seiragakis, and Dimitrios Petrakis. 2000. "Bleaching properties of alumina-pillared acid-activated montmorillonite." *Clays and clay minerals* 48(5):549-556. Accessed August 18, 2019. doi:10.1346/CCMN.2000.0480507.

Fatimah, Is, and M. Rapei. 2015. "The improvement of montmorillonite activity in 2-butanol dehydration by metal oxide pillarization." *Journal of Engineering Science and Technology* Special Issue on SOMCHE 2014 & RSCE 2014 Conference 74-82. Accessed August 4, 2019.

Fetter, Geolar, and Pedro Bosch. 2010. "Microwave effect on clay pillaring." In *Pillared clays and related catalysts,* edited by Antonio Gil·Sophia A. Korili·Raquel Trujillano and Miguel Angel Vicente,1-21. New York: Springer. Accessed August, 15, 2019. doi:10.1007/978-1-4419-6670-4_1.

Fripiat, J. J., and M. I. Cruz-Cumplido. 1974. "Clays as catalysts for natural processes." *Annual review of earth and planetary sciences* 2(1):239-256. Accessed June, 15, 2019. doi:10.1146/annurev.ea.02.050174.001323.

Furrer, Gerhard, Christian Ludwig, and Paul W. Schindler. 1992. "On the chemistry of the Keggin Al13 polymer: I. Acid-base properties." *Journal of colloid and interface science* 149(1):56-67. Accessed August 4, 2019. doi:10.1016/0021-9797(92)90391-X.

G.J. Churchman, W.P. Gates, B.K.G. Theng, G. Yuan 2006. Clays and Clay Minerals for Pollution Control." In: *Developments in Clay Science*, vol. 1, edited by Faïza Bergaya, Benny K.G. Theng, Gerhard Lagaly, 625-675, Elsevier. Accessed June, 26, 2019. doi:10.1016/S1572-4352(05)01020-2.

Galán, Emilio. 1996. "Properties and applications of palygorskite-sepiolite clays." *Clay minerals* 31(4):443-453. Accessed June, 16, 2019 doi:10.1180/claymin.1996.031.4.01.

Georgescu, Ana-Maria, Françoise Nardou, Valentin Zichil, and Ileana Denisa Nistor. 2018. "Adsorption of lead (II) ions from aqueous solutions onto Cr-pillared clays." *Applied Clay Science* 152:44-50. Accessed August 24, 2019. doi:10.1016/j.clay.2017.10.031.

Geyer, Roland, Jenna R. Jambeck, and Kara Lavender Law. 2017. "Production, use, and fate of all plastics ever made." *Science advances* 3(7)1-5. Accessed September 13, 2019. doi: 10.1126/sciadv.1700782.

Gil, A., A. Massinon, and Paul Grange. 1995. "Analysis and comparison of the microporosity in Al-, Zr-and Ti-pillared clays." *Microporous materials* 4(5):369-378. doi:10.1016/0927-6513(95)00021-Z.

Gil, A., and M. Montes. 1994."Effect of thermal treatment on microporous accessibility in aluminium pillared clays." *Journal of Materials Chemistry* 4(9):1491-1496. Accessed August 11, 2019. doi: 10.1039/JM9940401491.

Gil, A., S. A. Korili, and M. A. Vicente. 2008 "Recent advances in the control and characterization of the porous structure of pillared clay catalysts." *Catalysis Reviews* 50(2):153-221. Accessed June, 26, 2019.doi:10.1080/01614940802019383.

Gil, Antonio, Luis M. Gandia, and Miguel A. Vicente. 2000. "Recent advances in the synthesis and catalytic applications of pillared clays." *Catalysis Reviews* 42(1-2):145-212. doi:10.1081/CR-100100261.

Gil, Antonio, Sophia A. Korili, Raquel Trujillano, and Miguel Angel Vicente, 2010. *Pillared clays and related catalysts*, 1-517. New York: Springer Science & Business Media.

Gomes, Celso Figueiredo. 1988. *Argilas: O que são e para que servem.*Lisboa: Fundação Calouste.

Gong, Zhiqiang, Zhentong Wang, Zhenbo Wang, Peiwen Fang, and Fanzhi Meng. 2018. "Study on pyrolysis characteristics of tank oil sludge and pyrolysis char combustion." *Chemical Engineering Research and Design* 135:30-36. Accessed September 16, 2019. doi: 10.1016/j.cherd.2018.05.027.

González, Beatriz, Alba Pérez, Raquel Trujillano, Antonio Gil, and Miguel Vicente. 2017. "Microwave-assisted pillaring of a montmorillonite with Al-polycations in concentrated media." *Materials* 10(8): 886. Accessed August 4, 2019. doi:10.3390/ma10080886.

Gregg, S. J., and K. S. W. Sing. 1982. *Adsorption, surface area and porosity*. 2nd ed., 1-312. London: Academic Press.

Grim, Ralph E. 1962. *Applied Clay Mineralogy*. New York: McGraw Hill Press.

Guerra, D. L., V. P. Lemos, E. F. Fernandes, R. S. Angélica, and M. L. da Costa. 2005. "Influência da temperatura de formação do íon de Keggin no processo de pilarização de esmectitas."*Eclética Química* 30(4): 27-32. Accessed August, 15, 2019. http://www.scielo.br/pdf/%0D/eq/v30n4/27707.pdf [Influence of Keggin ion formation temperature on smectite pillaring process *Eclética Química* 30(4): 27-32.]

Guerra, D. L., V. P. Lemos, R. S. Angélica, and C. Airoldi. 2006. "Influência da razão Al/argila no processo de pilarização de esmectita."*Cerâmica* 52:200-206. Accessed August, 15, 2019. http://www.scielo.br/pdf/%0D/ce/v52n323/32090.pdf [Influence of the Al /clay ratio on the pillaring process of smectite *Cerâmica* 52:200-206.]

Gyftopoulou, M. E., M. Millan, A. V. Bridgwater, D. Dugwell, R. Kandiyoti, and J. A. Hriljac. 2005. "Pillared clays as catalysts for hydrocracking of heavy liquid fuels."*Applied Catalysis A: General* 282(1-2):205-214. Accessed August 30, 2019. doi:10.1016/j.apcata.2004.12.012.

Hendricks, Sterling B. 1929. "XVIII. Diffraction of X-Radiation from some Crystalline Aggregates."*Zeitschrift für Kristallographie-Crystalline Materials* 71(1-6):269-273. Accessed June, 16, 2019. doi:10.1524/zkri.1929.71.1.269.

Hendricks, Sterling B., and William H. Fry. 1930. "The results of X-ray and microscopical examinations of soil colloids." *Soil Science* 29(6):457-480. Accessed June, 16, 2019. doi: 10.1097/00010694-193006000-00003.

Herney-Ramirez, J., Miguel A. Vicente, and Luis M. Madeira. 2010. "Heterogeneous photo-Fenton oxidation with pillared clay-based catalysts for wastewater treatment: a review." *Applied Catalysis B: Environmental* 98(1-2):10-26. Accessed August 27, 2019. doi:10.1016/j.apcatb.2010.05.004.

Houdry, E., W. F. Burt, A. B. Pew, and W. A. Peters. 1938. "Catalytic processing by the Houdry process." *Nat. Petroleum News* 39:R570-R580.

https://dspace.uii.ac.id/bitstream/handle/123456789/2826/SOMCHE%202014_6_2015_074_082.pdf?sequence=1.

Hu, Guangji, Jianbing Li, and Guangming Zeng. 2013. "Recent development in the treatment of oily sludge from petroleum industry: a review." *Journal of hazardous materials* 261:470-490. Accessed September 19, 2019. doi.org/10.1016/j.jhazmat.2013.07.069.

Hunt, John M., R. Paul Philp, and Keith A. Kvenvolden. 2002. "Early developments in petroleum geochemistry." *Organic geochemistry* 33(9):1025-1052. Accessed September 13, 2019. doi:10.1016/S0146-6380(02)00056-6.

Inguanzo, M., A. Domınguez, J. A. Menéndez, C. G. Blanco, and J. J. Pis. 2002. "On the pyrolysis of sewage sludge: the influence of pyrolysis conditions on solid, liquid and gas fractions." *Journal of Analytical and Applied Pyrolysis* 63(1):209-222. Accessed October 8, 2019. doi: 10.1016/S0165-2370(01)00155-3.

Ismadji, Suryadi, Felycia Edi Soetaredjo, and Aning Ayucitra. 2015. "The Characterization of Clay Minerals and Adsorption Mechanism onto Clays." In *Clay Materials for Environmental Remediation*, 93-112. Berlin: Springer Briefs. doi:10.1007/978-3-319-16712-1_5.

Jafarinejad, S. 2017. "Pollution and wastes from Petroleum Industry". In Jafarinejad, S. *Petroleum waste treatment and pollution control*. 19-78. Butterworth-Heinemann.

Jambeck, Jenna R., Roland Geyer, Chris Wilcox, Theodore R. Siegler, Miriam Perryman, Anthony Andrady, Ramani Narayan, and Kara Lavender Law. 2015. "Plastic waste inputs from land into the ocean." *Science* 347(6223):768-771. Accessed September 13, 2019. doi:10.1126/science.1260352.

Jamradloedluk, Jindaporn, and Chaloenporn Lertsatitthanakorn. 2014. "Characterization and Utilization of Char Derived from Fast Pyrolysis of Plastic Wastes." *Procedia Engineering* 69:1437-1442. Accessed October 2, 2018. doi: 10.1016/j.proeng.2014.03.139.

Jia, Hanzhong, Song Zhao, Xiehong Zhou, Chengtun Qu, Daidi Fan, and Chuanyi Wang. 2017 "Low-temperature pyrolysis of oily sludge: roles of Fe/Al-pillared bentonites." *Archives of Environmental Protection* 43(3):82-90. Accessed September 25, 2019. doi:10.1515/aep-2017-0027.

Johns, William D. "Clay mineral catalysis and petroleum generation. 1979."*Annual Review of Earth and Planetary Sciences* 7(1):183-198. Accessed June, 15, 2019. doi:10.1146/annurev.ea.07.050179.001151.

Johns, William D., and Akira Shimoyama. 1972. "Clay minerals and petroleum-forming reactions during burial and diagenesis." *AAPG Bulletin* 56(11):2160-2167. Accessed June, 15, 2019. Alternative identifier OSTI ID: 7303752.

Joppert Jr, Ney, Alexsandro Araujo da Silva, and Mônica Regina da Costa Marques. 2015. "Enhanced diesel fuel fraction from waste high-density polyethylene and heavy gas oil pyrolysis using factorial design methodology." *Waste management* 36:166-176. Accessed September 13, 2019. doi.org/10.1016/j.wasman.2014.11.023.

Kausar, Abida, Munawar Iqbal, Anum Javed, Kiran Aftab, Haq Nawaz Bhatti, and Shazia Nouren. 2018 "Dyes adsorption using clay and modified clay: a review." *Journal of Molecular Liquids* 256:395-407. Accessed August 4, 2019. doi:10.1016/j.molliq.2018.02.034.

Keggin, J. F. 1934. "The structure and formula of 12-phosphotungstic acid." *Proceedings of the Royal Society of London. Series A, Containing Papers of a Mathematical and Physical Character* 144(851):75-100. Accessed August 4, 2019. doi:10.1098/rspa.1934.0035.

Kiran, N., E. Ekinci, and C. E. Snape. 2000. "Recyling of plastic wastes via pyrolysis." *Resources, Conservation and Recycling* 29(4):273-283. Accessed October 9, 2018. doi:10.1016/S0921-3449(00)00052-5.

Kiricsi, Imre, Árpád Molnár, István Pálinkó, Ágnes Fudala, and János B. Nagy. 1997. "Nanoscale redox catalysts: Cr-and Cr, Al-pillared layer clays: Characterization and catalytic activity." *Solid state ionics* 101:793-797. Accessed August 30, 2019. doi:10.1016/S0167-2738(97)00295-6.

Kloprogge, J. T. 1998. "Synthesis of smectites and porous pillared clay catalysts: A review." *Journal of Porous Materials* 5(1) :-41. Accessed August 18, 2019. doi:10.1023/A:1009625913781.

Kloprogge, J. Theo, Loc V. Duong, and Ray L. Frost. 2005. "A review of the synthesis and characterisation of pillared clays and related porous materials for cracking of vegetable oils to produce biofuels." *Environmental Geology* 47(7):967-981. Accessed August 4, 2019. doi:10.1007/s00254-005-1226-1.

Kumar, Sachin, Achyut K. Panda, and Raghubansh Kumar Singh. 2011. "A review on tertiary recycling of high-density polyethylene to fuel." *Resources, Conservation and Recycling* 55(11): 893-910. Accessed September 13, 2019. doi:10.1016/j.resconrec.2011.05.005.

Kurian, Manju, and S. Kavitha. 2016. "A Review on the Importance of Pillared Interlayered Clays in Green Chemical Catalysis." *IOSR Journal of Applied Chemistry* (1):47-54. Accessed August 11, 2019. https://pdfs.semanticscholar.org/60ad/bee15f03774c73fc33fb5ba85bae1ab5018f.pdf.

Lahav, N., U. Shani, and J. Shabtai. 1978. "Cross-linked smectites. I. Synthesis and properties of hydroxy-aluminum-montmorillonite." *Clays and Clay Minerals* 26, no. 2: 107-115. Accessed August 28, 2019. http://citeseerx.ist.psu.edu/viewdoc/download?doi=10.1.1.476.9149&rep=rep1&type=pdf.

Li, Dan, Chunshan Li, and Kenzi Suzuki. 2013. "Catalytic oxidation of VOCs over Al-and Fe-pillared montmorillonite." *Applied Clay Science* 77:56-60. Accessed August 4, 2019. doi:10.1016/j.clay.2013.02.027.

Li, Jingrong, Maocong Hu, Shufeng Zuo, and Xianqin Wang. 2018. "Catalytic combustion of volatile organic compounds on pillared interlayered clay (PILC)-based catalysts." *Current opinion in chemical engineering* 20:93-98. Accessed August 24, 2019. doi:10.1016/j.coche.2018.02.001.

Li, Kaixin, Junxi Lei, Guoan Yuan, Piyarat Weerachanchai, Jing-Yuan Wang, Jun Zhao, and Yanhui Yang. 2017. "Fe-, Ti-, Zr-and Al-pillared clays for efficient catalytic pyrolysis of mixed plastics." *Chemical Engineering Journal* 317:800-809. Accessed October 10, 2018. doi:10.1016/j.cej.2017.02.113.

Lin, Yan, Yanfen Liao, Zhaosheng Yu, Shiwen Fang, Yousheng Lin, Yunlong Fan, Xiaowei Peng, and Xiaoqian Ma. 2016. "Co-pyrolysis kinetics of sewage sludge and oil shale thermal decomposition using TGA–FTIR analysis." *Energy Conversion and Management* 118:345-352. Accessed October 8, 2019. doi:10.1016/j.enconman.2016.04.004.

Loeppert, R. H., and M. M. Mortland. 1979. "The influence of heat-stable intercalate on the rate of dehydroxylation of smectite." *Clays and Clay Minerals* 27(5):373-376. Accessed August 20, 2019. http://citeseerx.ist.psu.edu/viewdoc/download?doi=10.1.1.567.3221&rep=rep1&type=pdf.

López, A., I. De Marco, B. M. Caballero, M. F. Laresgoiti, A. Adrados, and A. Aranzabal. 2011. "Catalytic pyrolysis of plastic wastes with two different types of catalysts: ZSM-5 zeolite and Red Mud." *Applied Catalysis B: Environmental* 104(3-4):211-219. Accessed October 10, 2018. doi:10.1016/j.apcatb.2011.03.030.

Lopez, A., I. De Marco, B. M. Caballero, M. F. Laresgoiti, and A. Adrados. 2011a. "Influence of time and temperature on pyrolysis of plastic wastes in a semi-batch reactor." *Chemical Engineering Journal*173(1): 62-71. Accessed September 16, 2019. doi:10.1016/j.cej.2011.07.037.

Lopez, Gartzen, Maite Artetxe, Maider Amutio, Javier Bilbao, and Martin Olazar. 2017. "Thermochemical routes for the valorization of waste polyolefinic plastics to produce fuels and chemicals. A review." *Renewable and Sustainable Energy Reviews* 73: 346-368. Accessed September 16, 2019. doi:10.1016/j.rser.2017.01.142.

Luna, Fernando J., and Ulf Schuchardt. 1999. "Argilas pilarizadas–uma introdução."*Química Nova* 22(1):104-109. Accessed August, 16, 2019. doi:10.1590/S0100-40421999000100017 [An introduction to pillared clays *Química Nova* 22(1):104-109.]

Luo, Guohua, Tomohiko Suto, Satomi Yasu, and Kunio Kato. 2000. "Catalytic degradation of high density polyethylene and polypropylene into liquid fuel in a powder-particle fluidized bed." *Polymer Degradation and Stability* 70(1):97-102. Accessed September 25, 2019. doi:10.1016/S0141-3910(00)00095-1.

Ma, Lingya, Jianxi Zhu, Yunfei Xi, Runliang Zhu, Hongping He, Xiaoliang Liang, and Godwin A. Ayoko. 2015. "Simultaneous adsorption of Cd (II) and phosphate on Al_{13} pillared montmorillonite." *RSC Advances* 5(94):77227-77234. Accessed August 4, 2019. doi: 10.1039/c5ra15744g.

Ma, Wenchao, Bin Liu, Ruixue Zhang, Tianbao Gu, Xiang Ji, Lei Zhong, Guanyi Chen, Longlong Ma, Zhanjun Cheng, and Xiangping Li. 2018. "Co-upgrading of raw bio-oil with kitchen waste oil through fluid catalytic cracking (FCC)."*Applied energy* 217:233-240. Accessed September 22, 2019. doi:10.1016/j.apenergy.2018.02.036.

Manju, Kurian. 2005. "Catalytic activity of iron pillared montmorillonites for benzylation of benzene with benzyl chloride." *Bulletin of the Catalysis Society of India* 4:88-100. http://www.macollege.in/uploads/publishers_file/doc_166.pdf.

Marosz, Monika, Andrzej Kowalczyk, and Lucjan Chmielarz. 2019. "Modified vermiculites as effective catalysts for dehydration of methanol and ethanol." *Catalysis Today*. Accessed August 4, 2019. doi:10.1016/j.cattod.2019.07.003.

Mata, Guiomar, Raquel Trujillano, María Asunción Vicente, C. Belver, Marcos Fernández-García, S. A. Korili, and Arturo Gil. 2007 "Chromium–saponite clay catalysts: preparation, characterization and catalytic performance in propene oxidation." *Applied Catalysis A: General* 327(1): 1-12. Accessed August 29, 2019. doi:10.1016/j.apcata. 2007.04.021.

Melendi-Espina, Sonia, R. Alvarez, M. A. Diez, and M. D. Casal. 2015."Coal and plastic waste co-pyrolysis by thermal analysis–mass spectrometry." *Fuel processing technology* 137:351-358. Accessed October 10, 2019. doi:10.1016/j.fuproc. 2015.03.024.

Meng, Xianghai, Chunming Xu, Jinsen Gao, and Li Li. 2005. "Studies on catalytic pyrolysis of heavy oils: reaction behaviors and mechanistic pathways." *Applied Catalysis A: General* 294(2):168-176. Accessed September 25, 2019. doi: 10.1016/j.apcata.2005.07.033.

Miandad, R., M. A. Barakat, Asad S. Aburiazaiza, M. Rehan, and A. S. Nizami. 2016. "Catalytic pyrolysis of plastic waste: A review." *Process Safety and Environmental Protection* 102:822-838. Accessed September 13, 2019. doi:10.1016/j.psep. 2016.06.022.

Miandad, R., M. A. Barakat, M. Rehan, A. S. Aburiazaiza, I. M. I. Ismail, and A. S. Nizami. 2017. "Plastic waste to liquid oil through catalytic pyrolysis using natural and synthetic zeolite catalysts." *Waste management* 69:66-78. Accessed September 22, 2019. doi:10.1016/j.wasman.2017.08.032.

Milato, Jônatas V., Rodrigo J. França, and Mônica R. C. Marques. 2019. "Pyrolysis of oil sludge from the offshore petroleum industry: influence of different mesoporous zeolites catalysts to obtain paraffinic products." *Environmental technology* 1-10. Accessed September 22, 2019. doi:10.1080/09593330.2019.1650833.

Ming-Yuan, He, Liu Zhonghui, and Min Enze. 1988. "Acidic and hydrocarbon catalytic properties of pillared clay." *Catalysis today* 2(2-3): 321-338. Accessed August 18, 2019. doi:10.1016/0920-5861(88)85013-2.

Mishra, Trilochan, and Kulamani Parida. 1998a. "Transition metal pillared clay 4. A comparative study of textural, acidic and catalytic properties of chromia pillared montmorillonite and acid activated montmorillonite." *Applied Catalysis A: General* 166(1):123-133. Accessed August 30, 2019. doi:10.1016/S0926-860X(97)00247-0.

Mishra, Trilochan, and Kulamani Parida. 1998b "Transition metal oxide pillared clay: 5. Synthesis, characterisation and catalytic activity of iron–chromium mixed oxide pillared montmorillonite." *Applied Catalysis A: General* 174(1-2):91-98. Accessed August 30, 2019. doi:10.1016/S0926-860X(98)00154-9.

Mishra, Trilochan. 2010."Transition metal oxide-pillared clay catalyst: synthesis to application." In *Pillared Clays and Related Catalysts*, edited by Gil, Antonio, Sophia A. Korili, Raquel Trujillano, and Miguel Angel Vicente, 99-128. New York: Springer. doi:10.1007/978-1-4419-6670-4_5.

Miskolczi, Norbert, and Funda Ateş. 2016. "Thermo-catalytic co-pyrolysis of recovered heavy oil and municipal plastic wastes." *Journal of Analytical and Applied Pyrolysis* 117: 273-281. Accessed October 10, 2019. doi:10.1016/j.jaap.2015.11.005.

Miskolczi, Norbert, Chunfei Wu, and Paul T. Williams. 2017. "Pyrolysis of waste plastics using Catalysts: activated carbon, MCM-41 and HZSM-5." International *Journal of Chemical Engineering and Applications* 8(1):67-74. Accessed September 25, 2019. doi: 10.18178/ijcea.2017.8.1.632.

Mnasri-Ghnimi, S., and N. Frini-Srasra. 2019. "Removal of heavy metals from aqueous solutions by adsorption using single and mixed pillared clays." *Applied Clay Science* 179:105151. Accessed August 4, 2019. doi:10.1016/j.clay.2019.105151.

Mokaya, R., and W. Jones. 1995. "Pillared clays and pillared acid-activated clays: a comparative-study of physical, acidic, and catalytic properties." *Journal of Catalysis* 153(1):76-85. Accessed August 18, 2019. doi:10.1006/jcat.1995.1109.

Mokaya, Robert, William Jones, Mary E. Davies, and Mavis E. Whittle. 1993. "Chlorophyll adsorption by alumina-pillared acid activated clays." *Journal of the American Oil Chemists Society* 70(3): 241-244. Accessed August 18, 2019. doi:10.1007/BF02545302.

Mortland, M. M., and V. Berkheiser. 1976. "Triethylene diamine-clay complexes as matrices for adsorption and catalytic reactions." *Clays and Clay Minerals* 24(2):60-63. Accessed August 20, 2019. http://citeseerx.ist.psu.edu/viewdoc/download?doi=10.1.1.491.177&rep=rep1&type=pdf.

Noreña, Luis, Julia Aguilar, Violeta Mugica, Mirella Gutiérrez, and Miguel Torres. 2012. "Materials and Methods for the Chemical Catalytic Cracking of Plastic Waste." In *Material Recycling-Trends and Perspectives*, edited by Dimitris S. Achilas. IntechOpen.

Oakes, W. G., and R. B. Richards. 1949. "619. The thermal degradation of ethylene polymers." *Journal of the Chemical Society(Resumed).* 2929-2935.

Occelli, Mario L., J. Lynch, and J. V. Senders. 1987 "TEM analysis of pillared and delaminated hectorite catalysts." J. Catal.;(United States) 107(2):557-565. Accessed August 11, 2019. doi:10.1016/0021-9517(87)90319-8.

Önal, Eylem, Başak Burcu Uzun, and Ayşe Eren Pütün. 2014. "Bio-oil production via co-pyrolysis of almond shell as biomass and high density polyethylene." *Energy conversion and management* 78:704-710. Accessed October 10, 2019. doi:10.1016/j.enconman.2013.11.022.

Önenç, Sermin, Mihai Brebu, Cornelia Vasile, and Jale Yanik. 2012. "Copyrolysis of scrap tires with oily wastes." *Journal of analytical and applied pyrolysis* 94:184-189. Accessed October 2, 2018. doi:10.1016/j.jaap.2011.12.006.

Panda, Achyut K., and R. K. Singh. 2011. "Catalytic performances of kaoline and silica alumina in the thermal degradation of polypropylene." *Journal of Fuel Chemistry and Technology* 39(3):198-202. Accessed September 19, 2019. doi:10.1016/S1872-5813(11)60017-0.

Pandey, Prashant, and Vipin Kumar Saini. 2018. "Pillared interlayered clays for pollution remediation." In *Green adsorbents for pollutant removal*, edited by Crini G., Lichtfouse E., 353-376. Springer, Cham. doi:10.1007/978-3-319-92162-4_11.

Pánek, Petr, Bruno Kostura, Iveta Čepeláková, Ivan Koutník, and Tomáš Tomšej. 2014. "Pyrolysis of oil sludge with calcium-containing additive." *Journal of analytical and applied pyrolysis* 108:274-283. Accessed September 13, 2019. doi:10.1016/j.jaap.2014.04.005.

Paraschiv, Maria, Radu Kuncser, Mohand Tazerout, and Tudor Prisecaru. 2015. "New energy value chain through pyrolysis of hospital plastic waste." *Applied Thermal Engineering* 87:424-433. Accessed September 13, 2019. doi.org/10.1016/j.applthermaleng.2015.04.070.

Pauling, L. 1930. "The Structure of the Micas and Related Minerals." *Proceedings of the National Academy of Sciences of the United States of America* vol. 16(2):123-129. Accessed June, 16, 2019. doi:10.1073/pnas.16.2.123.

Pauling, Linus. 1930a. "The structure of the chlorites." *Proceedings of the National Academy of Sciences of the United States of America* 16(9):578-582. Accessed June, 16, 2019. doi: 10.1073/pnas.16.9.578.

Pawar Harshal, R., and M. Lawankar Shailendra. 2013. "Waste plastic pyrolysis oil alternative fuel for CI Engine–A Review". *Research Journal of Engineering Sciences* 2(2):26-30. Accessed September 13, 2019. https://pdfs.semanticscholar.org/29be/2ff2f16dae1548c457b7d97586118a568c8a.pdf.

Pergher, S. B. C., A. Corma, and Vicente Fornes. 1999. "Materiales laminares pilareados: preparación y propiedades."*Química nova* 22(5):693-709. Accessed August, 13, 2019. doi:10.1590/S0100-40421999000500013 [Pillared layered materials: preparation and properties *Química nova* 22(5):693-709.]

Pergher, Sibele Berenice Castellã, and Renato Sprung., R. 2005. "Pilarização de uma argila brasileira com poliidroxications de alumínio: preparação, caracterização e propriedades catalíticas."*Quimica Nova* 28(5):777-782. Accessed August 28, 2019. http://www.scielo.br/pdf/%0D/qn/v28n5/25899.pdf [Pillarization Of A Brazilian Clay With Aluminium Polyhydroxications: Preparation, Characterization And Catalytic Properties *Quimica Nova* 28(5):777-782.]

Pinnavaia, Thomas J., Ming Shin Tzou, and Steven D. Landau. 1985. "New chromia pillared clay catalysts." *Journal of the American Chemical Society* 107(16):4783-4785. Accessed August 12, 2019. doi:10.1021/ja00302a033.

Pires, J., and M. L. Pinto. 2010 "Pillared interlayered clays as adsorbents of gases and vapors." In *Pillared clays and related catalysts*, edited by Gil A., Korili S., Trujillano R., Vicente M., 23-42. New York: Springer. doi:10.1007/978-1-4419-6670-4_2.

Plastic Europe: Association of Plastic Manufacturer: Brussels, 2018. "Plastics: The Facts 2018." Accessed October 10, 2019. https://www.plasticseurope.org/en/resources/publications/619-plastics-facts-2018.

Poutsma, Marvin L. 2000. "Fundamental reactions of free radicals relevant to pyrolysis reactions." *Journal of Analytical and Applied Pyrolysis* 54(1-2):5-35. Accessed October 8, 2019. doi:10.1016/S0165-2370(99)00083-2.

Rana, Mohan S., Vicente Sámano, Jorge Ancheyta, and J. A. I. Diaz. 2007. "A review of recent advances on process technologies for upgrading of heavy oils and residua." *Fuel* 86(9):1216-1231. Accessed October 4, 2019. doi:10.1016/j.fuel.2006.08.004.

Ross, Clarence S., and Earl V. Shannon. 1925 "The chemical composition and optical properties of beidellite."*Journal of the Washington Academy of Sciences* 15(21):467-468.

Ross, Clarence S., and Earl V. Shannon. 1926. "The minerals of bentonite and related clays and their physical properties 1."*Journal of the American Ceramic Society* 9(2):77-96. Accessed June, 15, 2019. doi:10.1111/j.1151-2916.1926.tb18305.x.

Ross, Clarence S., and Paul F. Kerr. 1930. "The Kaolin Minerals" *Journal of the American Ceramic Society* 13(3):151-160. Accessed June, 15, 2019. doi:10.1111/j.1151-2916.1930.tb16556.x.

Ross, Clarence Samuel, and Paul Francis Kerr. 1931."The clay minerals and their identity." *Journal of Sedimentary Research* 1(1):55-65. Accessed June, 15, 2019. doi:10.1306/D4268DC9-2B26-11D7-8648000102C1865D.

Roulia, M. 2005. "Synthesis and characterization of novel chromium pillared clays." *Materials chemistry and physics* 91(2-3):281-288. Accessed August 12, 2019. doi:10.1016/j.matchemphys.2004.11.023.

Saladino, Raffaele, Claudia Crestini, Umberto Ciambecchini, Fabiana Ciciriello, Giovanna Costanzo, and Ernesto Di Mauro. 2004. "Synthesis and degradation of nucleobases and nucleic acids by formamide in the presence of montmorillonites." *ChemBioChem* 5(11):1558-1566. Accessed June, 15, 2019. doi:10.1002/cbic.200400119.

Santos, Everton, Bruna Rijo, Francisco Lemos, and M. A. N. D. A. Lemos. 2019. "A Catalytic Reactive Distillation Approach to High Density Polyethylene Pyrolysis–Part 1–Light Olefin Production." *Chemical Engineering Journal* 378:122077. Accessed September 25, 2019. doi:10.1016/j.cej.2019.122077.

Santos, P. S. 1989. *Ciência e Tecnologia das Argilas*, 2ª edição vol. 1: São Paulo: Edgard Blücher Ltda

Sassi, Halima, Gwendoline Lafaye, Hédi Ben Amor, Abdelaziz Gannouni, Mohamed Razak Jeday, and Jacques Barbier. 2018. "Wastewater treatment by catalytic wet air oxidation process over Al-Fe pillared clays synthesized using microwave irradiation." *Frontiers of Environmental Science & Engineering* 12(1):2. Accessed August 27, 2019. doi:10.1007/s11783-017-0971-1.

Sawhney, B. L. 1968. "Aluminum interlayers in layer silicates effect of OH/Al ratio of Al solution, time of reaction, and type of structure." *Clays and Clay Minerals* 16(2): 157-163. Accessed August 20, 2019. http://citeseerx.ist.psu.edu/viewdoc/download?doi=10.1.1.492.5015&rep=rep1&type=pdf.

Schoonheydt, Robert A., Tom Pinnavaia, Gerhard Lagaly, and Nick Gangas. 1999. "Pillared clays and pillared layered solids." *Pure and Applied Chemistry* 71(12):2367-

2371. Accessed August 15, 2019. https://webpages.ciencias.ulisboa.pt/~jpsilva/quersaber/PILCS.pdf.

Schubert, Teresa, Markus Lehner, Thomas Karner, Wolfgang Hofer, and Andreas Lechleitner. "Influence of reaction pressure on co-pyrolysis of LDPE and a heavy petroleum fraction. 2019. "*Fuel Processing Technology* 193:204-211. Accessed October 8, 2019. doi:10.1016/j.enconman.2016.04.004.

Sen, Sudip Kumar, and Sangeeta Raut. 2015. "Microbial degradation of low density polyethylene (LDPE): A review." *Journal of Environmental Chemical Engineering* 3(1):462-473. Accessed September 13, 2019. doi:10.1016/j.jece.2015.01.003.

Seo, Young-Hwa, and Dae-Hyun Shin. 2002. "Determination of paraffin and aromatic hydrocarbon type chemicals in liquid distillates produced from the pyrolysis process of waste plastics by isotope-dilution mass spectrometry." *Fuel* 81(16):2103-2112. Accessed October 12, 2018. doi:10.1016/S0016-2361(02)00197-7.

Serwicka, Ewa M., and Krzysztof Bahranowski. 2004. "Environmental catalysis by tailored materials derived from layered minerals." *Catalysis Today* 90(1-2):85-92. Accessed August 6, 2019. doi:10.1016/j.cattod.2004.04.012.

Sharma, Sumedha, and Aloke K. Ghoshal. 2010. "Study of kinetics of co-pyrolysis of coal and waste LDPE blends under argon atmosphere." *Fuel* 89(12):3943-3951. Accessed September 29, 2019. doi:10.1016/j.fuel.2010.06.033.

Sharuddin, Shafferina Dayana Anuar, Faisal Abnisa, Wan Mohd Ashri Wan Daud, and Mohamed Kheireddine Aroua. 2016. "A review on pyrolysis of plastic wastes." *Energy conversion and management* 115: 308-326. Accessed September 19, 2019. doi:10.1016/j.enconman.2016.02.037.

Siddiqui, Mohammad Nahid, and Halim Hamid Redhwi. 2009. "Catalytic coprocessing of waste plastics and petroleum residue into liquid fuel oils." *Journal of Analytical and Applied Pyrolysis* 86(1):141-147. Accessed October 10, 2019. doi:10.1016/j.jaap.2009.05.002.

Silva, D. C., A. A. Silva, C. F. Melo, and M. R. C. Marques. 2017. "Production of oil with potential energetic use by catalytic co-pyrolysis of oil sludge from offshore petroleum industry." *Journal of Analytical and Applied Pyrolysis* 124:290-297. Accessed October 6, 2018. doi:10.1016/j.jaap.2017.01.021.

Sogancioglu, Merve, Gulnare Ahmetli, and Esra Yel. 2017. "A Comparative Study on Waste Plastics Pyrolysis Liquid Products Quantity and Energy Recovery Potential." *Energy Procedia* 118: 221-226. Accessed September 19, 2019. doi: 10.1016/j.egypro.2017.07.020.

Srinivasan, Rajani. 2011. "Advances in application of natural clay and its composites in removal of biological, organic, and inorganic contaminants from drinking water." *Advances in Materials Science and Engineering* 1–17. Accessed June, 26, 2019. doi:10.1155/2011/872531.

Starshinova, V., Ya Abdelsalam, V. Gorelysheva, A. Oparkin, S. Gnevashev, A. Pyataev, R. Khamidullin, and E. Karalin. 2019. "Heterogenous catalysts based on Fe-pillared clays for catalytic aquathermolysis processes of heavy oil." In *IOP Conference Series: Earth and Environmental Science*, 282(1):012036. IOP Publishing. Accessed August 27, 2019. doi:10.1088/1755-1315/282/1/012036.

Stern, Robert, and Mark J. Jedrzejas. 2008. "Carbohydrate polymers at the center of life's origins: the importance of molecular processivity."*Chemical reviews* 108(12):5061-5085. Accessed June, 15, 2019. doi:10.1021/cr078240l.

Sterte, Johan. 1988. "Hydrothermal treatment of hydroxycation precursor solutions." *Catalysis Today* 2(2-3):219-231. Accessed August 4, 2019. doi:10.1016/0920-5861(88)85005-3.

Sychev, Mikhail, Vincent HJ de Beer, and Rutger A. van Santen. 1997. "Chromia and chromium sulfide pillared clays differing in pillar density." *Microporous materials* 8(5-6):255-265. Accessed August 30, 2019. doi:10.1016/S0927-6513(96)00086-7.

Takuma, Kazuhiko, Yoshio Uemichi, and Akimi Ayame. 2000. "Product distribution from catalytic degradation of polyethylene over H-gallosilicate."*Applied Catalysis A: General* 192(2):273-280. Accessed October 4, 2019. doi:10.1016/S0926-860X(99)00399-3.

Tan, Xue-Cai, Chun-Chun Zhu, Qing-Kun Liu, Tian-Yi Ma, Pei-Qing Yuan, Zhen-Min Cheng, and Wei-Kang Yuan. 2014. "Co-pyrolysis of heavy oil and low density polyethylene in the presence of supercritical water: the suppression of coke formation." *Fuel processing technology*118:49-54. Accessed October 10, 2019. doi:10.1016/j.fuproc.2013.08.007.

Tang, Xiaozhong, Wen-Qing Xu, Yan-Fei Shen, and Steven L. Suib. 1995. "Preparation and characterization of pillared gallium aluminum clays with enriched pillars." *Chemistry of materials* 7(1):102-110. Accessed August 11, 2019. doi:10.1021/cm00049a016.

Teixeira-Neto, E., and A. A. Teixeira-Neto. 2009. "Modificação química de argilas: desafios científicos e tecnológicos para obtenção de novos produtos com maior valor agregado."*Química Nova* 32(3):809-817. Accessed June, 24, 2019. doi:10.1590/S0100-40422009000300023 [Chemical modification of clays: scientific and technological challenges for obtaining new value-added products *Química Nova* 32(3): 809-817.]

Thomas, José Eduardo. 2004 "O Petróleo." In *Fundamentos de engenharia de petróleo* 2ed, José Eduardo Thomas (organizador) Rio de Janeiro: Interciência-PETROBRAS.

Tomul, Fatma, and Suna Balci. 2009. "Characterization of Al, Cr-pillared clays and CO oxidation." *Applied Clay Science* 43(1):13-20. doi:10.1016/j.clay.2008.07.006.

Tomul, Fatma. 2016. "The effect of ultrasonic treatment on iron–chromium pillared bentonite synthesis and catalytic wet peroxide oxidation of phenol." *Applied Clay Science* 120: 121-134. Accessed August 12, 2019. doi:10.1016/j.clay.2015.11.007.

Toranzo, R., M. A. Vicente, and M. A. Banares-Munoz. 1997. "Pillaring of a Saponite with Aluminum– Chromium Oligomers. Characterization of the Solids Obtained." *Chemistry of materials* 9(8): 1829-1836. Accessed August 29, 2019. doi:10.1021/cm970117v.

Tyagi, Vinay Kumar, and Shang-Lien Lo. 2013. "Sludge: a waste or renewable source for energy and resources recovery?." *Renewable and Sustainable Energy Reviews* 25: 708-728. Accessed September 16, 2019. doi:10.1016/j.rser.2013.05.029.

Tzou, M. S., and T. J. Pinnavaia. 1988 "Chromia pillared clays." *Catalysis today* 2(2-3): 243-259. Accessed August 29, 2019. doi:10.1016/0920-5861(88)85007-7.

Tzou, Ming-Shin 1983 "Clay catalysts pillared by metal hydroxypolymers." Ph.D.Thesis, Michigan State University.

Uçar, Suat, Ahmet R. Özkan, and Selhan Karagöz. 2016. "Co-pyrolysis of waste polyolefins with waste motor oil." *Journal of Analytical and Applied Pyrolysis* 119:233-241. Accessed October 10, 2019. doi:10.1016/j.jaap.2016.01.013.

Uddin, Mohammad Kashif. 2017. "A review on the adsorption of heavy metals by clay minerals, with special focus on the past decade." *Chemical Engineering Journal* 308:438-462. Accessed August 6, 2019. doi:10.1016/j.cej.2016.09.029.

Undri, Andrea, Luca Rosi, Marco Frediani, and Piero Frediani. 2014. "Efficient disposal of waste polyolefins through microwave assisted pyrolysis." *Fuel* 116:662-671. Accessed September 19, 2019. doi:10.1016/j.fuel.2013.08.037.

Uzoejinwa, Benjamin Bernard, Xiuhua He, Shuang Wang, Abd El-Fatah Abomohra, Yamin Hu, and Qian Wang. 2018. "Co-pyrolysis of biomass and waste plastics as a thermochemical conversion technology for high-grade biofuel production: Recent progress and future directions elsewhere worldwide." *Energy conversion and management* 163: 468-492. Accessed October 10, 2019. doi:10.1016/j.enconman. 2018.02.004.

Vaccari, A. 1999. "Clays and catalysis: a promising future." *Applied Clay Science* 14(4):161-198. Accessed June, 24, 2019. doi:10.1016/S0169-1317(98)00058-1.

Valle, Maria Letícia M., Maria José O. C. Guimarães, and Cristiane Sampaio. S 2004. "Degradation of polyolefins using zeolitic catalysts."*Polímeros* 14(1):17-21. Accessed September 13, 2019. doi:10.1590/S0104-14282004000100009.

Varma, Rajender S. 2002. "Clay and clay-supported reagents in organic synthesis." *Tetrahedron* 58(7):1235-1255. Accessed June, 24, 2019. https://www.thevespiary.org/rhodium/Rhodium/hive/hiveboard/picproxie_docs/000458467-Tetrahedron_58_2002_1235-1255.pdf.

Velde, Bruce B., and Alain Meunier. 2008. "Introduction" In *The origin of clay minerals in soils and weathered rocks*, 1-403. Springer Science & Business Media.

Vicente, M. A., A. Gil, and F. Bergaya. 2013. "Pillared clays and clay minerals." In *Developments in Clay Science*, vol. 5, edited by Faïza Bergaya, Gerhard Lagaly, 523-557. Elsevier. doi:10.1016/B978-0-08-098258-8.00017-1.

Volzone, C. 1995. "Hydroxy-chromium smectite: influence of Cr added." Clays and clay minerals 43(3):377-382. Accessed August 30, 2019. doi:10.1346/CCMN 1995.0430312.

Volzone, C., A. M. Cesio, R. M. Torres Sanchez, and E. Pereira. 1993. "Hydroxy-chromium smectite." *Clays and Clay Minerals* 41(6):702-706. Accessed August 29, 2019.doi:10.1346/CCMN.1993.04106.

Volzone, C., and A. M. Cesio. 2003. "Changes in OH-Cr-montmorillonite after heating in air and nitrogen atmospheres." *Materials chemistry and physics* 79(1):98-102. Accessed August 30, 2019. doi:10.1016/S0254-0584(02)00430-3.

Volzone, Cristina. 2001 "Pillaring of different smectite members by chromium species (Cr-PILCs)."*Microporous and mesoporous materials* 49(1-3):197-202. Accessed August 30, 2019. doi:10.1016/S1387-1811(01)00423-1.

Weckhuysen, Bert M., Israel E. Wachs, and Robert A. Schoonheydt. 1996. "Surface chemistry and spectroscopy of chromium in inorganic oxides." *Chemical reviews* 96(8): 3327-3350. Accessed August 29, 2019. doi:10.1021/cr940044o.

Xu, Menghan, Haifeng Liu, Hui Zhao, and Weifeng Li. 2014. "Effect of oily sludge on the rheological characteristics of coke-water slurry." *Fuel* 116:261-266. Accessed September 13, 2019. doi:10.1016/j.fuel.2013.07.114.

Yang, Penghui, Pei Zhou, Yuan Li, Chengtun Qu, and Ningsheng Zhang. 2018. "Recent development in pyrolytic catalysts of oil sludge." *Petroleum Science and Technology* 36(7): 520-524. Accessed September 16, 2019. doi:10.1080/10916466.2018.1431661.

Yao, Peng Quan, Lin Hua Zhu, Jin Yang, and Tian Si. 2013. "Isomerization of α-pinene over alumina-pillared montmorillonite." *Advanced Materials Research* 746:49-52. Accessed August 4, 2019. doi:10.4028/www.scientific.net/AMR.746.49.

Yuan, Peng, Hongping He, Faïza Bergaya, Daqing Wu, Qin Zhou, and Jianxi Zhu. 2006. "Synthesis and characterization of delaminated iron-pillared clay with meso–microporous structure." Microporous and Mesoporous Materials 88(1-3): 8-15. Accessed August 27, 2019. doi:10.1016/j.micromeso.2005.08.022.

Yun, Ju Byeong, Seong Ho Hwang, and Jin Ho Choe. 2000. "Trimeric chromium oxyformate route to chromia-pillared clay." *Bulletin of the Korean Chemical Society* 21(10):1049-1051. Accessed August 30, 2019. http://ocean.kisti.re.kr/downfile/volume/chemical/JCGMCS/2000/v21n10/JCGMCS_2000_v21n10_989.pdf

Yurdakoç, Mürüvvet, Mehmet Akçay, Yalçın Tonbul, Fatih Ok, and Kadir Yurdakoç. 2008. "Preparation and characterization of Cr-and Fe-pillared bentonites by using $CrCl_3$, $FeCl_3$, $Cr(acac)_3$ and $Fe(acac)_3$ as precursors." *Microporous and mesoporous materials* 111(1-3):211-218. Accessed August 27, 2019. doi:10.1016/j.micromeso.2007.07.032.

Zeng, Q. H., A. B. Yu, G. Q. Lu, and D. R. Paul. 2005. "Clay-based polymer nanocomposites: research and commercial development." *Journal of nanoscience and nanotechnology* 5(10):1574-1592. doi:10.1166/jnn.2005.411.

Zhang, Huiyan, Jianlong Nie, Rui Xiao, Baosheng Jin, Changqing Dong, and Guomin Xiao. 2014. "Catalytic co-pyrolysis of biomass and different plastics (polyethylene, polypropylene, and polystyrene) to improve hydrocarbon yield in a fluidized-bed reactor." *Energy & Fuels* 28(3):1940-1947. Accessed October 10, 2019. doi:10.1021/ef4019299.

Zhang, Xuesong, Hanwu Lei, Shulin Chen, and Joan Wu. 2016. "Catalytic co-pyrolysis of lignocellulosic biomass with polymers: a critical review." *Green Chemistry* 18(15): 4145-4169. Accessed September 22, 2019. doi: 10.1039/C6GC00911E.

Zheng, Chenggang, Manman Wang, Yongli Wang, and Zhiyong Huang. 2012. "Optimization of biosurfactant-mediated oil extraction from oil sludge."*Bioresource technology* 110:338-342. Accessed September 13, 2019. doi:10.1016/j.biortech. 2012.01.073.

Zhou, Qi, Jianxun Yang, Miaomiao Liu, Yang Liu, Stefanie Sarnat, and Jun Bi. 2018. "Toxicological Risk by Inhalation Exposure of Air Pollution Emitted from China's Municipal Solid Waste Incineration." *Environmental science & technology* 52(20):11490-11499. Accessed September 16, 2019. doi:10.1021/acs.est.8b03352.

Zhu, H. Y., Q. Ma, and GQ Max Lu. 1999. "Influence of heat treatments on the pore and adsorption characteristics of sodium doped alumina pillared bentonite." *Journal of Porous Materials* 6(2):135-142. Accessed August 11, 2019. doi: 10.1023/A: 1009683422478.

Zhu, Jianxi, Ke Wen, Ping Zhang, Yuebo Wang, Lingya Ma, Yunfei Xi, Runliang Zhu, Hongmei Liu, and Hongping He. 2017. "Keggin-Al30 pillared montmorillonite." *Microporous and Mesoporous Materials* 242: 256-263. Accessed August 4, 2019. doi:10.1016/j.micromeso.2017.01.039.

Zhu, Jianxi, Ke Wen, Yuebo Wang, Lingya Ma, Xiaoli Su, Runliang Zhu, Yunfei Xi, and Hongping He. 2018 "Superior thermal stability of Keggin-Al30 pillared montmorillonite: A comparative study with Keggin-Al13 pillared montmorillonite." *Microporous and Mesoporous Materials* 265:104-111. Accessed August 4, 2019. doi:10.1016/j.micromeso.2018.02.007.

Zubkov, S. A., L. M. Kustov, V. B. Kazansky, G. Fetter, D. Tichit, and F. Figueras. 1994. "Study of the nature of acid sites of montmorillonites pillared with aluminium and oligosilsesquioxane complex cations. 1. Brönsted acidity." *Clays and clay minerals* 42(4): 421-427. Accessed August 11, 2019. doi:10.1346/CCMN.1994.0420407.

Biographical Sketches

Monica Regina da Costa Marques

Affiliation: Universidade do Estado do Rio de Janeiro (UERJ)

Education: Organic Chemistry Doctor

Business Address: Street São Francisco Xavier, 524, Maracanã, RJ, Brazil.

Research and Professional Experience:

Titular Professor, Department of Organic Chemistry, Institute of Chemistry, Rio de Janeiro State University - UERJ, since 1995. Graduated in Chemistry and Technological Chemistry (UFRJ), Master of Science (UFRJ), PhD in Sciences (UFRJ) and conducted postdoctoral studies at the Professor Eloísa Mano Institute of Macromolecules (UFRJ) developing methodologies for analysis of polymer microstructure by nuclear magnetic resonance spectrometry. Undergraduate student of the Nutrition course at UERJ (since 2017). At UERJ, she was head of the department of Organic Chemistry (1998-2000), Vice-Coordinator of the Graduate in Chemistry (2004-2006, 2011-2012 and 2016-2018). Coordinator of the Graduate Program in Chemistry (2006-2008) and Coordinator of the Doctorate Program in Environment (2008-2010), Special Advisor of the Sub-Rectorate of Graduate Studies and Research (2008-2015). As a researcher, she is a UERJ Researcher (since 1998) and Research Leader at CNPq of the Environmental Technology Laboratory (link dgp.cnpq.br/dgp/espelhogrupo/6348267049447329) and coordinates the Fernanda Coutinho Analytical Center. She was coordinator of the Integrated Solid Waste Management Plan (PEGIRS) with the Secretariat of the Environment (2008-2009), researcher at INCT - National Institute of Oil and Gas and Red Thematic Glicocience in Salud. As a teacher and researcher, she guides Master and Doctoral students of the Postgraduate Courses in Chemistry and Postgraduate in Environment, both at UERJ, in the Graduation of Chemistry and Chemical Engineering, the disciplines of Organic Chemistry, Experimental Organic Chemistry, Methods applied to organic chemistry. Areas of Expertise: The area of expertise in scientific research is multidisciplinary (Environmental Sciences) involving the assessment of environmental impacts of industries and the development of treatment techniques to minimize these impacts, as well as the development of techniques for pollutant analysis in various environmental matrices. She is a reviewer of environmental and chemical journals and ad hoc consultant to CNPq, Capes, Faperj, FAPES and ARI. (California-USA).

Honors: Awarded by Walter Baptist Mors Medal for relevant contributions to Química Fluminense by the Brazilian Society of Chemicals - Rio Regional. CNPq Level 1D Productivity Fellow. FAPERJ Scientist of Our State

Publications from the Last 3 Years:

1. Nazareth, Monick, Mônica RC Marques, Marcia CA Leite, and Ítalo Braga Castro. 2019. "Commercial plastics claiming biodegradable status: Is this also accurate for marine environments?." *Journal of hazardous materials* 366: 714-722. Accessed August 11, 2019. doi:10.1016/j.jhazmat.2018.12.052.
2. Gomes, A. C. S. A., L. C. Costa, D. C. Brito, R. J. França, and Mônica RC Marques. 2019. "Development of a new ion-imprinted polymer (IIP) with Cd^{2+} ions based on divinylbenzene copolymers containing amidoxime groups." *Polymer Bulletin* 1-13. Accessed August 11, 2019. doi:10.1007/s00289-019-02842-8.
3. Milato, Jônatas V., Rodrigo J. França, and Mônica RC Marques. 2019."Pyrolysis of oil sludge from the offshore petroleum industry: influence of different mesoporous zeolites catalysts to obtain paraffinic products." *Environmental technology* 1-10. Accessed August 11, 2019. doi:10.1080/09593330.2019.1650833.
4. Oliveira, B. R.; Oliveira, R. N.; Scheid, C. M.; Ferraz, A. S. F. S.; Marques, M. R. C.; Calçada, L. A. 2018. "Effects Of PH And Solid Concentration On The Rheology Of Drilling Fluids Composed By Natural Clay, Water, And Nacmc. Brazilian Journal Of Petroleum And Gas". 12:99 – 106.
5. Soares, Andréia Silveira Freire, Mônica Regina Costa Marques, Luis Américo Calçada, Moacyr Nogueira Borges Filho, and Carmem Lúcia de Oliveira Petkowicz. 2018. "Interaction of blockers on drilling fluids rheology and its effects on sealing of fractures and prevention of filtrate invasion." *Journal of Petroleum Science and Engineering* 171: 260-270. Accessed August 14, 2019. doi:10.1016/j.petrol.2018.07.053.
6. Fernandes, A. S., J. L. Mazzei, H. Evangelista, M. R. C. Marques, E. R. A. Ferraz, and I. Felzenszwalb. 2018. "Protection against UV-induced oxidative stress and DNA damage by Amazon moss extracts." *Journal of Photochemistry and Photobiology B: Biology* 183: 331-341. Accessed August 14, 2019. doi:10.1016/j.jphotobiol.2018.04.038.
7. Felzenszwalb, Israel, Andreia da Silva Fernandes, Lara Barroso Brito, Gisele Augusto Rodrigues Oliveira, Paula Aquino Soeiro Silva, Maria Elena Arcanjo, Monica Regina da Costa Marques et al. 2018. "Toxicological evaluation of nail polish waste discarded in the environment." *Environmental Science and Pollution Research* 1-14. Accessed August 14, 2019. doi:10.1007/s11356-018-1880-y

8. Paula, Mariana S., Rodrigo PM Ribeiro, Alexandre A. Cerqueira, Fábio Merçon, Daniele Maia Bila, Louise da Cruz Felix, and Mônica RC Marques. 2018. "Use of reverse osmosis as a polish for the cationic surfactant after electro-oxidative treatment: Acute and chronic toxicity assessment." *Ecotoxicology and environmental safety* 163: 521-527. Accessed August 16, 2019. doi:10.1016/j.ecoenv.2018.07.112.
9. Oliveira, Alan, Alexandre Cerqueira, Paulo Souza, Michelle Rigo, Julieta Paiva, and Monica Marques. 2017. Evaluation of Electrochemical and Fenton Processes in the Treatment of Mature Landfill L eachates. *International Journal of Scientific Research in Science and Technology*. (3):285-293. Accessed August 16, 2019. https://pdfs.semanticscholar.org/4a04/893681b6e9b99bf6283173c17f1554c3515c .pdf.
10. Fernandes, A. S., F. V. C. Mello, S. Thode Filho, R. M. Carpes, J. G. Honório, M. R. C. Marques, I. Felzenszwalb, and E. R. A. Ferraz. 2017. "Impacts of discarded coffee waste on human and environmental health." *Ecotoxicology and environmental safety* 141: 30-36. Accessed August 16, 2019. doi:10.1016/j. ecoenv.2017.03.011.
11. Faillace, Jacyra Guimarães, Claudinei Fernandes de Melo, Soraia Peres Lima de Souza, and Mônica Regina da Costa Marques. 2017. "Production of light hydrocarbons from pyrolysis of heavy gas oil and high density polyethylene using pillared clays as catalysts." *Journal of Analytical and Applied Pyrolysis* 126: 70-76. Accessed August 16, 2019. doi:10.1016/j.jaap.2017.06.023.
12. Silva, D. C., A. A. Silva, C. F. Melo, and M. R. C. Marques. 2017."Production of oil with potential energetic use by catalytic co-pyrolysis of oil sludge from offshore petroleum industry." *Journal of Analytical and Applied Pyrolysis* 124: 290-297. Accessed August 19, 2019. doi:10.1016/j.jaap.2017.01.021.
13. Couto, Rafael Schirmer de Paula, Aline Faria Oliveira, Alcides Wagner Serpa Guarino, Daniel Vidal Perez, and Mônica Regina da Costa Marques. 2017. "Removal of ammonia nitrogen from distilled old landfill leachate by adsorption on raw and modified aluminosilicate."*Environmental technology* 38(7): 816-826. Accessed August 19, 2019. doi:10.1080/09593330.2016.1212935.
14. Melo, Claudinei F., Leandro A. Silva, Luciana C. Costa, and Monica RC Marques. 2017. "Synergistic Effect of Adsorption and Enzymatic Conversion in the Bisphenol-A Removal by Laccase Immobilized on Poly (glycidyl methacrylate-co-ethyleneglycol dimethacrylate)."*Journal of the Brazilian Chemical Society* 28(11): 2192-2201. Accessed August 19, 2019. doi:10.21577/0103-5053.2017 0070.

Ana Cristina da Silva Serra

Affiliation: Universidade do Estado do Rio de Janeiro (UERJ)

Education: Petroleum Systems Master

Business Address: Street São Francisco Xavier, 524 – Maracanã, RJ, Brazil.

Research and Professional Experience:

PhD student in Chemistry at Rio de Janeiro State University (UERJ), working in the area of waste oil treatment, catalysis and clays since 2019.

Graduated in Chemistry and Technological Chemistry (UFRJ), Master of Science in Petroleum Systems (UFRJ). She has experience in geochemistry as a researcher at the Petrobras Research and Development Center (CENPES) since 2003. At Cenpes she worked in the areas of petroleum geochemistry and surface geochemistry. Has experience in geochemical interpretation of piston-core and oil samples by liquid chromatography (LC), gas cromatography (GC), stable carbon isotopes ($\delta^{13}C$) saturated and aromatic biomarkers (GC-MS) and development of tools for onshore and offshore surface geochemistry exploration targets.

She was member of Brazilian Geochemical Society (SBGQ) as Director Advisor SBGQ (2014-2015). Member of Editorial Board of Scientific Journal *Geochimica Brasiliensis* (2007 -2009).

As a teacher, she worked in an oil and gas technical course teaching about oil and refining, instrumentation and control, safety, renewable energy and regulatory standards (2008-2009).

Jacyra Guimarães Faillace

Affiliation: Federal Institute of Education, Science and Technology of Rio de Janeiro (IFRJ)

Education: Chemistry Doctor

Business Address: Street Lúcio Tavares, 1045, Nilópolis - RJ, Brazil.

Research and Professional Experience:

She has a degree in Chemistry from the State University of Rio de Janeiro (1994), a master's degree in Inorganic Chemistry from the Federal University of Rio de Janeiro (2001) and a doctorate in Chemistry from the State University of Rio de Janeiro (2018). She is currently a professor at the Federal Institute of Education, Science and Technology of Rio de Janeiro. Has experience in Chemistry, focusing on Inorganic Chemistry, acting on the following subjects: general and inorganic chemistry teaching. As a teacher and researcher, she guides graduation students in Chemistry at IFRJ. In the Graduation of Chemistry she teaches the disciplines of general chemistry and inorganic chemistry

Publications from the Last 3 Years:

Faillace, Jacyra Guimarães, Claudinei Fernandes de Melo, Soraia Peres Lima de Souza, and Mônica Regina da Costa Marques. 2017. "Production of light hydrocarbons from pyrolysis of heavy gas oil and high density polyethylene using pillared clays as catalysts." *Journal of Analytical and Applied Pyrolysis* 126: 70-76. Accessed August 19, 2019. doi:10.1016/j.jaap.2017.06.023.

In: An Introduction to Aluminosilicates
Editor: Nero Regina Blevins
ISBN: 978-1-53617-250-8

Chapter 3

APPLICATION OF SYNTHETIC ALUMINOSILICATES IN ADHESIVE DRY CONSTRUCTION MIXTURES

Loganina Valentina Ivanovna
and Zhegera Kristina Vladimirovna
Penza State University of Architecture and Construction,
Department "Quality Management and Technology of
Construction Production," Penza, Russia

ABSTRACT

Information is giving on the possibility of using amorphous aluminosilicates as a modifying additive in the formulation of cement-based dry building mixtures (DBM) with the cement. The physicochemical properties of the additive are considered, the results of XRD and DTA analyzes are presented.

The regularities of the structure formation of the cement composite with the use of additives based on amorphous aluminosilicates in the formulation are established. He is additionally consist in the formation of calcium-sodium hydrosilicates and minerals of the zeolite group in reducing the amount of free lime and increasing the amount of chemically bound water reducing the total and capillary porosity and increasing the contraction and gel porosity.

It has been established that the introduction of an amorphous aluminosilicate-based additive into the formulation of adhesive DBM helps to accelerate the increase in strength of tile adhesive. A model of the strength of the cement composite has been compiled, allowing you to choose the optimal content of components in the formulation of the dry building mixture.

It was revealed that tile glue made on the basis of the developed formulation is crack resistant, resistant to slipping, and has high adhesive strength. The regularities of changes in the stress state of the adhesive layer from the action of temperatures for various climatic conditions of operation are established. The scope of the developed adhesive dry building mix is determined depending on the climatic conditions of use.

The stress distribution in the adhesive layer based on the developed composition of tile glue from the action of temperature in various climatic regions. It were found that the maximum values of tensile stresses σx occurring in the adhesive layer at the border with the substrate and at the border with the ceramic tile are observed in March.

It were revealed that in the center of the adhesive joint the maximum values of tangential stresses are observed in March (Penza) and in February (Yakutsk), and in the marginal zone in September (Penza) and in October (Yakutsk). For the conditions of Penza, the maximum values of shear stresses in the adhesive layer at the border with the substrate are observed in February, and at the border with ceramic tile in September. For the conditions of Yakutsk, the maximum values of tangential stresses in the adhesive layer are observed at the border with the substrate and at the border with ceramic tile in October. It was revealed that tile adhesive based on the developed adhesive DBM is crack resistant for Russian cities, located in a dry moisture zone and climatic subareas IIB.

The formulation of a cement-based adhesive dry building mix designed for gluing tiles to surfaces and containing Portland cement, quartz sand of the Ukhta deposit with the optimal ratio of fractions, the additive based on amorphous aluminosilicates, the plasticizing additive Kratasol PFM and the redispersible powder Neolith P 4400. Tile adhesive made on the basis of the developed DBM is characterized by the following indicators: adhesion strength upon tearing after 28 days of air-dry hardening is more than R_{adg}> 1,4 MPa, frost resistance grade F50, frost resistance grade of the contact zone F_{kz}50, water absorption with capillary suction W = 1.43 kg/m^2 H 0.5, tile sliding is less than 0.3 mm, shrink strain 0.029% - 0.034%.

1.1. Physical and Chemical Indicators of Additives Based on Synthetic Aluminosilicates for Dry Construction Mixtures

To expand the market of raw materials and reduce the supply of imported additives, we investigated the possibility of using dry construction mixtures of synthetic aluminosilicates obtained by precipitating aluminosilicates from sodium liquid glass with aluminum sulfate $Al_2 (SO_4)_3$ in the formulation.

The additives based on amorphous aluminosilicates were obtained by precipitating aluminosilicates from liquid sodium glass with a silicate module 2.8 by introducing a 15% solution of technical aluminum sulfate $Al_2(SO_4)_3$ followed by washing of the obtained Sediment with distilled water and its drying in a drying oven at a temperature t = (105 ± 5) ^{0}C to constant mass and grinding.

Physico-chemical indicators of mineral additive based on synthetic metallic silicates are presented in Table 1.1.

Table 1.1. Physico-chemical indicators of mineral supplements

№	Name of the indicator	Value of the indicator
1	Appearance	White powder
2	Activity, mg/g	350
3	Specific surface, S_{sp}, m^2/kg	688,6
4	Bulk density, kg/m^3	568,2
5	True density, kg/m^3	2140

The activity of mineral additives were determined by their solubility in a 20% KOH solution as follows: the initial mineral additive were ground to complete passage through sieve No. 008 and dried to constant weight in an oven for 2 hours at (105 ± 5)°C. The dried additive were placing in a glass beaker and poured with a 20% KOH solution. The resulting solution were placing for 3 h in a boiling water bath. After, the settled solution were filtering through a loose ashless filter. The precipitate in the glass were washing with hot distilled water to completely transferring the precipitate to the filter. The washed precipitate together with a funnel filter were dried to constant weight.

Figure 1.1. Microstructure of amorphous aluminosilicate.

It were established that the structure of the additive is represented mainly by particles of round shape with a size of 5.208–5.704 microns, but there are found parts of a flaky shape with a size of 7.13–8.56 microns. The specific surface of the particles, measured by the BET method, were S_{sp} = (68.6 ± 3.5) m^2/g. The chemical composition of the additive based on synthetic aluminosilicates is presented in Table 1.2 according to the spectra presented in Figure 1.2.

Analyzing the data in Table 1.2 revealed that predominate chemical elements O, Si, Na, S, and Al in chemical composition amorphous aluminosilicates - containing 60.69%, 31.26%, 24.23%, 18.69% and 8.29% respectively. The preponderance of these elements has a positive effect on the formation of cement stone structures with used synthetic additives.

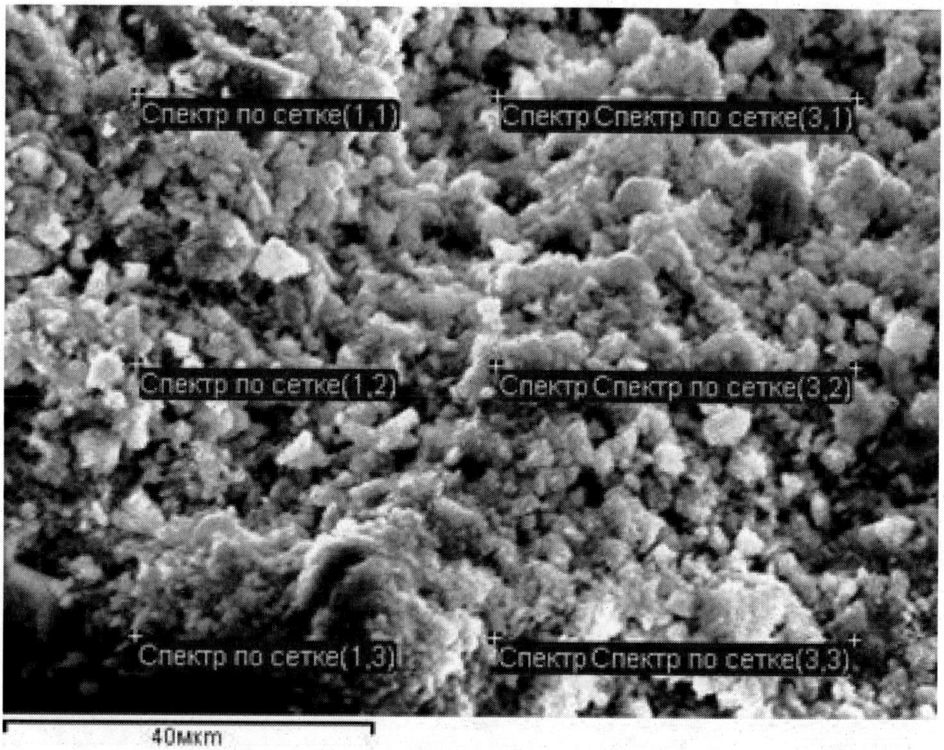

Figure 1.2. Microstructure of amorphous aluminosilicate.

Table 1.2. Chemical composition of the admixture

Spectrum (according to Figure 1.2)	Chemical elements additives weight, [%]				
	O	Na	Al	Si	S
Spectrum on the grid (1,1)	52,42	24,23	1,10	7,92	14,34
Spectrum on the grid (3,1)	57,71	8,61	2,59	31,09	-
Spectrum on the grid (1,2)	60,69	12,36	8,29	17,27	1,40
Спектр по сетке Spectrum on the grid (3,2)	60,23	8,67	3,15	27,95	-
Spectrum on the grid (1,3)	36,73	10,67	2,66	31,26	18,69
Spectrum on the grid (3,3)	60,00	10,22	1,75	27,35	0,68
Max	60,69	24,23	8,29	31,26	18,69
Min	36,73	8,61	1,10	7,92	0,68

1.2. STRUCTURING CEMENT COMPOSITION IN THE PRESENCE OF AN ADDITIVE BASED ON SYNTHETIC ALUMINOSILICATES

Portland cement Volsky PC 400 – D20 were used in the work with fine grinding (sieve No. 008) 93.6%, true density ρ_{true} = 3000 kg/m^3, bulk density ρ_{bulk} = 1300 kg/m^3, with specific surface S_{sp} = 340 m^2/kg (GOST 31108).

The introduction of additives based on synthetic aluminosilicates in the recipe of the cement composite leads to an acceleration of the setting time and an increase in the value of the normal density of the cement paste. The data obtained are given in Table 1.3.

Table 1.3. Change in the normal density and setting time of the cement paste depending on the content of the additive

Content of the additives, %, by weight of cement	The normal density of the cement test, %	Setting time	
		Beginning	End
-	28	2h 30 min	5h
10	34	50 min	1h 40 min
20	41	40 min	1h 30 min
30	43	20 min	1h 15 min

Analysis of the data (Table 1.3) indicates that the cement binder with additive synthetic aluminosilicates has a higher value of the normal density of cement paste (34–43% depending on the content of the additive). There is an acceleration of the setting time. In the cement paste without an additive, the beginning and end of the setting are, respectively, 2 h 30 min and 5 h, and in the cement binder with additive 20% synthetic aluminosilicates, respectively - 40 min and 1 h 30 min. With increasing content of the additive, setting time is accelerated. The presence in the recipe of cement paste synthetic aluminosilicates affects the structure formation of cement stone.

For confirm the obtained data, the nature of changes in the rheological properties of cement paste were studded. Carry the determination of plastic strength or shear stress out using a conical plastometer KP-3. For do this, a metal cone were immersing in the system under investigation at a certain interval, on which the load acted. The measurement were caring out systematically until the moment when the mixture turned into a cement stone. The plastic strength of cement paste Pm (MPa) were calculated by the formula

$$P_m = K_\alpha \frac{F}{h_m^2} \qquad 1.1)$$

where F - tapered load, grams;

h_m – depth of immersion of the cone in the mortar mixture;

K_α – coefficient depending on the value of the vertex angle of the cone; for the metal cone with an apex angle of 30° - K_α = 1,116;

$$K_\alpha = \frac{1}{\pi} \cdot \cos\left(\frac{\alpha}{2}\right) \cdot ctg\left(\frac{\alpha}{2}\right). \quad (1.2)$$

where α - cone angle at the apex (in its axial cross section).

Cement paste had same ratio water/cement, equal to W/C = 0.47. The research results are shown in Figure 1.3.

The analysis of plastograms (Figure 1.3), confirms previously obtained data that the introduction of synthetic aluminosilicate in the cement paste recipe leads to an earlier structure formation. So, the value of plastic strength of the control composition (without the use of synthetic aluminosilicates) at the age of 20 min from the moment of turning is = 0.0015 MPa (Figure 1.3, curve 1), and at the introduction of synthetic aluminosilicates in the amount of 10%, 20% and 30% by weight of the binder at the age of 20 minutes from the moment of mixing is = 0.011 MPa (Figure 1.3, curve 1), τ = 0.014 MPa, respectively (Figure 1.3, curve 2) and τ = 0.020 MPa (Figure 1.3, curve 3).

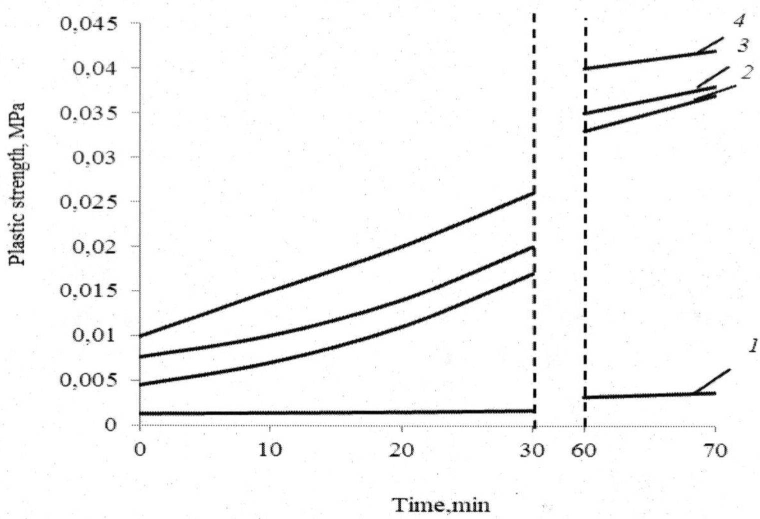

Figure 1.3. The change in the plastic strength of the cement paste: 1 - control sample (without the content of synthetic additives); 2 - content of synthetic additives 10% by weight of cement; 3 - content of synthetic additives 20% by weight of cement; 4 - the content of synthetic additives 30% by weight of cement.

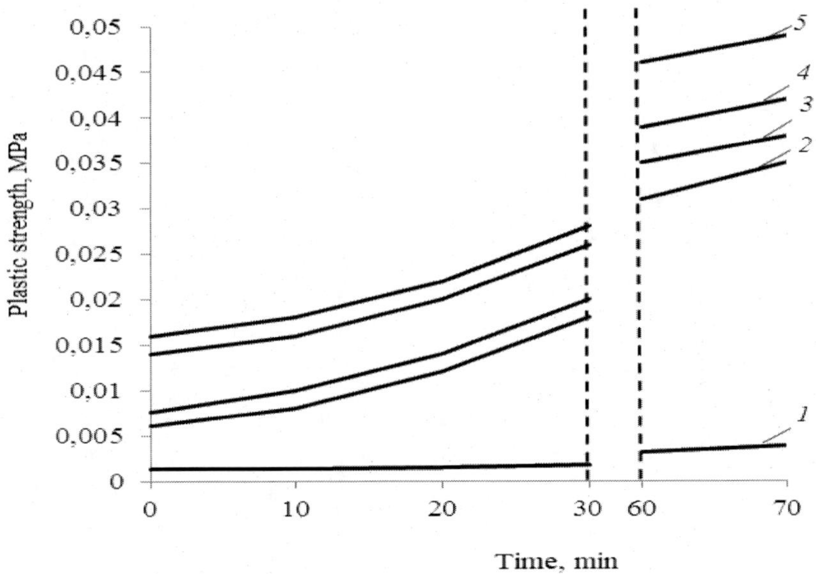

Figure 1.4. The change in the plastic strength of the cement paste: 1 - control (without additives); 2 - cement paste (the specific surface of the additive $S_{sp} = 1.03\ m^2/g$); 3 - cement paste (specific surface of the additive $S_{sp} = 0{,}69\ m^2/g$); 4 - cement paste (the specific surface of the additive $S_{sp} = 0.31\ m^2/g$);5 - cement paste (the specific surface of the additive $S_{sp} = 0.1\ m^2/g$).

The effect of the dispersity of the additive based on synthetic aluminosilicates on the change in the rheological properties of cement paste were studded. Volsky Portland cement with a specific surface S_{sp} equal to $S_{sp} = 0.34\ m^2/g$ were used. The content of the additive were 20% by weight of cement. The research results are shown in Figure 1.4.

The analysis of plastograms (Figure 1.4) indicates that the value of the specific surface of the synthetic additive affects the time of the formation of the cement paste. Therefore, the value of plastic strength of the cement paste (control composition) at the age of 20 minutes from the moment of mixing is $\tau = 0.0015$ MPa (Figure 1.4, curve 1), and with additive of synthetic aluminosilicates $\tau = 0.012–0.022$ MPa depending on the dispersity of the additive.

The obtained data of the rheological characteristics of the cement paste with additives based on synthetic aluminosilicates of different dispersion are confirmed by the results of the carried out analysis of the hardening kinetics of the cement stone (Figure 1.5). A synthetic additive were introducing into the cement paste recipe for 20% by weight of cement. The strength of cement stone were set in air-dry conditions at a temperature of (20 ± 2)^{0}C.

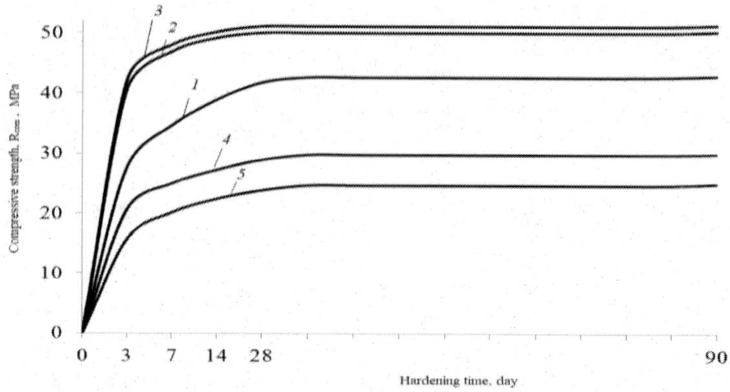

Figure 1.5. The kinetics of the strength of the samples in air-dry curing conditions: 1 - control (without the content of synthetic additives); 2 - the specific surface of the additive $S_{sp} = 1.03\ m^2/g$; 3 - the specific surface of the additive $S_{sp} = 0{,}69\ m^2/g$; 4 - the specific surface of the additive $S_{sp} = 0.31\ m^2/g$; 5 - the specific surface of the additive $S = 0.1\ m^2/g$.

Analysis of the experimental data (Figure 1.5) testifies that the introduction of an additive based on synthetic aluminosilicates with a specific surface $S_{sp} = 1.03\ m^2/g$ and $S_{sp} = 0.69\ m^2/g$ leads to an increase in compressive strength of cement samples at the age of 90 days of air-dry hardening by 13–20% compared with the control sample (without additives). Application in the formulation of cement stone additives based on synthetic aluminosilicates with a specific surface $S_{sp} = 0.31\ m^2/g$ and $S_{sp} = 0.1 m^2/g$ reduces the compressive strength of the samples on 35–41% compared with the control sample (without additives).

The formation of the structure of the cement stone is affecting by the conditions of curing samples. O confirm this, an analysis of the hardening kinetics of the cement stone under different hardening conditions: wet (Figure 1.6) and air-dry (Figure 1.7) were carried out. In the formulation is an additive based on synthetic aluminosilicates.

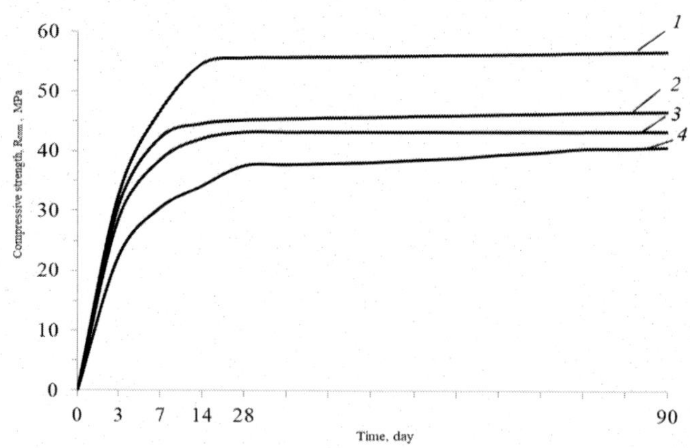

Figure 1.6. Hardening kinetics of cement samples in wet conditions: 1 - control (without the content of the additive); 2 - a sample with a content of 10% additive by weight of cement; 3 - sample containing 20% of the additive by weight of cement; 4 - sample containing 30% additive by weight of cement.

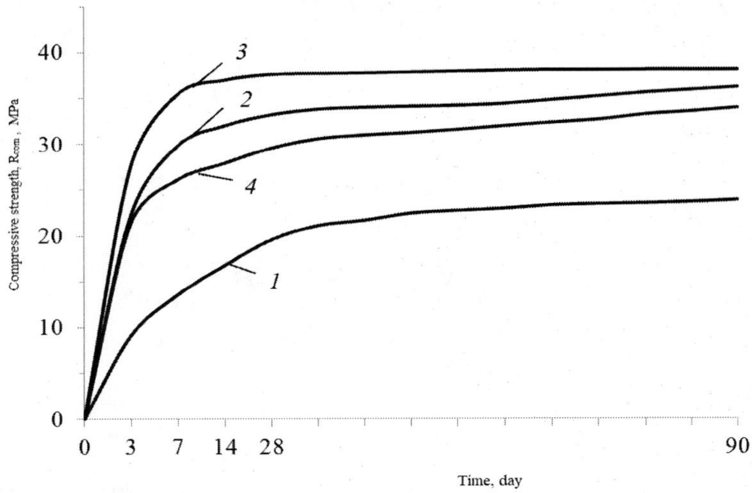

Figure 1.7. Hardening kinetics of cement samples in air-dry conditions: 1 - control (without the content of the additive); 2 - a sample with a content of 10% additive by weight of cement; 3 - sample containing 20% of the additive by weight of cement; 4 - sample containing 30% additive by weight of cement.

Analyzing the experimental data (Figure 1.6), it were established, that during hardening in wet conditions, the cement stone with additive based on synthetic aluminosilicates were introduced, has a lower strength than the control sample (without the additive). So, the compressing strength of a control sample at the age of 90 days were 56.7 MPa, and for samples with a content of synthetic aluminosilicates of 10–30% by weight of cement, respectively, 46.7–40.76 MPa.

Analysis of the experimental data (Figure 1.7) testifies that the compressive strength of a cement stone with additive based on synthetic aluminosilicates at of air-dry hardening in the age of 90 days is increasing by 40.2–52.7% depending on the content of the additive.

Analyzing Figure 1.6-1.7, it were revealed, that the hardening of the cement stone, with additive based on synthetic aluminosilicates is present, takes place in more favorable humidity conditions. Thus, it can be assumed, that the additive based on synthetic aluminosilicates has a water-holding capacity.

For confirm this the drying kinetics of cement samples were studied during hardening in air-dry conditions. The investigated samples were made with water/cement ratio W/C equal to W/C = 0.43. The results of the experiment are presented in Figure 1.8.

Analysis of Figure 1.8 showed that moisture cement stone evaporates intensively from samples of cement stone without an additive during air-dry hardening (Figure 1.8, curve 1) compared with samples with additive based on synthetic aluminosilicates (Figure 1.8, curves 2 - 5). Thus, for cement stone (without an additive), water content on the 20th day of hardening were 0.64%, and for cement stone with an additive content of 1%, 10%, 20% and 30% by weight of cement, respectively 1.99%; 2.52%; 2.45% and 2.75%. The data obtained confirm that the cement stone with additive of synthetic aluminosilicates has a more pronounced water-retaining effect than the control sample (without additives).

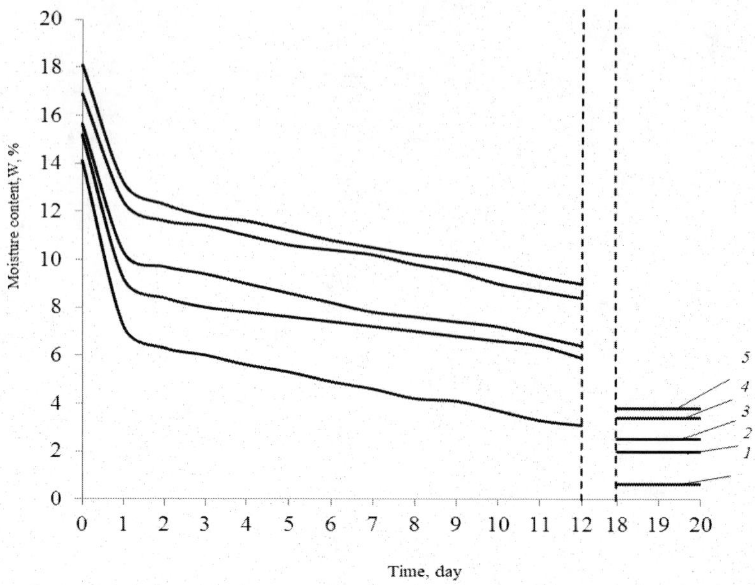

Figure 1.8. Kinetics of changes in the moisture content of cement samples during curing: 1 - control sample; 2 - the content of the additive is 1% by weight of cement; 3 - additive content of 10% by weight of cement; 4 - the content of the additive is 20% by weight of cement; 5 - the content of the additive is 30% by weight of cement.

For confirm the obtained results, the water-holding effect of the additive based on synthetic aluminosilicates in depending on its percentage (Figure 1.9).

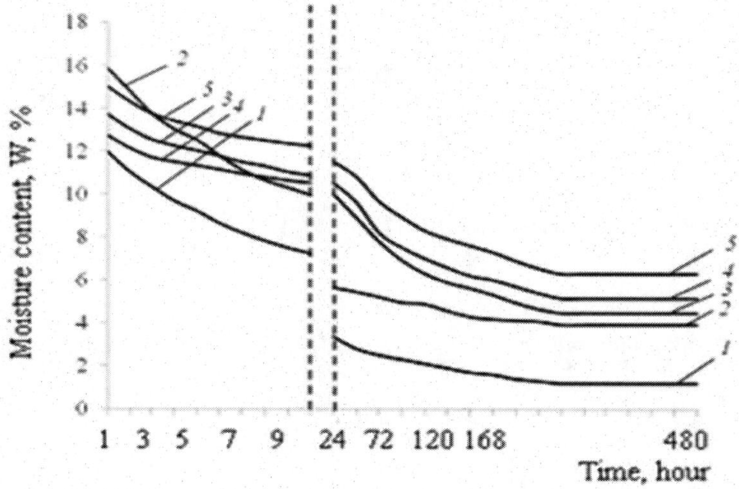

Figure 1.9. Drying kinetics of cement samples: 1 - control sample (without the content of the additive); 2 - content of synthetic aluminosilicates 1% by weight of cement; 3 - additive content of 10% by weight of cement; 4 - the content of the additive is 20% by weight of cement; 5 - the content of the additive is 30% by weight of cement.

Data analysis Figure 1.9 established that the additive based on synthetic aluminosilicates has a significant water-retaining effect compared with the control sample (without the content of the additive). So, in the cement stone without the additive content, the moisture content on the 20th day (480 h) of hardening were 1.2% (Figure 1.9, curve 1), and in the cement stone with an additive content of 1–30% accordingly, it were 3.89–6.3% depending on the content of the additive (Figure 1.9, curve 2–5).

For compare the water-retaining effect of the additive based on synthetic aluminosilicates, the additives FMC 2094 and Bermokool 425 methylcellulose were selected. The test results are shown in the Figure 1.10.

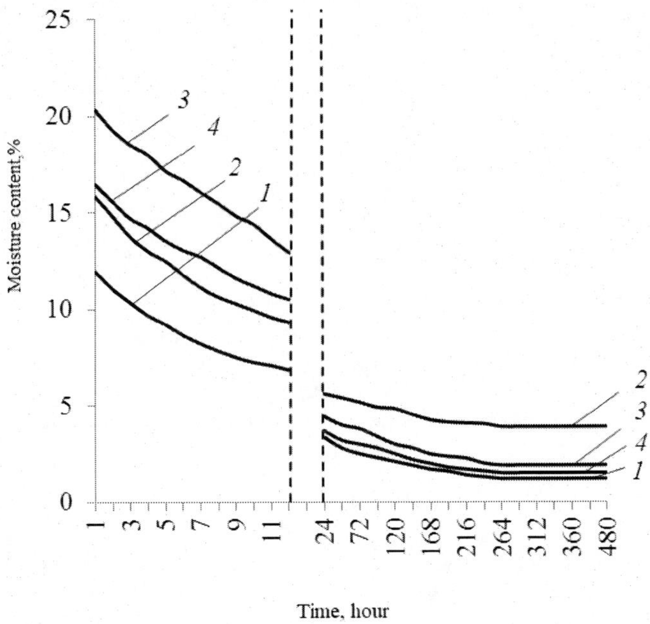

Figure 1.10. Drying kinetics of cement samples: 1 - control sample; 2 - content of synthetic aluminosilicates 1% by weight of cement; 3 - the content of the additive methylcellulose brand FMC 2094 - 1% by weight of cement; 4 - the content of the additive Bermocool 425 - 1% by weight of cement.

Data analysis Figure 1.10 shows that the additive based on synthetic aluminosilicates has the same water-retaining effect as methylcellulose brand FMC 2094 and Bermokool 425. So, in the cement stone with an additive based on synthetic aluminosilicates in an amount of 1% by weight of cement moisture content on the 20th day (480 h) hardening were 3.89% (Figure 1.10, curve 2), and for cement stone with a content of 1% methylcellulose brand FMC 2094 and 1% Bermokool 425, respectively correspondingly - 1.9% (Figure 1.10, curve 3) and 1.5% (Figure 1.10, curve 4).

The presence of an additive based on synthetic aluminosilicate in the formulation contributes to a change in the pH of the liquid phase of the cement paste (Figure 1.11). The kinetics of the change in the pH of the solution were measuring during the hardening of the images using an Aquilon pH meter. As can be seen from the data (Figure 1.11), the liquid phase of the cement test of the control sample is characterized by a higher initial pH value of pH 12.08, while the cement paste with the use of a synthetic additive in the amount of

10% and 20% is respectively pH = 11.88 and pH =11.50. This is obviously explained by the interaction of the amorphous phase of the synthetic additive with calcium hydroxide, which leads to a decrease in the concentration of the latter in the liquid phase of the cement paste, as well as by the chemical composition of the additive based on synthetic aluminosilicates.

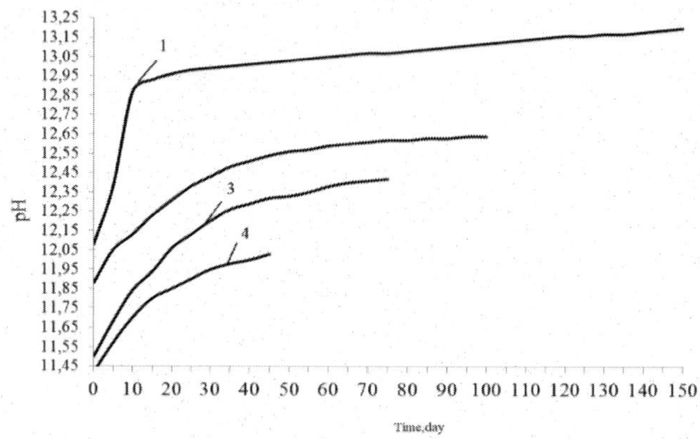

Figure 1.11. The change in pH of the liquid phase of the cement paste: 1 - control sample; 2 - the content of the additive is 10% by weight of cement; 3 - the content of the additive is 20% by weight of cement; 4 - the content of the additive is 30% by weight of cement.

The kinetics of heat release during hardening of the cement binder in the presence of an additive based on synthetic aluminosilicates were also evaluating. The kinetics of heat generation during the hardening of the samples were determined using a Dewar vessel. Cement paste were immersing in a vessel, hermetically closed with a stopper of an instrument and the change in temperature over time were recording. The end of the test took time when the temperature of cement paste did not change for 30 minutes. The research results are presented in Figure 1.12.

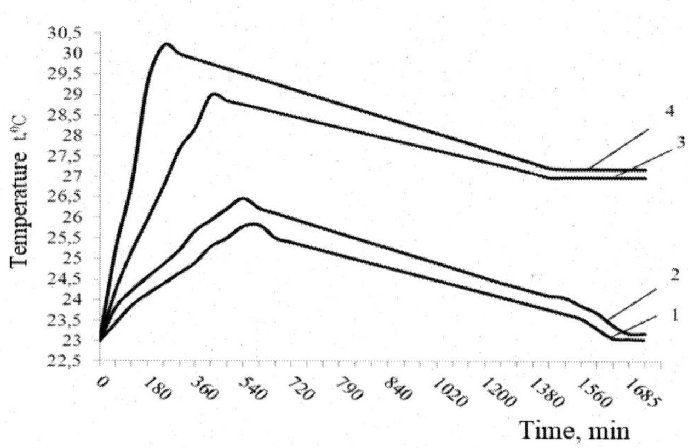

Figure 1.12. Heat dissipation kinetics during sample hardening: 1 - control sample; 2 - the content of the additive is 10% by weight of cement; 3 - the content of the additive is 20% by weight of cement; 4 - the content of the additive is 30% by weight of cement.

From the presented data (Figure 1.12) it can be seen that the maximum temperature in the process of hydration of the cement paste of the control composition is 26^0C during the time t = 565 min. The maximum temperature in the process of hydration of cement paste samples with additive based on synthetic aluminosilicates in the amount of 10%, 20% and 30% by weight of cement, is respectively 26.5°C for 530 minutes, 29°C for 420 minutes and $30,2^0C$ for 240 minutes.

For study the physicochemical processes of hardening of the cement stone, were holding X-ray phase and differential thermal analyzes (Figure 1.13 – 1.16).

The mineralogical structure of cement stone is represented by such hydrated formations as: portland (d, (A) = 6.57, 4.93, 3.11, 2.67, 2.62, 2.28), hydrosilicates CSH (d, (A) = 6.46, 5.74, 3.21), ettringite (d, (A) = 5.61, 4.704), calcium hydroaluminates (d, (A) = 5.30, 5.14, 4.44, 4.303, 3.50, 2.80, 2.30), SN(II) (d, (A) = 2.88, 2.00, 1.83), tobermorite (d, (A) = 4.89, 3.09, 3.03, 2.78), calcium hydroferrite (d, (A) = 4.50, 4.48, 2.70) and calcite CaCO3 (d, (A) = 3.88, 2.75, 2.204, 2.189, 2.088) (Figure 1.13).

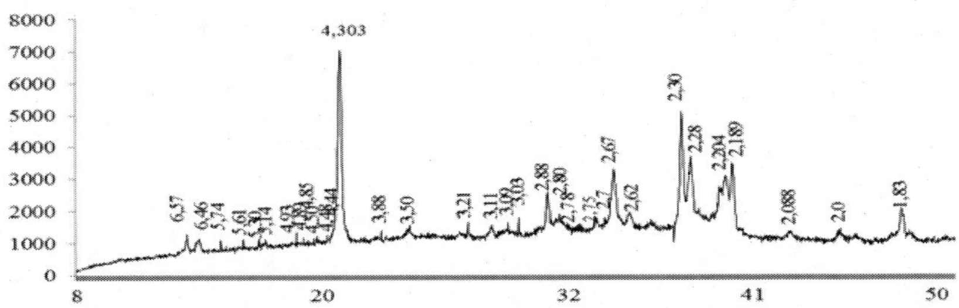

Figure 1.13. X-ray of cement stone.

At small angles (up to 20°), the following compounds are observed: portlandite (d (Å) = 7.32, 6.62, 6.28, 4.92, 4.63), hydrosilicates of CSH (d (Å) = 9.8, 6.46, 5.77), ettringite (d (Å) = 9.73, 7.00, 5.61, 4.704), calcium hydroaluminates (d (Å) = 10, 60, 8.20, 6.71, 6.34) calcium hydroferrite (d (Å) = 7.60, 5.46, 5.18, 4.50, 4.48) and tobermorite (d (Å) = 11.5, 6.00, 5.86) (Figure 1.14).

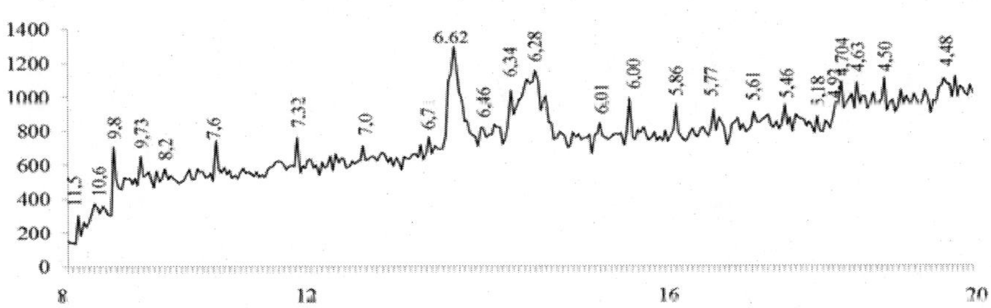

Figure 1.14. X-ray of cement stone at low angles.

The mineralogical composition of cement stone with additive based on amorphous aluminosilicates is represented portlandite (d(Å) = 4.93, 2.67, 2.28), hydrosilicates CSH (d(Å) = 6.62, 6.46, 6.42), ettringit (d(Å) = 4.704, 3.88), calcium hydroaluminates (d(Å) = 4.48, 4.303, 2.30), CSH(II) (d(Å) = 2.88, 2.00, 1.83), white (d(Å) = 2.67, 2.74, 2.61, 2.189), C_3A (d(Å) = 2.72, 2.204, 2.04), zeolite (d(Å) = 3.21, 3.37), $CaNaHSiO_4$ (d(Å) = 2.84, 2.23), aluminum hydrosilicate (d(Å) = 4.67, 3.45, 3.14), calcite $CaCO_3$ (d (Å) = 3.86, 2.42, 2.088, 1.86), natrolite (d(Å) = 6.53, 3.64, 3.18), $Ca_3Al_2[SiO_4]$ 1,53(OH) (d (Å) = 3.043, 2.72, 2.22), hydrosilicates sodium hydrates (d(Å) = 6.25, 4.099, 3.92) and hydroferrite (d(Å) = 4.5, 3.19, 1.7) (Figure 1.15).

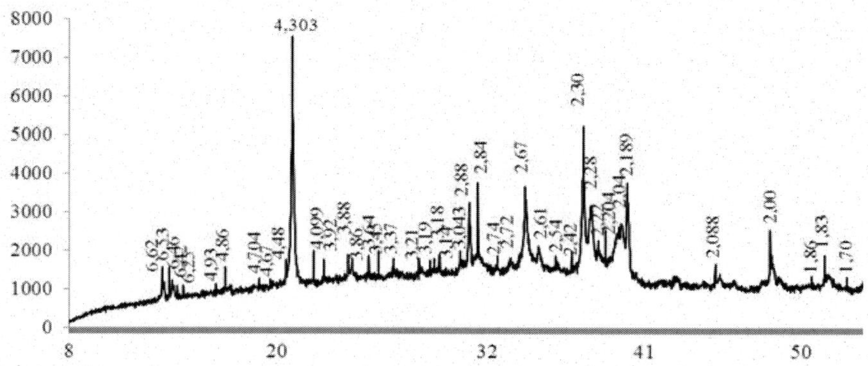

Figure 1.15. X-ray of cement stone using in the formulation of additives based on amorphous aluminosilicates.

At small angles (up to 20°), CSH hydrosilicates were identified. (d(Å) = 6.62, 6.46, 6.42), ettringite (d(Å) = 5.61, 5.62), CSH(II) (d(Å) = 8.67, 6.51), zeolite (d (Å) = 10.67, 7.54, 6.16, 5.48), calcium hydroaluminates (d(Å) = 8.2, 7.8, 5.14, 4.86, 4.50, 4.48), hydrosilicates sodium hydrates (d (Å) = 6.25) and natrolite (d (Å) = 6.53) (Figure 1.16).

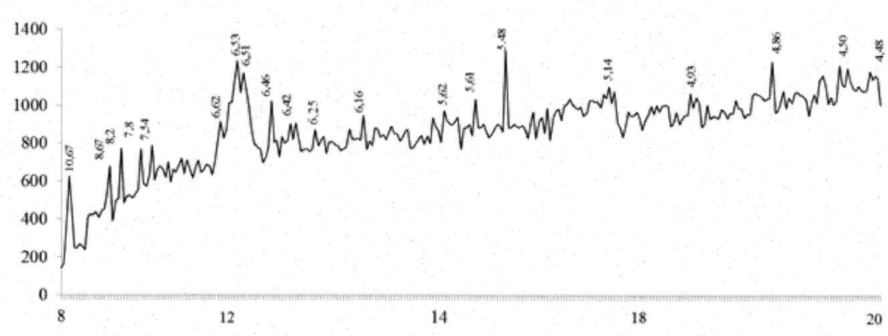

Figure 1.16. X-ray of cement stone with the use in the formulation of additives based on amorphous aluminosilicates at low angles.

Analyzing the obtained data, it can be argued that the cement stone with additive of amorphous aluminosilicates has more hydrate formations. It is further confirmed by differential thermal analysis (DTA) (Figure 1.17–1.18).

Thermogramms of the samples were obtained on a Q-1500 derivatograph. Software heating of the furnaces from 20 to 1000°C were carrying out with an electronic thermal heater at a rate of 20°C/min. Temperature is measuring with a platinum thermocouple with an accuracy of ± 5°C. The temperature difference between the studied substance and the standard, proportional to the thermal effect, were recording in the form of a DTA curve (sensitivity 500 μV). Simultaneously with the DTA curve, the weight loss curve and its derivative curve were recorded (sensitivity 500 μV). When conducting thermal analysis, the temperature range of the research were in the range from 30 to 1000°C, the rate of temperature rise in the analytical chamber were 3°C/min, and the duration of a single experiment were about 5 hours and 10 minutes.

Analysis of the thermograms (Figure 1.17 - 1.18) showed that in the sample of cement stone with synthetic additive, the endo effect at 150°C is more pronouncing than in the control sample, which indicates a greater presence of calcium hydrosulfate. The endoeffect at a temperature of 320°C shows, that a cubic hydroaluminate C_2AN_6 is forming in the cement stone. Due to the water-retaining properties of the additive based on synthetic aluminosilicates, hardening of the cement stone in its presence occurs under more favorable conditions. The decrease in the intensity of the endoeffect at a temperature of 510°C, indicating the dehydration of $Ca(OH)_2$, proves that there is a smaller amount of portlandite in the sample of cement stone with the use of a synthetic additive. Obviously, $Ca(OH)_2$ interacts with additive with education $CaNaHSiO_4$, $Na_{8.4}Ca_{3.2}Si_{36.4}Al_{11.6}O_{96}5H_2O$ and $Ca_3Al_2\,[SiO4]_{1.53}(OH)_{5.88}$.

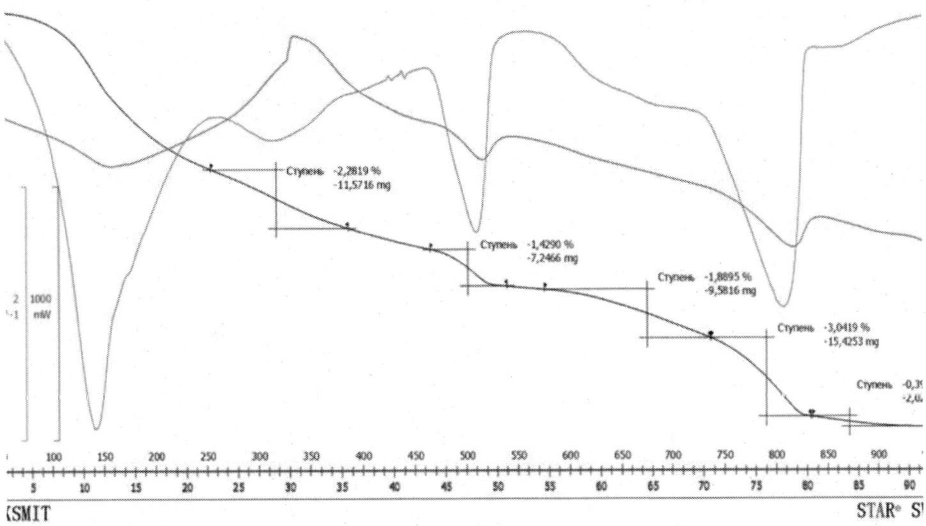

Figure 1.17. Thermogram of a control sample of cement stone.

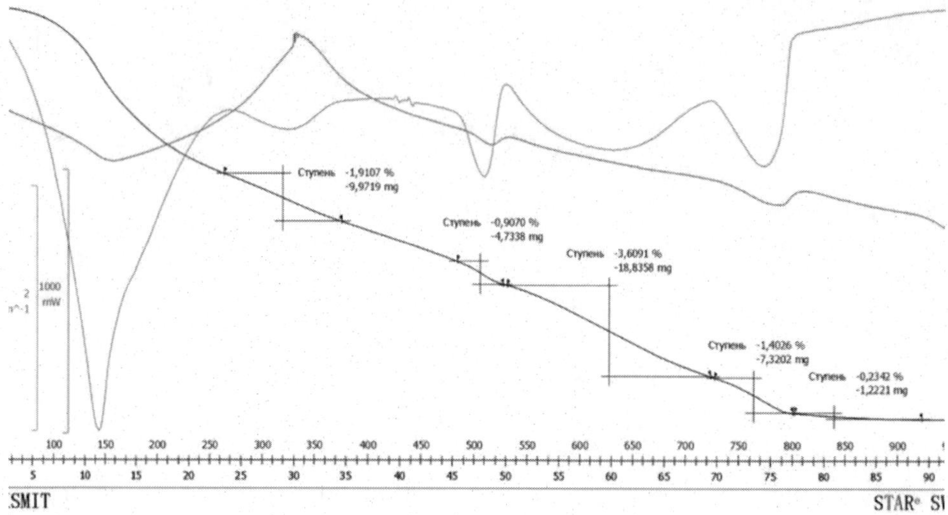

Figure 1.18. Thermogram of a cement stone sample with an additive based on synthetic zeolites used in the formulation.

For samples based on cement stone (control composition), a more pronounced endo-effect is observed at a temperature of 810°C due to $CaCO_3$ dissociation, compared to a cement stone with the addition of an additive based on synthetic aluminosilicates.

The exoeffect at a temperature of 550°C is pronounced in a sample of a cement stone with synthetic additive is present, which indicates the presence of a ten-water calcium hydroaluminate SAN_{10} in the composition of the cement stone.

The exoeffect at a temperature of 740°C, in the sample of cement stone using synthetic aluminosilicates, shows the presence of $5Ca_6SiO_29H_2O$ in the composition of tobermorite, turning into wollastonite.

The work also estimated the content of free and chemically bound water in the cement stone, depending on the percentage of the additive. The research results are summarized in Table 1.4.

Table 1.4. Content of free and chemically bound water in cement stone

Content additives (%), by weight of cement	The content of free water in the sample,%	The content of chemically bound water in the sample,%
–	7.3	14.5
10	6.7	15.2
20	6.1	17.0
30	3.3	18.3

As can be seen from the data (Table 1.4), in the cement stone with additive based on synthetic aluminosilicates, there is a decrease in the amount of free water and an increase in chemically bound water in comparison with the control sample. So, in the control composition, the content of free and chemically bound water is 7.3 and 14.5%, respectively, and for a sample with additive of 20%, is respectively 6.1 and 17.0%.

For further confirm the obtained results, we determined the amount of free lime in cement samples at the age of 28 days of air-dry hardening. A decrease in the amount of free lime $Ca(OH)_2$ in the cement stone, in the formulation of which there is an additive based on synthetic aluminosilicates, has been established. The content of free lime in the control sample were 13%, and in the sample with a content of 20% of the additive by weight of cement - 6.5%. The results obtained indicate the interaction of synthetic additives with free lime.

Considered topology structure of cement stone with the use of additive based on amorphous aluminosilicates. Preliminary studies found the optimal content of an additive of 20%. The ratio of cement and additives is a fraction 0.76:0.24 or 5:1 by weight. Counting the number of particles were calculating using the formulas:

$$n_c = \frac{6m_c}{\rho_c \pi d_c^3}, \quad (1.3)$$

$$n_a = \frac{6m_a}{\rho_a \pi d_a^3}. \quad (1.4)$$

where m_c and m_a are the masses of cement and synthetic additive in a mixed binder in fractions of a unit;

n_c and n_a – quantity of cement particles and additives based on amorphous aluminosilicates;

ρ_c and ρ_a – density of cement and synthetic additives.

The number of particles n were calculating by the formula:

$$n = \frac{V_{mat}}{V_{ball}}, \quad (1.5)$$

where V_{mat} – volume of material;

V_{ball} – volume of a particle.

The ratio of the amount of cement particles relative to the additive particles based on amorphous silica-alumina C were calculating by the formulas:

$$C = \frac{n_a}{n_ц} = \frac{m_a \rho_c d_c^3}{m_c \rho_a d_a^3}, \qquad (1.6)$$

or through a specific surface area:

$$C = \frac{m_a \rho_a^2 S_a^3}{m_c \rho_c^2 d_c^3}. \qquad (1.7)$$

where d_c^3 and d_a^3 – accordingly, the diameter of the particles of cement and additives;

$S_ц$ and $S_д$ – accordingly, the specific surface of cement and synthetic additives.

The calculation results are shown in Table 1.5. It has been established that with a specific cement surface S_{sp} = 340 m^2/kg and a specific surface of the additive based on synthetic aluminosilicates S_{sp} = 1029.8 m^2/kg, the ratio of particles is C = 4.5, and with a specific surface additive S_{sp} = 311,1m^2/kg - C = 0.1.

Table 1.5. The ratio of cement particles and additives based on synthetic aluminosilicates, depending on their dispersion

Size of specific surface, m^2/kg		The average particle diameter, microns		The amount of particles of synthetic additives per particle of cement
Cement	Synthetic additive	Cement	Synthetic additive	
340	1029,8	6	2,8	4,5
340	688,6	6	4,1	1,3
340	311,1	6	10,2	0,1

For the ratio C = 1.3 – 4.5 the distance from the cement particles to the particle of the additive based on amorphous aluminosilicates were calculated. The elementary cubic cell with the size of the cube *A* were choosing, and the volume of the cube were A^3. It were assuming that this ball contains one spherical particle.

This particle was placed in a cell in the form of ball sectors in 1/8 of the ball in each corner of the cube. In the free volume of the cell, not filled with spherical sectors of the cement matter, all the particles of the additive based on amorphous aluminosilicates were placed (Figure 1.19).

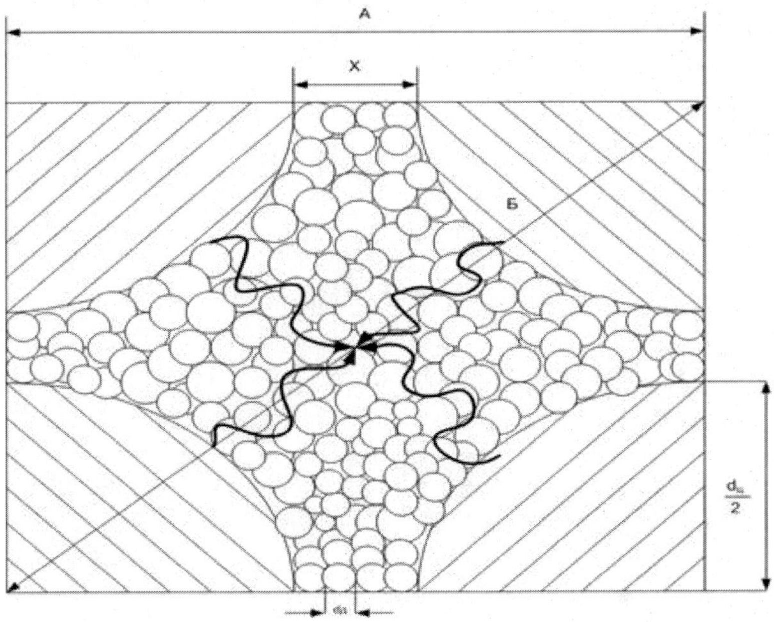

Figure 1.19. Topology of the Idealized structure of the cement binder with application in the formulation of synthetic additives and migration paths of the products of hydration of cement with monodisperse particles additive based on amorphous aluminosilicates.

Then the total volume of the unit cell will be composed of one of cement particles and the total volume of particles of synthetic supplements packed with varying degree of packing of the K_{pac}. In the resulting system cement composite binder (CCB) forms a dense hexagonal packing, in this regard, the coefficient of packing of equal CCB = 0.74. Given the K_{pac}, can be recorded for the full amount V_{cell}:

$$V_{\text{cell}} = A^3 = \frac{\pi d_c^3}{6} + \frac{\pi d_c^3}{6} CK_{pac}, \qquad (1.8)$$

here

$$A = \sqrt[3]{\frac{\pi}{6}(d_c^3 + d_c^3 CK_{pac})}. \qquad (1.9)$$

For the adopted characteristics of synthetic additives and cement, the calculated values $C = 4.5$; 1.3 and calculated $K_{pac} = 0.74$ it were obtained, respectively, $A = 5.3\times10^{-6}$ m, $A = 6.1\times10^{-6}$ m. Distance X between the surfaces of cement particles along the edge of the unit cell will be calculated by the formula:

$$X = A - d_c. \qquad (1.10)$$

Then it were calculated value of X for different values of A will be equal, respectively $X_1 = -0.7\times10^{-6}$ m and $X_2 = 0.1\times10^{-6}$ m.

Since the length of the diagonal B of the face of the cubic cell is $\sqrt{2A}$, and the diagonal of the cube $L = \sqrt{3A}$, then the distance between the surfaces of the particles Y along the diagonal of the face will be:

$$Y = \sqrt{2}A - d. \qquad (1.11)$$

The calculated values of Y for different values of A are equal to $Y_1 = 0.7\times10^{-6}$ m and $Y_2 = 1.7\times10^{-6}$ m.

The distance Z between the surfaces of the particles along the diagonal of the cube is:

$$Z = \sqrt{3}A - d. \qquad (1.12)$$

According the topological scheme (Figure 1.19) follows, that for binding of hydrated cements particles with the particles of the additive based on amorphous aluminosilicates in the air-dry conditions necessary migration of hydration products of cement binder to the particles of the additive based on amorphous aluminosilicates. Most remote from the surface zones are the center of the cube and the center of his edge.

The transfer of chemical products should be carried out at a distance from the surface equal to $t/2 = 0.8\times10^{-6}$ m and $t/2 = 1.9\times10^{-6}$ m. In such a percolation cluster system, the synthetic additive particles bound by the film water. The flow proceeds through the most permeable tetrahedral voids to the contact areas. The strength of the composition is ultimately determined by the strength of these contacts.

Table 1.6. The ratio of cement particles and additive based on amorphous aluminosilicates depending on their specific surface

Size specific surface, [m^2/кg]		The average particle diameter of, [mkm]		The number of particles of synthetic additives into one particle of cement
Cement	Synthetic additive	Cement	Synthetic additive	
340	1029,8	6	2,8	4,5
340	688,6	6	4,1	1,3
340	311,1	6	10,2	0,1

For the adopted characteristics of synthetic additives and cement, the calculated values $C = 4{,}5$; 1,3 and calculated $K_{pac} = 0{,}74$ we obtain, respectively, $A = 5{,}3{*}10^{-6}$ m, $A = 6{,}1{*}10^{-}$

^{6}m. Distance X between the surfaces of cement particles along the edge of the unit cell will be calculated by the formula:

$$X = A - d_ц \quad (1.13)$$

Then the calculated value of X for different values of A will be equal, respectively $X_1 = -0{,}7*10^{-6}$ m and $X_2 = 0{,}1*10^{-6}$ m.

Since the length of the diagonal B of the face of the cubic cell is $\sqrt{2A}$, and the diagonal of the cube $L = \sqrt{3A}$, then the distance between the surfaces of the particles Y along the diagonal of the face will be:

$$\mathrm{Y} = \sqrt{2}A - d. \quad (1.14)$$

The calculated values of Y for different values of A are equal to $Y_1 = 0{,}7*10^{-6}$ m and $Y_2 = 1{,}7*10^{-6}$ m.

The distance between the surfaces of the particles Z along the diagonal of the cube is:

$$\mathrm{Z} = \sqrt{3}A - d. \quad (1.15)$$

The calculated value of Z for the values of A is equal to $Z_1 = 1{,}6*10^{-6}$ m and $Z_2 = 2{,}8*10^{-6}$ m respectively.

Under such difficult conditions, the most probable mechanism for the formation of strength in time is obviously diffusion transfer of the binder ions due to osmosis and the subsequent reaction of the formation of a new phase on the surface of the additive particles. Thus, even a simplified consideration of the topology of the structure of composite binders makes it possible to more correctly explain the mechanism of formation of the strength of a cement composite using synthetic aluminosilicates.

Introduction to the formulation of the cement composite additives based on synthetic aluminosilicates leads to a change in its porosity (Table 1.6). The porosity of the cement stone were determined by the obtained values of the content of free and chemically bound water in the cement stone and its water-cement ratio. The calculation were carrying out according to the following formulas:

$$П_1 = \frac{В/Ц - 0{,}5 \cdot \alpha}{В/Ц + 0{,}32} \cdot 100 \quad , \quad (1.16)$$

$$П_2 = \frac{0,09 \cdot \alpha}{В/Ц + 0,32} \cdot 100$$,(1.17)

$$П_3 = \frac{0,2 \cdot \alpha}{В/Ц + 0,32} \cdot 100$$,(1.18)

$$П_{общ} = \frac{В/Ц - 0,21 \cdot \alpha}{В/Ц + 0,32} \cdot 100$$, (1.19)

where P_1 - is the capillary porosity, %;

P_2 - contraction porosity, %;

P_3 - gel porosity, %;

P_{gen} - total porosity, %;

W/C - water/cement ratio;

α - the degree of hydration of cement, determined by the formula

$$\alpha = 4 \cdot W_n$$, (1.20)

where W_n is the amount of chemically bound water.

Open porosity were determined by the following formula

$$P_{open} = \frac{m_s - m_{dr}}{V_{nat}} \cdot \frac{1}{P_{H_2O}} \cdot 100$$

$$П_{отк} = \frac{m_{нас} - m_{сух}}{V_{ест}} \cdot \frac{1}{P_{H_2O}} \cdot 100$$, (1.21)

where m_s - is the mass of the material in a saturated state, kg;

m_{dr} - is the mass of the material in the dry state, kg;

V_{nat} - natural volume of material, m^3;

P_{H_2O} - water density taken ρ = 1000 kg/m^3.

In the Table 1.7 presents the results of a study of the porosity of cement stone with the addition of an additive based on amorphous aluminosilicates.

Table 1.7. Change in the porosity value of cement samples depending on the content of the additive

Compositions	Porosity,%			
	The total	Capillary	Gel	Contractional
Control	41,2	18,7	15,5	7,0
10% Additives	40,3	16,7	16,3	7,3
20% Additives	38,3	12,0	18,1	8,2
30% Additives	37,3	9,0	19,5	8,8

The total and capillary porosity of the control samples is 1.1 and 1.6 times higher than the porosity of the sample with the additive (the amount of the additive is 20% of the cement mass), and the contractile and gel porosity is 1.2 times lower.

As can be seeing from the data given in Table 1.7, in cement stone with the use of synthetic aluminosilicates compared with the control sample, a decrease in total and capillary porosity and an increase in gel and contraction porosity are observed, which leads to an increase in the resistance of the cement composite. Thus, in the control sample, the total and capillary porosity is less than the porosity of the sample, using in the formulation 20% of the additive by weight of cement, respectively, by 6.8% and 35.8%, and the contraction and gel porosity is higher, respectively, by 17.1% and 16.8%.

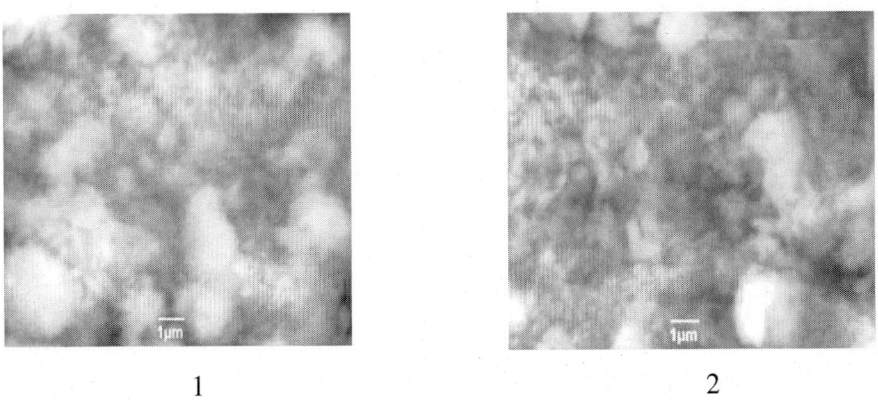

Figure 1.20. Microstructure of samples: 1 - control sample, 2 - sample with the addition of 10% by weight of cement.

In Figure 1.20 shows a photograph of the structure of cement stone, obtained using a scanning electron microscope of the type Electron Microscope JSM - 6390 LV of the company Schanning. The photographing mode were carried out under low vacuum conditions — 50 MPa at a magnification of 10,000 times.The results of the analysis of the microstructure of the sample with the additive based on synthetic aluminosilicates (Figure 1.20, image 2) showed that the structure is characterized by smaller particle sizes of the neoplasms.

The connection of the hydrosilicate fibers with the honeycomb grid of the synthesized aluminosilicates is observed. The microstructure of the control sample (Figure 1.20, image 1), in turn, is characterized by a large particle size, which are located at a considerable distance from each other.

2. Development of a Recipe for a Dry Adhesive Construction Mixture with the Application of Additive Based on Synthetic Alyumosilicates

2.1. Kinetics of Hardening Cement Composite in the Presence of an Additive Based on Synthetic Aluminosilicates

The kinetics of hardening of cement-sand samples were studied using an additive based on synthetic aluminosilicates (10% - 30% by weight of cement) and using imported modifying additive — methyl cellulose of the brand FMC 2094 (1% by weight of cement). Samples were made on the basis of sand from the Ukhta field and with the same water-cement ratio equal to W/C = 0.47. The results of the tests are presented in Figure 2.1 and 2.2.

In Figure 2.1 and 2.2 shows the kinetics curves of the strength of cement composites according to the results of the approximation of experimental data to determine the compressive strength of the samples under study. The experimental data were approximated using the Curve Expert 1.3 program. The dots in the figures indicate the experimental values of the compressive strength of cement composites. Based on the obtained results of approximation and experimental studies (Figure 2.1, 2.2), it follows that the introduction to the recipe of the cement-sand mortar of the additive based on synthetic aluminosilicates contributes to a more intensive structure formation of the cement composite than the use of in the formulation of the additive methylcellulose brand FMC 2094.

Curves, (Figure 2.1 and 2.2), are described by the following exponential equation:

$$y = a \cdot (1 - e^{-bx}), \quad (2.1)$$

where a is a constant that takes into account the maximum possible compressive strength;

b is the rate constant of structure formation;

x is the hardening time.

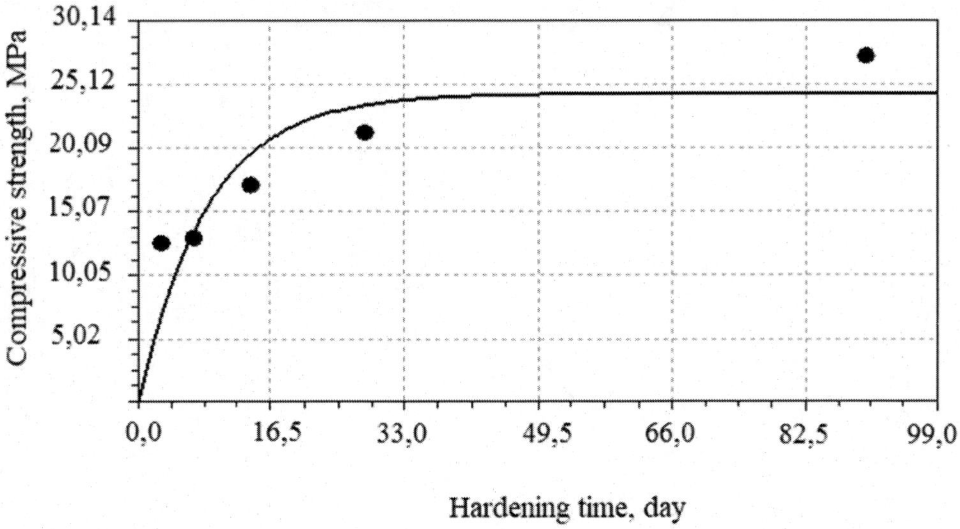

Figure 2.1. Strength kinetics of cement-sand composition using in the formulation of the additive methylcellulose brand FMC 2094 (1% by weight of cement).

In Table 2.2 shows the values of the constants *a* and *b* for the reduced equation describing the kinetics of curing of the cement composite.

Table 2.1. The value of the constants of the equation of the kinetics of strength

Modifying additive	Amount, %	*a*	*b*
Synthetic aluminosilicates	10	30,386	0,178
	20	34,674	0,221
	30	28,963	0,145
Methyl cellulose FMC 2094 brands	1	24,434	0,117

An analysis of the calculations shows that with the introduction in the formulation of a synthetic additive in the amount of 20% by weight of cement, the value of the rate constant for the structure formation is higher compared to the cement composite with additive of methylcellulose of the brand FMC 2094 (synthetic additive (10% and 30% by weight of cement).

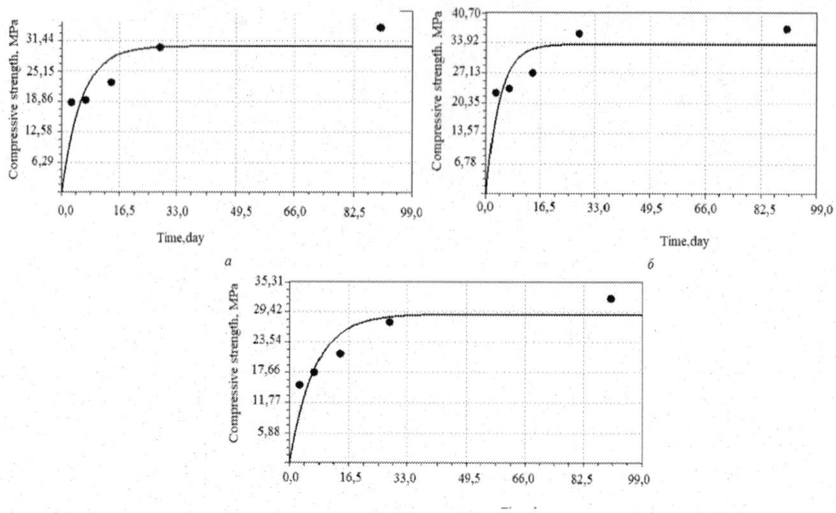

Figure 2.2. Kinetics of cement-sand strength set using the synthetic synthetic additive used in the formulation: a - 10% by weight of cement; b - 20% by weight of cement; in - 30% by weight of cement.

Therefore, the rate constant of structure formation b for cement composites with synthetic additive in the formulation (20% by weight of cement) were b = 0.221 MPa/day, and for compositions with methyl methylcellulose FMC 2094 (1% by weight cement) - 0,117MPa/day.

The kinetics of hardening of the cement-sand solution in the presence of an additive based on synthetic aluminosilicates and Neolith P 4400 powder has been evaluated. Samples were made with the addition of sand from the Ukhta field with a cement-sand ratio C:P = 1:2 and a water-cement ratio W/C = 0.47. Hardening of the samples took place in air-dry conditions (temperature $(18\pm2)^0$C, relative air humidity 60–70%). The research results are presented in Figure 2.3.

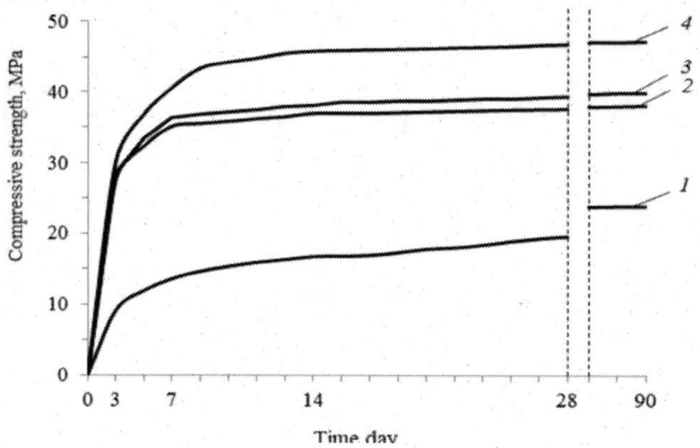

Figure 2.3. Hardening kinetics of cement composite: 1 - control (without the content of additives); 2 - a sample with a content of 20% synthetic additive by weight of cement; 3 - a sample with a content of 0.5% additive Neolith P 4400 and 20% synthetic additive; 4 – sample with 1% Neolith P 4400 additive and 20% synthetic additive.

Analysis of the experimental data (Figure 2.3) revealed that the introduction in cement-sand mortar of the redispersible powder Neolith P 4400 and a synthetic additive increases its compressive strength.

It has been established that the introduction of an additive based on synthetic aluminosilicates (20% by weight of cement) to the recipe of a cement-sand mortar increases the compressive strength of samples at the age of 90 days of air-dry hardening by 37.3% (Figure 2.3, curve 2) compared to with control samples (without additives).

Introduction to the recipe of a cement-sand mortar of a synthetic additive (20% by weight of cement) and Neolith P 4400 powder (0.5% by weight of cement) increases the compressive strength of samples at the age of 90 days by 39.9% (Figure 2.3, curve 3) compared with the control sample (without additives). The application in cement-sand solution of a synthetic additive (20% by weight of cement) and Neolith P 4400 powder (1% by weight of cement) in the formulation increases the compressive strength of samples at the age of 90 days by 49.4% (Figure 2.3, curve 4) compared with the control sample (without additives).

Thus, for the subsequent development of a dry building mixture recipe used as a tile adhesive, the optimal amount of Neolith P 4400 redispersible powder were chosen in an amount of 1% by weight of cement.

To create a model of the strength of the cement composite, the method of mathematical planning of the experiment were applying in work. A mathematical model of strength were compiling for a composition that includes: cement, an additive — the plasticizer Kratasol PFM and an additive based on synthetic aluminosilicates.

Table 2.2. Conditions for changing variables

Name	Coded designation	Variables		
		x_1	x_2	x_3
Lower level	-1	4886	10	0,7
Main level	0	6886	15	1,0
Upper level	+1	8886	20	1,3
Variation interval	Δ	2000	5	0,3

Three main factors affecting the strength of the cement composite are considered: the specific surface area of the additive based on synthetic aluminosilicates (x_1), the amount of additive content in the cement composite (x_2) and the content of the superplasticizer Kratasol PFM in the cement composite (x_3). The conditions for changing the variables and the planning matrix are given in Table 2.2 and 2.3.

Table 2.3. Planning matrix

Experience number	x_1	x_2	x_3
1	-1	-1	-1
2	1	-1	-1
3	-1	1	-1
4	1	1	-1
5	-1	-1	1
6	1	-1	1
7	-1	1	1
8	1	1	1

The homogeneity of the variances were checking by the Cochren criterion, the adequacy of the model were checking by the Fisher criterion and the significance of the coefficients by the Student criterion. After processing the obtained experimental data and excluding insignificant coefficients from the regression equation, the equation became as follows:

$$y = 33{,}388 + 2{,}2125x_1 + 3{,}525x_2 - 0{,}9875x_3 \, . \quad (2.2)$$

Equation (2.2) characterizes the dependence of the strength of the cement composite on the factors studied.

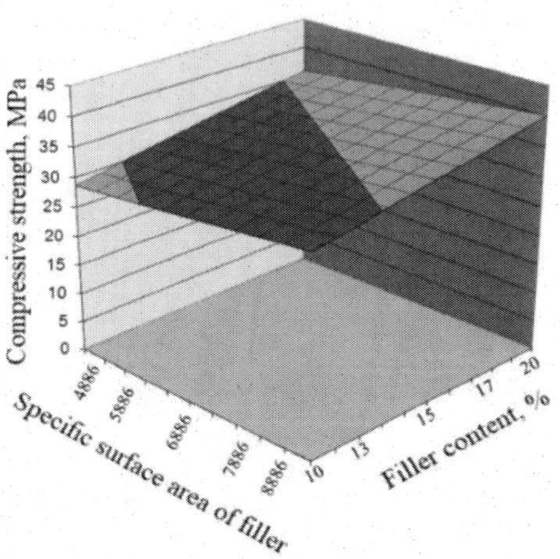

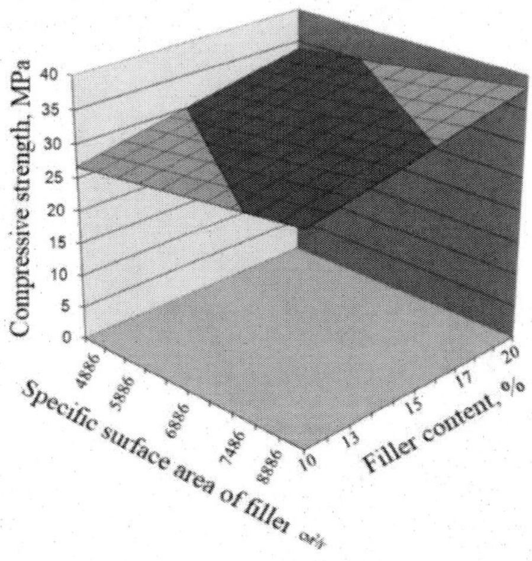

Figure 2.4. Strength of a cement composite stone depending from the specific surface SUD and the amount of synthetic additives: a - when the content of Kratasol PFM, equal to 0.5%; b - when the content of Kratasol PFM, equal to 1.5%.

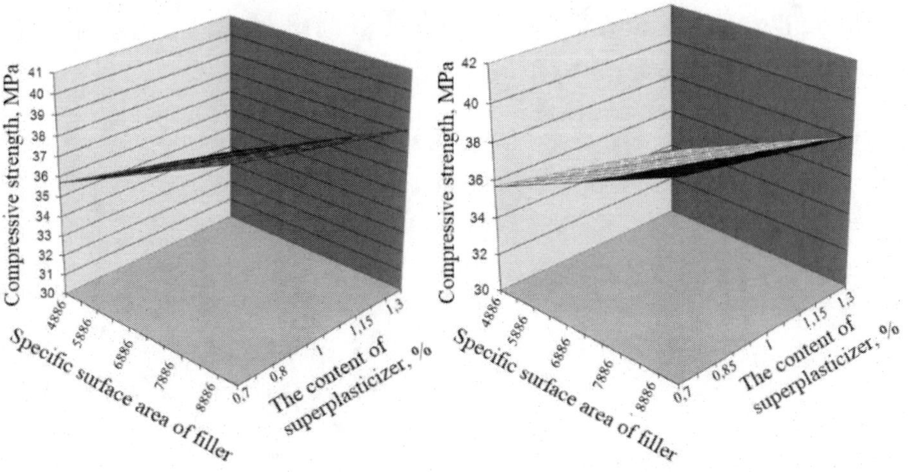

a b

Figure 2.5. The strength of the cement composite stone, depending on the specific surface S and the amount of content superplasticizer Kratasol PFM: a - with the introduction of synthetic aluminosilicates, equal to 10% by weight of the binder; b - with the introduction of synthetic aluminosilicates, equal to 20% by mass of binder.

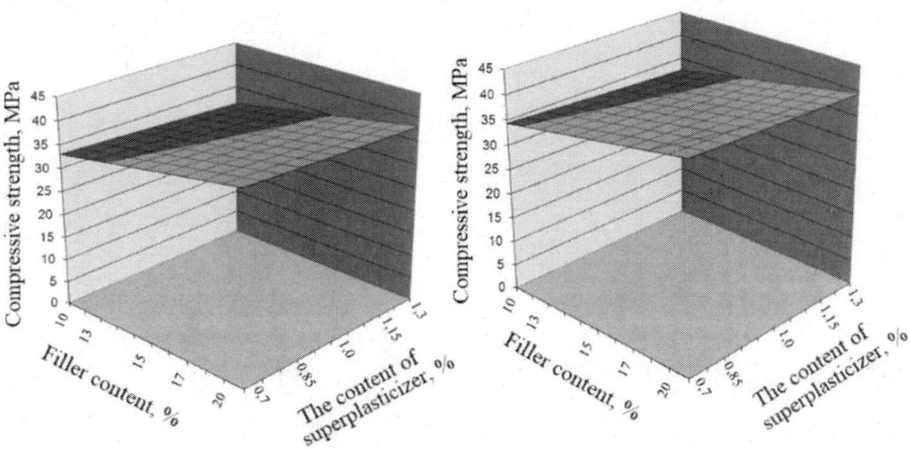

a b

Figure 2.6. Strength of a cement composite stone depending on the content of the synthetic additive and the amount of the content of the superplasticizer Kratasol PFM: a - with a specific surface S_{sp} equal to $S_{sp} = 4886\ cm^2/g$, b - with a specific surface area S_{sub} equal to $S_{sp} = 8886\ cm^2/g$.

The obtained coefficients of the regression equation and graphs (Figure 2.4–2.6) indicate that the increase in the strength can be achieved by increasing the specific surface of the additive and simultaneously reducing the content of the superplasticizer. The proposed model allows you to choose the optimal content of components in the recipe of the dry building mixture. The optimum content of synthetic aluminosilicates in the cement stone recipe is 20% with a specific surface $S_{sp} = 6886\ cm^2/g$.

2.2. Rheological and Technological Properties of Dry Building Mixtures with the Use of Additives Based on Synthetic Aluminosilicates

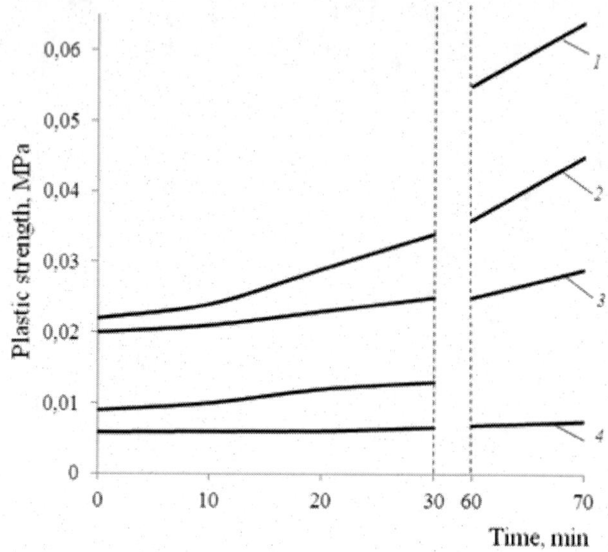

Figure 2.7. The change in plastic strength of the cement-sand mortar (C:S = 1:2): 1 - the content of synthetic additives 30% by weight of cement; 2 - the content of synthetic additives 20% by weight of cement; 3 - content of synthetic additives 10% by weight of cement; 4 - control sample (without the content of synthetic additives).

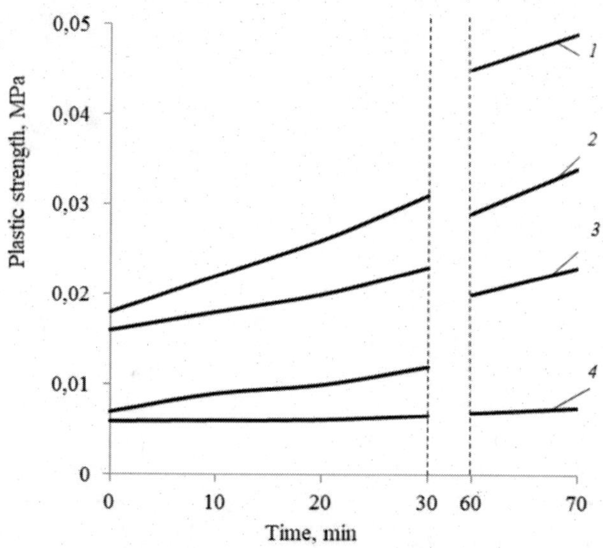

Figure 2.8. Change in plastic strength of cement-sand mortar (C:S = 1:3): 1 - the content of synthetic additives 30% by weight of cement; 2 - the content of synthetic additives 20% by weight of cement; 3 - content of synthetic additives 10% by weight of cement; 4 - control sample (without the content of synthetic additives).

The rheological properties of cement-sand compositions were evaluate by the plastic strength indicator determined using a KP-3 conical plastometer. The effect of synthetic zeolites on the kinetics of changes in the plastic strength of the compositions were evaluate. The control composition (without the additive) and the compositions with synthetic additive with a cement-sand ratio of 1: 2 (Figure 2.7) and a ratio of 1:3 (Figure 2.8) were studding. The studied compositions were made with the same water-cement ratio W/C = 0.7.

Analysis of plastograms (Figure 2.7) testifies that the introduction in cement-sand mortar (C:S = 1:2) of synthetic aluminosilicates accelerates the gain of plastic strength on the early stages of hardening. So, the value of the plastic strength of the cement-sand mortar (control composition) at the age of 20 minutes from the moment of mixing is $\tau = 0.0061$ MPa (Figure 2.7, curve 4), and with additive based on synthetic aluminosilicates in the amount of 10%, 20% and 30% by weight of the binder at the age of 20 minutes from the time of mixing, it is $\tau = 0.012$ MPa (Figure 2.7, curve 3), $\tau = 0.023$ MPa (Figure 3.10, curve 2) and $\tau = 0.029$ MPa (Figure 2.7, curve 1).

Analyzing plastograms (Figure 2.8) it were revealed that the introduction in cement-sand mortar (C:S = 1:3) accelerates the set of plastic strength of the solution in the early stages of hardening.

Thus, the value of the plastic strength of the cement-sand mortar (control composition) at the age of 20 min from the moment of mixing is $\tau = 0.0059$ MPa (Figure 2.8, curve 4), and with the introduction of an additive based on synthetic aluminosilicates in the amount of 10%, 20% and 30% of the mass of the binder at the age of 20 min from the moment of mixing is respectively $\tau = 0.01$ MPa (Figure 2.8, curve 3), $\tau = 0.02$ MPa (Figure 3.11, curve 2) and $\tau = 0.026$ MPa (Figure 2.8, curve 1).

Thus, from Figure 2.7 and 2.8 it were found that the introduction of additive in cement-sand mortar with C: S ratio = 1: 2 leads to an earlier set of plastic strength compared to cement-sand mortar with C: S ratio = 1: and synthetic additive. So, at the age of 20 minutes from the moment of mixing the mortar using a synthetic additive in the formulation (20% by weight of the binder), the plastic strength of the cement-sand mortar with the ratio C: S= 1: 2 were $\tau = 0.023$ MPa (Figure 2.7, curve 2) and with the ratio C:S= 1:3, respectively, $\tau = 0.02$ MPa (Figure 2.8, curve 2).

In this regard, in further developing the composition of a dry construction mixture a tile adhesive based on which has fast curing times it were decided to use a cement-sand mortar with a ratio C:S = 1:2.

For comparison, an assessment were made of the effect of the additive based on synthetic aluminosilicates and the addition of methylcellulose brand FMC 2094 on the change in the plastic strength of the cement-sand mortar. Samples were made on cement-sand mortar C: S= 1: 2 and with an equal water-cement ratio W/ C = 0.7. The test results are shown in Figure 2.9.

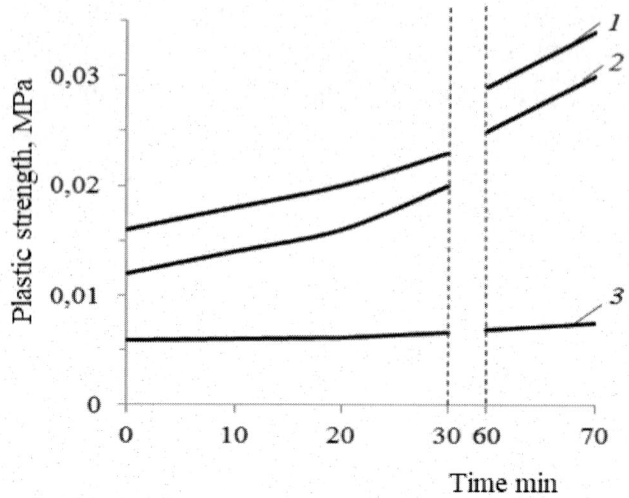

Figure 2.9. The change in the plastic strength of the cement-sand mortar with the use of additives in the formulation: 1 - synthetic additive (20% by weight of cement); 2 - methylcellulose brand FMC 2094 (1% by weight of cement); 3 - control composition (without additives).

An analysis of plastograms (Figure 2.9) showed that the introduction in cement-sand mortar additive of synthetic aluminosilicates accelerates the plastic strength of the mortar along with the import-modifying additive, methylcellulose of the FMC 2094 brand. Therefore, the plastic strength value of the cement-sand mortar with synthetic additive (20%) at the age of 20 min from the moment of mixing is $\tau = 0.02$ MPa (Figure 2.9, curve 1) and with additive methylcellulose brand FMC 2094 (1% by weight of cement - that) plastic strength were $\tau = 0,016$ MPa (Figure 2.9, curve 2).

The effect of Sika ViscoCrete 3180 and Kratasol PFM additives on the plastic strength value of a cement-sand sample were studied using an additive based on synthetic aluminosilicates (20% by weight of cement) depending on their percentage in the formulation. Sika ViscoCrete 3180 superplasticizer is a superplasticizer based on polycarboxylate esters for concrete mixtures with low exotherm. Additive Sika ViscoCrete 3180 is used as a highly effective plasticizer with a long time of preserving its mobility for any concrete mixes and meets the requirements of the technical specifications. Samples were made on cement-sand mortar C:S = 1:2 and with an equal water-cement ratio B/C = 0.7. The research results are presented in Figure 2.10 and 2.11.

Analysis of plastograms (Figure 2.10) revealed that with the introduction of additives Sika ViscoCrete 3180 and Kratasol PFM in the amount of 1% by weight of cement the plastic strength of the solution at the age of 20 minutes with the mixing momentum were $\tau = 0.015$ MPa (Figure 2.10, curve 2), $\tau = 0.012$ MPa (Figure 2.10, curve 3) and the plastic strength of the control composition were $\tau = 0.02$ MPa (Figure 2.10, curve 1).

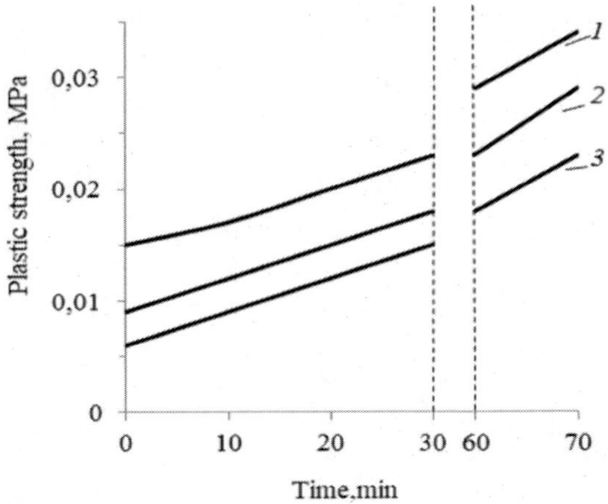

Figure 2.10. The change in the plastic strength of the cement-sand mortar with synthetic additive (20% by weight of cement) with the use of additives in the formulation: 1 - control composition (with the use of synthetic additives); 2 - Sika ViscoCrete 3180 (1% by weight of cement); 3 - Kratasol PFM (1% by weight of cement).

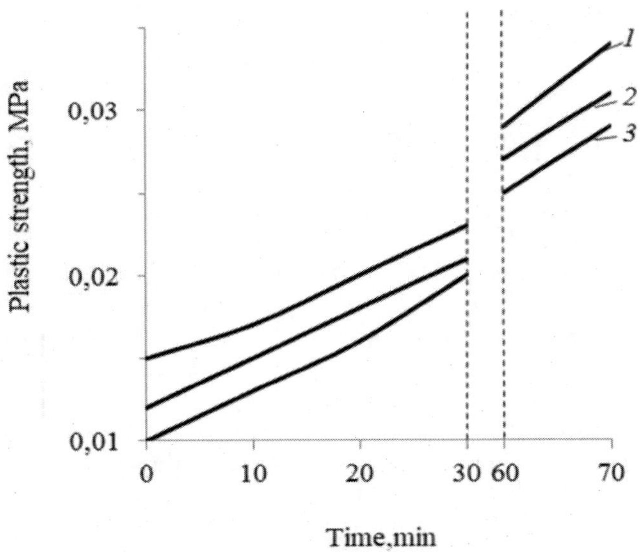

Figure 2.11. The change in the plastic strength of the cement-sand mortar with synthetic additive (20% by weight of cement) with the use of additives in the formulation: 1 - control composition (with the use of synthetic additives); 2 - Sika ViscoCrete 3180 (0.5% by weight of cement); 3 - Kratasol PFM (0.5% by weight of cement).

In Figure 2.11 presents the results of changing the plastic strength of the cement-sand mortar with synthetic additive (20% by weight of cement) and with additives Sika ViscoCrete 3180 and Kratasol PFM for 0.5% by weight of cement. Analysis of plastograms (Figure 2.11) revealed that the introduction in cement-sand mortar of synthetic aluminosilicates (20% by weight of cement) and additives Sika ViscoCrete 3180 and Kratasol PFM in the amount of 0.5% by weight of cement mortar plastic strength at the age of 20 min from the moment of mixing were τ = 0.018 MPa (Figure 2.11, curve 2) and τ = 0.016 MPa (Figure 2.11, curve 3), respectively.

In this case, the plastic strength of the control composition were τ = 0.02 MPa (Figure 2.11, curve 1). Analyzing the data in Figure 2.10 and Figure 2.11 found that the additive Kratasol PFM has a greater plasticizing effect on the cement-sand mortar using a synthetic additive in the formulation (20% by weight of cement) than the additive Sika ViscoCrete 3180.

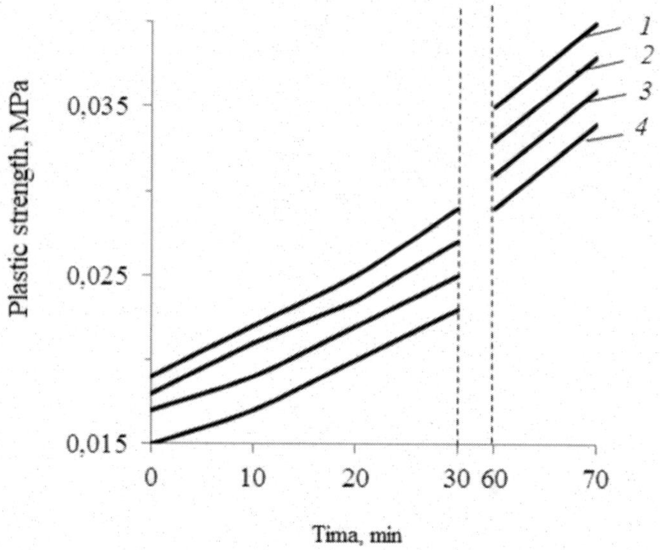

Figure 2.12. The change in the plastic strength of the cement-sand mortar with a synthetic additive (20% by weight of cement): 1 - the content of the redispersible powder Neolith P 4400 1% by weight of cement; 2 - the content of redispersible powder Neolith P 4400 0.8% from the mass of cement; 3 - the content of the redispersible powder Neolith P 4400 0.5% by weight of cement; 4 - control composition (using synthetic additives).

The effect of Neolith P 4400 redispersible powder on the value of plastic strength of a cement-sand mortar were studding with the use of an additive based on synthetic aluminosilicates (20% by weight of cement) in the formulation depending on its percentage. Samples were made on cement-sand mortar C:S= 1:2 and with an equal water-cement ratio W/C equal to W/C = 0.7. The results of the research are presented in Figure 2.12.

It were revealed that with the introduction of a synthetic additive in the amount of 20% by weight of cement into the cement-sand solution, the plastic strength of the solution at the age of 20 min after mixing were τ = 0.02 MPa (Figure 2.12, curve 4). Plastic strength of the cement-sand mortar using a synthetic additive (20% by weight of cement) and Neolith P 4400 redispersible powder in a volume of 0.5%; 0.8% and 1% at the age of 20 min from the moment of mixing were respectively $\tau = 0.022$ MPa (Figure 2.12, curve 3), $\tau = 0.023$ MPa (Figure 2.12, curve 2) and $\tau = 0.025$ MPa (Figure 2.12, curve 1).

Thus, for increase the plastic strength of the mortar based on the developed DBM it were decided to use the Neolith P 4400 redispersible powder in a volume of 1% by weight of cement.

The use in the formulation of tiling glue of additives based on synthetic aluminosilicates, plasticizer Kratasol PFM and redispersible powder Neolith P 4400 allows obtaining a mixture with rheological properties that can hold the finishing tile on a vertical surface, eliminating its slipping. In accordance with the requirements of DIN EN 1308, compositions based on dry mixes should be resistant to slipping when applied to vertical surfaces. In this regard, the work evaluated the stability of the developed composition to slipping.

The DBM recipe has been developing which is using as a tile glue for tiling the faces and interior walls of buildings. In the recipe including Volsky Portland brand M 400, sand from the Ukhtin location in the ratio of fractions 0.63-0.315:0.315-0.14, respectively, 80%:20% and a bulk density of 1538.2 kg/m^3 plasticizer Kratasol PFM, redispersible powder neolith p 4400 and addition based on synthetic aluminosilicates. The calculation of the resistance to slipping of the tile adhesive layer were carrying out. The essence of the calculation is as follows. A layer of adhesive glue thick δ and high h is bonded to the wall material (substrate) and holds the tile with a weight of P_2. Layer loaded with bulk forces P_1. At the boundary with the substrate and tile, shear stresses appear in the adhesive layer, respectively, τmax1 and τ_{max2}, which keep the layer under consideration in equilibrium (Figure 2.13).

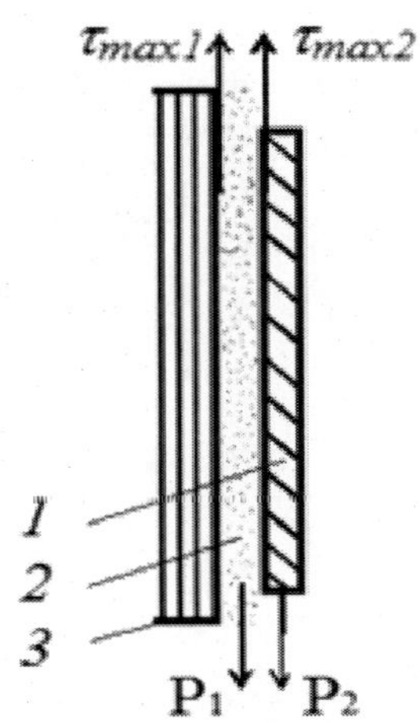

Figure 2.13 - Design scheme: 1 - wall tile; 2 - adhesive layer; 3 - vertical backer board.

To assess the condition of finding a layer of tile adhesive in equilibrium (the absence of «slipping»), you can use the system of equations:

$$\begin{cases} \tau_{max1} S = P_1 + P_2; \\ \tau_{max2} S = P_2. \end{cases} \quad (2.3)$$

where P_1 and P_2 are respectively the weight of the layer of tile adhesive and the weight of the tile, kg; S - the area of the adhesive layer, equal to the area of the tile, m^2.

To assess the condition of the adhesive layer in equilibrium (the absence of «slipping»), you can use the inequality:

$$|\tau_{max}| \leq \tau_{сд}. \quad (2.4)$$

τ_{cd} is limit stress of the shift τ_o, rheological characteristics and adhesion strength of the tile adhesive τ_{ad} summed up:

$$\tau_{сд} = \tau_0 + \tau_{ад} \quad (2.5)$$

Considering that at the initial moment of application of tile adhesive layer the magnitude of τ_{ad} is very small, the expression (2.6) can be written as:

$$\tau_{max} \leq \tau_0 \quad (2.6)$$

Volume forces (P_1), appearing in the adhesive layer, are calculated by the formula:

$$P_1 = \rho_1 \cdot V_1 \quad (2.7)$$

where ρ_1 stands for the density of the applied adhesive solution, kg/m^3;

V_1- amount of the applied adhesive, m^3.

Volume forces (P_2), appearing in wall tiles were calculated by the formula:

$$P_2 = \rho_2 \cdot V_2 \quad (2.8)$$

where ρ_2 is the density of facing tiles, kg/m^3;

V_2 - the volume of facing tiles, m^3.

Evaluation of frost resistance of the adhesive on the basis of the developed cement-based DBM with the use of an additive based on amorphous aluminosilicates were performed by alternately freezing and thawing of samples of a glue solution measuring 70mm×70mm×70mm after 28 days of air-dry hardening.

We have proposed an estimate of the shear strength of shear bonding with the help of the device GT 2.2.3 (Figure 2.14).

The mechanism of the GT 2.2.3 device is the creation of a horizontal shear load on the sample in the test installation based on the test composition. The force created by the reduction gear is transmitted to the movable carriage of the shear box and measured by a force sensor (Figure 2.14).

The test sample is placed in the carriage of the device in such a way that the substrate is in a small shear ring, and the test composition is in a large shear ring. The technological gap of 1 mm forms the cut-off area. During the test, as the tangential load applied to the lower shear ring increased, the value of the shear strains of the test composition relative to the substrate were recording.

Figure 2.14. Appearance of the device GT 2.2.3: 1 - hairpin; 2 - shearing mechanism; 3 - shear box.

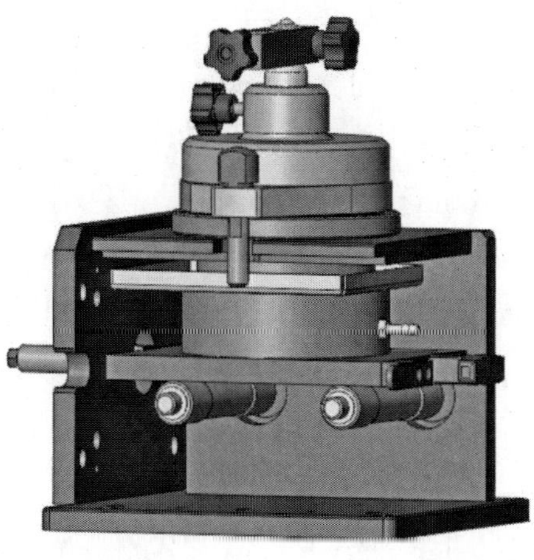

Figure 2.15. Shear box - view without side wall.

This technique and GT 2.2.3 instrument were used to determine the shear strength of the solvent adhesive layer relative to the substrate. A tile adhesive were used for the tests based on the developed formulation of the dry glue mixture. The developed adhesive formulation contains Portland cement, mineral aggregate (sand), plasticizer, polymer and mineral additive.

For the tests, cement bases of cylindrical shape were used, the geometric dimensions of which are 71.4 × 15 mm. On the bases, the test composition of a tile adhesive with a thickness of 5 mm were applying. The optimum shear rate of the solution adhesive layer is 0.2 mm/min. The samples were tested according to the scheme shown in Figure 3. For testing, bases were prepared on a cement-sand base of cylindrical shape, the geometric dimensions of which were 71.4 × 15 mm. After hardening, the test composition 10 mm thick were applying to the bases.

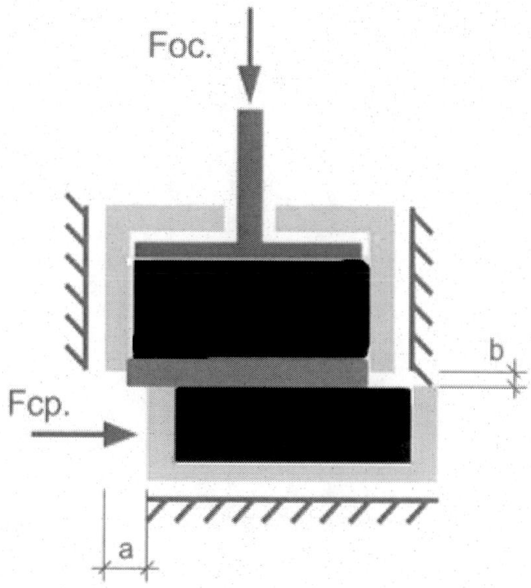

Figure 2.16. Plan of test of specimens for shear analysis.

The test sample were placing in the carriage of the device in such a way that the bases were in a small shear ring 2 and the test composition in a large shear ring 3 (Figure 2.16). The remaining volume of the large shear ring were filling with a gypsum test to ensure better fixation of the sample and an even distribution of the load on it. The technological gap of 1 mm forms the cut-off area. During the test, as the tangential load applied to the lower shear ring increased, the value of the shear strains of the test composition relative to the base were recording.

To measure the homogeneity of the data obtained during the experiment using the proposed technique, the coefficient of variation were determined. The shear strength values obtained during the series test were used for the calculation. The root-mean-square deviation when testing a series of samples is 3-6 kPa. The resulting coefficient of variation is equal to 1.5-2.6%, therefore, the variability of the variational series can be considered insignificant. The measurement error of the instrument itself is 0.5% of the measuring range.

Analysis of the data presented in the Table 2.4 testifies that the values of the tangential stresses τ_{max1} and τ_{max2} are less than the value of the maximum shear stress of the tile adhesive τ_0. So, the maximum tangential voltage τmax1 in the zone of contact with the substrate τ_{max1} is τ_{max1} = 0.00024MPa, and in the zone of contact with ceramic tile of size 0.3 × 0.3 × 0.005 m -τ_{max2} = 0.00015MPa. The ultimate shear stress of tile adhesive is τ_0 = 0.002MPa. Thus, the condition of the absence of slipping of ceramic tiles and tile glue is fulfilled.

Table 2.4. Resistance to creep of tile adhesive based on of the developed DBM

Tile type	Tile size, m	Shearing stresses, MPa		Limit stress of tile adhesive shift, τ_0, MPa	Condition of creep absence
		τ_{max1}	τ_{max2}		
Ceramic	0,1×0,1×0,005	0,00022	0,00013	0,002	satisfied
	0,3×0,3×0,005	0,00024	0,00015	0,002	satisfied
Ceramic granite	0,6×0,6×0,01	0,00033	0,00024	0,002	satisfied
	1,2×1,8×0,02	0,00060	0,00052	0,002	satisfied
Granite	0,3×0,6×0,015	0,00069	0,00042	0,002	satisfied

The analysis of the calculation results given in Table 2.4 shows that the condition of absence of creep is meeting when using other types of tiles as well. The results of these calculations are confirmed by experimental data. The experiment were to measure the maximum creep of a tile under its own weight with a caliper.

Maximum creep of a tile under its own weight were calculating as the difference between the two readings on calipers. The amount of creep of the adhesive layer based on the developed DBM formulation were 0.3 mm. The German standard DIN EN 1308 - 2007 sets 0.5 mm as the maximum tile creep ($\tau_{сп}$).

The obtained results of the calculations are confirmed by experimental data. The experiment were to measure the maximum slipping of a tile under the action of its own weight with a caliper. The slipping of ceramic tiles (fixed on the substrate by the tested composition of tile adhesives) were determined according to the European standard DIN EN 1308. The method consisted in measuring the maximum slipping of a tile under the action of its own weight. The test took place under normal conditions - temperature (23±2)°C relative air humidity (50±5)%, air circulation rate in the test area less than 0.2 m/s.

The steel ruler were fastening with clips to the upper end of the substrate so that its lower edge after placing the substrate in a vertical position passed horizontally. A protective tape 25 mm wide were gluing directly under the steel ruler, and then a thin layer of the tested adhesive glue were applying to the substrate with a spatula. Next, a thick layer of the test solution were applying to the substrate in such a way that it covered the lower edge of the protective tape, after which the solution were smoothed with a notched trowel perpendicular to the steel ruler. Immediately after smoothing the solution, a protective tape were removing with a notched trowel. Two stops with a width of 25 mm were applied to the steel ruler, after 2 minutes ceramic tiles were attached to the stop and loaded with a weight of (50±0.1)N. The distance between the ruler and the tile were measuring with a caliper at three points with an accuracy of ± 0.1mm. After (30±5)s weights and stops were removed and the substrate with the test composition and ceramic tiles applied in a vertical position were set. After (20 ± 2) minutes the distance were measuring in three previously measured points. The maximum slipping of the tile under the action of its own weight were calculated as the difference between the two readings of the caliper.

The amount of slipping of the adhesive layer based on the developed DBM recipe were 0.3 mm. The German standard DIN EN 1308 sets the permissible maximum slipping value of a tile that is $\tau_{sp.} \leq 0.5$ mm.

Thus, the results of the calculations and the data of the experimental studies suggest that the adhesive layer based on the developed composition of the DBM is resistant to slipping.

When finishing the surface with facing tiles, tile adhesive is using, which is applying with a layer of 3–10 mm. When applying a thin layer in the adhesive composition, there is a loss of moisture, which is accompanying by an increase in the stiffness of the mixture, complicating technological operations to level the finishing layer increasing the complexity of the work performed. In this regard, one of the important characteristics of the adhesive mixture is the water holding capacity that allows you to maintain the mobility of the mixture for a certain time.

Water retention were determined by testing the mortar layer 12 mm thick laid on blotting paper (sheets of blotting paper size of 150 mm × 150 mm TU 13-7308001-758-88).

Before the test, 10 sheets of blotter paper were weighed with an accuracy of up to ± 0,1g. The sheets of blotting paper were placed on a glass plate stacked on top of a pad of gauze fabric installed a metal ring and once again weighed. Carefully blended mortar mix laid flush with the edges of the metal rings, aligned, weighed and left for 10 min. Metal ring with a solution carefully removed together with gauze. Blotting paper were weighed with an accuracy of up to ± 0,1g.

Water-holding capacity of the mortar were measuring expressed as a percentage of water content in the sample before and after the experiment using:

$$V = \left(100 - \frac{m_2 - m_1}{m_4 - m_3} \cdot 100\right) \quad (2.9)$$

where m_1 - mass of blotting paper before test, g;
m_2 - mass of blotting paper after the test, g;
m_3 - mass installation without mortar, g;
m_4 - mass installations with mortar mixture, g.

Water-holding capacity of the mortar were determined twice for each sample solution mixture, and were calculated as the arithmetic mean of the two determinations, differing by no more than 20% of the lower value.

Table 2.5. The influence of the recipe DBM on the rheological and technological characteristics of tile glue

№	Composition	W/C	The water-holding capacity,%	Initial plastic strength τo, MPa×10^{-3}
1	Control composition (cement + sand)	0,7	93,3	5,9 ± 0,186
2	the same + synthetic additive (10% by weight of cement)	0,7	95,7	7,0 ± 0,155
3	the same + synthetic additive (20% by weight of cement)	0,7	96,6	15,2 ± 0,147
4	the same + synthetic additive (30% by weight of cement)	0,7	96,2	18,1 ± 0,179
5	the same + synthetic additive (10% by weight of cement) + Kratasol PFM (1% by weight of cement)	0,63	95,8	4,2 ± 0,098
6	the same + synthetic additive (20% by weight of cement) + Kratasol PFM (1% by weight of cement)	0,63	96,9	5,4 ± 0,115
7	the same + synthetic additive (30% by weight of cement) + Kratasol PFM (1% by weight of cement)	0,63	96,1	7,5 ± 0,088
8	the same + synthetic additive (10% by weight of cement) + Kratasol PFM (1% by weight of cement) + Neolith P 4400 (1% by weight of cement)	0,55	96,8	6,3 ± 0,104
9	the same + synthetic additive (20% by weight of cement) + Kratasol PFM (1% by weight of cement) + Neolith P 4400 (1% by weight of cement)	0,55	99,3	10,1 ± 0,093
10	the same + synthetic additive (30% by weight of cement) + Kratasol PFM (1% by weight of cement) + Neolith P 4400 (1% by weight of cement)	0,55	98,2	13,2 ± 0,113

In Table 2.5 shows the effect of additives on the basis of synthetic aluminosilicates plasticizing and modifying additives on the water retention capacity and the initial plastic strength of the compositions using sand from the Ukhta deposit with a ratio of 0.63 - 0.315: 0.315 - 0.16 respectively 80%: 20%.

The experimental data (Table 2.5), it is confirmed that dry mortar with the use of a synthetic additive redispersible Neolith P 4400 powder and additives Kratasol PFM in the formulation has a sufficient water retention capacity of 95.7–99.3%.

3. Operational Resistance of Adhesive Layer Based on Dry Construction Mixture

3.1. Evaluation of the Adhesion Strength of the Tile Adhesive Layer Using Synthetic Aluminosilicates in the Formulation

An important property of coatings based on dry construction mixtures is their adhesion strength to the surfaces of the finished elements and ceramic tiles. To assess the adhesive strength of the adhesive layer relative to the substrate and ceramic tile can using the value of adhesive strength. This is an indicator depending on the formulation of the adhesive composition, the type and condition of the substrate, the technology of deposition and the conditions for the formation of coatings.

Adhesive strength is summarized from:

$$R_{adg} = R_1 + R_2, \quad (3.1)$$

where R_1 is adhesion strength provided by the forces of physicochemical interaction between the substrate and the adhesive layer; R_2 is the adhesive strength provided by the forces of mechanical adhesion of the adhesive layer and the substrate.

For assess the adhesiveness of the adhesive layer based on the developed DBM the adhesion strength of adhesive compositions with the surface of cement-sand substrates were studied. A mixture with additive based on synthetic aluminosilicates and modifying additives, as well as a control formulation were studding.

For assess the strength of adhesion with the base, the method of tearing the punch (normal tear) were using according to the European standard DIN EN 1348. The method is based on determining the force of exfoliation of the stamp from the surface by separation. The samples were made, representing the substrate, on which the tested composition of tile adhesive were applied and ceramic tile (10×10×5)mm in size.

After 27 days of storage of the samples under normal conditions, the tiles with the help of epoxy glue (EDP-TU 0751-018-48284381-00) were gluing to cylindrical dies with a diameter of 20 mm. After another 24 hours of storage of the samples under normal conditions, the adhesion strength of the test solution with the base were determined. For this purpose, a load were applying to the stamp with a constant rate of increase (250±50) N/s. The adhesion strength of the finishing composition with the substrate were determined by the formula

$$R_{adg} = \frac{P}{F}, \quad (3.2)$$

where P is the total tensile force, H; F - gluing surface, mm^2.

The strength of adhesion with the base in a water-saturated state were determining by the method of tearing the punch (normal tear) according to the European standard DIN EN 1348. The test samples were kept for 7 days under normal conditions and then stored in water at room temperature. After 20 days the test specimens were removed from the water, wiped dry with a towel and glued to the tiles stamped cylinders with a diameter of 20 mm using epoxy glue (EDP-TU 0751-018-48284381-00). After 7 hours, the test specimens were again lowered into room temperature water. After 24 hours, the test specimens were taken out of the water and immediately tested to determine the strength of adhesion with the base.

During the tests, the cohesive nature of the separation of the ceramic tile were set at R_{kog} = 1.4 MPa (Figure 3.1). The separation of the tile from the tile adhesive during the tests did not occur.

Thus, it were found, what is tiled cement glue have adhesive strength higher than 1.4 MPa.

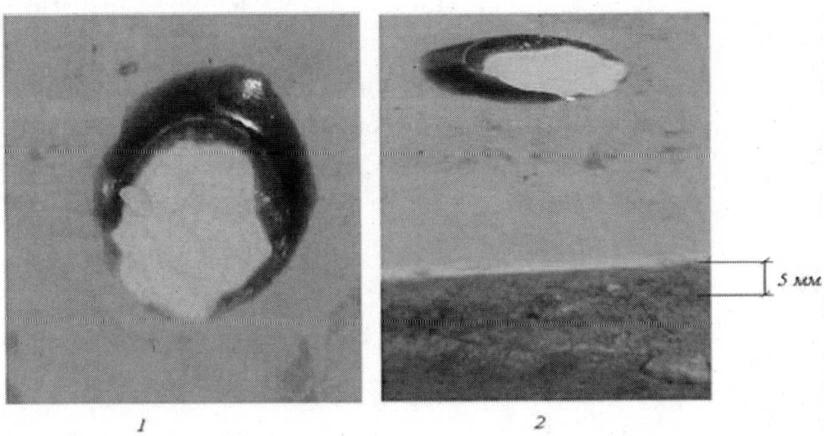

Figure 3.1. The nature of the separation washers.

It were revealing that tiled cement glue are characterizing by high adhesive strength. The greatest contribution to the increase in adhesion strength is made by the addition of Neolith P 4400, which disperses when the mixture is mixed with water and, as moisture evaporates, forms high-strength films on solid particles of the solution, increasing its adhesion to the cement-sand substrate. The adhesive strength of the adhesive layer with gypsum and polystyrene substrates were evaluating. The test results are presented in Table 3.1.

Table 3.1. Adhesion strength of the adhesive layer to tear

Substrate view	Adhesion strength, R_{adg}, МПа	
	Nature of separation	composition based on the developed mixture
Gypsum	on tile	more 1,4
Extruded polystyrene	on tile	more 1,4
Styrofoam ПСБ С-35 Polyfoam PSB S-35	on tile	more 1,4
Cement sand	on tile	more 1,4

As can be seen from the results (Table 3.1), tile adhesive, made on the basis of the developed formulation with synthetic aluminosilicates, has sufficient adhesion strength with different types of surface. So, the adhesion strength of a sample of tile adhesive with a gypsum substrate, extruded polystyrene PSB S-35 and a cement-sand substrate is more than $R_{adg}>1.4$ MPa. Comparing the results obtained with the adhesion strength requirements of the tile glue with the base we conclude that the adhesion strength of the developed formulation meets the conditions specified in the standard DIN EN 12004 where $R_{adg} \geq 0.5$ MPa.

During operation, the mortar layer also experiences shear loads relative to the substrate. Therefore, it is also necessary to carry out shear tests. At present, there is no actual method for determining the shear adhesion strength of mortar layers based on DBM. In accordance with GOST 14759 «Adhesives. The method for determining the shear strength» shear strength is determined on a testing machine that allows tensile testing. However, the technique involves the use of metal plates as substrates, and polymer compositions as glue. Of interest application, the method for determining shear resistance, which is using in testing soils. The technique provides for the use of the device GT 2.2.3 produced by LLC NPP Geotek. The principle of determining the shear load of the adhesive layer were to create in the test facility (GT 2.2.3 produced by OOO NPP Geotek), shown in Figure 3.2, horizontal shear load on the test specimen. Testing of samples were carried out according to the scheme shown in Figure 3.3.

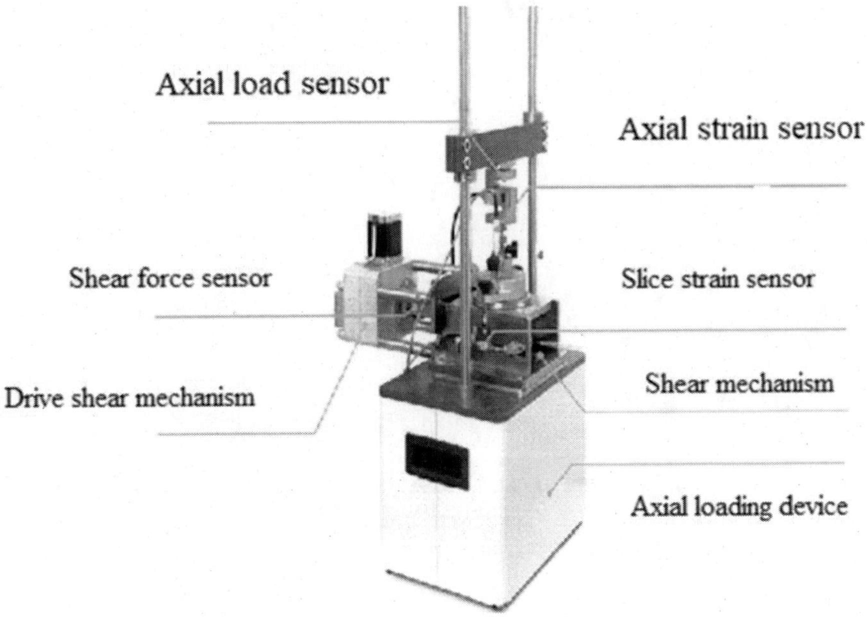

Figure 3.2. The appearance of the test unit GT 2.2.3.

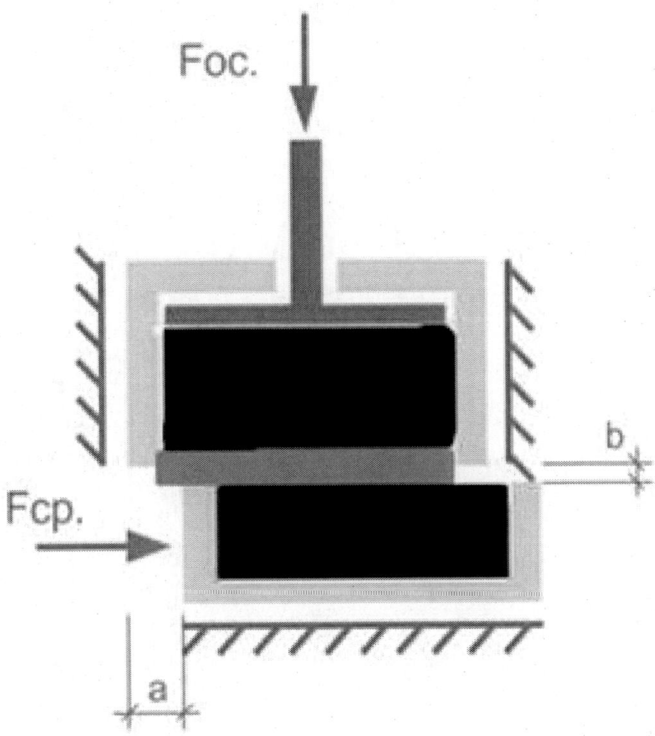

Figure 3.3. Plan of test of specimens for shear analysis.

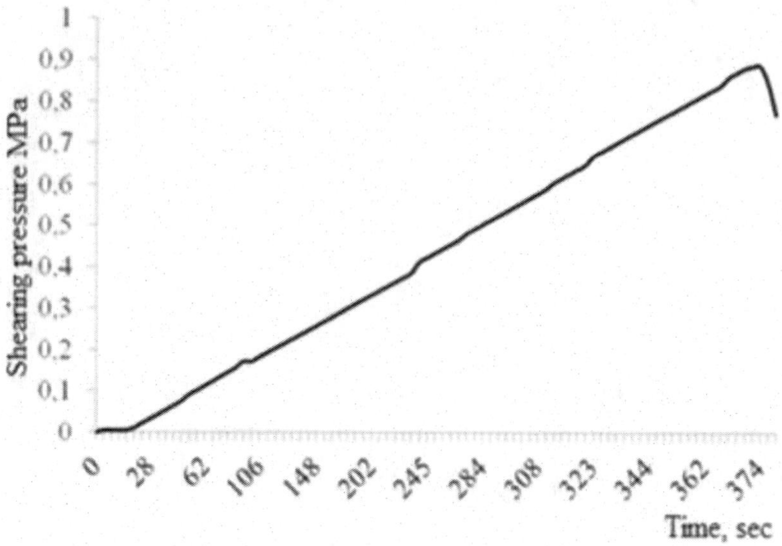

Figure 3.4. Shear-based cement slip shear test.

The analysis of the data showed that the adhesion strength on the shear of the solution layer on the basis of the developed formulation of the tile adhesive is 0.92 MPa, which meets the requirements imposed on the adhesion strength of the adhesive layer.

Analysis of the data presented in Figure 3.4 showed that the shear strength of the mortar layer based on the developed recipe of tile adhesive is $R_{sdv} = 0.92$ MPa.

3.2. Hydrophysical Properties of Tile Glue with Use in the Formulation of Additives Based on Synthetic Aluminosilicates

A capillary-porous structure, (determines the nature of the interaction of tile glue with moisture contained in an air-humid environment or in direct contact with water), is characteristic of a layer of tile glue based on the developed DBM recipe using synthetic aluminosilicates. Because of the interaction of the tile adhesive layer with moisture, its physic mechanical properties change. Therefore, it is necessary to investigate the interaction with moisture during the operation of the composition based on the developed tile adhesive formulation.

To assess the hydrophysical properties of the adhesive layer on the basis of the developed recipe of tile adhesive, their absorption kinetics and water absorption kinetics with capillary suction were studied. The kinetics of water absorption of finishing compositions based on dry mix were determined according to GOST 5802. Samples (predried to constant weight) were placed in a container filled with water. At the same time, the water temperature in the tank were (20 ± 2)°C. Samples were weighed with an interval of 1 hour on ordinary scales with an error of no more than 0.1%. When weighing, the samples taken out of the water were previously wiped out with a pressed damp cloth. The mass of water that leaked from the pores of the sample onto the weighing pan were included in the mass of the saturated sample. The test were carrying out until the results of two consecutive weighing differed by no more than 0.1%. Determined by water absorption by weight.

Water absorption of a separate sample by weight Wm in percent were determined with an error of up to 0.1% according to the formula

$$W_m = \frac{m_{\text{вл}} - m_{\text{сух}}}{m_{\text{сух}}} \cdot 100 \quad , \qquad (3.3)$$

where m_{dr} - the mass of the dried sample, g; m_w - mass of water-saturated sample.

Determination of water absorption with capillary suction took place according to GOST 31356. The water absorption at the capillary suction of the WcP of a sample dried to constant weight were determined by the amount of water absorbed by it at atmospheric pressure due to capillary and adsorption forces. The test were carried out on samples-beams, size 40 × 40 × 160 mm, aged 28 days at a temperature equal to, t = (20 ± 2)°C, according to the following mode: 2 days - storage of samples in the form; 5 days - storage of the decomposed samples at ambient air humidity (95 ± 5)% and, then, for 21 days, the samples were stored at ambient air humidity of (65 ± 5)%. After 28 days, the end faces of the test specimens were treated with a plaster grater to obtain a rough surface. These samples were dried to constant weight at a temperature equal to, t = (105 ± 5)°C. After that, the side faces of the samples were coated with a waterproof composition (molten paraffin) and weighed with an error of ± 0.01 g (mass m1). The linear dimensions of the wetted end face of the beams were measured with a caliper with an accuracy of ± 0.1 mm.

The prepared samples were placed an end face in a bath filled with water (temperature t = (20 ± 5)°C) on a mesh stand in such a way that the end of the sample were immersed in water by 5 mm - 10 mm. The water level in the bath were kept constant throughout the test. After 24 hours, the test samples were removed from the water, the excess water removed from the surface of the samples using a damp cloth and weighed with an error of ± 0.01 g (mass m^2). Water absorption at capillary suction W_{kp}, $kg/(m^2 \cdot h^{0.5})$, were determined by the formula

$$W_{\text{кп}} = K_w \cdot \frac{m_2 - m_1}{S} \quad , \qquad (3.4)$$

m_1 is the mass of the dry sample, kg;

m_2 is the mass of the sample after saturation with water, kg;

S is the area of the sample face wetted, m^2;

K_w – coefficient taking into account the time of sample saturation and equal, $\frac{1}{\sqrt{24}}$ $h^{-0.5}$.

The arithmetic mean value of the test results of three samples, rounded to 0.1 kg/(m^2 $h^{0.5}$), were taken as the test result.

When assessing the kinetics of water absorption of tile adhesive samples, it were established that during the first hour the most intensive water absorption occurs based on mass W_m (Figure 3.5). At the time of first ingress of moisture on the sample (1 hour), the dependence of moisture a quadratic parabola described content on the time of moistening:

$$W_m = f\left(\sqrt{\tau}\right). \quad (3.5)$$

Analysis of Figure 3.5 showed that the W_m water absorption of a tile adhesive sample based on the control composition (Figure 3.5, curve 1) after the 1st hour of water saturation were W_m = 14.6%, with additive based on synthetic aluminosilicates, respectively, Wm = 13.2% (Figure 3.5, curve 2). After the 2nd hour of water saturation, a gradual increase in water absorption values is observed. After 24 hours of water saturation, the water absorption of the control sample were W_m = 16.55%, and the sample with the use of synthetic aluminosilicates in the formulation - W_m = 15.03%.

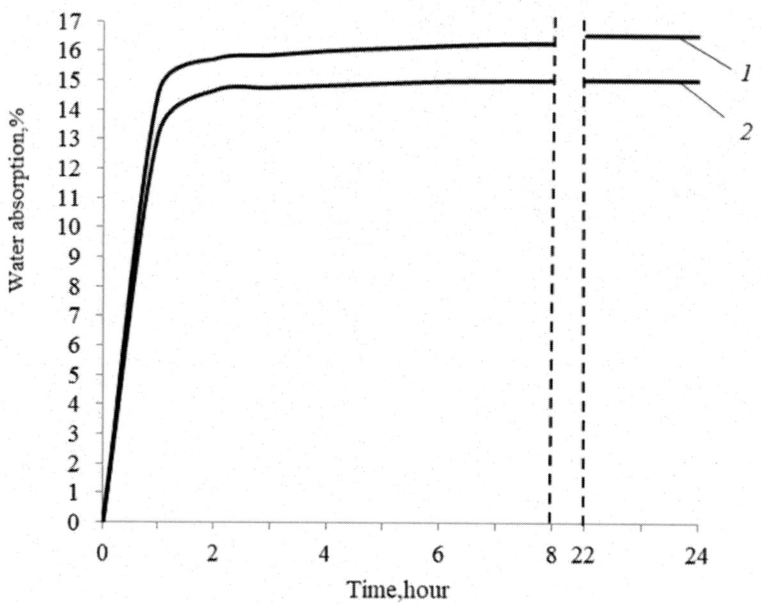

Figure 3.5. Water absorption kinetics by weight of tile adhesive samples: 1 –control composition; 2 - layer of tile glue using synthetic aluminosilicates in the formulation.

Thus, analyzing the data in Figure 3.5 it were revealed that tiled glue using synthetic aluminosilicates in the formulation has less water absorption than the control compound (without additives). This is due to the change in the nature of the pores of the cement stone when a synthetic additive is introduced into its formulation. To confirm the obtained data, an experiment were conducting to determine the water absorption at the capillary suction. When determining the water absorption at the capillary suction, a method were used, consisting in determining the change in the mass and volume of samples for 24 hours of direct contact with the aqueous medium. The results of the tests are presenting in Table 3.2.

Table 3.2. Water absorption of samples with capillary suction

Sample type	Water absorption value kg/m^2 h$^{0.5}$
Control sample	4,14 ± 0,056
Sample with the addition of synthetic aluminosilicates	1,43 ± 0,043

Analysis of the results obtained (Table 3.2) revealed that the value of the water absorption coefficient at capillary suction in the sample based on the developed formulation using synthetic aluminosilicates is much lower than that of the control sample and are respectively W_{cs} = 1.43 kg/m^2 h$^{0.5}$ and W_{cs} = 4.14 kg/m^2 h$^{0.5}$. Thus, the experiment performed confirms that the sample with the use of a synthetic additive in the formulation has a lower water absorption compared to the control sample (without the additive).

Estimated adhesion strength of the samples when exposed to moisture. The test results are shown in Figure 3.6.

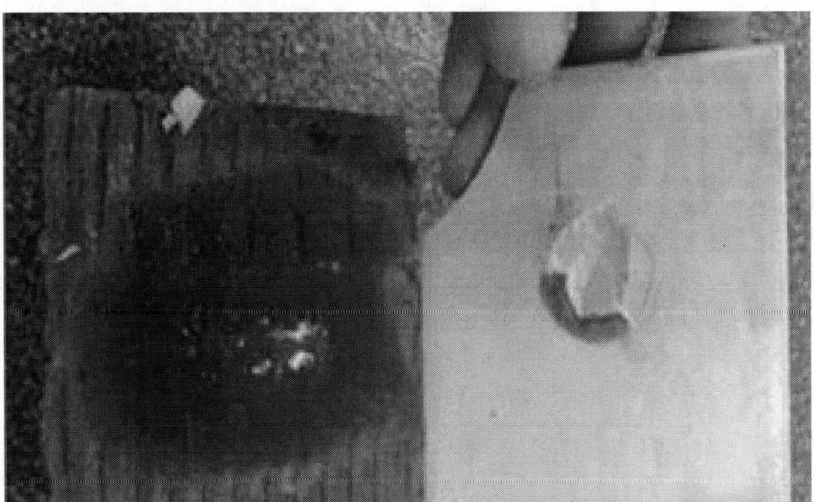

Figure 3.6. Adhesion strength with cement-sand substrate after aging in the aquatic environment.

Analysis of the results obtained (Figure 3.6) showed that the adhesion strength of the adhesive mortar based on the developed dry building mixture using synthetic aluminosilicates after 20 days of exposure to the aquatic environment is $R_{adg} = 1.1$ MPa, which meets the requirements specified in the DIN EN 12004 series standard - $R_{adg} \geq 0.5$ MPa.

3.3. Evaluation of the Properties of the Composite Based on Tile Adhesive in Relation to the Effects of Temperature

In order to confirm the possibility of using tile adhesives based on the developed dry mixture, were studied the frost resistance of the adhesive layer based on DBM. The ability of the adhesive layer not to be destroyed for a long time during repeated freezing and thawing in a water-saturated state is determined by the presence of reserve pores in its structure that are not filled with water.

Under the pressure of growing ice crystals during freezing, a portion of water is squeezed into the reserve pores. It helps to reduce the destructive effect of ice on the structural strength of the material. The destruction of the material in a water-saturated state with repeated freezing and thawing occurs only when all the reserve pores are filled with ice. The more conditionally closed porosity of the sample, the more cycles of alternate freezing and thawing is necessary to cause the sample to break.

Evaluation of frost resistance of the adhesive on the basis of the developed cement-based dry building mix with the use of an additive based on amorphous aluminosilicates in the formulation were carried out by alternately freezing and thawing on the samples with size 70 ×70 ×70mm, which were hardened 28 days on the air.

1 2

Figure 3.7. Samples before testing.

Figure 3.8. Samples of tile adhesive after testing.

As can be seen from Figure 3.7 and 3.8, samples of cement-based tile glue have subtle chalking, cracks on the surface visible to the naked eye make up to 5% of the surface, corrosion is absent. Thus, during testing of tile adhesive compositions based on the developed dry building mix formulation for frost resistance, it were established that the tile adhesive possesses a frost resistance mark F50 (Table 4.3) i.e., Can withstand 50 freeze-thaw cycles without changing the strength (within an acceptable error of 10%).

Table 3.3. Frost resistance of tile adhesive

Test result of samples, cycles							
0	5	10	15	25	35	50	75
43,2	42,83* / 0,16	42,70 / 0,47	42,60 / 0,70	42,07 / 1,94	39,86 / 7,78	39,36 / 8,24	38,43 / 10,42

* Above the line is the average strength of the tile adhesive (MPa), below the line is the change in the average strength of the tile adhesive (%).

Table 3.4. The frost resistance of tile adhesive contact zone

Test result of samples, cycles							
0	5	10	15	25	35	50	75
1,2	1,18* / 1,67	1,17 / 2,5	1,12 / 6,67	1,03 / 14,16	1,00 / 16,67	0,97 / 19,17	The tile abruption

* Above the line is the average adhesion strength of the tile adhesive (MPa), below the line is the change in the average bond strength of the tile adhesive (%).

Analysis of the data presented in Table 4.4 showed that the contact area of the tile adhesive with a cement-sand substrate and a ceramic one has sufficient frost resistance. The mark on the frost resistance of the contact layer were $F_{cz}50$.

The evaluation of the adhesion strength with the cement-sand substrate after the freeze-thaw cycles were carrying out. The test results are shown in Figure 3.9.

1 2

Figure 3.9. Samples after the test: 1 - thawing in the air; 2 - thawing in a humid environment.

The analysis of the obtained results showed that the separation occurs on the ceramic tile and is R_{kog} = 0.97 MPa.

Thus, the adhesion strength of the adhesive mortar based on the developed DBM with the use of synthetic aluminosilicates after freezing and thawing is R_{adg} 0.97 MPa, which satisfies the requirements specified in the DIN EN 12004 series - $R_{adg} \geq 0.5$ MPa.

3.4. Deformative Properties of Coatings Based on the Developed DBM

For assess the crack resistance of tile adhesives shrinkage deformations of the coating were measured during its curing. Measurement of shrinkage deformations of tile adhesive samples were carrying out using an IZA-2 optical comparator. Curing of the samples took place in air-dry conditions at a temperature of (20 ± 2) ° C and relative humidity of air φ = 50% - 55%. In Figure 3.10 is a graph showing the change in shrinkage deformations of the coating on the basis of a control composition (without the use of synthetic aluminosilicates) and a composition made on the basis of the DBM developed: Portland cement Mark 400 M, Ukhta sand in the ratio of fractions 0, 63 - 0,315: 0,315 - 0,14, respectively, 80%: 20%, Kratasol PFM in the amount of 1% by weight of cement, Neolith P 4400 in the amount of 1% by weight of cement and as a mineral additive - synthetic aluminosilicates in the amount of 20% by weight of cement.

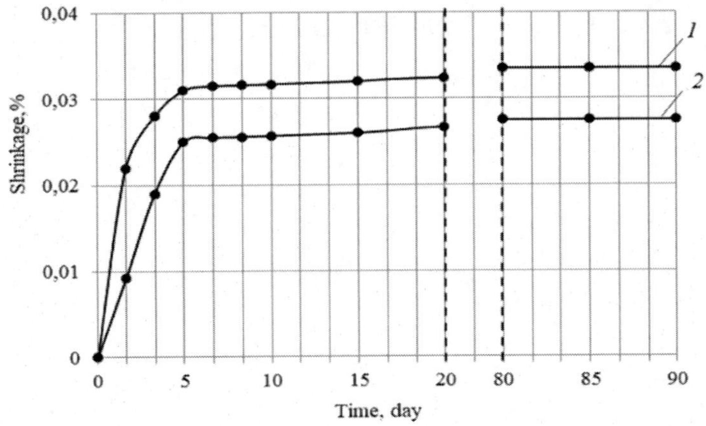

Figure 3.10. Changes in shrinkage deformation during hardening: 1 - control composition; 2 - developed composition.

Data analysis (Figure 3.10) shows that the most intensive growth of shrinkable deformations is observed within 5 days of air-dry hardening after which there is a noticeable stabilization of the values. So, the value of shrinkage deformation of the composition with the use of synthetic aluminum silicates for 5 days is $\varepsilon = 0.025\%$, and after 90 days - $\varepsilon = 0.028\%$ (see. Figure 3.10, curve 2). The values of the shrinkage deformations of the control composition on the 5th day of hardening were $\varepsilon = 0.031\%$, and after 90 days -$\varepsilon = 0.034\%$ (Figure 3.10, curve 1). The cracking of coatings based on the DBM occurs when the internal tensile stresses reach the cohesive strength value of the coating material. In this regard, the tensile strength of the coatings were determined using a tensile testing machine 5057-50 on samples after 90 days of air-dry hardening. Determination of tensile strength (cohesive strength) were carrying out on GOST 18299. The method is based on stretching the test specimen to rupture at a deformation rate of 1 mm/min. Samples of compositions with dimensions of 10×10×50mm were fixed in the clamps of a tensile machine. Its longitudinal axis were located in the direction of tension. The applied forces acted equally across the entire cross section of the sample. The tests were carried out at air temperature (20 ± 2)°C and relative air humidity of 60%.

The calculation of tensile strength were carrying out according to the results of testing at least four samples of each composition. The tensile strength R_{kog} (MPa) for each sample were calculated by the formula

$$R_{\text{kog}} = \frac{F_{Pi}}{S_{Oi}}, \quad (3.6)$$

where F_{Pi} is the tensile load at the moment of rupture, N;

S_{Oi} is the initial cross-sectional area of the sample, mm^2

The modulus of elasticity were calculated according to the stress – strain diagram using the tangent of the angle of inclination to the abscissa axis of the tangent (Z) drawn to the initial straight-line section of the diagram. The modulus of elasticity for each sample E_{el}, MPa, were calculated by the formula

$$E_{упр} = \frac{R'_{kogi}}{\varepsilon'_i} \cdot 100 \quad , \quad (3.7)$$

where R'_{kogi} -ultimate tensile strength at the moment of detachment of a tangent from the stress-strain diagram, MPa;

ε'_i -relative elongation at break, %.

The results of the experiments are presented in Figure 3.11.

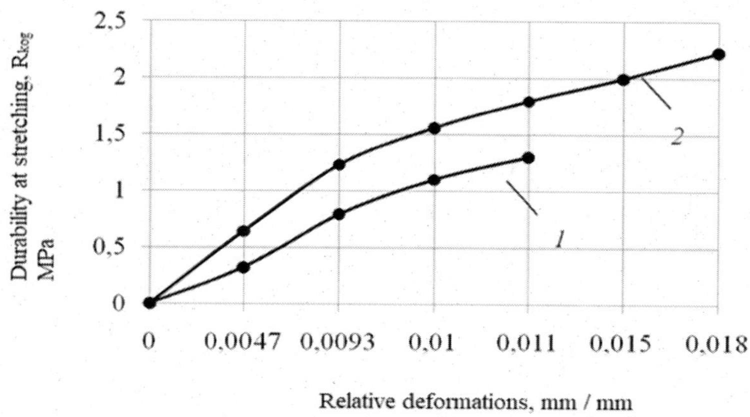

Figure 3.11. Change in relative deformations when stretching tile adhesive samples: 1 - control sample; 2 - sample using synthetic aluminosilicates.

Analysis of the data (Figure 3.11) revealed that the introduction of synthetic aluminosilicates into the recipe of tile adhesive allows to obtain a material characterized by tensile strength R_{kog} = 2.2MPa. The limiting extensibility ε_{lim} thus ε_{lim} =0,018 mm/mm. The destruction of control samples occurs at a relative deformation ε_{lim} equal to ε_{lim} = 0.011 mm/mm. The deformative properties of tile adhesives made based on the developed formula of dry glue mixes are presenting in Table 3.5.

Table 3.5. Deformative properties of the adhesive layer

№	Name of the indicator	Value of the indicator
1	The modulus of elasticity, MPa	128,6
2	relative deformations, ε, mm/mm	0,00028
3	limiting extensibility, ε_{lim}, mm/mm	0,018
4	$K_{тр} = \varepsilon/\varepsilon_{lim}$	0,015

The data obtained indicate sufficient shrinkage cracking of the adhesive layer on the basis of the developed recipe.

3.5. Evaluation of the Stress State of Tile Adhesive Based on Cement Using Synthetic Aluminosilicates

During operation tile adhesive is exposed to the external environment. Stresses occur inside the layer due to the action of rain, high relative humidity of the air, temperature changes, and also due to the diffusion of water vapor through the enclosing structure from the inner to the outer surface during the cold period of the year. It can lead to destruction layer of tile adhesive. Earlier studies established that to ensure the durability of coatings based on dry building mixtures it is necessary to ensure their crack resistance. It is knew that under the action of internal and external factors, tensile deformations ε_{def} occur inside the material, which may exceed the tensile strength of the material ε_{lim}.

Cracking of coatings based on DBM occurs, when the internal tensile stresses reach the cohesive strength value of the coating material.

$$\sigma = R_{kog} \quad (3.8)$$

or tensile strains are equal to ultimate stretch ability

$$\varepsilon_{деф} = \varepsilon_{пред}. \quad (3.9)$$

The crack resistance of the adhesive layer were evaluating based on the condition

$$\sigma < R_{kog}, \quad (3.10)$$

where σ is the internal stress, MPa

R_{kog} - cohesive strength of the adhesive layer, MPa

The stress state of the adhesive layer from the effects of temperatures as one of the factors of destruction were estimating. The calculation were carrying out for cities of Russia (located in different climatic regions). For the calculations, the cities of Penza and Yakutsk were selected (located in the same humidity zone 3 (dry) and in different climatic regions, respectively –II B and I A). The design scheme of the building envelope for Penza is presented in Figure 3.12 (a), for the city of Yakutsk is shown in Figure 3.12 (b).

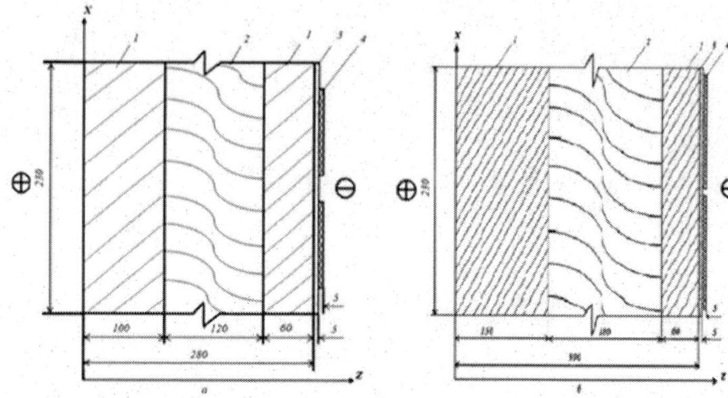

Figure 3.12. Constructive solution of the building envelope for the city of Penza (a) and for the city of Yakutsk (b): 1 - claydite; 2 - expanded polystyrene; 3 - a layer of tile adhesive cement based; 4 - ceramic tiles.

As a wall material in the design scheme of the enclosing structure (Figure 3.12) keramsit concrete with a density of ρ = 1600 kg/m^3 and thermal conductivity λ = 0.58 W/(m°C) is taken as a heat-insulating material - expanded polystyrene with a density of ρ = 40 kg/m^3 and a thermal conductivity of λ = 0.038 W/(m°C), as a covering material - ceramic tile, a density of ρ = 2600 kg/m^3 and a thermal conductivity of λ = 1.2 W/(m°C). Tile adhesive, made according to the developed recipe of DBM, has a coefficient of thermal conductivity λ = 0.58 W/(m°C).

During operation, in the adhesive layer, not only tangential stresses arise, but also tensile stresses. The calculation of tensile stresses is make for the most dangerous zones of the adhesive layer in the presented design - the lines OV and AS indicated in Figure 3.13.

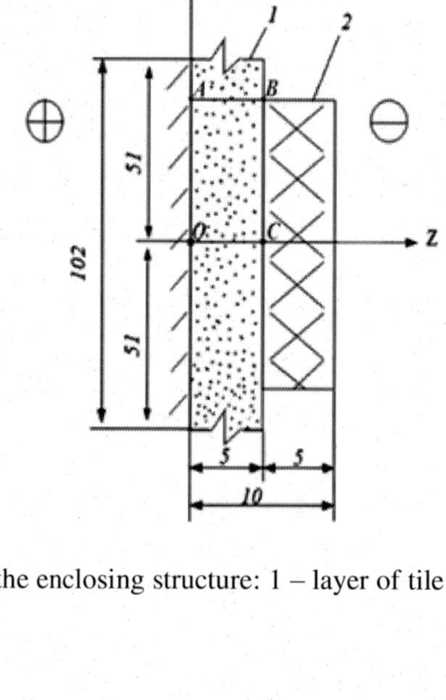

Figure 3.13. Design scheme of the enclosing structure: 1 – layer of tile adhesive; 2 – ceramic tile.

In Figure 3.14 and 3.15, respectively, are the values of tensile stresses arising in the adhesive layer at the border with expanded clay concrete and at the border with ceramic tiles during the year in Penza and in Yakutsk.

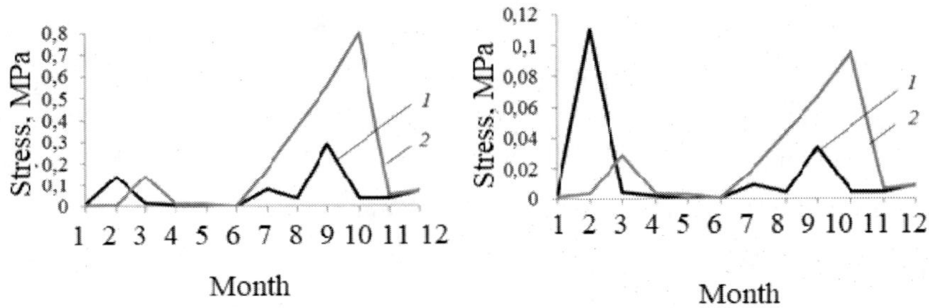

a b

Figure 3.14. Change of shear stresses σ_{xz} in the adhesive layer during the year Penza (curve 1), Yakutsk (curve 2): a – on the border with ceramic tiles (Figure 3.13, AC line); b – on the border with expanded clay concrete (Figure 3.13, line OB).

Revealed that the maximum shear stresses in the adhesive layer on the border with clay are observed in the month of February the composition they σ_{xz} = 0.11 MPa and in the month of September on the border with ceramic tile – σ_{xz} = 0.29 MPa (Penza, Figure 3.14, curve 1). In Yakutsk (Figure 3.14, curve 2) the maximum shear stresses in the adhesive layer are observed at the boundary with expanded clay and at the boundary with ceramic tiles in October respectively σ_{xz} = 0.096 MPa and σ_{xz} = 0.8 MPa.

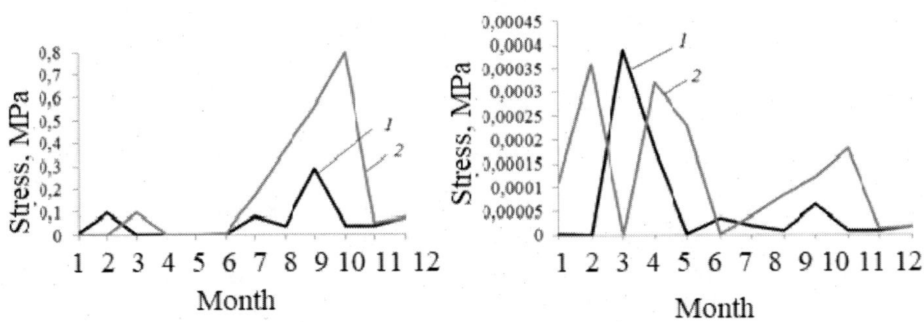

a b

Figure 3.15. Change of shear stresses σ_{xz} in the adhesive layer during the year Penza (curve 1), Yakutsk (curve 2): a – boundary zone (Figure 3.13, the line of BC); b – the center of the adhesive layer (Figure 3.13, the line OA).

In the center of the adhesive seam (Figure 3.13, line OA) for conditions propulsion Penza (Figure 3.15, curve 1) the maximum shear stresses are observed in the month of March and are σ_{xz} = 0.00039MPa and for the conditions of Yakutsk in the month of February – σ_{xz} = 0.00036MPa. In the marginal zone (Figure 3.13, the line of the sun) the maximum stresses are typical for the conditions of Penza in September components σ_{xz} = 0.29MPa and for the conditions of Yakutsk – in October components σ_{xz} = 0.8MPa.

Comparison of the obtained values of the maximum tangential stresses σ_{xz} arising in the adhesive layer with the value of adhesive strength of the cement-based tile adhesive equal to R_{adg} > 1.4 MPa showed that the shear stresses arising in the adhesive layer do not exceed the values of adhesive strength i.e., the adhesive seam based on the developed composition of the DBM struts to exfoliation.

In Figure 4.16 and 4.17, respectively, are the values of tensile stresses arising in the adhesive layer at the border with expanded clay concrete and at the border with ceramic tiles during the year in Penza and in Yakutsk.

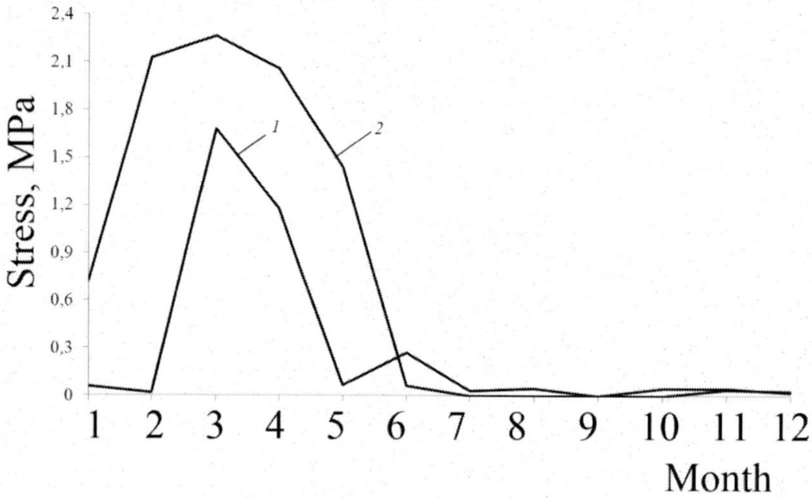

Figure 3.16. The change in tensile stresses σ_x in the adhesive layer at the border with concrete during the year (line OB, Figure 3.13): 1 - Penza; 2 – Yakutsk.

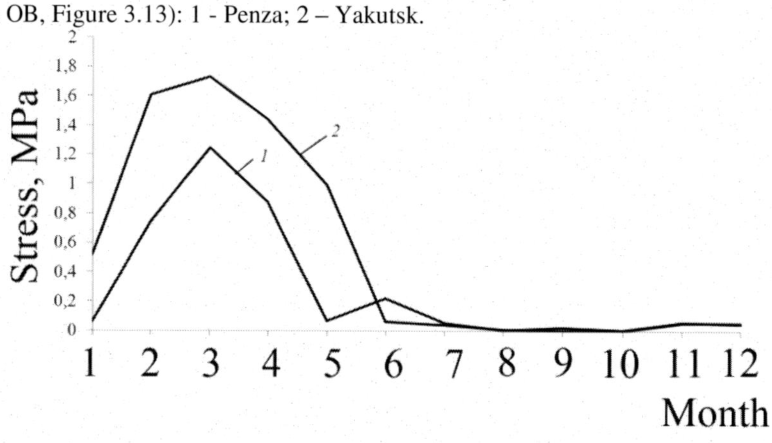

Figure 3.17. The change in tensile stresses σ_x in the adhesive layer on the border with ceramic tiles during the year (line AC, Figure 3.13): 1 - Penza; 2 – Yakutsk.

Analysis of the data presented in Figures 3.16 and 3.17 showed that the maximum tensile stresses σ_x arising in the adhesive layer at the border with claydite-concrete (Figure 3.16) and at the border with ceramic tile (Figure 3.17) are observing in March.

The values of the maximum tensile stresses σ_x of the adhesive layer on the border with expanded clay concrete in the city of Penza and in the city of Yakutsk are equal respectively $\sigma_x = 1.68$MPa and $\sigma_x = 2.3$MPa. The values of the maximum tensile stresses σ_x arising in the adhesive layer on the border with ceramic tiles in the city of Penza and in the city of Yakutsk, respectively, are - $\sigma_x = 1.24$MPa and $\sigma_x = 1.73$MPa.

Comparison of the obtained maximum tensile stresses arising in the adhesive layer in Yakutsk with the cohesive strength value of cement-based tile adhesive equal to R_{kog} = 2.2MPa showed that the condition of formula is not satisfied. The values of internal stresses are greater than the cohesive strength $\sigma_x > R_{kog}$.

Thus, the developed recipe of tile glue is not recommended to be used in the city of Yakutsk and in cities in the humidity zone - 3 (dry) and climatic subdistricts IA.

The results of the calculations allow us to state that the adhesive layer based on the developed recipe for dry adhesive construction mixture is crack resistant for the city of Penza and the cities of Russia located in the humidity zone - 3 (dry) and climatic subareas IIB. To confirm the obtained results, we consider tensile and compressive stresses arising in the adhesive layer for the conditions of the climate of the city of Penza.

In Figure 3.18 shows the values of tensile stresses over the year for the city of Penza. It were revealed that the maximum tensile stress along the X axis in the zone of contact of the adhesive layer with claydite-concrete (line S, Figure 3.13) is observed in March and is $\sigma_x = 1.68$ MPa. The maximum tensile stress along the X axis in the zone of contact of the adhesive layer with ceramic tile (line AC, Figure 3.13) occurs in the month of March and is $\sigma_x = 1.24$ Mpa.

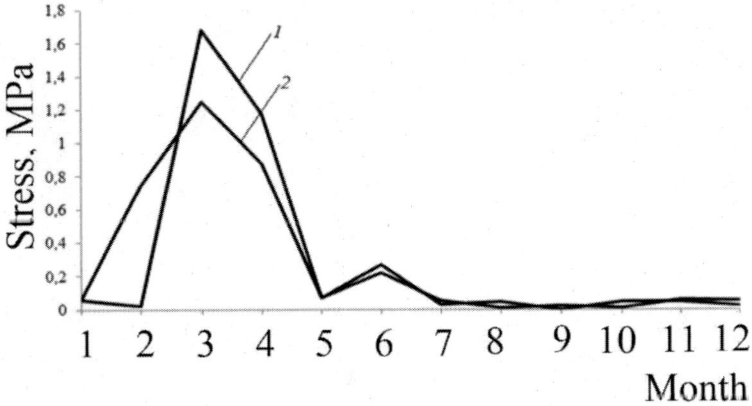

Figure 3.18. The change in tensile stresses σ_x in the adhesive layer during the year (Penza): 1 - on the border with expanded clay concrete; 2 - on the border with ceramic tiles.

In Figure 3.19 shows the stress values occurring over the thickness of the adhesive layer (Z-axis) during the year in the city of Penza.

The maximum tensile stress occurs in the center of the adhesive layer (OA line, Figure 3.13) in the month of September amounting to $\sigma_z = 0.02$ MPa. In the boundary zone of contact of the adhesive layer along the Z axis (line BC, Figure 3.13) tensile and compressive stresses appear. The maximum tensile stress along the Z axis of the marginal zone of contact of the adhesive layer on the grain with ceramic tiles arises in September and is $\sigma_z = 0.24$ MPa. The maximum compressive stress of the edge contact zone through the thickness of the adhesive layer at the border with ceramic tile occurs in the month of March and is $\sigma_z = -0.49$ MPa.

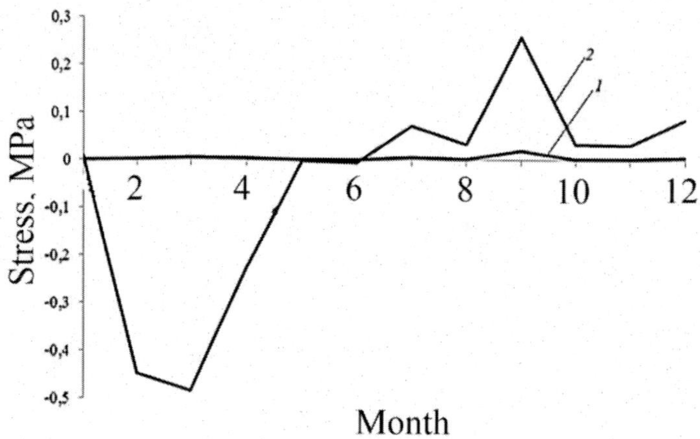

Figure 3.19. Change of maximum stresses σ_z in the adhesive layer during the year (Penza): 1 - the center of the adhesive layer; 2 - edge zone of the adhesive layer.

In Figure 3.20 shows the values of tensile stresses along the prosthesis of the adhesive layer characteristic for the month of March in the city of Penza.

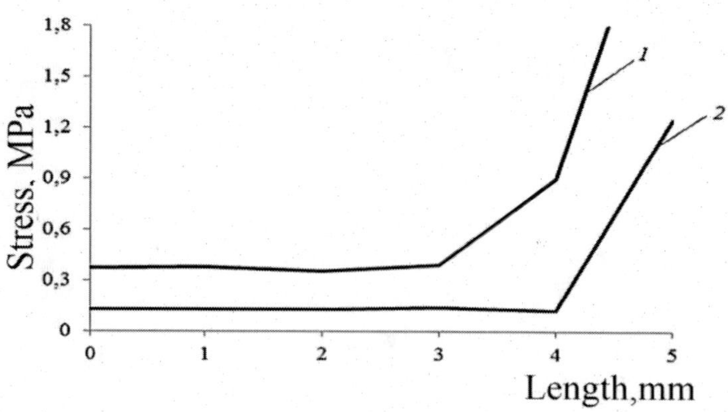

Figure 3.20. The change in tensile stresses σx along the length of the adhesive layer in March (Penza): 1 - on the border with expanded clay concrete (line OB, Figure 3.13); 2 - on the border with ceramic tiles (line AC, Figure 3.13).

As can be seen from the data (Figure 3.20) the maximum tensile stresses σ_x occur in the end zone of the adhesive layer at the boundary of the adhesive layer with claydite-concrete (OB line, Figure 3.13) and constitute σ_x = 1.68 MPa and adhesive layer with ceramic tiles (line AC, Figure 3.13) - σ_x = 1.24 MPa.

In Figure 3.21 shows the values of tensile and compressive stresses arising through the thickness of the adhesive layer characteristic of the month of September in the city of Penza.

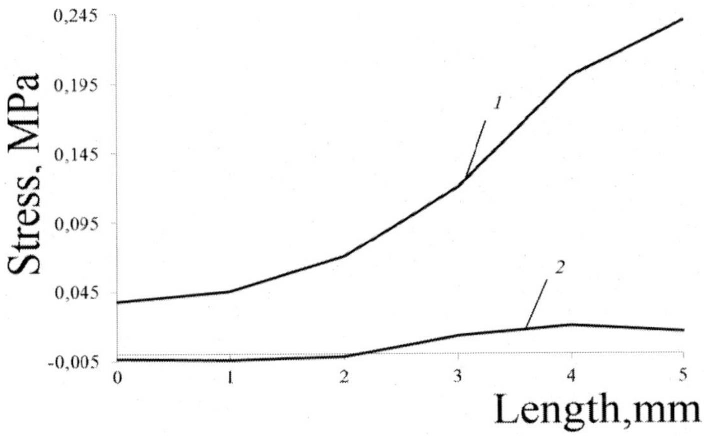

Figure 3.21. Stress variation σ_z across the thickness of the adhesive layer in the month of September (Penza): 1 - in the center of the adhesive layer (OA line, Figure 3.13); 2 - in the end zone of the adhesive layer (BC line, Figure 3.13).

Analysis of the experimental data presented in Figure 3.21 showed that the maximum tensile stress σ_z occurs in the center of the tile adhesive layer (OA line, Figure 3.13) at the border with the ceramic tile and is σ_z = 0.016 MPa and the maximum compressive stress in the center of the adhesive layer on the border with concrete is σ_z = –0.0042MPa. In the end zone of the adhesive layer (line BC, rice 3.13) the maximum tensile stress occurs in the zone of contact of the adhesive layer with ceramic tiles and is σ_z = 0.24 MPa.

In Figure 3.22 shows the values of compressive stresses appearing through the thickness of the edge zone of contact of the adhesive layer (line BC, Figure 4.13), characteristic for the month of March in the city of Penza. Analysis of the data obtained (Figure 3.22) showed that the maximum compressive stress occurs in the zone of contact of the adhesive layer with ceramic tiles (point C, Figure 3.13) and is σ_z = –0.4 MPa.

The obtained values of the maximum stresses were compared with the values of cohesive and adhesive strength of tile adhesive on a cement basis, respectively, equal to R_{kog} = 2.2 MPa and R_{adg}> 1.4 MPa.

Thus, the presented results of the calculations suggest that the adhesive layer using synthetic zeolites in the formulation is crack resistant when used in cities in the humidity zone - 3 (dry) and climatic subareas II B.

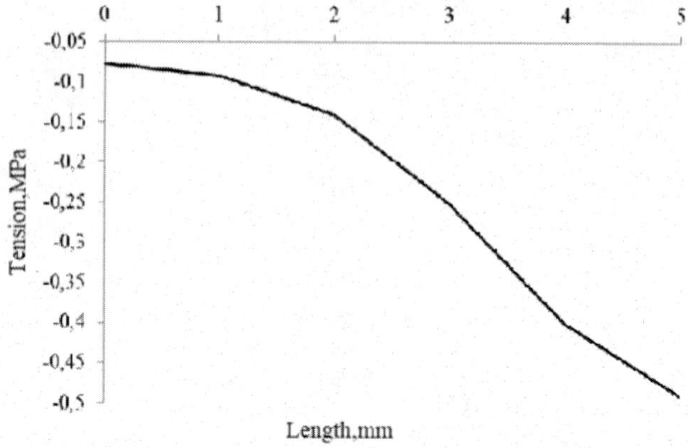

Figure 3.22. The change in compressive stress σ_z across the thickness of the edge zone of the adhesive layer in March (Penza).

Additionally, the crack resistance were estimating by the conventional coefficient of crack resistance K_{tr}, defined as the ratio of flexural strength to compressive strength by the formula

$$K_{tr} = \frac{R_b}{R_c} \quad (3.7)$$

The results of the calculation found that the samples of tile adhesives have sufficient crack resistance since the value of the conventional coefficient of crack resistance K_{tr} is $K_{tr} = 0.06$.

In Tab. 3.7 shows the main technological and operational properties of tile adhesives based on the developed recipes with the use of synthetic aluminosilicates.

Table 3.7. Technological and operational properties of tile adhesive based on the developed DBM

Name of the indicator	Unit of measurement	Indicator value		
		developed	prototype 1	prototype 2
The bulk density of the dry mixture, r, kg/m^3	kg/m^3	1800	1670	1500
Mixing time	min	3	4	4
Time correction, min	min	20	30	20
Viability during storage in open containers	min	100	180	120
Workability		good	good	good
Recommended thickness of one layer	mm	3 – 5	до 10	до 45
mConsumption of the composition when applied with a layer thickness of 5 mm	kg/m^2	3,2	6,0	4,2
The water-holding capacity of the solution, %	%	97,8– 99,3	95,0–97,0	96,2–98,8

Name of the indicator	Unit of measurement	Indicator value		
		developed	prototype 1	prototype 2
Tile sliding, mm	mm	0,3	0,5	0,5
Adhesion strength, R_{adh}				
after keeping in an air-dry environment	MPa	more 1,4	1,1	0,8
after aging in the aquatic environment	MPa	more 1,1	0,9	1,0
after 50 freeze-thaw cycles of the moisture-saturated sample	MPa	0,97	0,7	0,9
Shear bond strength	MPa	0,92	0,6	0,7
Mark on frost resistance	Mark	*F*50	*F*50	*F*50
Frost contact area	Mark	$F_{ca}50$	$F_{ca}50$	$F_{ca}50$
Water absorption with capillary suction	$kg/m^2 \cdot h^{0,5}$	1,43	2,00	1,78
Shrinkage, ε	%	0,028–0,034	0,030–0,040	0,029–0,038
Operating temperature	C	-50°C to +70°C	-50°C to +50°C	-50°C to +70°C
Warranty period of storage in undamaged packaging in dry rooms	month	12	12	12

As a prototype 1, UNIS -2000 tile adhesive were chosen on a cement basis (UNIS group of companies) and as a prototype 2, CM 11 tile adhesive (produced by Ceresit).

REFERENCES

[1] Bobryshev, A. A. Finishing adhesive solutions based on dry blends using complex powder polymer additives [Text]: author. dis. ... Cand. tech. Sciences: 05.23.05/A. A. Bobryshev. - Penza, 2003. - 22 p.

[2] Galperina, T. Ya. Application of zeolitized rocks of the Shivyr ¬tuy field in cement production [Text]/T. Ya. Galaprina, L. A. Vertopralova, I. A. Solovyov, F. I. Lyshov//*Cement,* 1992. - № 4. - p. 79-83.

[3] Golubev, V. I. New products on the market of additives for dry building mixes and concrete [Text]/V. I. Golubev, P. G. Vasilik//*Construction materials.* - 2006. - № 3. - p. 24-25.

[4] Gorchakov, G. I. Increasing the frost resistance of concrete [Text]/G. I. Gorchakov, M. M. Kapkin, B. G. Skramtaev– M.: Stroyizdat, 1965. - 190 p.

[5] GOST 310.5-88. Cements. Method of determining heat generation [Text]. - Moscow: USSR Ministry of Industry Building Materials, 2003 - 6c.

[6] GOST 427-75. Metal measuring rulers. Technical conditions [Text]. - M.: Standardinform, 2007. - 7 p.

[7] GOST 5382-91. Cements and materials for cement production. Methods of chemical analysis [Text]. - M.: Standards Publishing House, 1991. - 58 p.

[8] GOST 5802-86. Building solutions. Test methods [Text]. - M.: Standardinform, 2010. - 16 p.

[9] GOST 8735-88. Sand for construction work. Test methods [Text]. - M.: Standardinform, 2006. - 26 p.

[10] GOST 8736-2014. Sand for construction work. Technical conditions [Text]. - M.: Standardinform, 2015. - 11 p.

[11] GOST 11109-90. Gauze household cotton. General technical conditions [Text]. - M.: Gos.kom. USSR on product quality management and standards, 1991. - 6 p.

[12] GOST 12966-85. Aluminum sulfate technical purified. Technical conditions (with Changes No. 1, 2) [Text]. - M.: Standards Publishing House, 1999. - 12 p.

[13] GOST 13078-81. Sodium liquid glass. Technical conditions (with Changes N 1, 2) [Text]. - M.: Publishing house of standards, 1989. - 15 p.

[14] GOST 14759-69. Adhesives. Method for determining shear strength [Text]. - M.: Publishing house of standards, 1970. - 14s.

[15] GOST 18299-72. Paint materials. Method for determining tensile strength, elongation at rupture and modulus of elasticity [Text]. - M.: Publishing house of standards, 1989. - 10 p.

[16] GOST 22685-89. Forms for the manufacture of control samples of concrete. Technical conditions [Text]. - M.: Standardinform, 1990. - 11 p.

[17] GOST 24452-80. Concretes. Methods for determining the prism strength, elastic modulus and Poisson's ratio [Text]. - M.: Standartinform, 1982. - 14 p.

[18] GOST 31108-2003. Cements general construction. Technical conditions [Text]. - M.: TsPPP, 2004. - 26 p.

[19] GOST 31356-07. Mixes dry construction on cement knitting. Test methods [Text]. - M.: MNTKS, 2008. - 16 p.

[20] GOST 31357-2007. Mixes dry construction on cement knitting. General technical conditions [Text]. - M.: Standardinform, 2008. - 12 p.

[21] Danilov, A. M. Mathematical modeling of the structure and properties of new generation materials [Text]/A. M. Danilov, I. A. Garkina//*Review of Applied and Industrial Mathematics.* - 2008. - Vol. 1 (issue 5). - p. 876–877.

[22] Doroshenko, Yu. M. Structure formation processes and properties of cement stone with polymer modifiers [Text]/Yu. M. Doroshenko, J. I. Shanaev//15 Szilikatip. esszilikattund. konf. (12-16 jun., 1989): SILICONF R.l. - Budapest, 1989. - p. 273-276.

[23] Zhegera, K. V. Method of evaluation of the adhesion strength of the mortar layer on the basis of dry construction mixtures [Text]/K. V. Jegere, I. S. Pushkina, D. A. Ryzhov, A. A. zhivaev//*Regional architecture and construction.* – 2015. –№3. – P. 64-67.

[24] Ilinsky, V. M. Thermal physics (fencing structures and microclimate of buildings) [Text]/V. M. Ilyinsky. - M.: Higher School, 1974. - 320 p.

[25] Illichenko, O. T. Calculations of the thermal state of structures [Text]/OT. Ilyichenko. - Kharkov: Vishcha school, 1979. - 168 p.

[26] Kalashnikov, V. I. Glachnoglakovyh construction materials [Text]/V. I. Kalashnikov [et al.]. - Penza: PGASA, 2000. - 207 p.

[27] MECELLOSE® 2015 Product Catalog. - 1 p. [Electronic resource]: CJSC "EuroChem-1" [website]. - Access mode: http://chem.eurohim.ru/catalog/DryMortar Additives/Group30/sGroup1009/ProductList.html.

[28] Karjakina, M. I. Physical and chemical bases of the processes of formation and aging of coatings [Text]/M. I. Karyakin - M.: *Chemistry,* 1970. - 215 p.

[29] Kozlov, V. V. Dry construction mixtures [Text]/V. V. Kozlov. - M.: DIA, 2000. - 96s.

[30] BERMOCOLL CCA 425 technical characteristics sheet [Electronic resource]: AkzoNobel Functionalchemicals [site]. - Access mode: http://deltachem.ru/d/401 471/d/cca-425.pdf.

[31] Loganina, V.I. Additive based on synthesized aluminosilicates for cement systems [Text]/V. I. Loganina, I. V. Zhernovsky, M. A. Sadovnikova, K. V. Zhegera//*East European journal of advanced technologies.* – 2013. – Vol. 5. – № 6 (65). – P. 8 -11.

[32] Loganina, V. I. Formation of strength of cement composition in the presence of synthesized aluminosilicates [Text]/V. I. Loganina, K. V. Zhegera//Bulletin of SUSU. A series of Construction and architecture. – 2015. – Vol. 15. – №2. – P. 43 – 46.

[33] Loganina, V. I. Evaluation of the stress state of the adhesive layer on the basis of dry construction mixtures with use of the synthesized aluminosilicates [Text]/V. I. Loganina, M. V., Ariskin, A. V. Karpov, K. V. Zhegera//Bulletin of civil engineers. – 2015. – №3 (50). – P. 163 – 166.

[34] Loganina, V. I. Estimation of fracture toughness of the finishing layer on the basis of dry adhesive mixture with the use of the synthesized aluminosilicates [Text]/V. I. Loganina, M. V., Ariskin, A. V. Karpov, K. V. Zhegera//Building materials. – 2015. – № 10 – P. 86.

[35] Loganina, V. I. Estimation of fracture toughness of the finishing plaster layer on the basis of dry construction mixtures [Text]/V. I. Loganina, M. V., Ariskin, E. R. Akzhighitova, N. And. Petukhova//Izvestiya vuzov. Construction. – 2013. – № 1. P. AFR.

[36] Loganina V. I., Zhegera Ch. V., Grintsova O. V. and Ivashchenko J. G.. Structure Formation of Cement Stone in the Presence of Additive on Amorphous Silica-Alumina. *Journal of Engineering and Applied Sciences,* 2016.№ 11: pp. 1439-1443. DOI: 10.3923/jeasci.2016.1439.1443.

[37] Loganina V. I., Zhegera Ch. V. Evaluation of the creep resistance of the adhesive based on dry mix//AIP Conference Proceedings 1809,020031(2010);doi: 10.1063/1.4975446http:dx.doi.org. 10.1063/1.4975446.

[38] Loganina V. I., Sadovnikova M. A. and Skachkov Yu. P. The rheological properties of the finishing lime composition with additive based on synthetic aluminosilicates ARPN *Journal of Engineering and Applied Sciences* Vol.12, N 4, 2017. p.1-4.

[39] Loganina V. I., Sadovnikova M. A.. Structure formation of limy composites with the additive based on synthetic aluminosilicates//*J. Ponte* - Mar 2017 - Volume 73 – N3 p.77-80. DOI: 10.21506/j.ponte.2017.3.8.

[40] Loganina V. I., Zhegera Ch. V. The use of amorphous silica-alumina-based additive in the adhesive dry mixes of building materials *International Journal of Built Environment* Vol. 5.1 January 2018.

[41] Loganina V. I., Zhegera K. V. The Topology Structure of Cement Stope with the Use of Compounding Additive Based on Amorphous Aluminosilicates. *Solid State Phenomena,* vol.279.pp.271-275, doi:10.4028/www.schientific.net/SSP 279.271.

[42] Pat. 2132545. Russian Federation, G01N3/24, E02D1/00 Shift device [Text]/G. G. Boldyrev (RU), OV Khryanina (RU) patent holder Penza State Architectural Institute - 4 p.

[43] Pat. EN 2463108. USA, the Composition of amorphous aluminosilicate and method for producing and using such a composition [Text]/C. G. Michelle (US), R. K. Ackermann (US), D.E. Smigal (US), J. A. Van Ven (NL), patent holder for Shell International Research Maatshappiy B.V. (NL) - 7 s.

[44] Rehbinder, P. A. On the method of immersion of a cone for the characteristics of the structural-mechanical properties of plastic-viscous bodies [Text]/P. A. Rebinder, N. A. Semenenko//Reports of the USSR Academy of Sciences. - 1949. - T. LXIV. - № 6. - p. 835-838.

[45] Construction rules: SP 131.13330.2012. Construction Climate: updated version of SNIP 23-01-99 * [Text]. - M.: Rosstandart, 2012. - 113s.

[46] Aiello, R. Molecular sieve zeolites/R. Aiello, C. Collela, R. Sersale//*Advances Chem. Ser.* – Washington. 1971. – № 101. – p. 51.

[47] Broussard, L. J. *Am. Chem. Soc.*/L. Broussard, Shoemker, D. P.–1960. – V. 82. – № 5.– p. 1041.

[48] DIN EN 1308-2007. Solutions and adhesives for ceramic tiles. Determination of creep [Text]: per.s nem./German edition. – M.: STANDARTINFORM, 2010. – 10 p.

[49] DIN EN 1348:2007. Solutions and adhesives for ceramic tiles. Determination of the strength of adhesion to the base of cement-containing solutions for internal and external works [Text]: TRANS./German edition. – M.: STANDARTINFORM, 2010 – 10 p.

[50] DIN EN 12004 (2012 – 09). Solutions and adhesives for ceramic tiles. Requirements, conformity assessment, classification and designation [Text]: TRANS. with it. – M.: STANDARTINFORM, 2012. – 29p.

[51] Mumpton, F. A. La rocamagica: Uses of natural zeolites in agriculture and industry/F. A. Mumpton – PNAS, 1999. – V. 96. – №. 7. – Pp.3463 – 3470, doi:10.1073/pnas.96.7.3463.
[52] Ross, H. Praxis-Hand buch Putz/H. Ross, F. Stahl//Rudolf Muller. – Koln, 2003.

In: An Introduction to Aluminosilicates
Editor: Nero Regina Blevins
ISBN: 978-1-53617-250-8

Chapter 4

SURFACE MODIFICATION OF ALUMINOSILICATE BY AMPHIPHILES: FABRICATION OF ULTRATHIN FILMS

Shuhei Hirayama, Yoshinori Abiko and Atuhiro Fujimori*

Graduate School of Science and Engineering, Faculty of Engineering, Saitama University, Sakura-ku, Saitama, Japan

ABSTRACT

In the first half of this chapter, it is investigated the molecular orientation and surface morphology of organized molecular films with regard to solid-state structures for organo-modified aluminosilicates (montmorillonites) by surface pressure-area (π–A) isotherm, in-plane and out-of plane X-ray diffraction (XRD), and atomic force microscopy (AFM). From the results of out-of plane XRD, formation of highly ordered layer structure was confirmed in these clay Langmuir–Blodgett (LB) film. Further, two-dimensional lattice of long alkyl chain of organo-modified parts packed hexagonally or orthorhombically in the films. Surface morphology of Z-type monolayers on solid of organo-modified aluminosilicates indicates heterogeneous modification ratio in montmorillonite surface by AFM observation. In addition, monolayer behavior on the water surface and mesoscopic morphological formation on solid of mixed films of organo-modified aluminosilicate and poly-(L-lactide) (PLLA) were investigated by π-A isotherm and AFM. Collapsed surface pressures are independent on the mixed ratio, and indicate almost constant value in their isotherms. This tendency is a peculiarity of immiscible system. From the result of AFM observation, phase separated structure was confirmed in mesoscopic scales. This phase separated morphology remarkably varied with mixed ratio.

In the second half of this chapter, ultrathin films of regularly adsorbed biological molecules have been fabricated by means of the modified LB method using an organo-

* Corresponding Author Email: fujimori@apc.saitama-u.ac.jp.

modified aluminosilicate. In this combined method, clay LB films play a template role in the formation of the lysozyme thin layer. Chemisorption of biological molecules to the anionic montmorillonite surface was confirmed by a comparison between infrared spectra of multilayers of organo-clay and of chemisorbed lysozyme. The surface morphology of these monolayers was observed by atomic force microscopy. In the case of organo-modified aluminosilicate, it is also indicated that chemisorbed enzyme activity is possible to maintain until 160°C at attempting to evaluation of the heat-resistance. The maintaining of enzyme activity at high temperature region is possible together with the maintaining of three-dimensional steric structure of enzyme. This maintaining of steric structure of enzyme is based on the formation of aggregate of adsorbed enzyme induced by epitaxial-growth from aluminosilicate surface and the construction of sandwiched structure of enzyme between organo-aluminosilicates.

Keywords: surface modification; montmorillonite; organized molecular films; surface coverage; molecular orientation

1. Formation and Structure of Langmuir–Blodgett Films of Organo-Modified Aluminosilicate with High Surface Coverage

1.1. Introduction

One of the aims of modern materials research is to construct organic/inorganic nanostructured materials [1, 2]. Currently, developing alternatives for endangered resources and materials to detect environmental contamination is at the forefront of material science research. In recent years, the exhaustion of deposits and rising market prices of rare metals and rare earth elements have become topics of public concern all over the world.

Scientists are urgently looking for ways to substitute petroleum-derived resources and reduce greenhouse gas emissions. To address this problem, we investigated the use of aluminosilicate [3, 4] as an alternative material. Aluminosilicate is comprised of ubiquitous elements, such as atoms with higher ranked Clarke numbers, such as O (ranked first), Si (ranked second), and Al (ranked third). Furthermore, aluminosilicate materials do not emit greenhouse CO_2 gas by combustion, whereas general polymer products burn with the emission of CO_2.

In addition, because aluminosilicate clays, such as natural montmorillonite, are post-petroleum mineral resources, next generation eco-materials could potentially be developed. For example, biochips implanted in the human body function based on their phase-separated structures formed at mesoscopic levels.

Aluminosilicate clays are unique in the sense that they consist of negatively charged aluminosilicate layers that are held together by exchangeable interlayer cations. These interlayer cations enable the clay to undergo ion exchange, enabling the modification of their surfaces by cationic surfactants that intercalate the interlayer spaces. Thus, many alkylammonium-modified layered silicates (organo-modified clays) have been developed and are now widely used to form various nanocomposite materials with enhanced properties [5].

Over the past four decades, over 70,000 papers have been published in the academic field of clay materials [6]. From 1987, almost all highly cited papers have been related to polymer/clay nanocomposites [7]. Organo-modified clay composites have been extensively investigated in material science [8].

For example, modified clays have been recently shown to be powerful compatibilization agents of polymer blends [9] that behave in a synergistic manner with standard flame retardant formulations, rendering a large class of self-extinguishing polymers [10]. In order to achieve enhanced mechanical and thermal properties, clays have been exfoliated in polymer matrices [10-12].

Thin films of clay minerals have been studied for the development of modified electrodes, sensors, photochromic devices, and nonlinear optical devices. The Langmuir–Blodgett (LB) method [13-15] is one of the most useful techniques used to prepare ultrathin films. Over the past four decades, approximately 140 papers have been reported in the field of clay LB films [16, 17]. The LB method forms floating molecular monolayers at air/water interfaces, which are deposited on a solid substrate in a layer-by-layer manner. Ultrathin films of clay minerals prepared using the LB method have been reported, and clay LB films have recently been hybridized with alkylammonium cations. Smectite clay is exfoliated into single negatively charged clay sheets containing low concentrations of aqueous suspension.

When a solution of an amphiphilic alkylammonium salt is spread onto the clay suspension, the negatively charged clay platelets in the suspension are adsorbed onto a floating monolayer of alkylammonium cations at the air/clay suspension interface. The hybrid monolayers of the clay platelets and ammonium cations formed at the interface can be transferred onto a solid surface to fabricate hybrid multilayers. Interestingly, by hybridizing with clay platelets, a floating monolayer of water-soluble alkylammonium cations can be stabilized at the air–suspension interface. The surface pressure–area (π–A) isotherm of the hybrid monolayer is dependent on the clay concentration in the suspension. As the concentration increases, the isotherm shifts to the side of the layer with the largest area, suggesting that the structure and properties of hybrid monolayers and multilayers vary based upon the clay concentration in the suspension.

In addition, because organo-modified montmorillonites are post-petroleum resources, these materials could potentially be used to form next generation eco-materials (e.g., biochips).

The patterning surface used to construct the biological high density memory device is formed by a phase separation between a biodegradable polymer and natural mineral. Previous studies have used modification methods to form organo-modified montmorillonites with extremely high surface coverage and localized negative charges on the bottom layer at oil/water interfaces [18-21]. In this work, floating monolayers and multilayered films on the solids of preformed organo-modified aluminosilicate with high surface coverage were constructed using the LB method. The molecular arrangement and surface morphology of the films were estimated by π–A isotherm, out-of-plane and in-plane X-ray diffraction (XRD), and atomic force microscopy (AFM). In addition, the miscibility of organic polymer and organo-modified clays with surface coverage are discussed.

1.2. Experimental Section

1.2.1. Materials

Natural Na^+-montmorillonite (Figure 1(a)) was supplied by Kunimine Co., Ltd. with a cation exchange capacity of 108.6 meq/100 g. The organophilic clay was prepared using cation exchange reactions of natural clay (aqueous dispersion) with 0.1% aqueous solution of dimethyl dioctadecyl (DMDO) ammonium, dimethyl phenyl octadecyl (DMPO), methyl hydroxyl ethyl octadecyl (MHEO), and tetramethyl octadecyl (TMO) chloride ((A–D) in Figure 1(b), respectively). Organo-clays with low surface coverage (commercial grade) [19] were formed in clay suspensions at 50°C by the addition of four types of quaternary ammonium cations with stirring, whereas those with high surface coverage were formed at the oil/water interface (chloroform solutions of quaternary ammonium chlorides/clay suspensions) [18].

These procedures are illustrated in Figure 2. The high surface coverage modification method is similar to that of interfacial polycondensation. A schematic illustration of the process of cation exchange reaction and structure of the resulting organo-modified clay is shown on right side of Figure 1(a). Poly(l-lactide) (PLLA) pellets with a high L-lactide content (<99%) were supplied by Unitika Co., Ltd. The weight-average molecular weight (Mw), number-average molecular weight (Mn), and the polydispersity index (Mw/Mn) were determined to be 20×10^4, 10×10^4, and 1.94, respectively. In addition, poly(ethylene succinate) (PESu) and poly(ε-caprolactone) (PCL) were used as amphiphilic biodegradable polymers (Figure 1(c)).

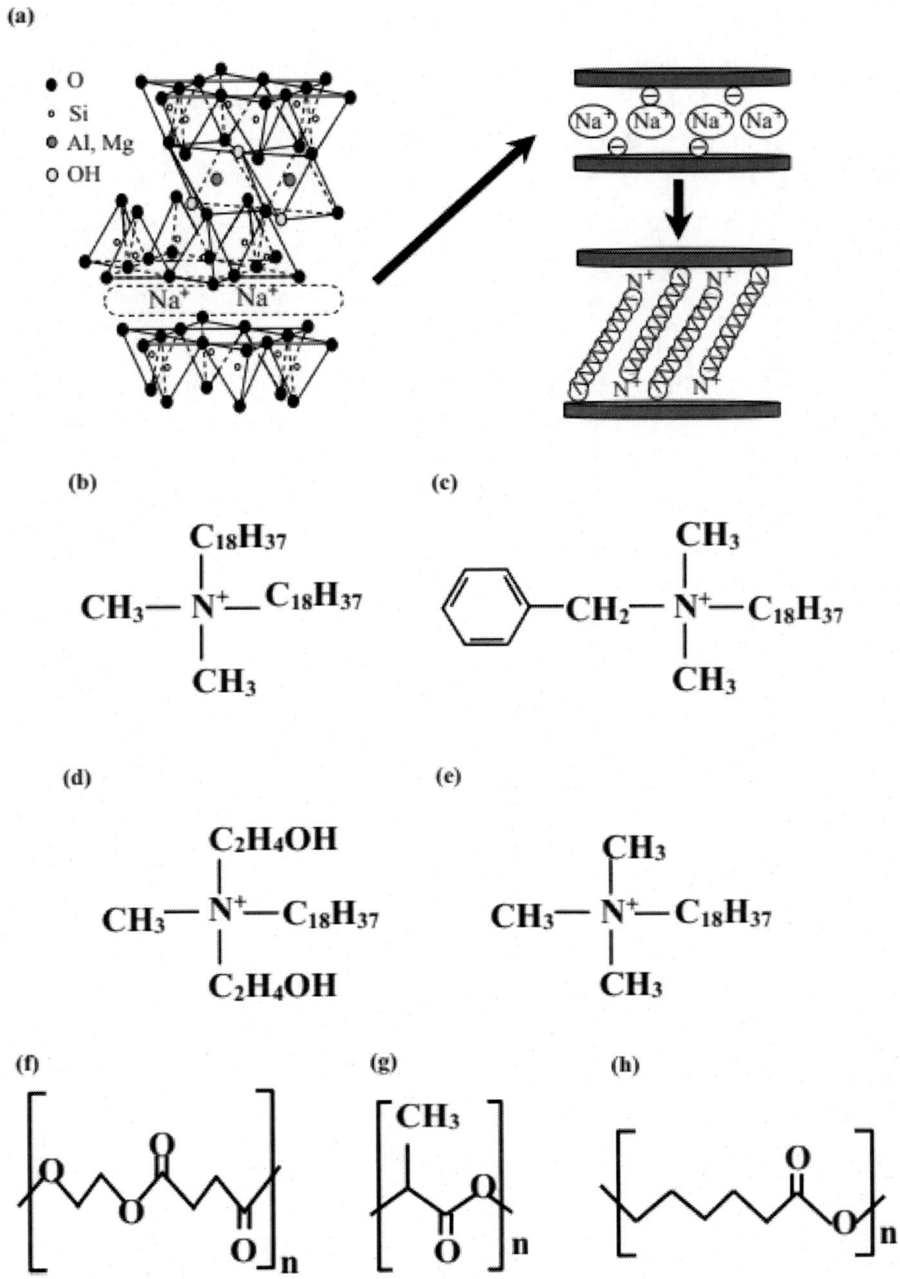

Figure 1. (a) Chemical structure of montmorillonite and schematic illustration of organo-modification. (b) Chemical structures of the following quaternary ammonium cations as organo-modified agents: (A) DMDO, (B) DMPO, (C) MHEO, and (D) TMO. (c) Chemical structures of the following amphiphilic biodegradable polymers used in this chapter: (E) PLLA, (F) PESu, and (G) PCL.

1.2.2. Measurements

Organo-modified clays were spread from toluene solutions (approximately 10^{-4} M) onto buffer solutions prepared using distilled water (approximately 18 MΩ·cm). PLLA,

PESu, and PCL were spread from chloroform solutions onto distilled water. Because these biodegradable polymer monolayers collapsed under 10 m Nm^{-1}, organo-modified clays and biodegradable polymers mixed monolayers were transferred onto mica substrates at 15°C below 5 m Nm^{-1} using the LB method [22]. The π–A isotherms were measured using a USI-3-22 film balance (USI Co., Ltd.) at 15°C.

These monolayers were transferred onto glass (XRD samples) or mica (AFM samples) substrates at 15°C. The structures of the organo-modified clays in bulk and layer structures of organized molecular films were characterized using a Rint-Ultima diffractometer (Rigaku Co., Ltd.).

However, in the case of organized molecular films, the out-of-plane X-ray diffraction method was used. Ni-filtered Cu Kα radiation (wavelength $\lambda = 0.154$ nm) was generated at 40 kV and 40 mA. The diffracted X-ray beam was monochromatized using a pyrographite monochromatic system and monitored by a scintillation counter. The in-plane spacing of the two-dimensional lattice of the films was determined using an in-plane X-ray diffractometer for two-dimensional geometrical arrangements [23, 24] (Bruker AXS, MXP-BX; Cu Kα radiation generated at 40 kV and 40 mA; instrument was specially made to order) equipped with a parabolic graded multilayer mirror.

The surface morphologies of the transferred monolayers were observed by tapping mode AFM (Seiko Instrument, SPA300 with SPI-3800 probe station) utilizing microfabricated rectangular Si cantilevers with integrated pyramidal tips and a constant force of 1.7 Nm^{-1}.

1.3. Results and Discussion

Figure 3 shows AFM images of organo-modified montmorillonites with low and high surface coverage. Organo-modification ratios were observed to significantly improve using the modification method at the oil/water interface [18]. The average organo-modification ratios determined by low and high surface coverage thermogravimetric analyses were 30% and 70%, respectively.

High surface coverage materials were only modified on one side of the clay surface. In this case, a modified surface corresponds to the exposed plane of an organo-modified agents/chloroform solution. Organo-modified clays with low coverage formed an irregular surface with a low density of alkyl chains. In contrast, organo-clays formed at the chloroform/distilled water interface exhibited regular morphology of high density alkylating agents on their surface.

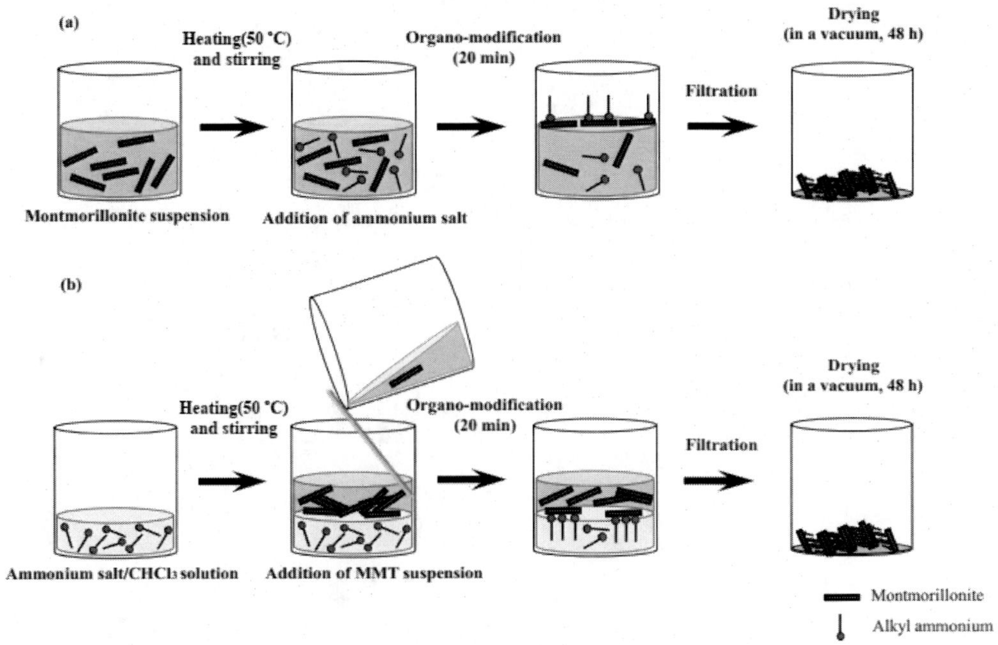

Figure 2. Modification procedure at oil/water interface.

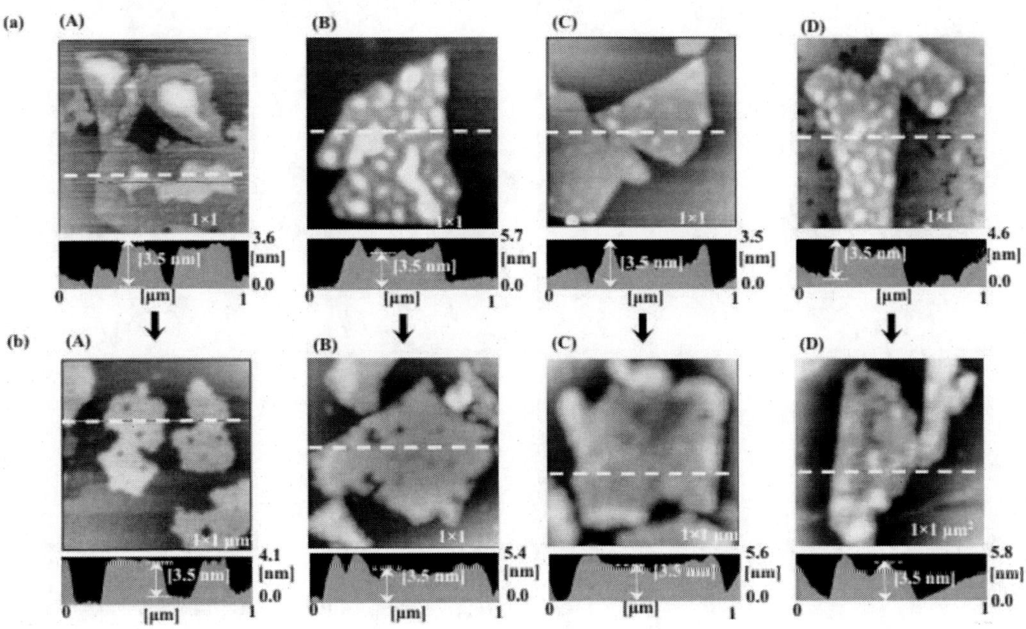

Figure 3. AFM images of organo-modified montmorillonites with: (a) low and (b) high surface coverage by (A) DMDO, (B) DMPO, (C) MHEO, and (D) TMO agents.

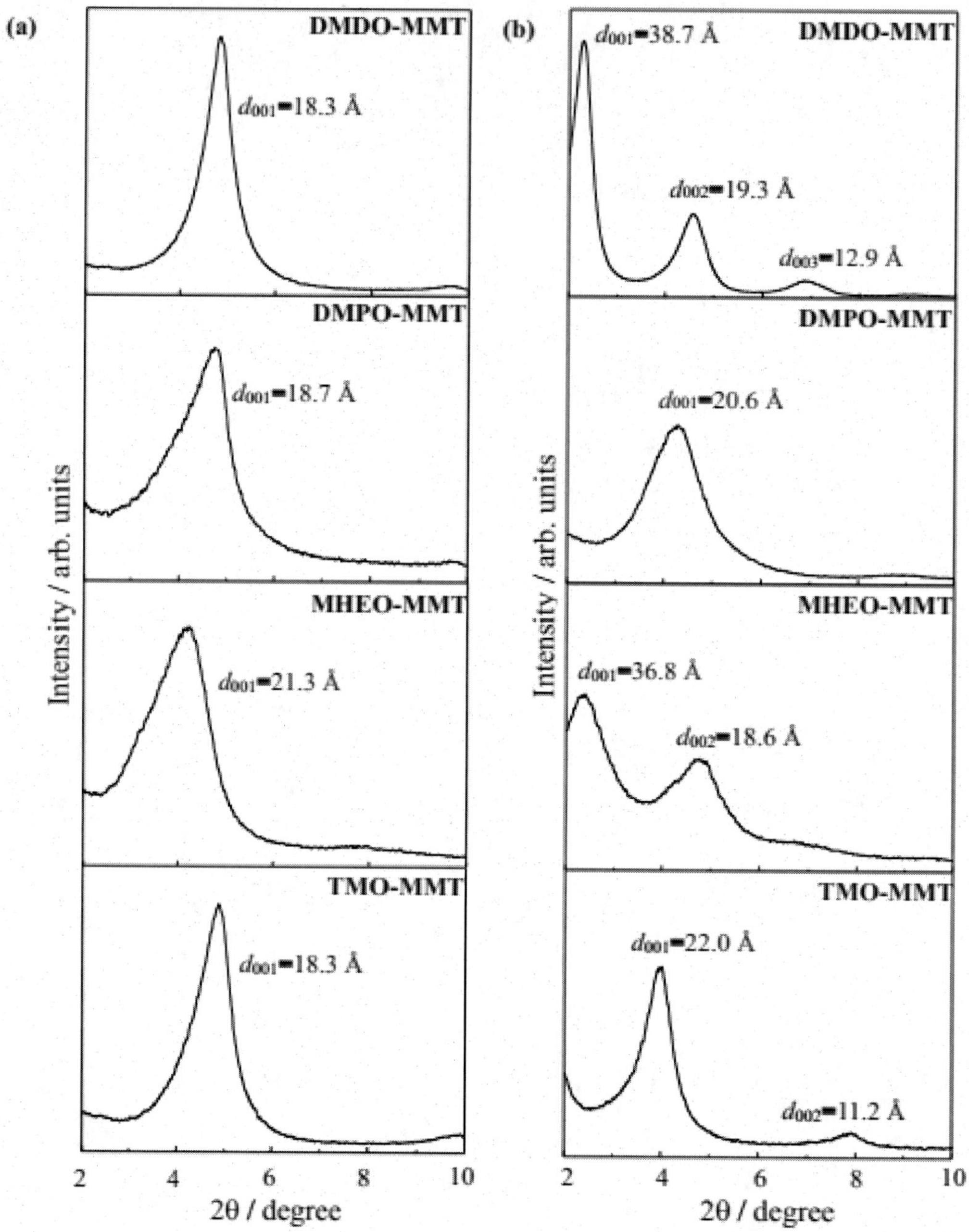

Figure 4. Powder X-ray diffraction profiles of organo-modified montmorillonites with: (a) low and (b) high surface coverage.

Figure 4 shows the X-ray diffraction profiles of these organo-clays in bulk powder, and Table 1 lists the d-spacing values estimated by powder XRD of organo-modified montmorillonites with low and high surface coverage. Almost, all organo-clays displayed an extension of interlayer spacing (d_{001}), which corresponded to their organo-modification ratios. The height of montmorillonite clay corresponds to 1.0 nm [3, 25]. Based on the results shown in Figure 4(a), the interlayer spacing of four types of organo-clays with low surface coverage is approximately 2.0 nm. Hence, we speculate that the organo-modified parts are lodged between the clay galleries. The interlayer spacing of clay galleries in DMDO-clays exhibited the most significant extension of 2.0 nm such that lodged alkyl chains are in a normal arrangement and the surface-modification ratio increased.

Table 1. The *d*-spacing values estimated by powder XRD in bulk and out-of-plane XRD of LB multilayers for organo-modified montmorillonites with (a) low and (b) high surface coverage

bulk	2θ (degree)	d-Spacing (Å)	hkl
DMDO-MMT	4.84	18.3	001
DMPO-MMT	4.76	18.7	001
MHEO-MMT	4.20	21.3	001
TMO-MMT	4.84	18.3	001
LB film	2θ (degree)	d-Spacing (Å)	hkl
DMDO-MMT	2.82	31.3	002
	4.24	20.8	003
	6.98	12.7	005
DMPO-MMT	1.44	61.4	001
	4.90	18.0	003
MHEO-MMT	1.66	53.2	001
	4.95	17.9	003
TMO-MMT	1.66	53.2	001
	4.60	19.2	003
DMDO-MMT	2.28	38.7	001
	4.58	19.3	002
	6.86	12.9	003
DMPO-MMT	4.28	20.6	001
MHEO-MMT	2.40	36.8	001
	4.76	18.6	002
TMO-MMT	4.02	22.0	001
	7.92	11.2	002
LB film	2θ (degree)	d-Spacing (Å)	hkl
DMDO-MMT	1.45	58.9	001
	4.52	19.5	003
DMPO-MMT	1.48	59.7	001
	4.90	18.0	003
MHEO-MMT	1.42	62.2	001
	5.08	17.4	003
TMO-MMT	1.45	58.9	001
	4.48	19.7	003

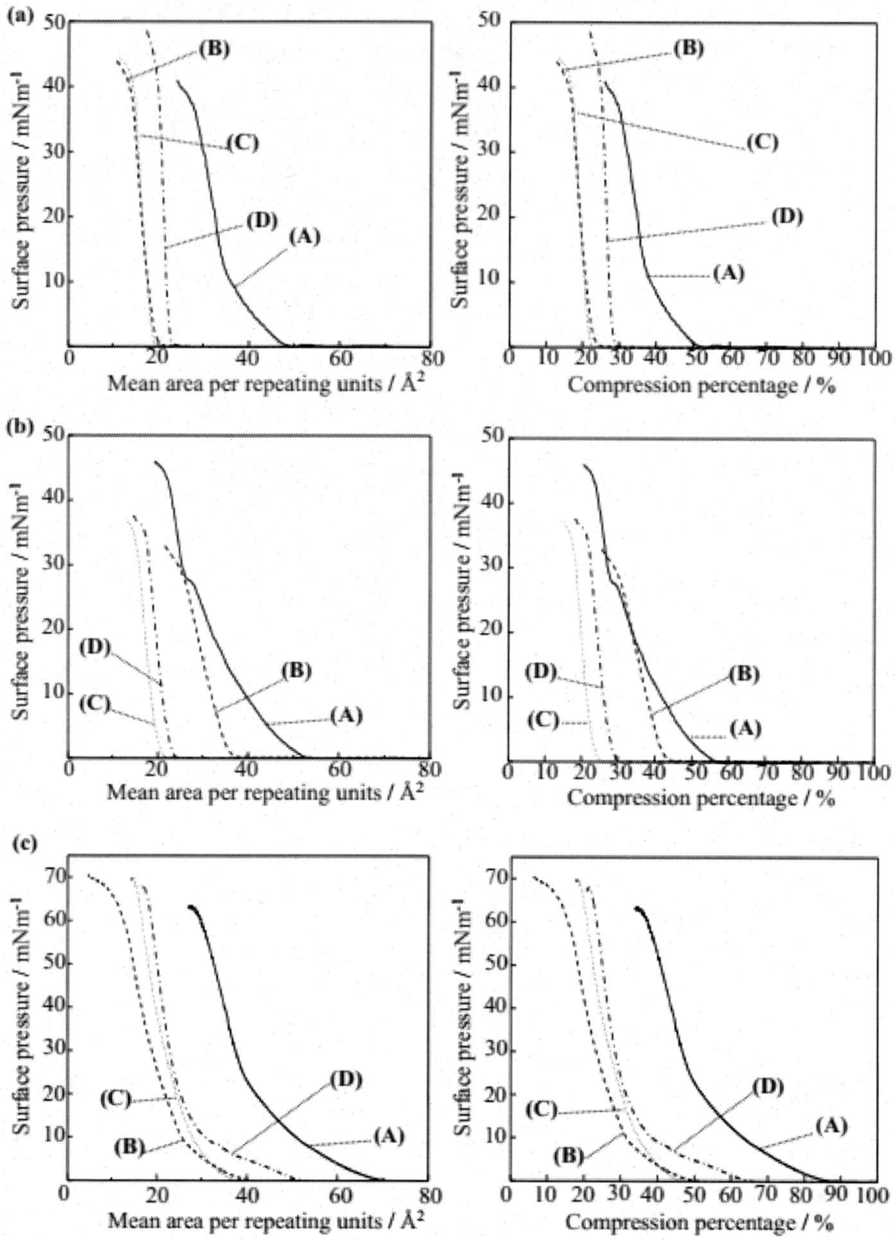

Figure 5. π–A isotherms of monolayers on the water surface of organo-modified montmorillonites with: (a) low and (b) high surface coverage. π–A curves of monolayers on buffer solution containing Na^+ ions of organo-clay with high surface coverage (c). The horizontal axis is converted to compression percentage on the right side of the figure.

Figures 5(a) and (b) show the π–A isotherms of the monolayers formed on the water surface of organo-modified montmorillonites with low and high surface coverage. The horizontal axis is converted to compression percentage shown on the right sides of these figures. Materials with irregular modification ratios did not appear to be suitable repeating units. The monolayers on the water surface of organo-clays formed extremely condensed films regardless of the ratio of surface-modification. In addition, the monolayer of organo-modified montmorillonites with high surface coverage could not be transferred onto a solid substrate. This phenomenon may be the result of both the delocalization of negative charge on the bottom layer of the montmorillonite and strong bonding to the oxonium ion in the subphase. Figure 5(c) shows π–A curves of monolayers on buffer solution containing Na+ ions of organo-clay with high surface coverage. In this case, these monolayers can be transferred to the substrate due to the disconnection of negative charge on the bottom layer of clay-H_3O^+ bonding.

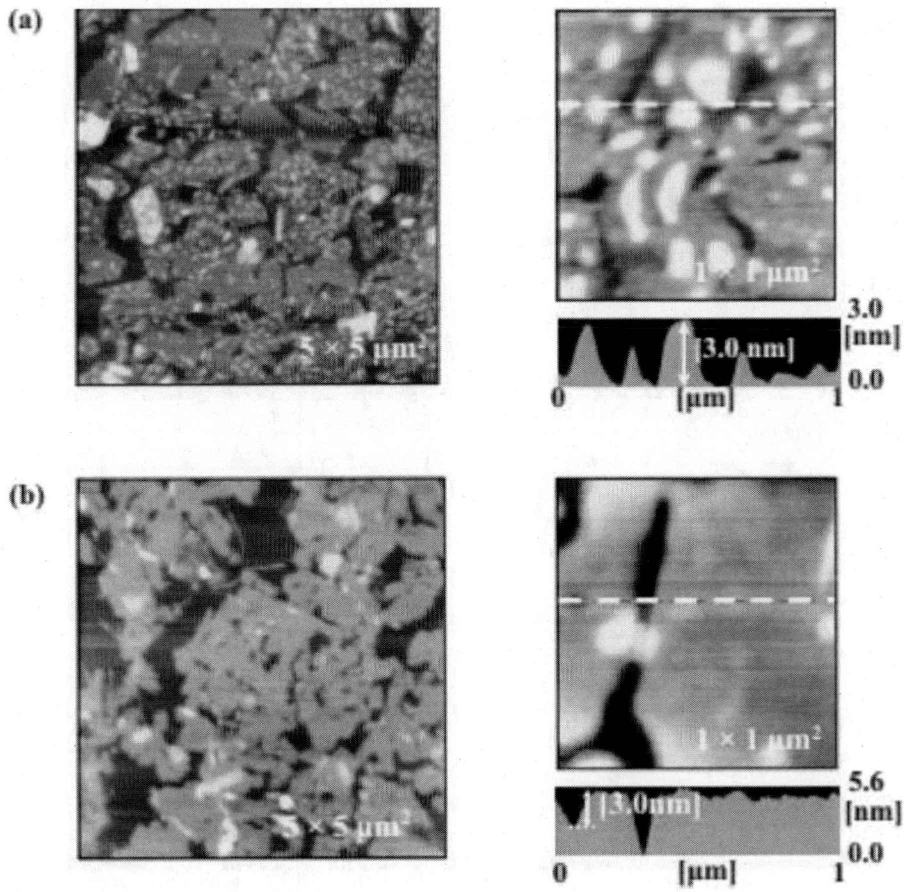

Figure 6. AFM images of Z-type monolayers of organo-modified montmorillonites with: (a) low and (b) high surface coverage.

Figure 6 shows AFM images of Z-type monolayers of organo-modified montmorillonites with low and high surface coverage. These results confirm the apparent difference between the surface-modification ratios is clear, whereas both organo-clays formed condensed monolayers at the air/water interface regardless of the ratio of surface-modification.

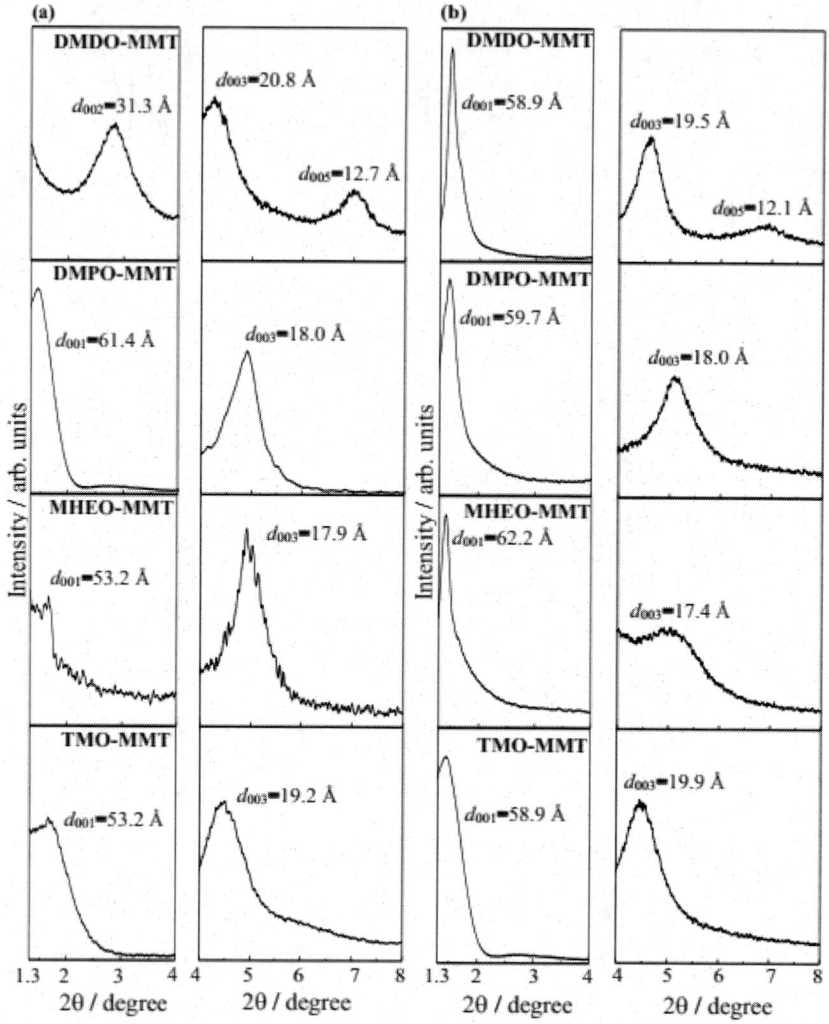

Figure 7. Out-of-plane XRD profiles of LB multilayers of organo-modified montmorillonites with: (a) low and (b) high surface coverage.

Figure 7 shows out-of-plane XRD profiles of LB multilayers for organo-modified clays with both low and high surface coverage, and Table 1 also lists the (00*l*) *d*-spacing values estimated by the out-of-plane XRD of organo-modified montmorillonites with low and high surface coverage. Almost, all multilayers formed highly developed layer structures. Notably, the diffraction profiles of LB multilayers of four types of organo-clays with high coverage indicate sharper peaks by almost same degree. The first order diffraction peaks of every material also appeared at similar positions of approximately 60 Å, indicating a perpendicular orientation to the surface. In contrast, MHEO- and TMO-modified montmorillonites with low surface coverage exhibited relatively low intensities and full width at half maximum (FWHM) of the (001) diffraction peaks in their profiles. These results indicate a decrease in the size of crystallites formed by hydrophobic chains along the ab-plane with surface coverage. In the case of DMPO-modified montmorillonites with low coverage, the peak width of (001) reflection becomes slight broader than that of high coverage. Only DMBO-modified montmorillonites with low coverage did not show (001) diffraction peaks, whereas the second, third, and fifth reflection peaks were clearly validated. This phenomenon appears to be a result of the extinction rule [26] related to the phase of diffraction. Thus, the improvement of the surface-modification ratio qualitatively enhances the regularity of layer structures along the c-axis, whereas remarkable differences are not shown in the case of DMBO- and DMDO-modified montmorillonites.

Figure 8 shows in-plane XRD profiles of LB multilayers for organo-modified clays with both low and high surface coverage. With the exception of DMPO-modified clay with bulky phenyl groups, these profiles indicate changes in double peaks of 4.1 and 4.5 Å in low coverage samples to single peaks of 4.5 Å in high coverage samples.

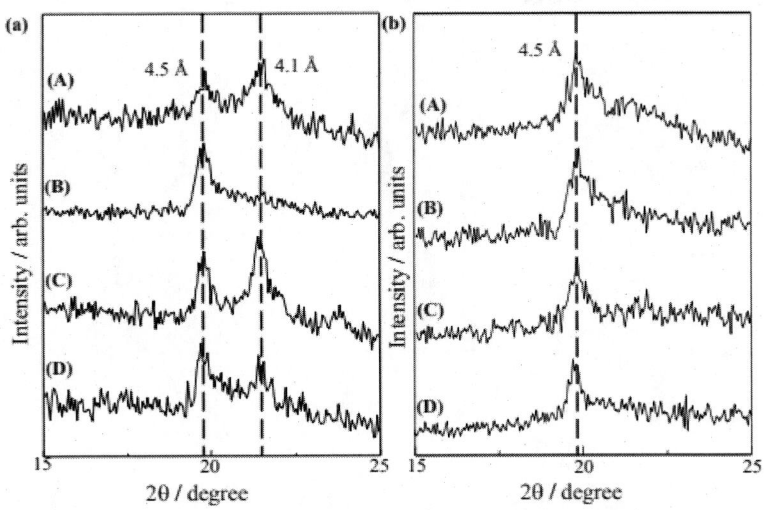

Figure 8. In-plane XRD profiles of LB multilayers of organo-modified clays with: (a) low and (b) high surface coverage by (A) DMDO, (B) DMPO, (C) MHEO, and (D) TMO agents.

Figure 9 shows a schematic of the molecular arrangement and packing models of long alkyl chains in several organo-clay films estimated by out-of-plane and in-plane XRD. As mentioned above, MHEO- and TMO-modified clays with an –OH group and single long alkyl chain, respectively, indicate low regularity and a tilted arrangement at 42° in their LB film. All high coverage materials form normal arrangements with the substrate and hexagonal sub-cell in the *ab*-plane. Low coverage DMBO- and DMDO-modified clays also form normal arrangements that are highly ordered. However, the former forms orthorhombic packing, which corresponds to other low coverage materials, and is later packed hexagonally. We speculate that the transition of packing modes with modification ratios is strongly related to distribution of surface charge. Essentially, two-dimensional hexagonal packing results in a similar structure to that of two-dimensional orthorhombic packing. To attain a high adsorption density by a quaternary ammonium cation, the packing mode appears to change to a more isotropic mode. With the exception of DMPO-modified clay, long alkyl chains of low coverage samples form two-dimensional orthorhombic packing in multilayers. Hydrocarbons of high surface coverage materials packed hexagonally in their LB films.

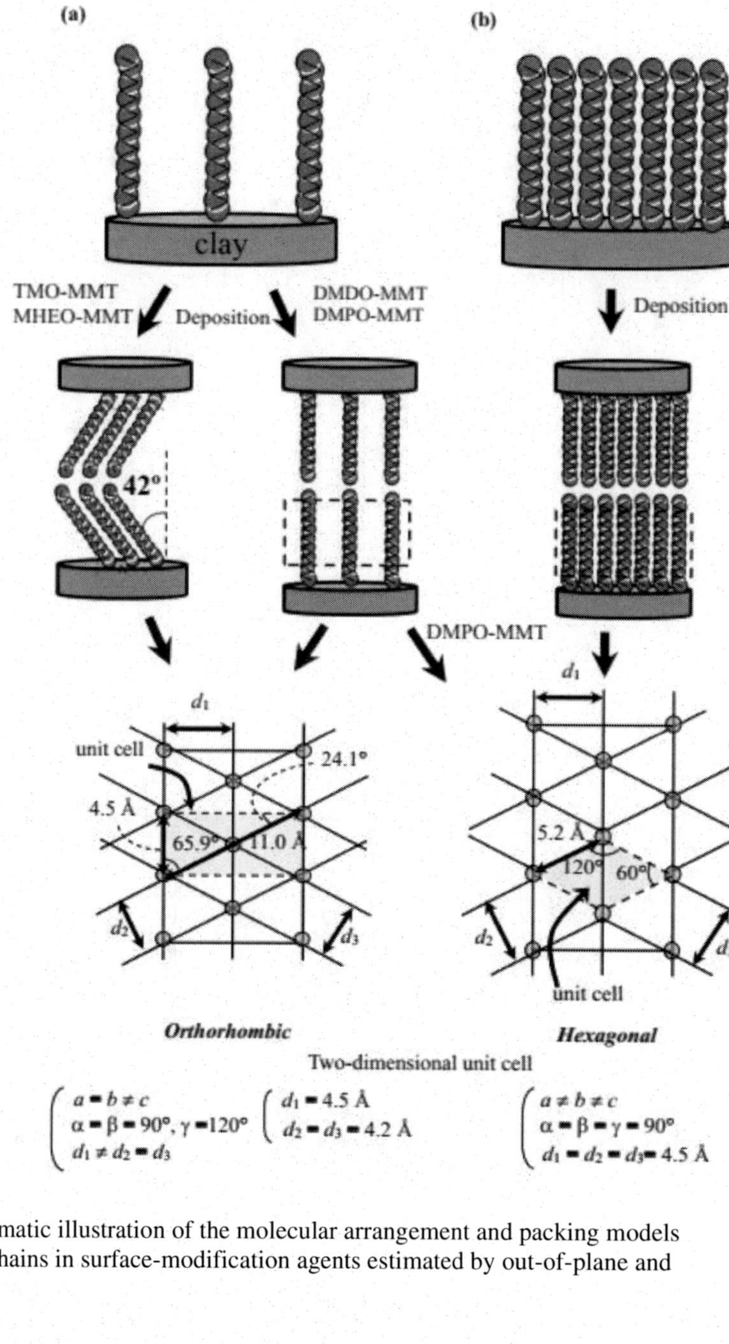

Figure 9. Schematic illustration of the molecular arrangement and packing models of long alkyl chains in surface-modification agents estimated by out-of-plane and in-plane XRD.

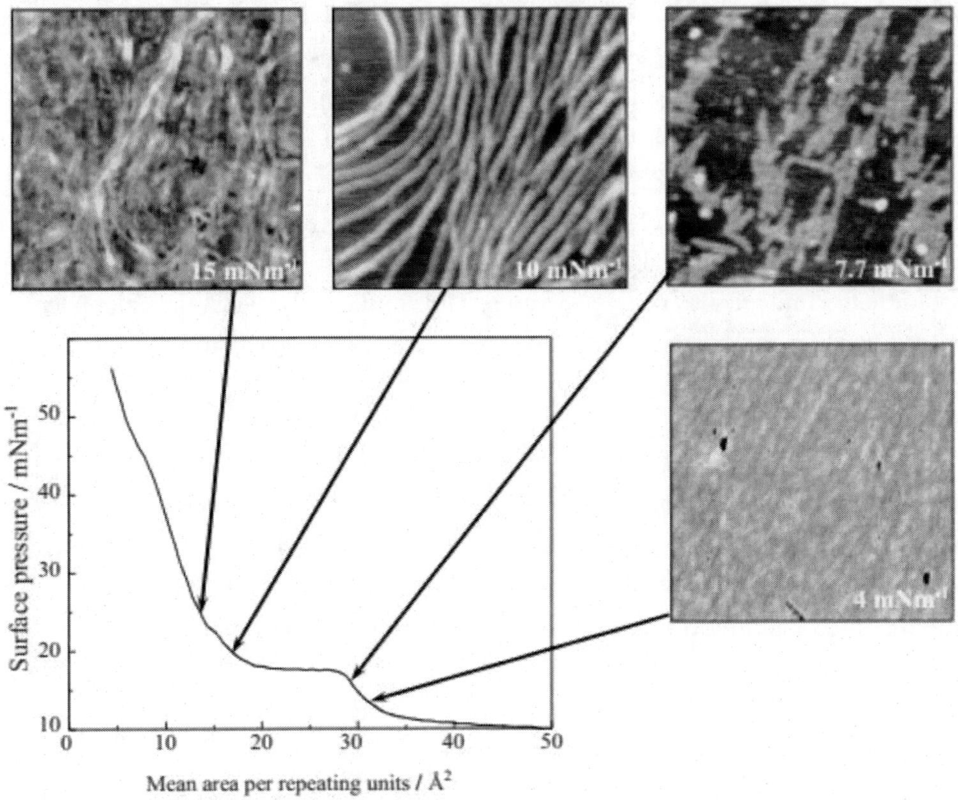

Figure 10. π–A isotherms of monolayers of PLLA on the water surface and AFM images of monolayers transferred at corresponding surface pressures.

Figure 10 shows the π–A isotherms of monolayers of PLLA on the water surface as well as AFM images of monolayers transferred at corresponding surface pressures [19]. PLLA monolayers appear to correspond to mixed monolayers with organo-modified clays. Based on the appearance of the plateau region in the π–A isotherm, the phase transition behavior from monolayers to three-dimensional crystals was confirmed. This plateau indicates that an equilibrium state at the interface exists and corresponds to the phase transition on the water surface.

Although this behavior generally corresponds to a collapse of monolayers and piles of film, these straight chain-type polymers may form a three-dimensional lamellae structure (made by the folded chain unit) at the plateau region in the isotherm. The morphological formation of organized molecular films of PLLA is similar to a bundle-like texture. Figure 10 demonstrates that when the surface pressures related to two-dimensional phase transitions are chosen, the monolayers are transferred onto solid substrates at each surface pressure.

Extrapolated images of the isotherm correspond to AFM results of their monolayers on solid substrates. The lamellar features are parallel to the LB film deposition direction. The orientation of the lamellae suggests that shape-persistent features behave like rigid-rod polymers. According to the pseudo-orthorhombic unit cell of this polymer, each 10^3 helix has a linear dimension of 2.78 nm.

Thus, a single chain of PLLA will be 50 nm long if the entire chain is in a perfect 10^3-helical conformation [27]. A value of 50 nm corresponds well to the bundle thickness estimated by the line profiles of AFM results. The similarity between the calculated chain length and lamellar dimension from AFM images suggests that individual PLLA chains exist as 10^3 helices, which persist over nearly the entire length of the chain with disordered chain ends.

Figure 11(a) shows π–A isotherms of DMDO-modified montmorillonite with low coverage: PLLA-mixed monolayers on the water surface in several mixed ratios. Furthermore, the inset of Figure 11(a) shows a plot of the collapsed surface pressure versus a mixed ratio with DMPO-MMT in this system. In the case of the mixed monolayers of DMDO-modified montmorillonites with low coverage: PLLA, the transition point is almost independent of the mixed ratio and exhibits a constant value in their corresponding isotherms.

Thus, this is immiscible system, and phase separation may occur in the two-dimensional film plane at macroscopic scales. On the other hand, Figure 11(b) shows π–A isotherms of DMDO-modified montmorillonites with high surface coverage: PLLA-mixed monolayers on the water surface at several mixed ratios. Based on classical π–A isotherms [13, 28], continuous changes in collapsed surface pressures and limited mixed ratios indicate ideal macroscopic miscibility. Although clear continuous changes were not observed, the point of collapsed surface pressure gradually diffused. Thus, we considered the formation of an apparently miscible state at the mesoscopic or microscopic level and tried to directly observe the film surface.

Figure 12 shows AFM images of mixed monolayers of PLLA- and DMDO-modified clays with low and high surface coverage. The functionality of high modification ratios improved the miscibility of the organic materials. These figures indicate the difference in the miscibility of amphiphilic polymers. The organo-modified clay separated from the PLLA phase in mixed monolayers. On the other hand, mixed monolayers of PLLA and organo-modified clay with high surface coverage formed apparent miscible structures at mesoscopic levels. That is to say, they form an apparently miscible system at the mesoscopic scale.

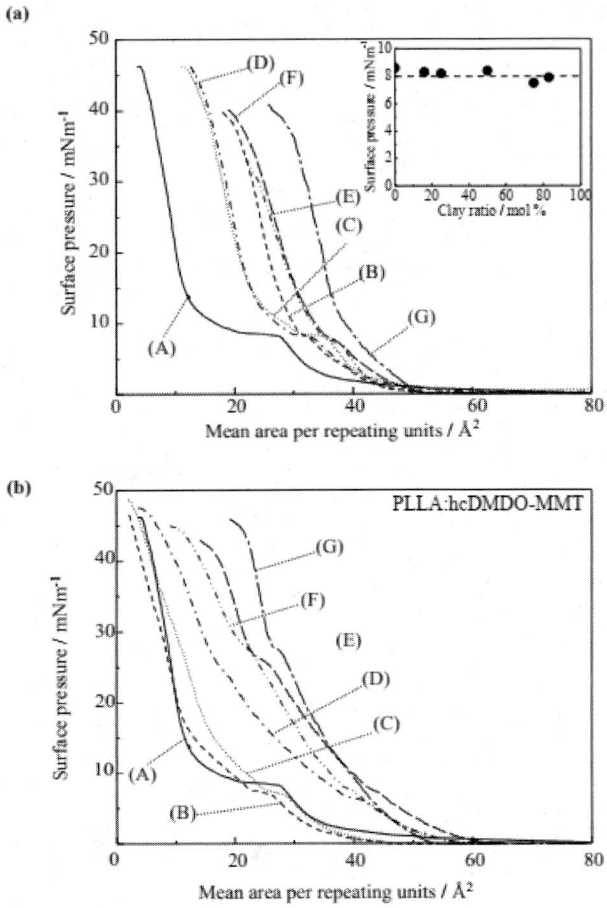

Figure 11. π–A isotherms of mixed monolayers of PLLA and organo-modified montmorillonites with (a) low and (b) high surface coverage.

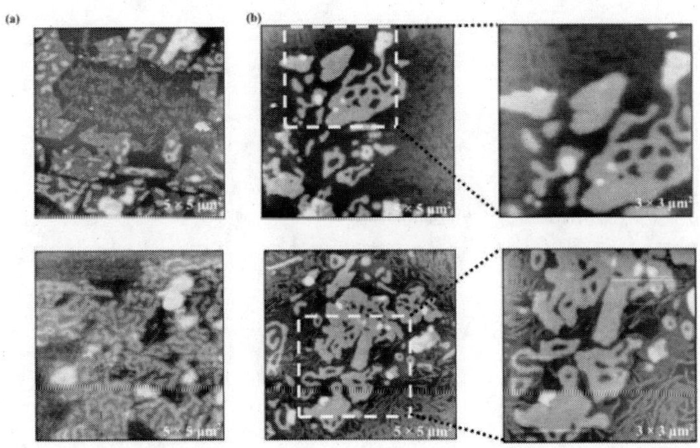

Figure 12. AFM images of mixed monolayers of PLLA- and DMDO-modified montmorillonites with (a) low and (b) high surface coverage. The following mixed content of PLLA : DMDO-modified montmorillonites was used: (A) 1:0, (B) 5:1, (C) 3:1, (D) 1:1, (E) 1:3, (F) 1:5, and (G) 0:1.

Figure 13 shows π–A isotherms of mixed monolayers on the water surface as well as AFM images of PESu, PLLA, and PCL on mica with DMDO-modified montmorillonites with high surface coverage. Furthermore, Zisman plots estimated by the measurement of the contact angles to the monolayers of biodegradable polymers are also shown. Using this plot, the values of surface free energy were calculated to be 20.4 (PESu), 21.2 (PLLA), and 23.1 (PCL) mJ/m2, respectively. In the case of PESu, the amount of phase-separated structures was significant as the polymers had fibril textures. On the other hand, the morphing of mixed monolayers of PCL [29]: organo-clay with high coverage was not possible. The miscibility of the organic polymer with organo-modified material in the two-dimensional plane appears to be closely related to the surface free energy of the polymer. The surface-modification ratios of inorganic materials were observed to significantly influence the construction of miscible states of organic/inorganic nanostructured materials.

1.4. Conclusion

Two-dimensional nanocomposites [30] of organo-modified aluminosilicate with low and high surface coverage were constructed by means of the Langmuir–Blodgett method. In this chapter, we developed an effective organo-modification method on the surfaces of natural aluminosilicates at organic solvent/distilled water interfaces. Using the LB method, corresponding multilayer films formed highly ordered layer structures along the c-axis.

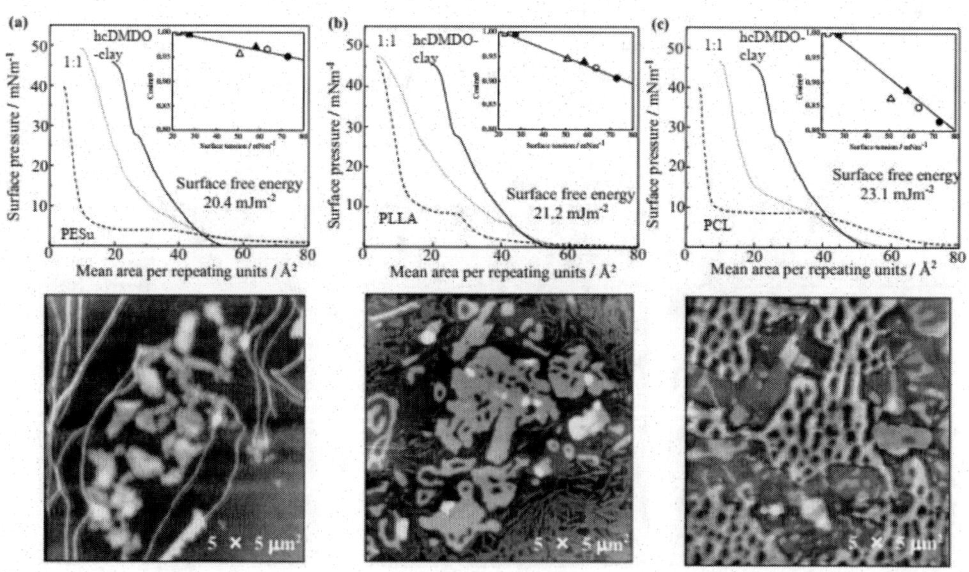

Figure 13. π–A Isotherms of mixed monolayers on the water surface and AFM images of (a) PESu, (b) PLLA, and (c) PCL on mica with DMDO-modified montmorillonites with high surface coverage. The inset shows Zisman plots estimated by the measurement of the contact angle to monolayer of biodegradable polymers.

Notably, high surface coverage materials displayed large crystalline sized of alkyl chains, which was estimated by the peak widths of out-of-plane XRD profiles. In-plane packing of these films closely packs hexagonal or orthorhombic systems. The morphology of mixed monolayers with amphiphilic PLLA monomolecular films changed drastically with modification ratios. In other words, the miscibility of these films with organic materials was significantly improved by using the modification method at oil/water interfaces.

2. Improvement of Thermal Stability of Enzyme via Immobilization on Langmuir–Blodgett Films of Organo-Modified Aluminosilicate with High Coverage

2.1. Introduction

An enzyme is a biocatalyst with several advantageous properties, including extremely high catalytic activity, substrate-, stereo-, regio-, and reaction-specificity, and a low environmental impact [31]. These properties have been widely exploited in advanced devices such as biosensors, bioreactors, or in enzymatic fuel cells [32]. Enzyme immobilization on a solid support offers several advantages over catalysis methods that employ free enzymes [33, 34]. Inorganic supports can enhance the enzyme stability, facilitate separation, may be reused, and permit continuous operation. Various techniques have been employed to immobilize the enzyme, including self-assembly [35], a sol–gel process, and a LB technique [36-39]. Among these methods, the LB technique takes advantage of preparing ultrathin films with precisely controlled thickness and molecular orientation.

Recently, we reported the preparation of a new type of bionanohybrid materials based on organo-modified clay minerals. Clay minerals are large surface materials and display various types of external and internal surfaces that are generally viable to modify [40]. In the field of biocatalysis, clay minerals have excellent biocompatibility, structural stability, and adjustable surface properties [41], thus proteins could be a good catalyst when they are supported on clay.

Layered silicates, clay minerals, have played important roles in modern technology [42, 43]. These materials have received considerable attention because of their potential use in numerous technical applications [44]. Ranging from nanocomposite materials to biomedical and personal-care products [6]. In addition, aluminosilicate clays are composed of ubiquitous elements whose abundance is characterized by high atomic Clarke numbers (the first: O, the second: Si, and the third: Al).

Clays are unique in the sense that they consist of negatively charged aluminosilicate layers kept together by exchangeable interlayer cations [45]. These enable ion exchange of the clays which, in turn, makes it possible to modify their surfaces by intercalation of a cationic surfactant into the interlayer spaces [46]. Thin films of clay minerals have been studied for applications in, among others, modified electrodes, (bio)sensors, photochromic devices, and nonlinear optical devices [30, 47, 48].

The LB method is one of the most useful techniques for preparing ultrathin films [16, 17, 49, 50]. In this method, floating molecular monolayers formed at an air–water interface are deposited on a solid substrate in a layer-by-layer way. Ultrathin films of clay minerals prepared by the LB method have been reported [51]. Recently, clay LB films have been hybridized with alkylammonium cations. Smectite clay is exfoliated into single clay sheets with negative charges in an aqueous suspension at low concentration. When a solution of an amphiphilic alkylammonium salt is spread onto the clay suspension, the negatively charged clay platelets in the suspension are adsorbed onto a floating monolayer of alkylammonium cations at the air–clay suspension interface [52]. The hybrid monolayers of clay platelets and ammonium cations formed at the interface can be transferred onto a solid surface to fabricate a hybrid multilayer. Interestingly, a floating monolayer of water-soluble alkylammonium cations is stabilized at the air–suspension interface by hybridization with the clay platelets. The surface π–A isotherm of the hybrid monolayer is dependent on the clay concentration in the suspension: at higher concentrations, the isotherm shifts to the larger area side. This suggests that the structures and properties of hybrid mono- and multi-layers could be varied, depending on the clay concentration in the suspension.

Previously, several authors have reported the formation of organo-modified montmorillonites with extremely high surface coverage and localized negative charges at the bottom by a modified LB method at an oil–water interface [19-21, 23, 53]. By using the LB method for the fabrication of films, regular, two-dimensional arrays of proteins, such as enzymes, can be controlled [40]. In recent years, attempts have been made to immobilize functional groups, and functional element immobilization in biomolecules has generated growing interest as a powerful tool in bioelectronics research [18]. Improved stability and heat resistance of the enzyme as a result of enzyme immobilization has been confirmed [54, 55].

Thus, in use in harsh environments, the catalytic properties of enzyme molecules can be excellent and, when applied in low-energy-consuming bioreactors with low environmental impact, ecologically beneficial production of fine chemicals can be expected [56-62]. In this work, floating monolayers and multilayer films of pre-formed organo-modified montmorillonites with a high coverage and localized negative charges at the bottom were constructed by the LB method on top of a solid substrate. A buffer solution containing lysozyme [63] as the cationic biomolecule was used as a subphase. The organo-modified aluminosilicate LB films play a template role in the formation of adsorbed lysozyme arrays. The LB multilayers of organo-modified MMTs with adsorbed enzymes were characterized by π–A isotherms, infrared (IR) spectroscopy, XRD, and AFM and fluorescence spectra.

2.2. Experimental Section

Natural Na^+-montmorillonite [Figure 14(a)] with a cation-exchange capacity of 108.6 meq/100 g was kindly supplied by the Kunimine Co. Montmorillonite are nanoparticles of layered mineral silicates. MMT consists of ~1 nm thick aluminosilicate layer surface-substituted with metal cations and the diameter is 0.1–1 μm. Isoelectric point of montmorillonite is pH 8. The organophilic clay was prepared by cation-exchange reactions of natural clay (aqueous dispersion) with a 0.1% aqueous solution of DMDO ammonium chloride [Figure 14(b)]. Molecular weight of montmorillonite per unit (monovalent anionic unit) and dimethyl dioctadecyl ammonium is 1078.3 and 551.0. In the case of assumption of single molecule adsorption of monovalent cationic organo-modified agent per one monovalent anionic site on the montmorillonite surface, molecular weight per 1:1 unit corresponds approximately 1629.

Quaternary ammonium cations, based on long alkyl chains, were used to form the hydrophobic parts of the clays. In order to estimate the hydrophobicity and the formation ability of ordered films, an organo-modification agent with two alkyl chains (DMDO) was used [53]. In this chapter, organo-clays with high surface coverage were formed at the oil/water interface (chloroform solutions of quaternary ammonium chlorides/clay suspensions). By the reaction at the interface, ammonium salt densely adsorbed on one side of the clay-platelets. These procedures are illustrated in Figure 2. Organo-modified clay with low surface coverage is corresponding to 30% coverage by surface modification ratio by organo-modified agent (Figure 2(a)). On the other hand, organo-modified clay with high surface coverage is corresponding to 70% coverage by surface modification ratio by organo-modified agent (Figure 2(b)). At the organo-modification in the hot water as conventional method, hydrophobic organo-modified clay particles aggregate each other into the reaction solution, as a result, modification rate is reduced. On the other hand, since the aggregation of the organo-modified clay is suppressed in the reaction solution by applying the modification method at oil/water interface, modification ratio is improved. Because it is considered that these are a chemical reaction at the interface, this organo-modification on only the one side occurs in preference.

Lysozyme was used as the water-soluble biological cationic enzyme (Figure 14(c), isoelectric point: pH 11). Chicken egg white lysozyme purchased from Wako Pure Chem. The stock solution of lysozyme of 10 mg·L^{-1} was prepared in 1 mM phosphate buffer at pH 7 (potassium dihydrogenphosphate and disodium hydrogenphosphate) and stored at 4°C. By varying the amount of each salt, a range of buffers can be prepared that buffer well between pH 5.8 and pH 8.0. Phosphates have a very high buffering capacity and are highly soluble in water. Fluorescein-modified casein and protease were purchased from Takara Bio Inc. The stock solution of fluorescein-modified casein of 20 mg·L^{-1} and protease of 10 mg·L^{-1} was prepared in 1 mM phosphate buffer at pH 7 (potassium dihydrogenphosphate and disodium hydrogenphosphate) and stored at 4°C. The organo-modified clays were spread out of a toluene solution (approximately 10^{-4} M) onto a buffer solution composed of distilled water (approximately 18 MΩ·cm) containing lysozyme molecules. If a spreading solution at this concentration have been prepared, formation of organo-modified clay monolayer of single layer exfoliation is possible.

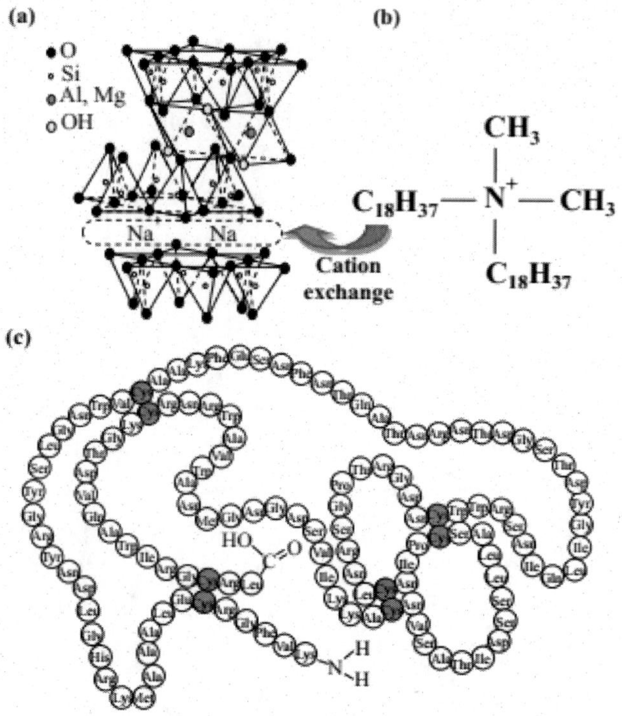

Figure 14. Structure of (a) montmorillonite, (b) the dimethyl dioctadecyl ammonium cation serving as organo-modifying agent, and (c) lysozyme used in this chapter.

The π–A isotherms were measured by means of a USI-3-22 film balance at 15°C. The value of compression speed corresponds 4.8 $cm^2 \cdot min^{-1}$ (2.5 Å $molecule^{-1} \cdot min^{-1}$). The monolayers were transferred onto CaF_2 (IR samples) or mica (AFM samples) or glass substrates with Ferric stearate monolayer as hydrophobic underlayer (out-of-plane XRD) at 15°C using the LB and a horizontal lifting method. IR spectra were measured using a 2000 spectrometer system (Parkin-Elmer Co., Ltd).

The structures of the organo-modified montmorillonites in bulk form and the layered structures of aligned molecular films were characterized with a Rint-Ultima diffractometer (in the case of aligned molecular films, the out-of-plane XRD method was used). Ni-filtered Cu Kα radiation (wavelength $\lambda = 0.154$ nm) was generated at a voltage of 40 kV and a current of 40 mA. The diffracted X-ray beam was monochromatized by a pyrographite monochromator system and monitored by a scintillation counter. The surface morphologies of the transferred monolayers were observed with a scanning probe microscope in tapping mode, using microfabricated rectangular Si single-crystal cantilevers with integrated pyramidal tips at a constant force of 1.7 Nm^{-1}. Fluorescence spectra were measured using a FP-6500 spectrofluorometer (Jasco Co., Ltd). The excitation wavelength is 485 nm. The fluorescence measurements have been performed using 4.5 mL quartz cuvettes and a bandwidth of 5 nm at excitation and emission.

2.3. Results and Discussion

Figure 15(a) shows schematic illustration of LB multilayers of organo-modified clay adsorbed lysozyme. Figure 15(b) shows the surface pressure–area (π–A) isotherm of organo-modified clay and their adsorbed lysozyme. In the case of monolayer of organo-clay, pH 7, 1.0 mM potassium buffer solution (potassium bicarbonate and potassium chloride) was used. Anion is localized on the hydrophilic surface of organo-clay. The monolayer of organo-modified montmorillonites with low surface coverage could be transferred onto a solid substrate. On the other hand, the monolayer of organo-modified montmorillonites with high surface coverage could not be transferred onto a solid substrate. This phenomenon may be the result of both the delocalization of negative charge on the bottom layer of the montmorillonite and strong bonding to the oxonium ion in the subphase. When buffer solution containing Na^+ ions were used, these monolayers can be transferred to the substrate due to the disconnection of negative charge on the bottom layer of clay–H_3O^+ bonding. In the case of lysozyme-adsorbed monolayer, the tendency of the π–A isotherm to expand toward larger areas was confirmed.

Therefore, this result supports adsorption of lysozyme to monolayer of organo-modified clay on the water surface. For the information of the characteristics of the change on isotherms, it finds that a new liquid expanded phase has appeared by adsorption of enzyme. Through this liquid expanded phase, isotherm shifts to liquid condensed phase via a gradual two-dimensional transition, but it does not indicate a solid phase. It is analyzed that reducing the condensed degree of the corresponding Langmuir monolayer by an adsorption of lysozyme molecule to the bottom of monolayer of organo-clay, and formation of liquid expanded phase having an almost twice value of the original limiting area has induced.

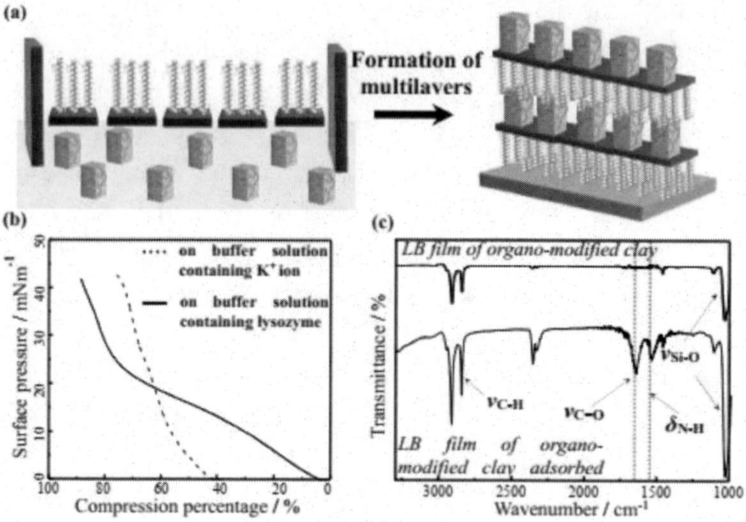

Figure 15. (a) Fabrication of multilayers of organo-modified aluminosilicate chemisorbed biomolecules and (b) *π–A* isotherms of monolayers for organo-modified montmorillonite on the phosphate buffer surface containing biological molecules and (c) IR spectra of multilayers of organo-modified aluminosilicates adsorbed by lysozyme (10 mNm^{-1}, 20 layers).

These monolayers of organo-modified montmorillonites and their adsorbed enzyme has been transferred on CaF_2 solid substrate by Langmuir–Blodgett method at 20 times. The characteristics of the IR spectra of corresponding LB films are shown in Figure 15(c) reveal the existence of clear C–H stretching and Si–O stretching vibrations. These bands originate from the long alkyl chains of quaternary ammonium reagent on the clay surface and the silicate layers itself [64, 65]. In Figure 15(c) of spectra of organo-modified clay LB film adsorbed lysozyme, bands of $\nu_{C=O}$ and δ_{N-H} are clearly confirmed. It seems that these bands come from carboxyl and amino groups of the amido bond in adsorbed lysozyme. Therefore, this result seems to confirm the adsorption of a biological enzyme onto monolayers of organo-modified montmorillonites.

Figure 16 shows the morphological characterization of LB film of organo-modified clay and that adsorbed to lysozyme transferred at high surface pressure by AFM observations. At first, monolayer on the water surface of organo-modified clay had prepared by the adding dropwise solution of organo-clay on the water subphase phase containing lysozyme. Lysozyme have adsorbed to the bottom of monolayer of organo-modified clay by electrostatic interaction. The monolayer of organo-modified clay adsorbed lysozyme has been compressed and transferred to the solid substrate. The films shown in Figures 16(a) and (b) correspond images of X-type monolayers transferred by the horizontal lifting method. These images show films exposed a surface of either hydrophilic clay or adsorbed enzymes. Comparing these images, the most important characteristic is the difference in height. Neat organo-clay monolayers have an alkylated part of thickness 2.5 nm and a montmorillonite part of thickness 1.0 nm [3, 6]. The height information presented in Figure 16(a) is in good agreement with these values. This experimental result indicates a normal orientation of the monolayers of organo-modified montmorillonite on a solid substrate. As evidenced from Figure 16(b), the height of a lysozyme-adsorbed film corresponds to 13.0 nm. The height of a lysozyme molecule along the long axis is 4.5 nm [66]. Therefore 13.0 nm correspond to twice for length of lysozyme on monolayer of organo-modified clay. Further bimolecular layer is adsorbed onto the montmorillonite surface which corresponds parallel orientation to the surface normal. This speculation is supported by the results of out-of plane XRD at low angles. Figure 16(c) shows a schematic model of a film composed of a double layer of adsorbed lysozyme on top of an organo-modified aluminosilicate monolayer, deposited on a solid substrate. The interaction between the adsorbed cationic enzyme and montmorillonite corresponds to an electrostatic force. In addition, the origin for the formation of a lysozyme double-layer structure may be attributed to the $–COO^{-}\cdots{+}H_3N–$ interaction. Indeed, the adsorption conditions of an organo-montmorillonite monolayer to a cationic enzyme at the air–water interface are pH 7, 30 $mN{\cdot}m^{-1}$ in the isotherm, and 15°C subphase temperature. In particular, a transferring of high surface pressure of the film facilitates the formation of piled up adsorbant molecule on the film.

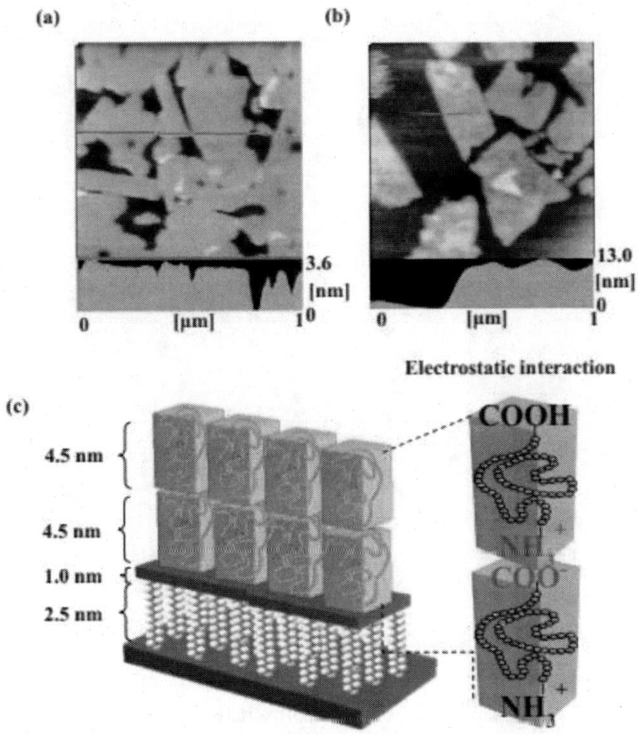

Figure 16. AFM images of (a) X-type monolayers of organo-modified aluminosilicates and (b) lysozyme-adsorbed monolayers (pH 7, 30 mNm^{-1}, 15 °C) (c) schematic illustration of a film composed of a double layer of adsorbed lysozyme on top of an organo-modified aluminosilicate monolayer, deposited on a solid substrate.

Figure 17 shows surface morphological changes in X-type monolayers of organo-modified MMT with lysozyme, which is depending on the transferring surface pressure and the pH value. In the case of a low transferring surface pressure (10 $mN \cdot m^{-1}$), a low-density adsorption of a single-layered lysozyme structure on the organo-MMT monolayer is observed.

Furthermore, the amount of adsorbed enzyme increases with higher acidic conditions. Model of Figure 17 have only showed the increasing of the amount of adsorption of lysozyme, comparing to the pH 7 condition of subphase. These experimental findings are due to the isoelectric point of lysozyme of pH 11. Previously, it has been reported that densely packed lysozyme molecules were observed by scanning tunneling microscopy (STM) with molecular resolution [67].

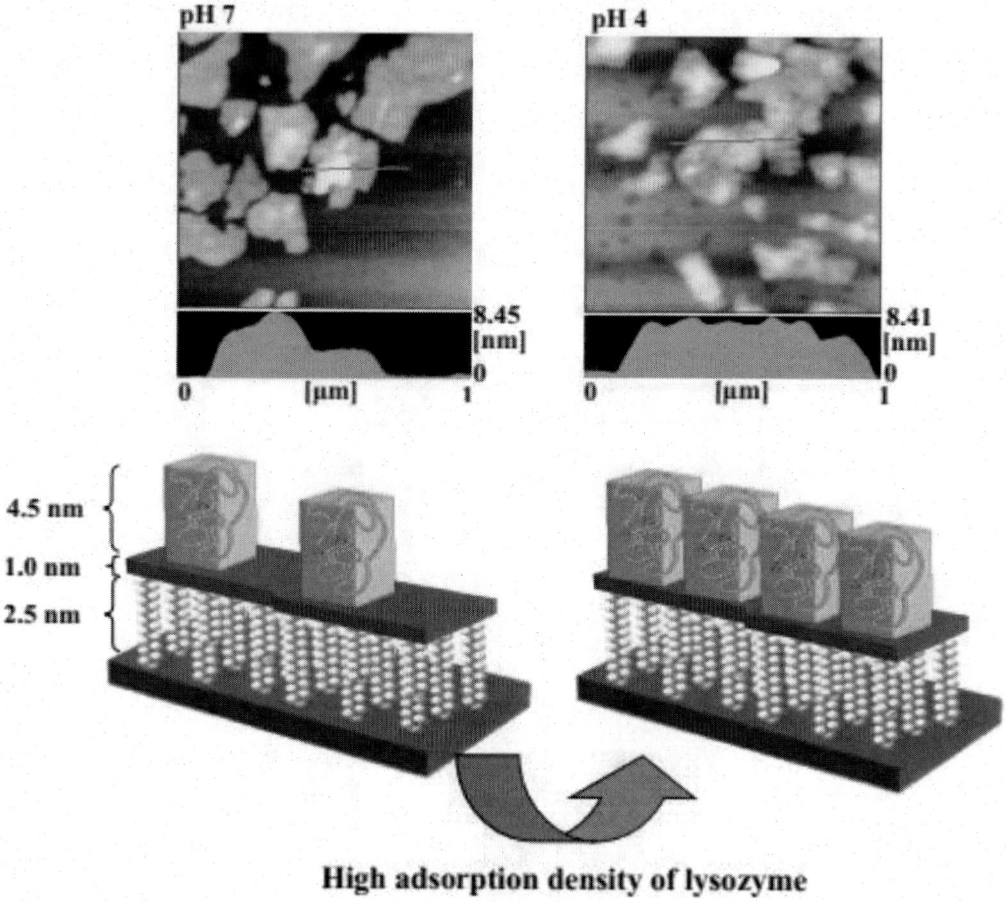

Figure 17. AFM images of X-type monolayers of organo-modified aluminosilicates adsorbed by lysozyme at pH 4 (left) and pH 7 (right) (10 $mN\ m^{-1}$, 15°C).

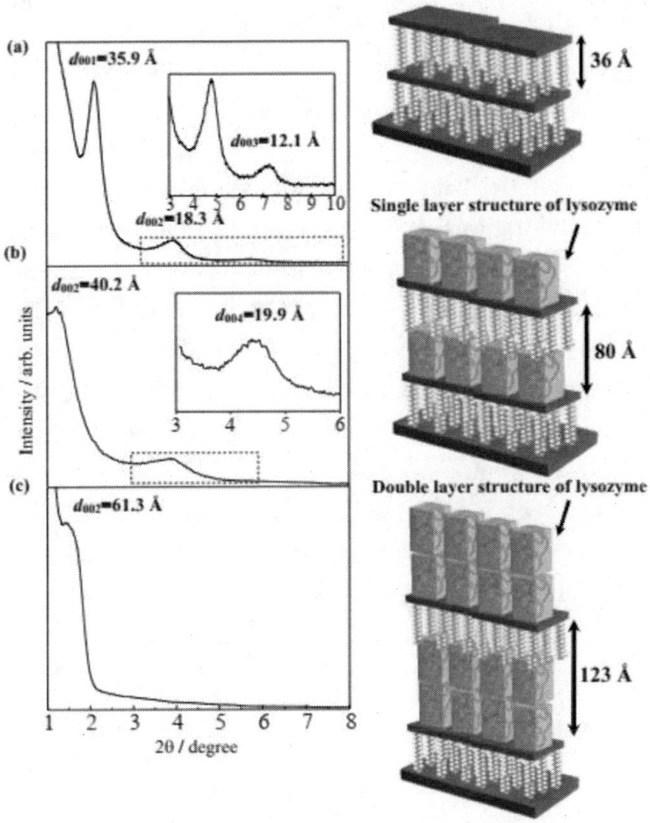

Figure 18. Out-of-plane XRD profiles for multilayers of (a) organo-modified montmorillonite, (b) with an additional chemisorbed lysozyme layer at a low (10 mNm^{-1}) and (c) at a high (30 mNm^{-1}) surface pressure.

Figure 18 shows the results of out-of-plane XRD profiles of multilayers transferred by the horizontal lifting method. In the case of neat organo-modified montmorillonite, reflections up to the third order are confirmed, which imply formation of a highly ordered layer sequence as shown in Figure 18(a). After lysozyme adsorption at low pressure (10 $mN{\cdot}m^{-1}$) under otherwise identical conditions [Figure 18(b)], higher-order reflections up to the fourth order can be confirmed in the profile of aluminosilicate LB film with enzyme. A single-layered adsorption is confirmed at the low surface pressure transferring of monolayer to solid substrate. In this case, interlayer spacing indicates 8 nm, which is well-corresponded to result of AFM observation. When the adsorbed lysozyme bilayer was prepared at high pressure (30 $mN{\cdot}m^{-1}$) on top of the aluminosilicate monolayer, a second-order reflection is observed [Figure 18(b)]. Since corresponding height of the film indicates 13 nm in the case of high surface pressure transferring as the Figure 16 and interlayer spacing indicates 12 nm in the case of multilayer formation at high surface pressure as the Figure 18, it finds that lysozyme form a bilayer structure. However, such a supramolecular-like structure generally loses its otherwise inherent activity for hydrolysis to saccharide. Therefore, control of the amount of adsorbed molecules, the adsorbate density, and the adsorption process are all important factors in the construction of new nanostructured biomaterials. Motivated by attempts to develop novel tissues of layered enzymes for biomedical applications, as described above, we further explored the physical properties at high temperatures.

Temperature-dependent IR measurements of lysozyme-adsorbed multilayers reveal stretching and bending vibrations of Si–O, C = O, and NH, derived from organo-aluminosilicate and lysozyme, evident at each individual temperature [Figure 19(a)]. Thus, the presence of biomolecules adsorbed on top of structured clay monolayers at high temperatures, up to 250°C, was confirmed. In addition, the fact that the third and fourth reflection could be confirmed at 25°C [Figure 19(b)], an interlayer distance of about 8 nm in the layered enzyme tissue was determined on the basis of the out-of-plane XRD measurements. The presence of a secondary reflection is confirmed upon increasing temperature up to 250°C, with a corresponding interlayer distance of about 7 nm [Figure 19(b)]. Therefore, it can be seen that the layered structure of the multilayers is comprised of single-layered lysozyme oriented perpendicular to the surface of the structured clay layer, and degradation of the biomolecules is suppressed because of the highly ordered, layered arrangement of the enzymes, even at high temperatures.

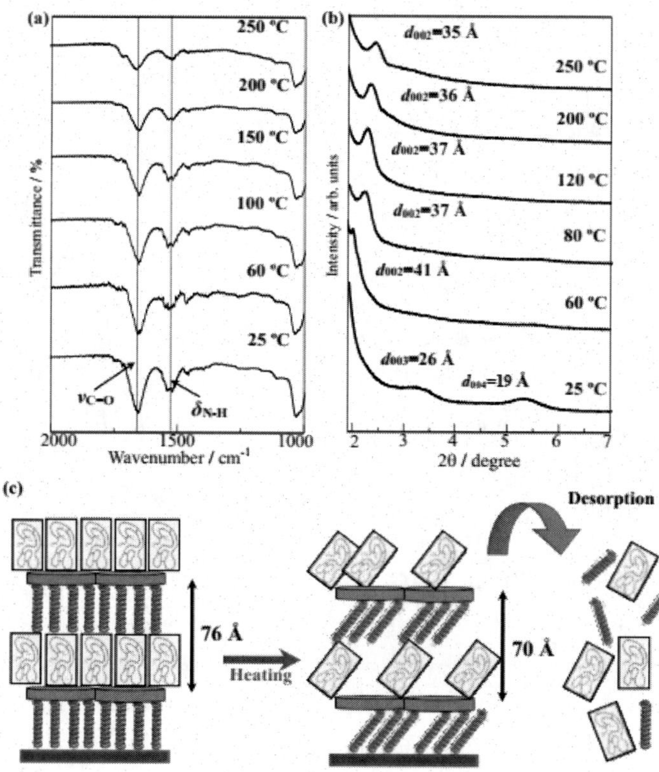

Figure 19. Lysozyme-adsorbed films at various high temperatures: (a) IR spectra and (b) out-of-plane XRD profiles (10 mNm^{-1}, 20 layers).

As a further important aspect, we investigated the fluorescence properties and the protease activity of the protease-adsorbed multilayers. The interaction between montmorillonite and protease is mainly electrostatic interaction. Organo-modified aluminosilicate adsorbed protease was immersed in fluorescein-labeled casein solution after annealing at 100°, 120°, and 160° during 10 min, respectively.

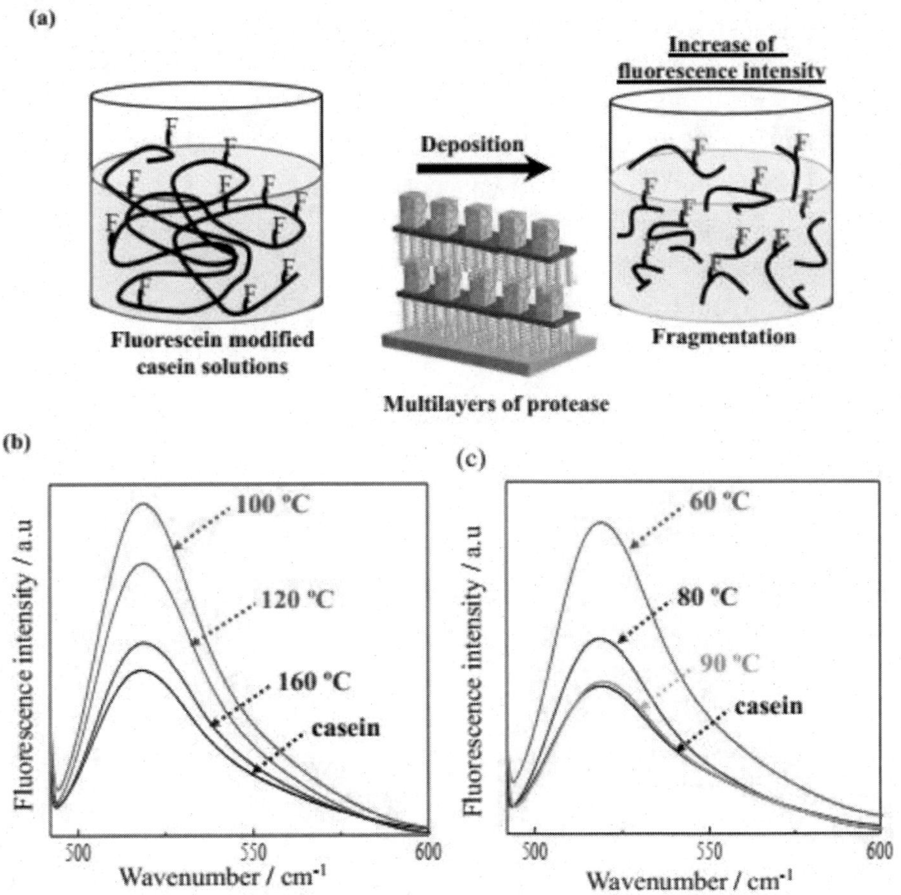

Figure 20. (a) Schematic representation of the fragmentation process and the accompanying increase in fluorescence intensity, including the protease activity of casein fluorescence spectra of casein and (b) casein/protease multilayers, (c) protease solution after annealing of the films at different temperatures.

Fluorescein-labeled casein is native casein that has been labeled using a large molar excess of fluorescein isothiocyanate (FITC). In self-quenching, the fluorescent molecule quenches its own fluorescence because of close proximity of identical molecules at high concentration. Upon keeping the protease activity, the casein is broken apart and the self-quenching is removed and an increase in fluorescence is observed (Figure 20). In order to determine whether the activity of the adsorbed protease molecules is maintained, the resulting fluorescence spectrum was measured.

Protease-adsorbed multilayers were incubated for 30 min in fluorescein-labeled casein solution, and after, fluorescence spectra were measured from 492 to 600 nm with 485 nm excitation. The fluorescence spectra of casein and casein/protease multilayers at different temperatures are shown in Figure 20. It is found that the fluorescence spectra show an overall similar behavior. The fluorescence intensity increased most strongly at 100°C and, therefore, the enzyme activity is retained most strongly at this temperature. Moreover, a decrease in the fluorescence intensity was observed at 120°C and 160°C, but, likewise, the enzyme activity was maintained (Figure 20). On the other hand, protease solution was annealed at 60°, 80°, 90° during 10 min, and putted into fluorescein-labeled casein solution. Figure 20(c) indicate enzyme activity was lost at 90°. These results support that the maintaining of enzyme activity at high temperature region is possible together with the maintaining of three-dimensional steric structure of enzyme. This maintaining of steric structure of enzyme is based on the formation of aggregate of adsorbed enzyme induced by epitaxial-growth from aluminosilicate surface and the construction of sandwiched structure of enzyme between organo-aluminosilicates.

2.4. Conclusion

Structured molecular films of adsorbed enzymes have been fabricated by means of a modified LB method using organo-modified aluminosilicate. In this method, clay monolayers play a template role in the formation of lysozyme arrays. The monolayer behavior on the buffer solution containing lysozyme was monitored by π–A isotherms. Adsorption of enzymes onto the anionic silicate surface could be confirmed by a comparison of the IR spectra of multilayers of organo-clay and those of lysozyme-adsorbed films. The IR bands of lysozyme were clearly confirmed in the spectra of enzyme-adsorbed multilayers. The difference between the heights of the organo-aluminosilicate monolayers and their lysozyme-adsorbed homolog films indicates an enzyme thickness of about 9 nm. This value corresponds to almost twice the length of the lysozyme molecule. Hence, the formation of a bilayered structure of lysozyme on top of the clay surface could be demonstrated in this system. In the case of organo-modified aluminosilicate, it is also indicated that chemisorbed enzyme activity is possible to maintain until 160°C at attempting to evaluation of the heat-resistance. The maintaining of enzyme activity at high temperature region is possible together with the maintaining of three-dimensional steric structure of enzyme. This maintaining of steric structure of enzyme is based on the formation of aggregate of adsorbed enzyme induced by epitaxial-growth from aluminosilicate surface and the construction of sandwiched structure of enzyme between organo-aluminosilicates.

ACKNOWLEDGMENTS

The studies in this chapter was supported by JSPS KAKENHI Scientific Research on Innovative Areas "MSF Materials Science (Grant Numbers JP 19H05118)," and JSPS KAKENHI Grant Numbers (C) JP 17K05988.

REFERENCES

[1] J. H. Fendler, A. Ulman (Ed.), *Thin Films: Organic Thin Films and Surfaces – Directions for the Nineties*, vol. 20, Academic Press, San Diego (1995) pp. 11-40.

[2] G. Ozin, *Adv. Mater.*, **4** (1993) 612.

[3] F. Cavani, F. Trifiro, A. Vaccari, *Catal. Today*, **11** (1991) 173.

[4] K. Takada, H. Sakurai, E. Takayama-Muromachi, F. Izumi, R. A. Dilanian, T. Sasaki, *Nature*, **422** (2003) 53.

[5] R. A. Vaia, H. Ishii, E. P. Giannelis, *Chem. Mater.*, **5** (1993) 1694.

[6] M. Alexandre, P. Dubois, *Mater. Sci. Eng. Rep.*, **28** (2000) 1.

[7] Y. Fukushima, S. Inagaki, *J. Inclus. Phenom.*, **5** (1987) 473.

[8] T. J. Pimavaia, J.-R. Butruille, G. Alberti, T. Bein (Eds.), *Comprehensive Supramolecular Chemistry*, vol. 7, Elsevier, Oxford (1996) (Chapter 7).

[9] J. W. Lee, Y. T. Lim, O. O. Park, *Polym. Bull.*, **45** (2000) 191.

[10] M. Si, J. Hefter, A. Song, M. H. Rafailovich, J. C. Sokolov, *J. Adhesion Sci. Technol.*, **19** (2005) 1459.

[11] M. Y. Gelfer, C. Burger, B. Chu, B. S. Hsiao, A. D. Drozdov, M. Si, M. Rafailovich, B. B. Sauer, J. W. Gilman, *Macromolecules*, **38** (2005) 3765.

[12] X. Wang, Y. Gao, K. Mao, G. Xue, T. Chen, J. Zhu, B. Li, P. Sun, Q. Jin, D. Ding, *Macromolecules*, **39** (2006) 6653.

[13] G. L. Gaines Jr., *Insoluble Monolayers at Liquid Gas Interfaces*, Wiley, New York (1966).

[14] G. G. Roberts, *Langmuir–Blodgett Films*, Plenum Press, London (1990). (b) M. C. Petty, *Langmuir–Blodgett Films*, Cambridge Univ. Press, New York (1996).

[15] Ulman, *An Introduction to Ultrathin Organic Films: from Langmuir–Blodgett to Self-Assembly*, Academic Press (1991).

[16] T. Yamamoto, Y. Umemura, O. Sato, Y. Einaga, *J. Am. Chem. Soc.*, **127** (2005) 16065. (b) T. Yamamoto, N. Saso, Y. Umemura, Y. Einaga, *J. Am. Chem. Soc.*, **131** (2009) 13196.

[17] H. Tetsuka, H. Takashima, K. Ikegami, H. Nanjo, T. Ebina, F. Mizukami, *Chem. Mater.*, **21** (2009) 21.

[18] K. Kurosaka, M. Kubota, A. Fujimori, Japan Patent, Japan Patent No. 2011-063475, and No. 2013-112567 (2009, 2011).

[19] Fujimori, J. Kusaka, R. Nomura, *Polym. Eng. Sci.*, 51 (2011) 1099.

[20] Fujimori, *SPE Plastic Res. Online*, (2011) 1, DOI: 10.1002/spepro.003663.

[21] Fujimori, J. Kusaka, M. Kubota, K. Kurosaka, *RFP Int.*, 6 (2011) 216.

[22] S. L. Ni, W. Yin, M. K. Ferguson–McPherson, S. K. Satija, J. R. Morris, A. R. Esker, *Langmuir*, **22** (2006) 5969. (b) S. L. Ni, W. Lee, B. B. Li, A. R. Esker, *Langmuir*, **22** (2006) 3672.

[23] Fujimori, Y. Sugita, H. Nakahara, E. Ito, M. Hara, N. Matsuie, K. Kanai, Y. Ouchi, K. Seki, *Chem. Phys. Lett.*, **387** (2004) 345.

[24] Fujimori, T. Araki, H. Nakahara, E. Ito, M. Hara, H. Ishii, Y. Ouchi, K. Seki, *Chem. Phys. Lett.*, **349** (2001) 6.

[25] P. H. Nam, M. Kaneko, N. Ninomiya, A. Fujimori, T. Masuko, *Polymer*, **46** (2005) 7403.

[26] H. P. Klug, L. E. Alexander, *X-ray Diffraction Procedures*, John Wiley and Sons, New York (1974).

[27] T. Miyata, T. Masuko, *Polymer*, **8** (1997) 4003.

[28] D. J. Qian, H. Nakahara, K. Fukuda, *Langmuir*, **11** (1995) 4491.

[29] T. Miyata, S. Shibata, T. Masuko, M. Mochizuki, *Sen-i Gakkaishi*, **54** (1998) 219.

[30] Y. Lvov, K. Ariga, I. Ichinose, T. Kunitake, *Langmuir*, **12** (1996) 3038.

[31] A. M. Klibanov, *Science*, **219** (1983) 722.

[32] J. A. Cracknell, K. A. Vincent, F. A. Armstrong, *Chem. Rev.*, **108** (2008) 2439.

[33] Wang, G. Zhou, Y. Li, N. Lu, H. Song, L. Zhang, *Colloids Surf., A*, **406** (2012) 75.

[34] P. A. Russo, M. M. L. Riberio Carrott, P. A. M. Mourao, P. J. M. Carotto, *Colloids Surf., A*, **386** (2011) 25.

[35] H. Ohnuki, T. Saiki, A. Kusakari, H. Endo, M. Ichihara, M. Izumi, *Langmuir*, **23** (2007) 4675.

[36] M. Sriyudthsak, H. Yamagishi, T. Moriizumi, *Thin Solid Films*, **160** (1988) 463.

[37] S. C. Sun, P. H. Hosi, D. J. Harrison, *Langmuir*, **7** (1991) 727.

[38] K. Wan, J. M. Chovelon, N. Jaffrezic-Renault, *Talanta*, **52** (2000) 663.

[39] Fujimori, M. Taguchi, S. Arai, *Colloids Surf., A*, **443** (2014) 432.

[40] S. Arai, J. Kusaka, M. Kubota, K. Kurosaka, A. Fujimori, *Chem. Lett.*, **41** (2012) 1181.

[41] G. Lagaly, S. Ziesmer, *Adv. Colloid Interface Sci.*, **100** (2003) 105.

[42] P. C. LaBarton, Z. Wang, T. Pinnavaia, *J. Appl. Clay Sci.*, **15** (1999) 11.

[43] K. Takada, H. Sakurai, E. Takayama-Muromachi, F. Izumi, R. A. Dilanian, T. Sasaki, *Nature*, 422 (2003) 53.

[44] M. Osada, T. Sasaki, *Adv. Mater.*, 24 (2012) 210.

[45] R. Ma, T. Sasaki, *Adv. Mater.*, **22** (2010) 5082.

[46] A. K. Mishra, S. Allauddin, R. Narayan, T. M. Aminabhavi, K. V. S. N. Raju, *Ceram. Int.*, **38** (2012) 929.
[47] J. H. Fendler, *Chem. Mater.*, **8** (1996) 1616.
[48] G. Schulz-Ekloff, D. Wohrle, B. van Duffel, R. A. Schoonheydt, *Microporous Mesoporous Mater.*, **51** (2002) 91.
[49] T. Yamamoto, N. Saso, Y. Umemura, Y. Einaga, *J. Am. Chem. Soc.*, **131** (2009) 13196.
[50] M. Muramatsu, K. Akatsuka, Y. Ebina, K. Wang, T. Sasaki, T. Ishida, K. Miyano, M. Haga, *Langmuir*, **21** (2005) 6590.
[51] Fujimori, S. Arai, J. Kusaka, M. Kubota, K. Kurosaka, *J. Colloid Interf. Sci.*, **392** (2013) 256.
[52] K. Ikegami, H. Tetsuka, Y. Hoshi, T. Ebina, H. Takashima, *Langmuir*, **26** (2010) 2514.
[53] Walcarius, Electroanalytical, 10 (1998), pp. 1217-1235.
[54] Y. Bai, H. Yang, W. W. Yang, Y. C. Li, C. Q. Sun, *Sens. Actuators, B*, **124** (2007) 179.
[55] W. Bemd, D. Margrarita, A. Pilar, R. Eduardo, *ACS Appl. Mater. Interfaces*, **3** (2011) 4339.
[56] A. J. Patil, S. J. Mann, *Mater. Chem.*, **18** (2008) 4605.
[57] E. Ruiz-Hitzky, P. Aranda, M. Darder, M. Ogawa, *Chem. Soc. Rev.*, **40** (2011) 801.
[58] E. Ruiz-Hitzky, P. Aranda, M. Darder, G. Rytwo, *J. Mater. Chem.*, **20** (2010) 9306.
[59] Mousty, *Anal. Bioanal. Chem.*, **396** (2010) 315.
[60] C. F. Cal, U. Bakowsky, E. Rytting, A. K. Schaper, T. Kissel, *Eur. J. Pharm. Biopharm.*, **69** (2008) 31.
[61] J. H. Choy, S. J. Choi, J. M. Oh, T. Park, *Appl. Clay Sci.*, **36** (2007) 122.
[62] M. Kawase, Y. Hayashi, F. Kinoshita, E. Y. Yamato, J. Miyazaki, J. Yamakawa, T. Ishida, M. Tamura, K. Yagi, *Biol. Pharm. Bull.*, **27** (2004) 2049.
[63] W. F. Jones, J. M. Wiencek, P. A. Darcy, *J. Cryst. Growth*, 232 (2001) 221.
[64] S. Miao, H. Leeman, S. D. Feyterb, R. A. Schoonheydt, *J. Mater. Chem.*, **20** (2010) 698.
[65] Vinu, M. Miyahara, K. Ariga, *J. Phys. Chem. B*, **109** (2005) 6436.
[66] Rouwendal, W. Norde, *J. Colloid Interf. Sci.*, **139** (1990) 169.
[67] L. Haggerty, B. A. Watson, M. A. Barteau, A. M. Lenhoff, *J. Vac. Sci. Technol., B*, **9** (1991) 1219.

In: An Introduction to Aluminosilicates
Editor: Nero Regina Blevins
ISBN: 978-1-53617-250-8

Chapter 5

AMORPHOUS MESOPOROUS ALUMINOSILICATES IN THE SYNTHESIS OF N-HETEROCYCLIC COMPOUNDS

Nellya G. Grigorieva[1,2,*], Marat R. Agliullin[1,2], Sergey V. Bubennov[1], Svetlana A. Kostyleva[1], Nadezhda A. Filippova[1], Vera R. Bikbaeva[2] and Boris I. Kutepov[1,2]

[1]Institute of Petrochemistry and Catalysis,
Russian Academy of Sciences,
Ufa, Bashkortostan, Russian Federation
[2]Ufa State Petrochemical Oil Engineering University,
Ufa, Bashkortostan, Russia

ABSTRACT

A new method *without using templates* synthesis of amorphous mesoporous aluminosilicates with acidic properties and a narrow pore size distribution in the range of 2-7 nm has been developed. The physicochemical properties of samples of amorphous mesoporous ASM aluminosilicate with a different Si/Al molar ratio (40, 80, 120) were studied.

The catalytic properties of samples of amorphous mesoporous aluminosilicate ASM with different Si/Al molar ratios (40, 80, 120) were studied in the synthesis of practically important pyridines (by the interaction of C_2–C_5 alcohols with formaldehyde and ammonia, cyclocondensation of acetaldehyde and propionic aldehyde with ammonia), dialkylquinolines and alkyltetrahydroquinolines (by reaction of aniline with C_3, C_4 aldehydes) and alkyldihydroquinolines (by interaction of aniline with ketones, acetone and

* Corresponding Author's Email: ngg-ink@mail.ru.

acetophenone). It is found that mesoporous aluminosilicate ASM sample with a molar ratio of Si/Al = 40, which has the highest acidity among the studied samples, exhibits the highest activity and selectivity in these reactions.

Keywords: mesoporous aluminosilicates ASM, cyclocondensation, carbonyl compounds, alcohols, ammonia, amines, pyridines, quinolines, hydroquinolines, tetrahydroquinolines, benzodiazepines, benzimidazoles, sol-gel synthesis, heterogeneous catalysis

INTRODUCTION

Recently, mesoporous aluminosilicates are increasingly used in the catalytic transformations of the acid-base type [1-5]. The most studied among mesoporous mesostructured aluminosilicates are Al-MCM-41 and Al-SBA-15 [6-9], the synthesis of which is difficult and multi-stage, requires the use of expensive templates.

A more affordable way to prepare catalytically active mesoporous aluminosilicates may be sol-gel synthesis using silicon and aluminum alkoxides [10].

Currently, sol–gel technology has become an integral feature of the modern practice of producing new materials with unique properties [11-19]. It is the basis of the latest technologies for the preparation of high-temperature superconducting oxides [13], ceramic ultrafiltration membranes [14], optical [15] and anticorrosive coatings [16], new biomaterials [17], porous sorbents or carriers [18, 19]. Furthermore, sol–gel synthesis is a powerful tool for controlling the formation of active sites and the porous structure of various heterogeneous catalysts at the stage of sol–gel processing [20, 21].

Sol–gel technology is widely used in the production of amorphous aluminosilicates that are part of the modern catalysts for cracking, hydrocracking, hydroisomerization of *n*-paraffins, biofuel processing, pyridine synthesis and many other important processes of oil refining and petrochemical industry [22-26]. Generally, industrially produced amorphous aluminosilicates are micro- or micro-mesoporous materials with a predominance of micropores. The presence of micropores reduces the diffusion rates of the reacting molecules and products of the catalytic transformations, resulting in a rapid deactivation of the catalyst. In addition, such materials are ineffective in the catalytic transformations of bulk molecules because of the transport limitations due to the parameters of their porous structure. Solutions for all of these problems will probably require preparation of aluminosilicates with a mesoporous structure, ensuring efficient diffusion of the reactants to the active sites and the reverse diffusion of the resulting reaction products.

Traditional methods for the synthesis of mesoporous aluminosilicates are based on the use of various templates, which are expensive and toxic materials.

We have developed a new method for the free-of-template synthesis of amorphous mesoporous aluminosilicates with acidic properties [10, 27]. The synthesized samples of mesoporous aluminosilicates were studied as catalysts for cyclocondensation reactions, by which various practically important nitrogen-containing heterocyclic compounds - derivatives of pyridine, quinoline, benzimidazole, and benzodiazepine - were obtained.

Many methods have been developed for the synthesis of N-heterocyclic compounds. Industrially important methods for producing N-heterocyclic compounds consist mainly in the condensation of various carbonyl compounds with ammonia or amines under the action of acids or bases [28]. These processes are characterized by the formation of a significant amount of by-products. The target products are formed unselectively, therefore the creation of highly selective heterogeneous catalytic methods for the production of N-heterocyclic compounds is an important and topical task.

In recent years, zeolite catalysts have been actively used for the synthesis of N-heterocyclic compounds [29], which provide many advantages in the conversion of harmful environmentally hazardous cyclocondensation processes of carbonyl compounds with ammonia into green chemical production. However, it should be noted that the significant disadvantage of zeolites is their rapid deactivation due to the blocking of the micropores of the crystal lattice by the reaction products and coke. In addition, the microporous structure of the zeolite limits or hinders the synthesis of bulk molecules with a diameter of about 1 nm or greater.

A new approach to the heterogeneous catalytic synthesis of N-heterocyclic compounds is the use of mesoporous materials as catalysts.

Next, the results of studying the catalytic properties of samples of amorphous mesoporous ASM aluminosilicate with various Si/Al molar ratios (40, 80, 120) in the synthesis of pyridine and alkyl pyridines, alkyl quinolines, alkyl tetrahydroquinolines, alkyl dihydroquinolines, alkyl (aryl) benzimidazoles, and benzodiazepines.

Sol-Gel Synthesis and Physico-Chemical Properties of Amorphous Mesoporous Aluminosilicates

Mesoporous aluminosilicates with Si/Al = 40, 80, 120 atomic ratio have been prepared by sol–gel synthesis at variable pH [10]. In a typical synthesis in the first step (synthesis in an acidic medium), 14 g of tetraethylorthosilicate and 0.16–0.64 g (0.16 g for Si/Al = 120, 0.32 g for Si/Al = 80, 0.64 g for Si/Al = 40) of aluminum nitrate were added to 12 ml of water and 28 ml of ethanol. The initial solution with pH 3 had been kept in a thermostat at 58–60°C for 20–30 h until the gel point was reached. Then, in the second step (synthesis in an alkaline medium), 6.3 g of aqueous ammonia solution was added to the gel with vigorous stirring until pH 10 was reached, followed by kept at 25°C for 24 h. Then, based on the results, all prepared gels were subjected to stepwise thermal treatment at 50°C (12 h), 150°C (24 h) and 650°C for 5 h. To investigate the catalytic properties of aluminosilicates, they were crushed to fine powders of the fractional composition from 20 to 100 μm. The samples with Si/Al ratio of 40, 80 and 120 are indicated as ASM-40, ASM-80 and ASM-120, respectively.

The chemical composition of the prepared aluminosilicates was analyzed on a EDX 720/900HS Shimadzu X-ray fluorescent spectrometer. The analysis results show that the Si/Al atomic ratio in all calcined aluminosilicate samples is close to the Si/Al ratio in the initial solution (Table 1).

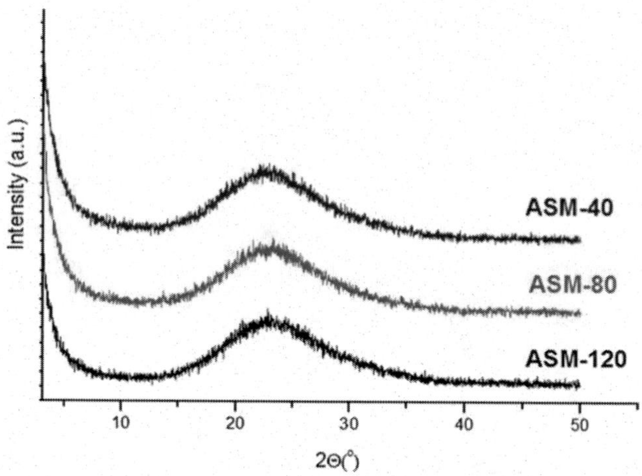

Figure 1. Powder X-ray diffraction of aluminosilicates. Reprinted with permission from ref. [27]. Copyright 2018 Springer Nature Switzerland AG.

Table 1. Chemical compositions of aluminosilicates

Sample	Chemical compositions	
	Initial compositions	Calcined aluminosilicates
ASM-40	$Si_{40.11}Al_{1.00}$	$Si_{40.07}Al_{1.00}$
ASM-80	$Si_{80.09}Al_{1.00}$	$Si_{80.10}Al_{1.00}$
ASM-120	$Si_{120.12}Al_{1.00}$	$Si_{120.22}Al_{1.00}$

[Reprinted with permission from ref. [27]. Copyright 2018 Springer Nature Switzerland AG.].

Diffraction patterns were recorded using Rigaku Ultima IV diffractometer in the monochromatic Cu Kα emission within the 2θ angle range of 3°–50° in 0.5°C/min increments and the 20 s integration time at each point. Figure 1 shows the X-ray patterns of the synthesized samples. For all samples, a halo at around 2θ = 20–30 is observed, indicating the formation of an exclusively amorphous material.

It is known [30] that Bronsted acid sites are formed while the aluminum atoms are introduced into the silicate framework in the aluminosilicates. Moreover, the aluminum atoms are characterized by a tetrahedral environment in oxygen with a signal in the ^{27}Al MAS NMR spectra from 50 to 60 ppm. Part of the aluminum atoms that did not enter the silicate framework can form the extrastructural Al_2O_3 microphases, with the signal in ^{27}Al MAS NMR spectra from 0 to 20 ppm.

Figure 2 shows the ^{27}Al NMR spectra of the synthesized samples (spectra were obtained using Avance-400 "Bruker" spectrometer equipped with the multi-nuclear sensor in the simple single pulse experiment under the ~ 104 Hz magic angle spinning of the samples in zirconium dioxide rotors). It is seen that on all spectra there is a main signal from 50 to 60 ppm and a weak signal from 0 to 10 ppm. The results show a high proportion of aluminum atoms included in the silicate framework with the predominant formation of the BAS (Bronsted Acid Sites). With an increase in the Si/Al ratio in aluminosilicates, the share of the signal decreases from 0 to 10 ppm and a fraction of the signal increases from 50 to 60 ppm, pointing to a better introduction of aluminum atoms with a lower Si/Al ratio.

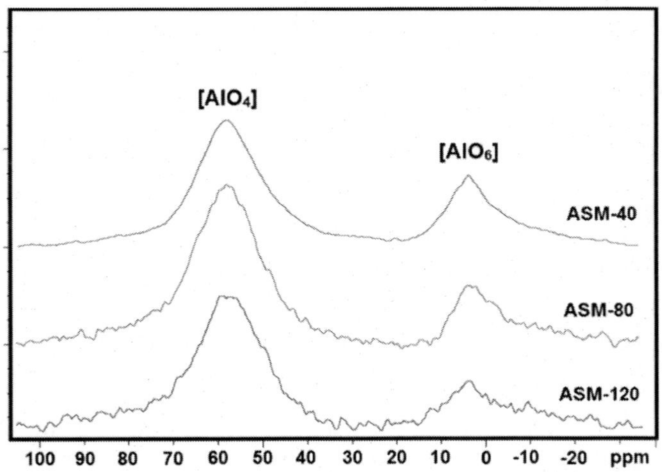

Figure 2. ^{27}Al MAS NMR spectra of aluminosilicates. Reprinted with permission from ref [27]. Copyright 2018 Springer Nature Switzerland AG.

Figure 3 shows nitrogen adsorption–desorption isotherms and pore size distribution for ASM-40, ASM-80 and ASM-120 samples. We can see that there is an isotherm of type IV with a hysteresis loop of type H^1 according to the IUPAC classification, characteristic of mesoporous materials, for all samples. In the pressure region ranging from 0.98 to 1.00, a sharp rise indicating the presence of macropores is observed.

Further, all samples exhibit a narrow pore size distribution in the range of 2–5 nm (Figure 3), and the change in the Si/Al ratio has practically no effect on the value of this parameter.

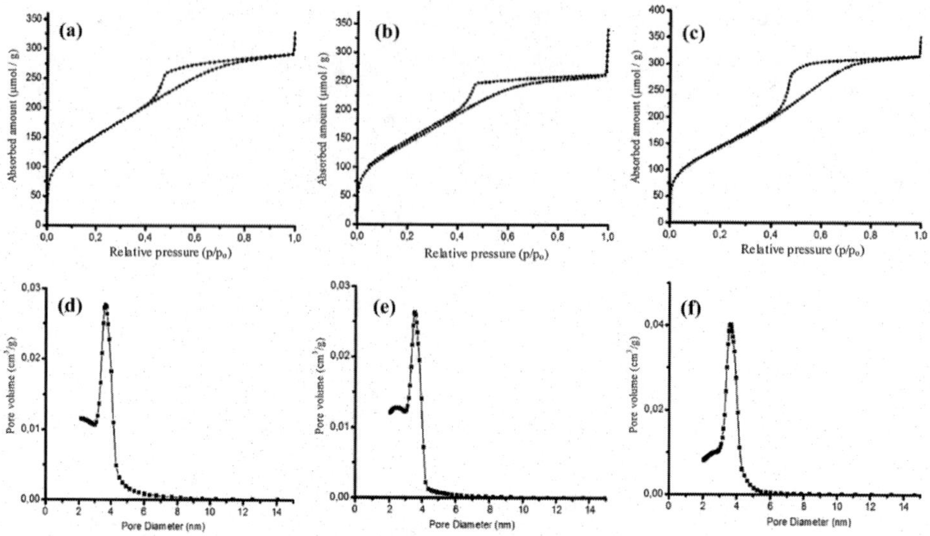

Figure 3. Nitrogen sorption isotherms and pore size distribution for aluminosilicates: a ASM-40; b ASM-80; c ASM-120; d ASM-40; e ASM-80; f ASM-120. Reprinted with permission from ref [27]. Copyright 2018 Springer Nature Switzerland AG.

Table 2 provides the characteristics of the porous structure by the adsorption–desorption data of nitrogen. Apparently, the ASM-40 sample has a microporous volume of 0.05 cm^3/g, a mesopore volume of 0.5 cm^3/g and BET specific surface of 670 m^2/g. The ASM-120 sample has a specific surface area of 520 m^2/g, and the micro- and mesopore volumes are 0.05 and 0.80 cm^3/g, respectively.

Thus, an increase of the Si/Al ratio in aluminosilicates results in an insignificant decrease in the specific surface and an increase in the mesopore volume are observed.

It is known that the porous structure of the materials obtained by the sol–gel method is largely dependent on the size, the shape of the primary particles of the sol and their links to each other [21]. Figure 4a shows the particle size distribution of the sol obtained for the ASM-80 gel before the gel point at the first step of the synthesis (in acidic medium). We can see that the size of primary particles ranges from 4 to 6 nm, similar results are observed for the ASM-40 and ASM-120 samples (Table 2).

Table 2. Effect of Si/Al on the characteristics of the porous structure

Sample	Si/Al	A_{BET}[a] (m^2/g)	V_{micro}[b] (cm^3/g)	V_{meso}[c] (cm^3/g)	Particle size[d] (nm)	Particle size[e] (nm)
ASM-40	40	670	0.05	0.50	4 - 6	10 - 15
ASM-80	80	640	0.05	0.71	4 - 6	10 - 15
ASM-120	120	520	0.05	0.80	4 - 6	10 - 15

[a] Surface area calculated by the Brunauer-Emmett-Teller (BET) method;

[b]micropore volume;

[c] mesopore volume;

[d]Determined by SAXX; [e]Determined by TEM.

[Reprinted with permission from ref. [27]. Copyright 2018 Springer Nature Switzerland AG].

Apparently, because of a relatively equal value of pH ~ 3, for all initial solutions in the first step of synthesis, there are no large differences in supersaturations in the nucleation centers, resulting in a relatively equal size of the primary particles of the sol.

Figures 4b–d show pictures for ASM-40, ASM-80 and ASM-120 samples obtained by transmission electron microscopy. One can see that the porous structure for all samples is framed by spherical particles with a size of 10–15 nm, which are not densely packed together. The voids formed between the particles are mesopores. It should be noted that particles with a size of 10–15 nm consist of smaller particles with a size of 4–6 nm. Thus, mesopores in these materials are formed between particles of similar size, which is the main reason for the formation of a narrow mesopore size distribution.

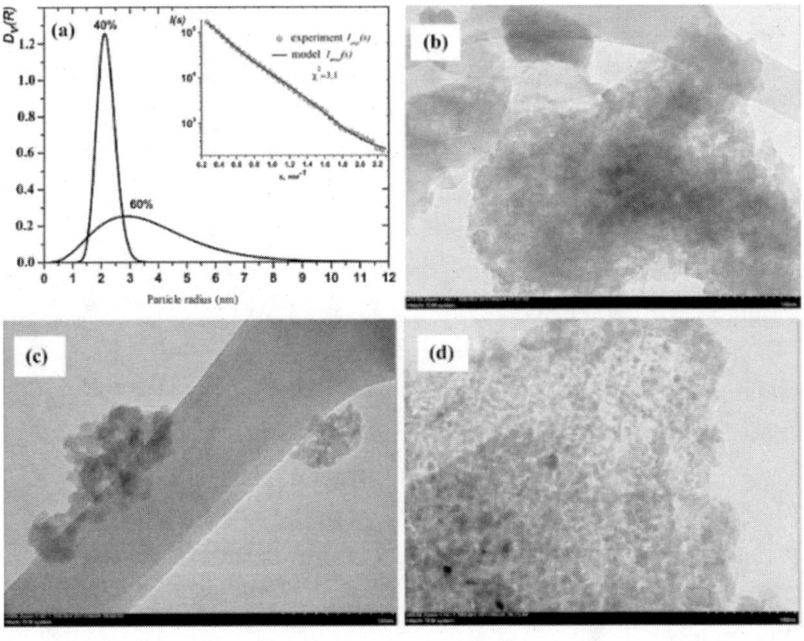

Figure 4. The particle size distribution of the sol for gel in the gel point (**a**). TEM images of aluminosilicates: ASM-40 (b); ASM-80 (c); ASM-120 (d). Reprinted with permission from ref. [27]. Copyright 2018 Springer Nature Switzerland AG.

Table 3. Type and concentration of LAS and BAS according to FTIR spectroscopy of adsorbed co and TPD NH_3

Type of acid sites	Lewis acid sites			Brønsted acid sites		TPD ΣNH_3
ν (cm^{-1})	ν_{co} = 2231-2223	ν_{co} = 2212-2210	ν_{co} = 2202- 2190	ν_{OH} = 3615	ν_{OH} = 3725-3740	150-550°C
The strength of sites		-	-	$\Delta\nu_{OH}$ = 300	$\Delta\nu_{OH}$ = 270-320	-
Sample	Concentration (μmol / g)					
ASM-40	78	19	65	8	42	554
ASM-80	58	11	50	5	23	427
ASM-120	34	8	32	3	12	324

[Reprinted with permission from ref. [27]. Copyright 2018 Springer Nature Switzerland AG].

Table 3 shows the TPD NH_3 data for the ASM-40, ASM-80 and ASM-120 samples. As shown, increasing Si/Al ratio reduces the total acidity values, which is common for all aluminosilicate systems [30]. The noteworthy feature here is that the total acidity values for the ASM-40 and ASM-80 samples are relatively equal, although the aluminum content in ASM-40 is twice as high as in ASM-80. The results obtained are due to the fact that, because of the limited possibility of introducing aluminum atoms into the silicate framework the amount of the latter in ASM-40 and ASM-80 samples is relatively equal. The remaining aluminum atoms in the ASM-40 sample are present, presumably in the form of clusters of aluminum oxide.

Figure 5a shows the IR spectrum of the ASM-80 sample in the valence vibrations region of OH groups. In the spectrum of the sample, absorption bands (a.b.) of six types of hydroxyl groups are pointed out.

The weak broad band at 3400–3650 cm^{-1} is classified as hydroxyl groups perturbed by hydrogen bonding with neighboring OH groups [31]. By analogy with zeolites, a.b. 3615–3620 cm^{-1} may be classified as acidic bridged hydroxyl groups SI–O(H)–Al. It is difficult to classify a.b. 3650 cm^{-1}. Presumably, it can belong to the category of acidic bridged hydroxyl groups SI–O (H)–Al in large pores or to Al–O(H)–Al groups [32]. The bands 3615–3620 and 3650 cm^{-1} in the presented spectrum are not clearly defined, but are clearly distinguishable in the difference spectra after adsorption of CO (Figure 5b). The triplet of a.b. 3730, 3742 and 3747 cm^{-1} characterizes the terminal Si–OH groups, and the first two points can be classified as Si–OH groups near surface defects [31]. The 3730 cm^{-1} signal presumably refers to the vibrations of the acidic Si–OH groups located in the immediate vicinity of the LAS formed by three-coordinated Al or Si atoms (Si–O(H)… $Al^{3+)}$ [32, 33]. A signal of 3742 cm^{-1} is observed in the spectrum of the sample after vacuum processing at room temperature, which suggests the non-acidity of these groups. The 3787 cm^{-1} band characterizes terminal Al–OH groups; similar absorption bands are observed for ASM-40 and ASM-120 samples.

It is known that a CO probe molecule, which is a weak base, can be used to evaluate the strength and concentration of Brunsted acid sites (BAS) of aluminosilicates. Adsorption of CO at 77 K is accompanied by a shift of a.b. OH-groups to the low-frequency region (ΔνOHOH…CO) caused by the perturbation of valence vibrations of OH groups due to the formation of a hydrogen bond with CO. At that, the larger the magnitude of the shift, the stronger the center is shifts of a.b. OH groups in the formation of the hydrogen bond and the corresponding a.b. adsorbed CO for samples of aluminosilicates are presented in Table 3.

The BAS concentration is detailed in Table 3. As expected, the ASM-40 sample exhibits the highest BAS concentration, as the Si/Al ratio increases, BAS concentrations decrease.

According to CO adsorption data (Figure 5c), 5 types of Lewis acid sites are observed on the surface of all samples. In the spectrum of the samples, there are a.b. at 2223 and

2231 cm^{-1}, characteristic of CO complexes with Al^{3+} ions in pentahedron environment being specific structure defects of aluminosilicates and zeolites; a.b. at 2210–2212 cm^{-1}, a.b. at 2202–2200 and 2190 cm^{-1}, which are characteristic of CO complex with Al^{3+} ions, apparently, clusters of aluminum oxide.

The types and concentration of LAS are presented in Table 3. In fact, with increasing Si/Al ratio, the concentration of LAS decreases, as the aluminum content in the material is reduced.

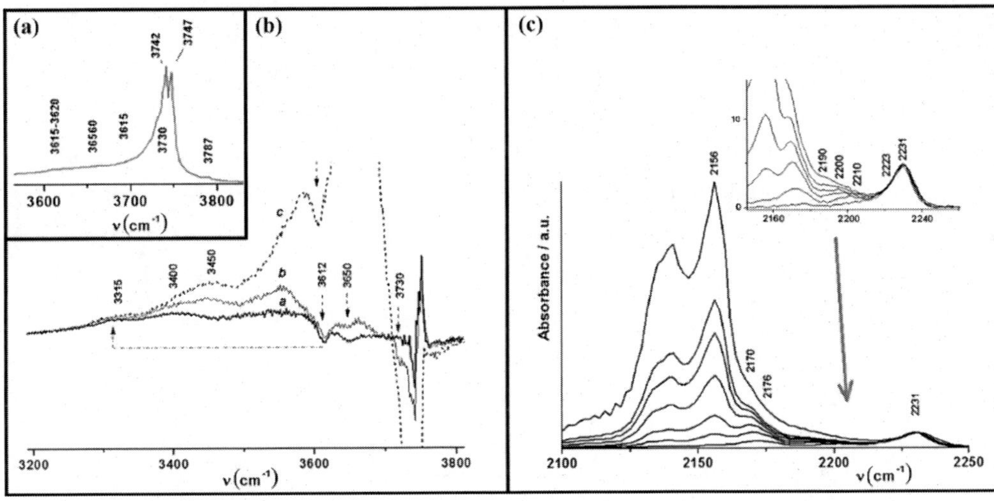

Figure 5. (a) FTIR spectra of aluminosilicate ASM-80; (b) IR difference spectra in the OH-region for the ASM-80 sample before and after the adsorption: (a) 0.2; (b) 0.3; (c) 5 TORR CO; c IR difference spectra carbonyl regions between the initial aluminosilicate ASM-80 and those with increased dosage of adsorbed CO from 0.1 (*bottom curve*) to 4 mbar (*top curve*) at liquid nitrogen temperature. Reprinted with permission from ref. [27]. Copyright 2018 Springer Nature Switzerland AG.

3. Aluminosylicates in Synthes of N-Heterocyclic Compounds

3.1. Synthesis of Pyridine and Alkylpyridines

Pyridines are the most demanded group among nitrogen-containing heterocyclic compounds. Polymer, chemical, pharmaceutical and other industries are consumers of pyridines. Over 150 years after the discovery of pyridine, hundreds of effective medicines have been found and synthesized [34], and the search for new forms of the pyridine compounds continues [35, 36]. Alkyl pyridines, specifically, α-picoline (2-methylpyridine), γ-picoline (4- methylpyridine), and 5-ethyl-2-methylpyridine, are used as a raw material for the production of 2-vinylpyridine, 4-vinylpyridine, 2-methyl-5-vinylpyridine, on the basis of which latexes for tire cord impregnation, ion exchange resins, and photographic materials are made [37-39]. Alkyl pyridines are also raw materials for the production of chemical plant protection products against weeds, pests, and various diseases [40, 41]. Extractants, inhibitors of metal corrosion, solvents, surfactants, rubber vulcanization accelerators, and other products were synthesized on the basis of substituted pyridines [28, 42].

The most common synthetic route for the production of various pyridines -the reaction of aldehydes or ketones with ammonia. This reaction was first studied in detail by A. E. Chichibabin in 1924 and since then been studied extensively for industrial manufacturing because of cheap access to raw materials. The reaction is usually carried out 350-550°C and a space velocity of 500-1000 h^{-1} in the presence of a solid acid catalyst, e.g., silica-alumina, zeolites.

We synthesized pyridine and its methyl derivatives by the reaction of ethanol with formaldehyde and ammonia first carried out by Kulkarni under the action of zeolites ZSM-5 [43]. Both on zeolite catalysts and on amorphous mesoporous aluminosilicate ASM samples, the main products of this reaction are pyridine (**1**), picolines (2-, 3-, and 4-methylpyridines) (**2**), and lutidines (dimethylpyridines) (**3**) (Figure 6).

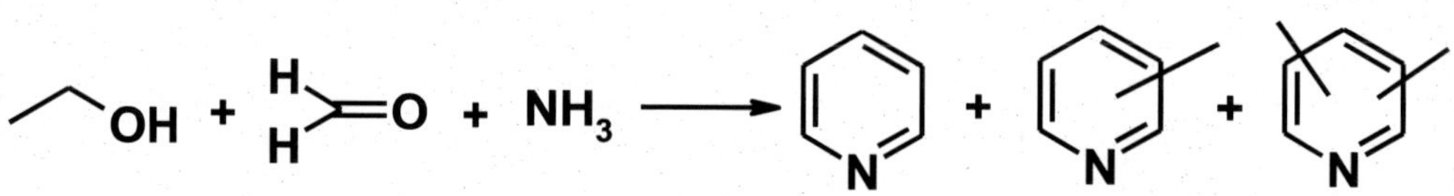

Figure 6. Synthesis of pyridines by the interaction of ethanol, formaldehyde and ammonia.

Table 4. Synthesis of pyridines over mesoporous aluminosilicates ASM by the reaction of ethanol, formaldehyde, and ammonia

Catalyst	T, °C	Alcohol conversion, %	Selectivity, %				Reference
			Pyri-dine (I)	Pico-lines (II)	Lutidines (III)	Heavy products	
ASM-40	300	50	19	57	18	6	[44]
ASM-80	300	30	20	56	22	2	[44]
ASM-120	300	10	24	49	25	2	[44]
ASM-40	400	80	18	47	24	11	[44]
H-ZSM-5** (Si/Al = 150)	420	72.5	53	32	15	Not determined	[43]
Pb-ZSM-5**	400	44.9	53	32	15	Not determined	[43]

Reaction conditions: molar ratio C_2H_5OH : CH2O : NH3 = 1.0 : 0.8 : 1.5, w = 7 h–1.

**w = 0.5 h–1.

In the composition of picolines, 3-methylpyridine accounts for 80%. In the composition of lutidines, 3,5-isomer prevails. Also, compounds that are heavier than dimethylpyridines are formed.

The results of the study of the catalytic properties of mesoporous aluminosilicates ASM (with molar ratios of Si/Al = 40, 80, and 120), differing in acid characteristics in the reaction of ethanol, formaldehyde, and ammonia, are given in Table 4 [44].

It was found that the ASM-40 sample is the most active in the reaction. The conversion of ethanol on mesoporous aluminosilicate samples decreases in the series ASM-40 > ASM-80 > ASM-120, which is associated with a decrease in the concentration of acid sites due to a decrease in the fraction of aluminum atoms in the aluminosilicate framework. The data obtained show that the catalytic systems ASM are active in the studied reaction only if they contain a certain number of strong acid sites. A decrease in the BAS concentration below a certain level leads to the loss of catalyst activity, which cannot be compensated by the accessibility of sites.

In the reaction products formed under the action of the studied catalysts, picolines (mainly 3-methylpyridine) have the highest content: 49–57%. Picolines are formed most selectively on the samples of mesoporous aluminosilicates with Si/Al molar ratios of 40 and 80. An increase in the ratio of framework atoms of Si/Al aluminosilicate samples from 40 to 120 leads to an increase in the selectivity to pyridine and lutidines, and the selectivity to picolines and heavy compounds decreases.

We compared the activity and selectivity of mesoporous aluminosilicates ASM with the properties of microporous catalysts: H-ZSM-5 and Pb-ZSM-5 zeolites [43]. It can be seen from the data presented in Table 4 that the conversion of ethanol at similar temperatures (400 and 420°C) on ASM-40 sample is slightly higher than on the most active zeolite H-ZSM-5. Taking into account that experiments on the ASM-40 samples were carried out at a higher feed space velocity of (7 h^{-1}) than on H-ZSM-5 and Pb-ZSM-5 zeolites (0.5 h^{-1}) and that alcohol conversion increases with a decrease in *w*, we can speak quite confidently about the high activity of mesoporous aluminosilicates ASM in the synthesis of pyridines by the reaction of ethanol, formaldehyde, and ammonia.

When comparing the composition of products formed on zeolite catalysts and amorphous aluminosilicates, it is clear that the reaction mass obtained on mesoporous aluminosilicate is rich in methylpyridines: picolines and lutidines. The pyridine: picolines: lutidines ratio in the products synthesized in the presence of H-ZSM-5 and Pb-ZSM-5 zeolites is 1 : 0.6 : 0.3, while for the ASM-40 sample, this ratio is 1 : 3 : 1.

The presented data convincingly show the applicability of amorphous mesoporous aluminosilicates with acidic properties in the synthesis of pyridines by the reaction of ethanol, formaldehyde, and ammonia. The activity of ASM-40 mesoporous aluminosilicate sample is not inferior to the H-ZSM-5 and Pb-ZSM-5 zeolite catalysts, and the total selectivity of the formation of picolines and lutidines on it reaches 75–78%.

The reaction of other alcohols (*n*-propanol and *n*-butanol) with formaldehyde and ammonia occurs in the presence of mesoporous aluminosilicates ASM to form mainly 3,5-dialkylpyridines (Figure 7).

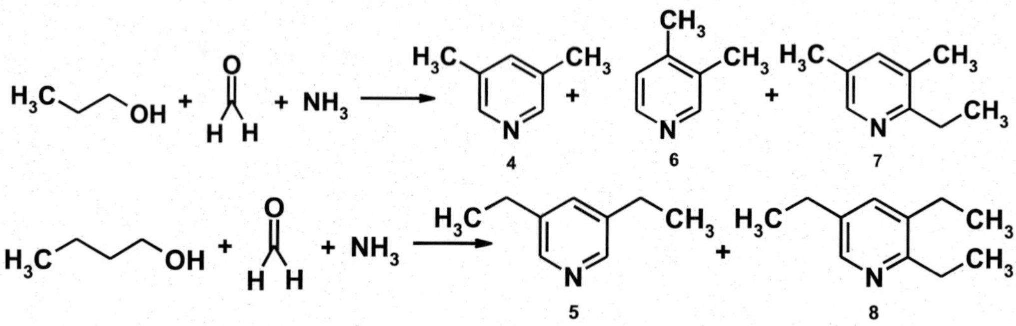

Figure 7. Synthesis of 3,5-dialkylpyridines over mesoporous aluminosilicate catalysts.

In the case of *n*-propanol, 3,5-lutidine (**4**) is the main product of the reaction. In the case of *n*-butanol, 3,5-diethylpyridine (**6**) is the main product. In addition to dialkylpyridines (**4**) and (**6**), light compounds are also present in the reaction products, among which both linear condensation products of alcohol, formaldehyde, and ammonia (imines, propyl, and butylamines, acetal) and 3-methylpyridine are identified. In the reaction with *n*-propanol, 3,4-lutidine (**5**) and a mixture of trialkylpyridines with a predominant content of 2-ethyl-3,5-dimethylpyridine (**7**) are formed along with 3,5-lutidine (**4**). In the reaction with *n*-butanol, the main by-product is 2-propyl-3,5-diethylpyridine (**8**).

Data presented in Table 5 show that the ASM-40 sample has a high selectivity in the synthesis of 3,5-dialkylpyridines. In the reaction of *n*-propanol with formaldehyde and ammonia at 300°C, the selectivity to 3,5-lutidine is 97% with an alcohol conversion of 25%.

As the temperature increases to 400°C, alcohol conversion increases to 70%, but the amount of by-products in the reaction mass increases, as a result of which the selectivity to dialkylpyridine decreases. Similar dependences were observed in the interaction of *n*-butanol with formaldehyde and ammonia: with increasing temperature from 300 to 400°C, the conversion of alcohol increases from 22 to 62%, and the selectivity to 3,5-diethylpyridine drops from 86 to 63%.

Further elongation of the hydrocarbon chain of the alcohol (*n*-pentanol) worsens the reaction performance: even at 400°C, the conversion of alcohol does not exceed 20%, and a complex mixture of products is formed, in which the content of 3,5-dipropylpyridine is 40%.

Table 5. Synthesis of 3,5-dialkylpyridines over ASM-40 mesoporous aluminosilicate and zeolite catalysts

Catalyst	Alco-hol	T,°C	Alcohol conversion, %	Selectivity, %				Reference
				light products	3,5-di-alkylpyri dines	other di- and trialkyl-pyridines	heavy products	
ASM-40	n-propanol	300	25	3	97	–	–	[44]
ASM-40	n-propanol	350	45	5	84	11	–	[44]
ASM-40	n-propanol	400	70	6	79	15	–	[44]
ASM-40	n-butanol	300	22	3	86	11	–	[44]
ASM-40	n-butanol	400	62	11	63	26	–	[44]
ASM-40	n-pentanol	400	20	25	40	25	10	[44]
HY	n-propanol	400	16	11	63	23	3	[44]
HY	n-buta-nol	400	14	30	41	28	1	[44]
La-ZSM-5**	n-propanol	400	88.6	25.1	72.2	2.7	Not deter-mined	[45]
H-ZSM-5***	n-butanol	400	91.2	12	69	19	Not deter-mined	[46]

Dashes mean that the corresponding products are not formed.

*Reaction conditions: molar ratio ROH : CH_2O : NH3 = 1.0 : 0.8 : 3.0; 400°C, w = 7 h–1.

**C3H7OH : CH_2O : CH3OH : NH3 = 1.0 : 0.4 : 0.9 : 6, 400°C, w = 0.5 h–1.

***C4H9OH : CH_2O : NH3 = 1 : 1 : 6, 400°C, w = 0.5 h–1.

Comparison of the catalytic properties of the ASM-40 sample and microporous HY zeolite with a SiO_2/Al_2O_3 molar ratio of 5.0 (our data) in the interaction of n-propanol/n-butanol with formaldehyde and ammonia showed that the selectivity to dialkylpyridines and the alcohol conversion are higher on the ASM-40 sample (Table 5). The appearance of a large number of by-products on the HY zeolite can be associated with a significant concentration of nonuniform active sites in it, including strong Brønsted acid sites.

When comparing the results obtained in this work with the literature data, it can be seen (Table 5) that mesoporous aluminosilicate ASM is close to the La-ZSM-5 zeolite (the reaction of *n*-propanol with formaldehyde, methanol, and ammonia, [45]) and the H-ZSM-5 zeolite (the interaction of *n*-butanol with formaldehyde and ammonia [46]) on the selectivity to dialkylpyridins (**4**) and (**5**), but its activity is lower.

Synthesis of 2-methyl-5-ethylpyridine. One of the most important industrial reactions is the interaction of acetaldehyde with ammonia to form 2-methyl-5-ethylpyridine (**10**), which is the main raw material in the production of nicotinic acid and 2-methyl-5-vinylpyridine (Figure 8).

It is known that in industry, methyl ethyl pyridine (**10**) is produced by the reaction of paraldehyde (acetaldehyde trimer) with ammonia in the liquid phase at 200–250°C and 5–10 MPa in the presence of ammonium salts (e.g., ammonium acetate) or acetic acid. The yield of methyl ethyl pyridine in this reaction reaches 60% [28].

The reaction of acetaldehyde with ammonia on the ASM-40 sample occurs with the formation of predominantly methyl ethyl pyridine (**10**), the selectivity to which under these conditions is 88% with an acetaldehyde conversion of 60%.

2-Methylpyridine (**9**) is found in a small amount. On the microporous HY zeolite, with which we compared the ASM-40 sample, the conversion of acetaldehyde is higher: 69%. This is due to the higher acidity of the HY catalyst (Table 6). Apparently, for the same reasons the selectivity of HY zeolite is 1.8 times lower than that on the ASM-40 sample and a significant amount of heavy compounds appear, which are different di- and trialkylpyridines. A comparison of our results with literature data shows that the yield of methyl ethyl pyridine (**10**) (53%) is close to that observed in industrial units.

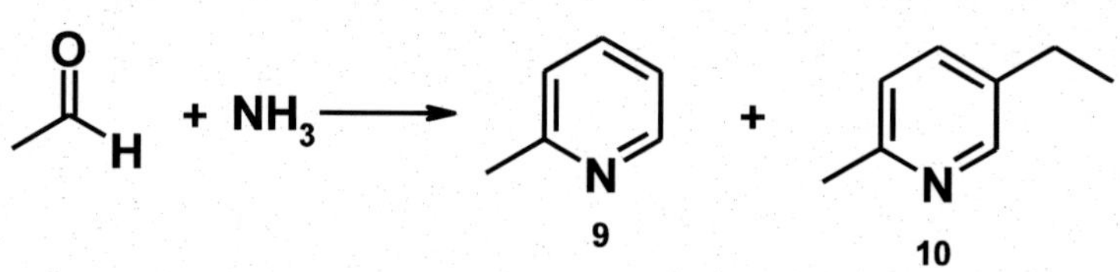

Figure 8. The interaction of acetaldehyde with ammonia over the ASM-40 and H-Y zeolite samples.

Table 6. The interaction of acetaldehyde with ammonia over aluminosilicate ASM-40

Sample	Conversion of acetaldehyde, %	Selectivity, %		
		2-Methyl-Pyridine (9)	2-methyl-5-ethylpyridine (10)	«heavy»
ASM-40	60	3	88	9
H-Y	69	5	55	40

Acetaldehyde: NH3 = 1:3, 150 oC, 10% wt. of catalyst.

The reaction of propionic aldehyde with ammonia leads to the formation of a mixture of mono-, di- and trialkylpyridines, among which 2-ethyl-3,5-dimethylpyridine (**7**) (Figure 9) is the main product.

In addition to 2-ethyl-3,5-dimethylpyridine (7), the following products were identified: imines (designated as light products), pyridine (**1**), 3- and 4-methylpyridines, dimethylpyridines, 3,5-dimethyl-4-ethylpyridine, and compounds heavier than 4-ethyl-3,5-dimethylpyridine (**11**). The conversion of propionic aldehyde and the composition of products formed on various samples of zeolite catalysts are given in Table 7.

Under conditions studied (150–300°C, 7 h^{-1}), the conversion of propionic aldehyde on the ASM-40 sample reaches 70–91%, the selectivity to 2-ethyl-3,5-dimethylpyridine (**7**) is 40–41%, and the amount of heavy compounds does not exceed 10%. The highest selectivity to 2- and 4-ethyl-3,5-dimethylpyridines (**7**) and (**11**) was observed at 300°C (their overall selectivity was 57%).

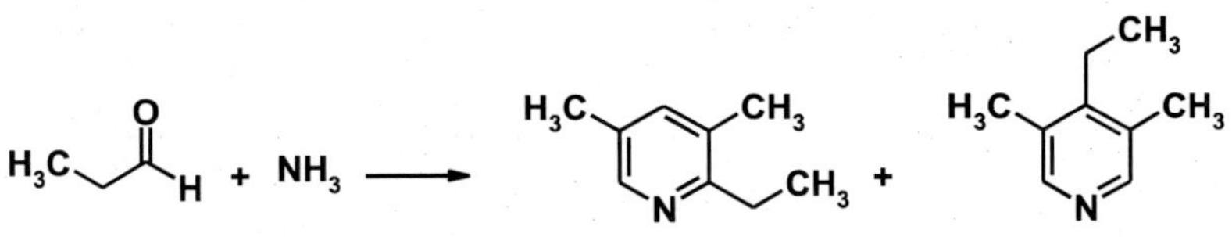

7

11

Figure 9. Synthesis of 2-ethyl-3,5-dimethylpyridine in the presence of mesoporous catalysts.

Table 7. Synthesis of 2-ethyl-3,5-dimethylpyridine in the presence of ASM aluminosilicate and zeolite catalysts

Reaction condition		Con	Selectivity (%)					
Sample	T (°C)	Version of propanal (%)	«Light»	**(1, 2)**	2,6-Lytidine	(**7**)	(**11**)	«Heavy»
H-Y	300	85	26	8	-	43	8	15
ASM-40	150	70	23	15	10	40	9	3
ASM-40	300	91	24	4	5	41	16	10

[a] molar ratio of propanal:ammonia = 1: 3; 200-350°C; 7 h^{-1}.

On microporous HY zeolite at 300°C, aldehyde conversion is lower (85%) and the reaction occurs with the formation of a larger amount of heavy compounds, which is apparently associated with the increased acidity of HY zeolite.

It was of interests to compare the results obtained with those already known. It turned out that the reaction of propionic aldehyde with ammonia under the action of Pb-ZSM-5 zeolite mentioned in [45] takes place at 400°C and 0.5 h^{-1} with an aldehyde conversion of 69.5%. The authors of [45] describe the product composition as a mixture of 2,6-lutidine (10%), 3,5-lutidine (15%), pyridine, and collidines (75%). Information on the selective formation of 2-ethyl-3,5-dimethylpyridine (**7**) is missing.

3.2. Synthesis of Quinoline Derivatives

Quinoline derivatives serve as important intermediates in the synthesis of biologically active compounds and herbicides [47]. A number of antimalarial [48, 49], antibacterial, antiseptic, anti-tuberculosis drugs [50], anticancer drugs [51] based on quinolines are known. The quinoline fragment is found in some commonly used antibiotics [52].

Most often, for the synthesis of quinolines, the Skraup, Döbner–Von Miller, Friedländer, and Combes methods are used with acid or base catalysts. Most of the proposed methods are characterized by several stages, severe reaction conditions, and low yields of the target products.

We attempted to synthesize dialkylquinolines in the presence of the ASM-40 sample by the reaction of aniline with aldehydes (propionic and butyric). It was found that the products of this reaction are N-alkylanilines (**12a**, **12b**), 2,3-dialkylquinolines (**13a**, **13b**), and 2,3-dialkyl-*N*-phenyl-1,2,3,4-tetrahydroquinolin-4-amines (**14a**, **14b**) (Figure 10). In addition to compounds (**12**)–(**14**), aldehyde condensation products, dialkyl dihydroquinolines, and heavy compounds are present in the reaction mass. They are indicated in Figure 11, like "others."

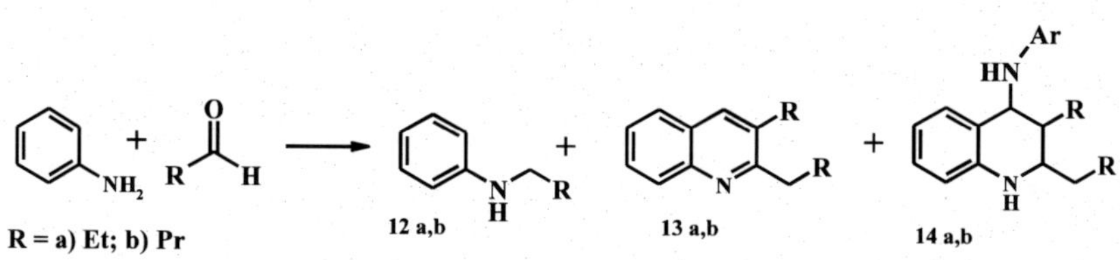

Figure 10. Synthesis of dialkylquinolines by the reaction of aniline with C_3-C_4 aldehydes over zeolite catalysts and mesoporous aluminosilicates.

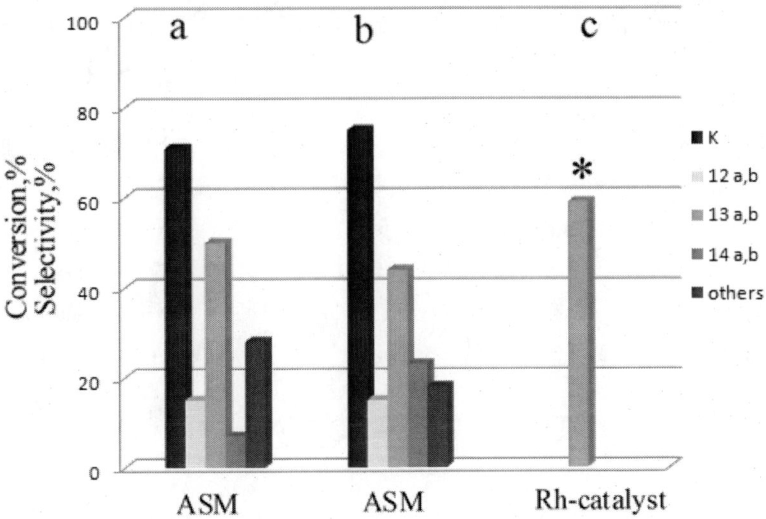

Figure 11. Interaction of aniline with aldehydes (C_3-C_4) in the presence of mesoporous aluminosilicate ASM-40. a)ASM, propanal; b)ASM, butabal; c) Rh(norbornadiene)Cl_2, propanal [53] (molar ratio of aniline:aldehyde = 1:2; 160°C, 20% wt. of the catalyst, the solvent - chlorobenzene). K- conversion of aniline, The selectivity of the formation of N-alkylanilines (12a, b), 2,3-dialkylquinolines (13a, b) and 2,3-dialkyl-N-phenyl-1,2,3,4-tetrahydroquinolin-4-amines (14a, b), respectively (12), (13) and (14). *-Yield.

Under the conditions studied, the conversion of aniline upon interaction with propionic and butyric aldehydes is rather high: 71 and 75%, respectively (Figure 11).

In the composition of the reaction products with propionic aldehyde, the main compound is dialkylquinoline (**13a**), the selectivity to which reaches 50%. Two main products, dialkylquinoline (**13b**) and tetrahydroquinoline (**14b**), were found in a ratio of about 2 : 1 in the reaction of aniline with butyric aldehyde. The selectivity to N-alkylanilines (**12a**, **12b**) in both reactions was 15%.

A comparison of the yield of dialkylquinoline (**13a**) obtained in the presence of the ASM-40 sample (39%) and [Rh(norbornadiene)Cl]$_2$ (59%) [53] showed that Rh catalytic systems are more effective in the reaction of aniline with propionic aldehyde.

The dihydroquinoline cycle is a compound in the structure of many natural and synthetic compounds with biological activity [54,55]. For example, compounds based on 2,2,4-substituted 1,2-dihydroquinolines exhibit antibacterial, antidiabetic, or anti-inflammatory properties [56-58]. In addition, derivatives of dihydroquinolines are inhibitors of lipid peroxidation, HMG-CoA reductase, bile acid transporter inhibitors, progesterone agonists and antagonists [59]. The use of 1,2-dihydroquinoline oligomers (di-, tri- and tetramers) as antioxidants of rubbers is known [60].

One of the simplest and most convenient methods for producing dihydroquinolines is acid-catalyzed interaction of aniline with ketones [59-64]. The reaction of aniline with acetone in the presence of ASM-40 mesoporous aluminosilicate proceeds with the formation of 2,2,4-trimethyl-1,2-dihydroquinoline (**15**) and N-phenyl-2-propanimine (**16**) (Figure 12).

The reaction proceeds with high conversion of aniline and selectivity of the formation of dihydroquinoline **15** to 49% over the amorphous aluminosilicate ASM-40 [65]. The main product of the reaction of aniline with acetone over zeolite catalyst H-Y is N-phenyl-2-propanimine **16**, which is due to the manifestation of the molecular sieve effect of the zeolite crystal framework.

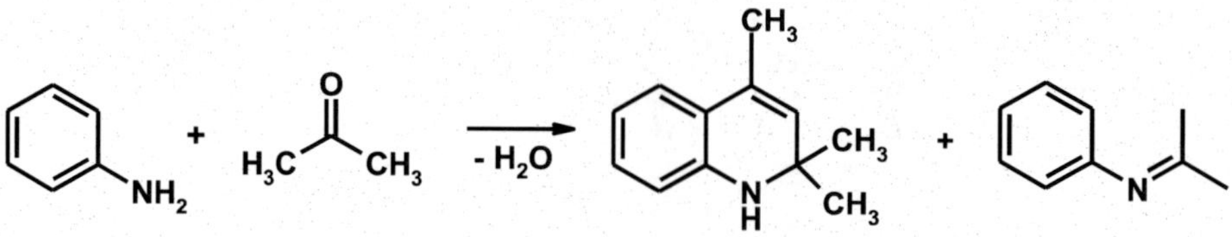

Figure 12. The interaction of aniline with acetone in the presence of aluminosilicates.

Table 8. Interaction of aniline with acetone in the presence of ASM mesoporous aluminosilicate and zeolites

Catalyst	Aniline conversion, %	Selectivity of formation, %		
		15	16	Other products
ASM-40	85	49	47	4
H-Y	11	7	88	5

Molar ratio of aniline: acetone = 1: 5; 10% catalyst; 230 °C; 24 h.

Obviously, only linear condensation of aniline with acetone can take place in the micropores of zeolites, leading to the formation of imine (**16**). The formation of bulky molecule of dihydroquinoline **15** inside the channels or cavities of the microcrystalline framework of zeolite catalysts is difficult and it is possible, apparently, only on surface acid centers. It should be noted the low activity of zeolite H-Y, in the presence of which the conversion of aniline in 24 hours does not exceed 11%. The reason for the low activity of zeolite H-Y can not only be sought in steric hindrance for the diffusion and interaction of reagents. The blocking of large cavities by the formed molecules of dihydroquinoline **15** can be also considered when one bears in mind that the diameter of the entrance windows to a large cavity is smaller than the diameter of the cavity itself (0.75 and 1.2 nm, respectively) [66].

The reaction of aniline with acetophenone on amorphous aluminosilicates occurs with the formation of three main products: (E)-*N*-(1-phenylethylidene)aniline (**17**), (1E, 2E)-*N*-(1,3-diphenylbut-2-en-1-ylidene) aniline (**18**) and 2-methyl-2,4-diphenyl-1,2-dihydroquinoline (**19**) (Figure 13).

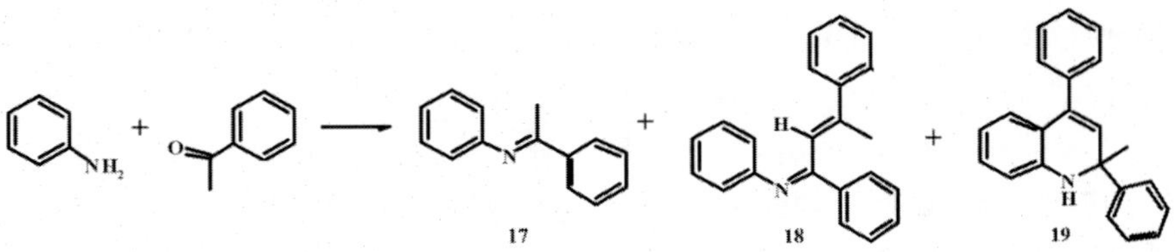

Figure 13. The interaction of aniline with acetophenone in the presence of ASM aluminosilicate and H-Y zeolite.

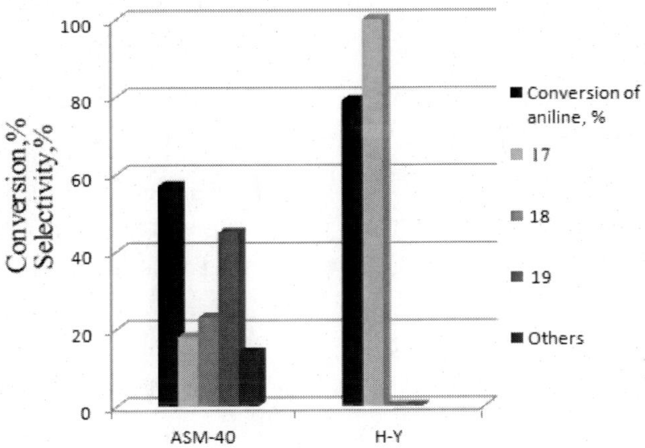

Figure 14. Interaction of aniline with acetophenone in the presence of ASM-40 mesoporous aluminosilicate and zeolite catalyst (molar ratio of aniline: acetophenone = 1: 2; 30% wt. catalyst; 130° C; 5 h).

Also, acetophenone aldol condensation can occur with the formation of 2(Z, E)-1,3-diphenylbut-2-ene-1-one and the trimerization of acetophenone in the presence of aniline with the formation of 1,3,5-triarylbenzene, indicated in Table 7 as "others."

In the reaction of aniline with acetophenone, the ASM-40 sample is less active than in the reaction with acetone, as evidenced by a lower conversion of aniline (57 and 85%, respectively). The selectivity to dihydroquinoline (**19**) is 45%. The imines (**17**) and (**18**) formed are the products of the interaction of aniline with both the initial acetophenone and acetophenone dimer, which are formed in parallel in the reaction mass under the conditions studied.

For comparison, in Figure 14 the results obtained in the presence of a microporous catalyst H-Y are presented. It is seen that both imine (**18**) and dihydroquinoline (**19**) are not formed on zeolite H-Y. The only product in this case is the Schiff base (**17**), which is formed as a result of the condensation of aniline with acetophenone.

The results obtained indicate a significant dependence of the activity and selectivity of the studied catalysts on their textural characteristics. Since most of the active centers in zeolites (strongly acidic structural OH groups) are known to be located inside channels or cavities, their accessibility to reacting molecules is the main factor determining catalytic properties. In the mesopores of the ASM-40 sample, steric hindrances to the formation of more bulky molecules than imine (**17**) are absent. Therefore, not only linear condensation of aniline with acetophenone, but also other reactions take place: condensation of aniline with acetophenone dimer and the subsequent heterocyclization of compound (**18**) leading to the formation of dihydroquinoline (**19**).

3.3. Synthesis of Benzimidazoles

Benzimidazole derivatives have a wide spectrum of pharmacological and biochemical activity, and they are used to obtain a number of vital pharmaceuticals with antibacterial, anti-inflammatory, antiulcer, antiviral, proto-tumor, hypotensive, analgesic, anti-spasmolytic properties [67, 68]. The use of benzimidazoles in the manufacture of corrosion inhibitors [69], agrochemicals [70], sensitizers of photographic materials [71, 72], chemosensors [73], optical brighteners [74], electrolytes in solar batteries [75], and single-component ferroelectrics [76] is known.

The most common method for the synthesis of benzimidazole derivatives is the condensation of o-aryl diamines with aldehydes or carboxylic acids in the presence of homogeneous and heterogeneous acid catalysts. We have studied the catalytic properties of the mesoporous aluminosilicate ASM-40 and microporous catalyst - zeolite H-Y in the reaction of o-phenylenediamine with p-chlorobenzaldehyde. It was established that, depending on the conditions, imine (**20**), monosubstituted benzimidazole (**21,** MBI) and di-substituted benzimidazole (**22,** DBI) may be present in the reaction products, as shown in Figure 15.

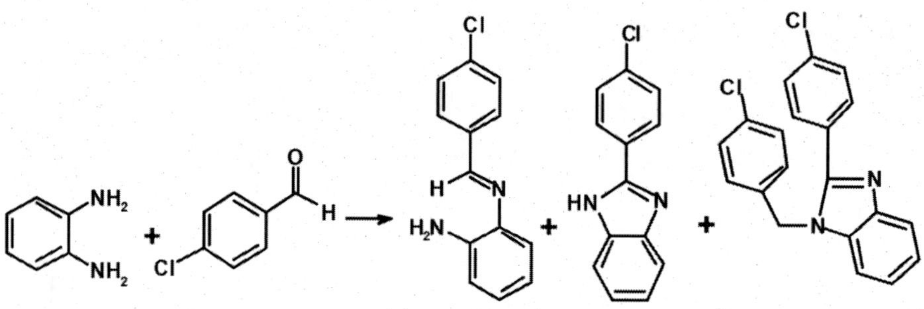

20 21 22

Figure 15. Reaction of o-phenylenediamine with para-chlorobenzaldehyde.

The most selective reaction takes place in the presence of mesoporous aluminosilicate ASM-40. So, under the indicated conditions (Table 9), the only reaction product is monosubstituted benzimidazole (**21**), while the conversion of o-phenylenediamine is 93%.

Table 9. Synthesis of benzimidazoles over ASM aluminosilicate and H-Y zeolite

Catalyst	o-phenylenediamine conversion, %	Selectivity of formation,%		
		(21)	(22)	(23)
ASM-40	93	-	100	-
H-Y	98	31	63	6

Molar ratio of o-phenylenediamine: aldehyde = 1: 1.5, 10 wt.% catalyst, 40° C, ethanol, 5 h.

On a microporous sample of H-Y with high conversion of diamine, the selectivity of formation monosubstituted product **21** is significantly lower - 63%. In addition to compound **21**, the imine (**20**) and disubstituted benzimidazole (**22**) are contained in the reaction mass.

3.4. Synthesis of 2,2,4-trimethyl-2,3-dihydro-1H-1,5-benzodiazepine

1,5-Benzodiazepines present a family of important nitrogen-containing heterocycles. Due to the wide spectrum of biological activity, benzodiazepines are the basis of many pharmacologically active compounds, which are used as anti-inflammatory, anticonvulsant, sedative and hypnotic drugs [77-79]. Over the past decade, they have been actively used to treat neuropsychiatric disorders, cancer, viral infections, and cardiovascular diseases.

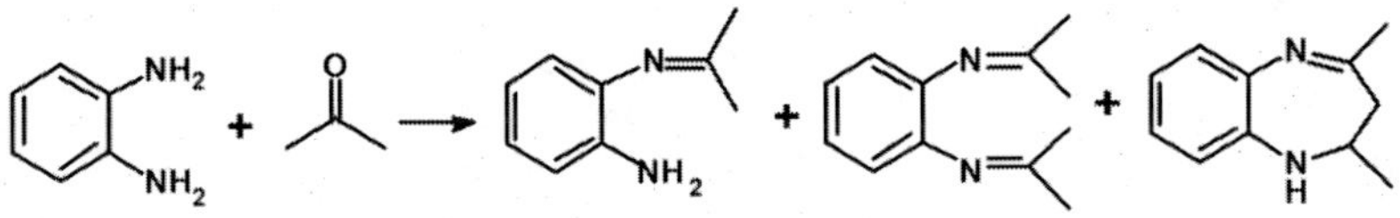

23 24 25

Figure 16. The reaction of 1,2-phenylenediamine with acetone.

In addition, derivatives of 1,5-benzodiazepines are used as dyes for acrylic fibers in photograms [80] and as synthons for other condensed ring systems, such as triazolo-, oxadiazolo-, oxazino- or furanobenzodiazepines [81].

The main method for producing 1,5-benzodiazepines is the acid-catalyzed condensation of 1,2-phenylenediamine with ketones.

In the presence of the catalysts studied by us - mesoporous aluminosilicate ASM-40 and zeolite HY, the reaction of o-phenylenediamine with acetone proceeds with the formation of monoimine (**23**), diimine (**24**), 2,2,4-trimethyl-2,3-dihydro-1H-1, 5-benzodiazepine (**25**) shown in Figure 16.

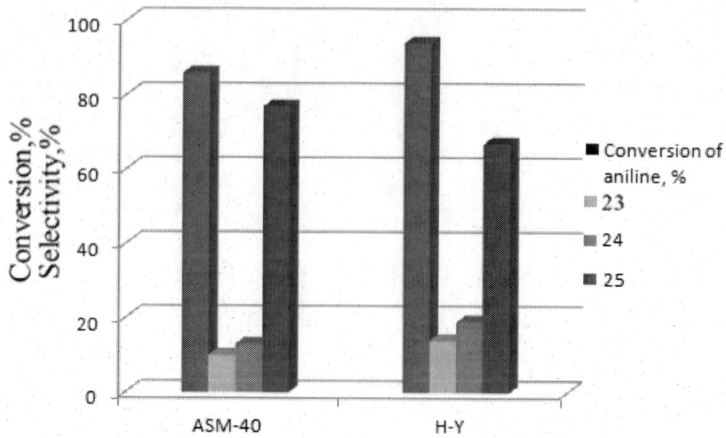

Figure 17. Synthesis of 1,5-benzodiazepine on aluminosilicate ASM-40 and zeolite H-Y (molar ratio of o-phenylenediamine: acetone = 1: 5, 10 wt.% catalyst, 50°C, methanol, 5 h).

A comparison of the selectivity of the action of the mesoporous aluminosilicate ASM-40 and zeolite HY shows (Figure 17) a higher efficiency of the aluminosilicate ASM-40 in the synthesis of 1,5-benzodiazepine (**25**): the selectivity of compound (**25**) over ASM-40 reaches 77%, herewith selectivity is 67% over zeolite HY. The observed small difference in the conversion of o-phenylenediamine (86% on ASM-40 and 94% on H-Y) is apparently due to the fact that the concentration of acid sites contained in ASM-40 is significantly lower than in the zeolite catalyst H-Y.

CONCLUSION

As a result of studying the catalytic properties of amorphous mesoporous aluminosilicates ASM with different Si/Al molar ratios (40, 80, and 160) in the synthesis of basic N-heterocyclic compounds (pyridines, quinolines, benzimidazoles, benzodiazepines) it was found that the ASM-40 sample (Si/Al = 40) it is most effective in the studied cyclocondensation reactions of alcohols/carbonyl compounds with ammonia/amines. This sample has the maximum concentration of acid sites.

The interaction of ethanol with formaldehyde and ammonia in the presence of aluminosilicate ASM-40 sample takes place with an alcohol conversion of 50–80% at 300–400°C. The reaction mass obtained on samples of mesoporous aluminosilicates is characterized by a high content of methylpyridines (up to 57% picolin and 24% lutidines) and a low content of heavy compounds.

The main product of the reaction of *n*-propanol with formaldehyde and ammonia over the ASM-40 sample is 3.5-lutidine, which is formed with a selectivity of up to 97% at a conversion of 25% (300°C). The interaction of *n*-butanol with formaldehyde and ammonia over HY zeolite led to the synthesis of 3,5-diethylpyridine with a selectivity of 86% at a conversion of 22% (300°C).

The aluminosilicate ASM-40 sample enables the formation of 5-ethyl-2-methylpyridine in the reaction of acetaldehyde with ammonia with a selectivity of 88% at an aldehyde conversion of 60%. The interaction of propionic aldehyde with ammonia over the aluminosilicate ASM-40 sample occurs with the formation of 2- and 4-ethyl-3,5-dimethylpyridines, the overall selectivity to which is 57% at an aldehyde conversion of 91%.

Aniline reacts with aldehydes (C_3, C_4) over ASM-40 with the formation of 2,3-dialkylquinolines and 2,3-dialkyltetrahydroquinolines with an overall selectivity of 57–67%.

It was found that the main products of the reaction of aniline with ketones (acetone and acetophenone) in the presence of aluminosilicate ASM-40 are alkyl(aryl)-1,2-dihydroquinolines (**15**, **20**), the selectivity of which is 45-49%. In the presence of microporous zeolite H-Y, the formation of dihydroquinoline bulky molecule is difficult, therefore, the main reaction products are Schiff bases.

The high activity and selectivity of ASM-40 aluminosilicate in the synthesis of benzimidazoles by the reaction of phenylenediamine with aldehydes (p-chlorobenzaldehyde, hexanal) have been established. The catalyst provides 97% yield of monosubstituted benzimidazoles.

The reaction of o-phenylenediamine with acetone over ASM-40 aluminosilicate proceeds with the formation of 2,2,4-trimethyl-2,3-dihydro-1H-1, 5-benzodiazepine, the selectivity of which reaches 77%.

The high activity and selectivity of the mesoporous aluminosilicate ASM in the synthesis of bulky molecules of alkylpyridines and alkylquinolines compared to microporous zeolites is achieved due to the presence of mesopores in its structure, which increase the accessibility of acid sites. Such a mesoporous structure makes it possible to reduce the diffusion hindrances for the movement of reactant molecules and reaction products, and to reduce pore blocking. In addition, in the mesopores, the concentration of reacting molecules may be higher, which leads to an increase in the number of chemical interaction acts of the reactants and an increase in their conversion. Another factor favoring the selective formation of target products may be the presence of a small amount of acid sites of aluminosilicate ASM with a lower strength than in zeolite catalysts, which prevents the occurrence of side reactions with the formation of significant amounts of heavy compounds and coke leading to catalyst deactivation.

REFERENCES

[1] Derouane, E. G. (2006). *Catalysts for Fine Chemical Synthesis: Microporous and Mesoporous Solid Catalysts*. Wiley, Chichester, 4:1-258.

[2] Tüysüz, H. and Schüth, F. (2012). Ordered mesoporous materials as catalysts. *Adv. Catal.,* 55:127-239.

[3] Trong, On. D., Desplantier-Giscard, D., Danumah, C., and Kaliaguine, S. (2001). Perspectives in catalytic applications of mesostructured materials. *Appl. Catal. A Gen.* 222:299-357.

[4] Carmo, Jr. A. C., de Souza, L. K. C., da Costa, C. E. F., Longo, E., Zamian Jose R. and da Rocha Filho G. N. (2009). Production of biodiesel by esterification of palmitic acid over mesoporous aluminosilicate Al-MCM-41. *Fuel,* 88(3):461-468.

[5] Corma, A. (1997). From Microporous to Mesoporous Molecular Sieve Materials and Their Use in Catalysis, *Chem. Rev.* 97:2373-2420.

[6] Chen X., Huang L., Ding G. and Li Q. (1997). Characterization and catalytic performance of mesoporous molecular sieves Al-MCM-41 materials. *Catal. Lett.,* 44(1-2):123-128.

[7] Corma, A., Fornes, V., Navarro, M. T. and Perez-Pariente, J. (1994). Acidity and Stability of MCM-41 Crystalline Aluminosilicates. *J. Catal.,* 148(2):569-574.

[8] Bhange, P., Bhange, Deu S., Pradhan, S. and Ramaswamy, V. (2011). Direct synthesis of well-ordered mesoporous Al-SBA-15 and its correlation with the catalytic activity. *Appl. Catal. A Gen.,* 400(1-2):176-184.

[9] Kumaran, G. M., Garg, S., Soni, K., Kumar, M., Gupta, J. K., Sharma, L. D., Rama Rao, K. S., and Dhar, G. M. (2008). Synthesis and characterization of acidic properties of Al-SBA-15 materials with varying Si/Al ratios. *Microporous Mesoporous Mater*, 114:103-109.

[10] Agliullin, M. R., Danilova, I. G., Faizullin, A. V., Amarantov, S. V., Bubennov, S. V., Prosochkina, T. R., Grigor'eva, N. G., Paukshtis, E. A. and Kutepov, B. I. (2016). Sol-gel synthesis of mesoporous aluminosilicates with a narrow pore size distribution and catalytic activity thereof in the oligomerization of dec-1-ene. *Microporous Mesoporous Mater,* 230:118-127.

[11] Kumar, A., Yadav, N., Bhatt, M., Mishra, N. K., Chaudhary, P. and Singh, R. (2015) Sol–gel derived nanomaterials and it's applications: a review. *Res J Chem Sci,* 5:98-105.

[12] Norfazilah, W., and Ismail, W. (2016) Sol–gel technology for innovative fabric finishing. *J Sol Gel Sci Technol*, 78:698-707.

[13] Kakihana, M. (1996). Solgel preparation of high temperature superconducting oxides. *J Sol Gel Sci Technol,* 6:7–55.

[14] Barb, C. J., Cassidy, D. J., Triani, G., Latella, B. A., Mitchell, D. R. G. and Bartlett, J. R. (2003). Low temperature bonding of ceramics by sol–gel processing. *J Sol Gel Sci Technol,* 26:1145–1150.

[15] Reisfeld, R. and Saraidarov, T. (2006). Innovative materials based on sol–gel technology. *Optic Mater*, 28:64–70.

[16] Wang, D. and Bierwagen, G. P. (2009). Sol–gel coatings on metals for corrosion protection, *Prog. Org. Coat,* 64:327-338.

[17] Owens, G. J., Singh, R. K., Foroutan, F., Alqaysi, M., Han, C-M., Mahapatra, C., Kim, H. W. and Knowles, J. C. (2016). Sol-gel based materials for biomedical applications. *Prog Mater Sci,* 77:1-79.

[18] Beltrarn-Osuna, A. A. and Perilla, J. E. (2016). Colloidal and spherical mesoporous silica particles: synthesis and new technologies for delivery applications. *J Sol Gel Sci Technol,* 77:480-496.

[19] Guo, X., Zhang, Q., Ding, X., Shen, Q., Wu, C., Zhang, L. and Yang, H. (2016). Synthesis and application of several sol–gel-derived materials via sol–gel process combining with other technologies. *J Sol Gel Sci Technol,* 79:328-358.

[20] Miller, J. B. and Ko, E. I. (1997). Control of mixed oxide textural and acidic properties by the sol-gel method. *Catal Today,* 35:269–292.

[21] Brinker, C. J. and Scherer, G. W. (1990). Sol-Gel Science, the physics and chemistry of sol–gel processing. Academic press Inc., Boston, 908.

[22] Chuang, K. C., Young, G. W. and Benslay, R. M. (1992). Advanced fluid catalytic cracking technology. In: *AlChE symposium series*, American Institute of Chemical Engineers, New York, 291:30.

[23] Ward, J. W. (1983). Design and preparation of hydrocracking catalysts. *Stud Surf Sci Catal.,* 16:587-618.

[24] Deldari, H. (2005). Suitable catalysts for hydroisomerization of long-chain normal paraffins. *Appl Catal A,* 293:1–10.

[25] Twaiq, F. A., Mohamed, A. R. and Bhatia, S. (2003). Liquid hydrocarbon fuels from palm oil by catalytic cracking over aluminosilicate mesoporous catalysts with various Si/Al ratios. *Microporous Mesoporous Mater,* 64:95–107.

[26] Grigor'eva, N. G., Filippova, N. A., Agliullin, M. R., Kutepov, B. I. and Narender, N (2017) Crystalline and amorphous aluminosilicates with different pore structures for the synthesis of pyridines. *J Chem Res,* 41:253–261.

[27] Agliullin, M. R., Talzi, V. P., Filippova N. A., Bikbaeva, V. R., Bubennov, S. V., Prosochkina T. R., Grigorieva, N. G., Narender Nama and Kutepov, B. I. (2018). Two-step sol–gel synthesis of mesoporous aluminosilicates: highly efficient catalysts for the preparation of 3,5-dialkylpyridines, *Applied Petrochemical Research,* 8:141–151.

[28] Schimizu, S., Watanabe, N., Kataoka, T., Shoji, T., Abe, N., Morishita, S. and Ichimura, H. (2012). Ullmann's Encyclopedia of Industrial Chemistry, Wiley-VCH Verlag GmbH & Co. KGaA. Weinheim, 1-34.

[29] Krishna, M. K. and Narender, N. V. V. (2012). Synthesis of *N*-heterocyclic compounds over zeolite molecular sieve catalysts: an approach towards green chemistry. *Catal. Sci. Technol.,* 2: 471-487.

[30] Niwa, M., Katada, N. and Okumura, K. (2010): Characterization and design of zeolite catalysts. Solid acidity, shape selectivity and loading properties. *Springer Series in Materials Science,* Amsterdam 1-184.

[31] Bordiga, S., Ugliendo, P., Damin, A. and Lamberti, C. (2001). Hydroxyls nests in defective silicalites and strained structures derived upon dehydroxylation: vibrational properties and theoretical modeling. *Top Catal*, 15:43–52.

[32] Trombetta, M., Busca, G., Rossini S. and Piccoli, V. (1998). FTIR studies on light olefin skeletal isomerization catalysis: III. Surface acidity and activity of amorphous and crystalline catalysts belonging to the SiO_2–Al_2O_3 system. *J Catal,* 179:581–596.

[33] Paukshtis, E. A. and Yurchenko, E. N. (1983). Study of the acid–base properties of heterogeneous catalysts by infrared spectroscopy. *Russ Chem Rev.*, 52:242-258.

[34] Taylor, A. P., Robinson, R. P., Fobian, Y. M., Blakemore, D. C., Jones, L. H. and Fadeyi, O. (2016). Modern advances in heterocyclic chemistry in drug discovery. *Organic & Biomolecular Chemistry,* 14(28):6611–6637.

[35] Altaf A. A., Shahzad A., Gul Z., Rasool N., Badshah A., Lal B. and Khan E. A (2015). Review on the Medicinal Importance of Pyridine Derivatives. *J. Drug Design & Med. Chem.,* 1(1):1-11.

[36] El-Sayed, H. A., Moustafa, A. H., El-Torky, A. E. and Abd El-Salam, E. A. (2017). A series of pyridines and pyridine based sulfa-drugs as antimicrobial agents: Design, synthesis and antimicrobial activity. *Russ. J. General Chem.*, 87(10):2401-2408.

[37] Weissermel, K. and Arpe, H.-J., Industrial organic chemistry, (2003). *Science,* 491.

[38] Abramovitch, R. A. Pyridine and Its Derivatives, Supplement, Chemistry of heterocyclic compounds: A Series of Monographs, (2009). John Wiley & Sons, 188: 451.

[39] Dick, J. S. and Rader, Ch. P. (2014). Raw Materials Supply Chain for Rubber Products: Overview of the Global Use of Raw Materials, Polymers, Compounding Ingredients, and Chemical Intermediates. Hanser Publishers, Munich XVI: 584.

[40] Kirk-Othmer Encyclopedia of Chemical Technology (2002). 4th Edition. John Wiley & Sons, Inc., New York, 20:582-602.

[41] Satyanarayana, T., Bhavdish Narain Johri and Prakash Anil. (2012). Microorganisms in Environmental Management: Microbes and Environment. *Springer Science & Business Media*, 819.

[42] Henry, G. D. (2004). De novo synthesis of substituted pyridines. *Tetrahedron*, 60:6043-6061.

[43] Kulkarni, S. J., Ramachandra, Rao R., Subrahmanyanm, M. and Rama Rao A. V. (1994). Synthesis of pyridine and picolines from ethanol over modified ZSM-5 catalysts. *Appl. Catal.,* 113:1-7.

[44] Grigor'eva, N. G., Agliullin, M. R., Kostyleva, S. A., Bubennov, S. V., Bikbaeva, V. R., Gataullin, A. R., Filippova, N. A., Kutepov, B. I. and Narender, N. (2019). Mesoporous aluminosilicates in the synthesis of N-heterocyclic compounds. *Kinetics and Catalysis*, 60(2):1-11.

[45] Rao R. R., Srinivas N., Kulkarni S. J., Subrahmanyam M. and Raghavan K. V. (1997). A new route for the synthesis of 3,5-lutidine over modified ZSM-5 catalysts. *Appl. Catal.:A: General.,* 161:37-42.

[46] Chandrakala, M., Rao, R. R., Kulkarni, S. J., Subrahmanyam, M. and Raghavan, K. V. (1998). Selective preparation of 3,5-diethylpyridine from n-butanol, formaldehyde and ammonia over modified ZSM-5 catalysts. *Indian J. Chem.*, 37:323-327.

[47] Marella, A., Tanwar, O. P., Saha, R., Ali, M. R., Srivastava, S., Akhter, M., Shaquiquzzaman, M. and Alam, M. M. (2013). Quinoline: A versatile heterocyclic. *Saudi Pharm J,* 21(1):1-12.

[48] Golden, E. B., Cho, H-Y., Hofman, F. M., Louie, S. G., and Chen, T. C. (2015). Quinoline-based antimalarial drugs: a novel class of autophagy inhibitors. *Neurosurgical Focus,* 38(3):12.

[49] Bawa, S., Kumar, S., Drabu, S. and Kumar, R. (2010). Structural modifications of quinoline-based antimalarial agents: Recent developments. *J. Pharm. Bioapllied Sci,* 2(2):64-71.

[50] Eswaran, S., Adhikari, A. V., Chowdhury, I. H., Pal, N. K., and Thomas, K. D. (2010). New quinoline derivatives: synthesis and investigation of antibacterial and antituberculosis properties. *European J. Med. Chem.,* 45(8):3374.

[51] Jain, S., Chandra, V., Kumar, Jain P., Pathak, K., Pathak, D. and Vaidya, A. (2016). Comprehensive Review on Current Developments of Quinoline-Based Anticancer Agents, *Arabian J. Chem.* 1-96.

[52] Kitagawa, W. and Tamura, T. (2008). A quinoline antibiotic from rhodoccus erythropolis JCM 6824, *The Journal of Antibiotics*, 61: 680–682.

[53] Watanabe, Y., Yamamoto, M., Shim, S. C., Mitsudo, T. and Takegami, Y. (1979). The rhodium catalyzed N-heterocyclization. The synthesis of quinolines from aminoarenes and aldehydes. *Chem. Lett.,* 8(8):1025.

[54] Balayer, A., Sévenet, T., Schaller, H., Hamid, A., Hadi, A. Chiaroni A., Riche C. and Païs, M. (1993). Dihydroquinoline-type alkaloids from *Bhesa Paniculata*, Celastraceae. *Natural Product Lellers*, 2(1):61-67.

[55] Abe, F., Yamauchi, T., Shibuya, H., Kitagawa I. and Yamashita, M. M. (1998). Indole alkaloids from the leaves of alstonia villosa in sunbawa (alstonia 6). *Chem. Pharm. Bull,* 46(8):1235–1238.

[56] Johnson, J. V., Rauckman, B. S., Baccanari, D. P. and Roth, B. (1989). 2,4-Diamino-5-benzylpyrimidines and analogs as antibacterial agents. 12.1,2-Dihydroquinolylmethyl analogs with high activity and specificity for bacterial dihydrofolate reductase. *Med. Chem.*, 32(8):1942-1949.

[57] Dillard, R. D., Pravey, D. E. and Benslay, D. N. (1973). Synthesis and antiinflammatory activity of some 2,2-dimethyl-1,2-dihydroquinolines. *J. Med. Chem.*, 16(3): 251-253.

[58] Theoclitou M.-E. and Robinson L. A. (2002). Novel facile synthesis of 2,2,4 substituted 1,2-dihydroquinolines via a modified Skraup reaction. *Tetrahedron Letters*, 43(21):3907–3910.

[59] Liu, Y., Gao Q., L. Liu and S. Li. (2013). Investigated on the Rubber Antioxidant 2,2,4-Trimethyl-1,2-dihydroquinoline Polymer. *Asian J. Chem.,* 25(6):2956-2958.

[60] Liu, Y., Gao, G., Liu L. and Shi, G. (2011). Study on the Industrial Process of Rubber Anti-oxidant RD. *J. Korean Chem. Soc.,* 55(5): 830-834.

[61] Kamakshi R. and Reddy B. S. R. (2007). Synthesis of 2,2,4-dihydroquinolines using heteropolyacid as a catalyst. *Catal. Commun.*, 8(5):825–828.

[62] Kundu D., Kundu S. K., Majee A. and Hajra A. (2008). A Facile synthesis of 2,2,4-trisubstituted-1,2-dihydroquinolines catalyzed by zinc triflate under solvent-free conditions. *J. Chin. Chem. Soc.*, (55):1186-1190.

[63] Yadav J. S., Reddy B. V. S., Premalatha K. and Murty M. S. R. (2007). $Bi(OTf)_3$-catalyzed condensation of 2,2-DMP with aromatic amines: A rapid synthesis of 2,2,4-trimethyl-1,2-dihydroquinolines. *J. Mol. Catal. A: Chem.,* 271:161–163.

[64] Durgadas, S., Chatare, V. K., Mukkanti, K. and Pal, S. (2010). Ceric Ammonium Nitrate: An Efficient Catalyst for One-Pot Synthesis of 2,2,4-Trimethyl-1,2-dihydroquinolines. *Lett. Org. Chem.*, 7(4):306-310.

[65] Grigorieva, N. G., Filippova, N. A., Gataulin, A. R., Bubennov, S. V., Agliullin, M. R., Kutepov, B. I. and Narender Nama. (2017). Aluminosilicates with different pores structure in the synthesis of 2,2,4-trimethyl-1-2-dihydroquinoline and N-phenyl-2-propanimine. *Russian Chemical Bulletin, Int. Ed.*, 66(11):2115-2121.

[66] Baerlocher, Ch., McCusker, L. B. and Olson, D. H. (2007). *Atlas of Zeolite Framework Types*, Amsterdam: Elsevier.

[67] Hossain, M. and Kumar, N. A. (2018). A Review on Heterocyclic: Synthesis and Their Application in Medicinal Chemistry of Imidazole Moiety. *Science Journal of Chemistry,* 6(5): 83-94.

[68] Shatha Ibrahim Alaqeel. (2017). Synthetic approaches to benzimidazoles from O-phenylenediamine: A literature review. *Journal of Saudi Chemical Society,* 21(2):229-237.

[69] Obot, I. B. and Ubong M. Edouk. (2017). Benzimidazole: Small planar molecule with diverse anti-corrosion potentials. *Journal of Molecular Liquids,* 246:66–90.

[70] Zamani, L., Mirjalili, B. B. F., Zomorodian, K., Namazian, M., Khabnadideh, S. and Mirzaei, E. F. (2014). Synthesis of benzimidazoles in the presence of nano-$TiCl_4$ center dot SiO_2 as antifungal agents and tautomerism theoretical study of some products. *Farmacia, 62:*459-474.

[71] US Pat. 3369897, (1968). Karl-Heinz M., Rolf, P. Photographic materials containing pyrazolo-[1, 5alpha]-benzimidazole color couplers.

[72] Govardhana Babu Bodedla, Thomas, J. K. R., Fan, M. S. and Kuo Chuan Ho (2016). Bi-anchoring Organic Dyes that Contain Benzimidazole Branches for Dye-Sensitized Solar Cells: Effects of π Spacer and Peripheral Donor Groups. *Chemystry an Asian J,* 11(18):2564-2577.

[73] Zhiqian Guo, Na Ri Song, Jong Hun Moon, Myounwoo Kim, Eun Jin Jun, Jiyoung Choi, Jin Yong Lee, Christopher W. Bielawski, Jonathan L. Sessler and Juyoung Yoon. (2012). A Benzobisimidazolium-Based Fluorescent and Colorimetric Chemosensor for CO_2. *J. Am. Chem. Soc*, 134:17846−17849.

[74] Rajadhyaksha, D. D. and Rangnekar, D. W. (1986). Synthesis of pyrazolo[4′,3′:5,6]pyrido[1,2-a]benzimidazole derivatives and study of their fluorescent properties. *Chem. Technology and Biotechnology*, 36(7): 300-304.

[75] Scheers, J., Johansson, P., Szczeciński, P., Wieczorek, W., Armand, M. and Jacobsson, P. (2010). Benzimidazole and imidazole lithium salts for battery electrolytes. *Journal of Power Sources*, 195(18):6081–6087.

[76] Balashova, E. V., Krichevtsov, B., Svinarev, F., Zaitseva, N. V. and Pankova, G. A. (2018). Structural and Dielectric Properties of Organic Ferroelectric 2-Methylbenzimidazole. *Journal of Surface Investigation X-ray Synchrotron and Neutron Techniques*, 12(2):233-239.

[77] Aastha, P Navneet, K Anushu, A Pratima and S Dharma, K. (2013). 1, 5-Benzodiazepines: Overview of properties and synthetic aspects. *Res J Chem Sci.*, (3):90-103.

[78] Bariwal, J. B., Upadhyay, K. D., Manvar, A. T., Trivedi, J. C., Singh, J. S. and Jain K. S. (2008). 1, 5-Benzothiazepine, a versatile pharmacophore: a review. *Eur J Med Chem.* 43(11):2279-2290.

[79] Salve, P. S. and Mali, D. S. (2013). 1,5-benzodiazepine: A versatile pharmacophore. *Int. J. Pharma Bio. Sci*., 43(11): 345-370.

[80] Rachedine, Kaoua Norah, Bennamane Saliha, Bakhta, Sihame, Benadji Cherifa, Rabia Bellara and Nedjar-Kolli (2010). Synthesis of Substituted 1,4-Diazepines and 1,5-Benzodiazepines using an Efficient Heteropolyacid-Catalyzed Procedure. *Molecules*, 16: 92-99.

[81] Essaber, M., Baouid, Hasnaoui, A., Benharret, A. and Lavergne, J-P. (1998). Synthesis of New Tri- and Tetraheterocyclic Systems: 1,3-Dipolar Cycloaddition of

Nitrilimines on 2,7-Dimethyl-4-Phenyl-3H-1,5-Benzodiazepine. *Synth. Commun.*, 28(22):4097-4104.

In: An Introduction to Aluminosilicates
Editor: Nero Regina Blevins
ISBN: 978-1-53617-250-8

Chapter 6

COMPOSITES OF BENTONITE, CHITOSAN AND POLYACRYLAMIDE: PREPARATION, CHARACTERIZATION, AND APPLICATION IN UREASE IMMOBILIZATION

Demet Baybaş[1,*], Serhat Öner[1] and Ulvi Ulusoy[2]
[1]Cumhuriyet University, Department of Biochemistry, Sivas, Turkey
[2]Cumhuriyet University, Department of Chemistry, Sivas, Turkey

ABSTRACT

The objectives of this work were the preparation, characterization, and investigation in urease immobilization of composite adsorbents polyacrylamide–bentonite (PAAmB), polyacrylamide–chitosan (PAAmCh), and polyacrylamide–chitosan–bentonite (PAAmChB) produced from polyacrylamide as a hydrogel, bentonite as an aluminosilicate mineral, and chitosan as a polysaccharide. The formation of the composites was proved by FT-IR, XRD, DSC, and SEM analysis.

The effect of concentration, time, and temperature on urease immobilization on the composites was investigated by conductometry. The monolayer adsorption capacities obtained from the Langmuir model were 93 mg g^{-1} for PAAmB, 174 mg g^{-1} for PAAmCh and 95 mg g^{-1} for PAAmChB. The immobilization was time-dependent with pseudo-second-order kinetics, but it was not temperature dependent.

The activity dependence of urease immobilized on the composites to the substrate (urea) concentration, pH and temperature was evaluated with reference to the activity measured for free urease. The V_{max} and K_m values derived from the Michaelis–Menten graphs were 11 units mg^{-1} and 128 mM for the free enzyme, 11 units mg^{-1} and 243 mM for PAAmB, 20 units mg^{-1} and 104 mM for PAAmCh, and 19 units mg^{-1} and 102 mM for

* Corresponding author: E-mail: dbaybas@cumhuriyet.edu.tr.

PAAmChB. The optimum pH and temperature were determined for the composites and free enzymes.

Keywords: bentonite, chitosan, polyacrylamide, composite, urease immobilization

1. INTRODUCTION

Urea, also known as carbamide, is a stable compound in aqueous solutions and its autohydrolysis half-life is 3.6 years at 30°C [Zeinali and Lenjannezhadian, 2018]. It is the waste product of amino acid metabolism, being produced in the liver in mammals and excreted from kidneys in the urine. Ethyl carbamate (urethane; EC), which naturally occurs in all fermented foods and beverages, is spontaneously formed by the reaction between urea and ethanol [Andrich et al., 2010]. Synthetic urea is used in the production of fertilizers, various plastics, resins, adhesives, solvents, medicines, and cleaning products owing to its high nitrogen content. It is also found in medicines and cosmetic products used for the treatment of dry skin [Krajewska, 2009a].

Urea from urea-using industries, fertilized soils, animal urine, and sewage contaminates the environment. Urea is generally known to have low ecotoxicity, but ammonia formed as a result of urea hydrolysis leads to various effects, such as toxicity, alkalinity, and gas emission. Moreover, eutrophication and contamination of underground water are among its long-term effects. The worldwide urea production is 10^8 tonnes year^{-1} and more than 90% is used in fertilizer production. In the production, 0.2% to 2% urea waste is formed during the purification and recovery stages following urea synthesis. The urea content of these wastes before being discharged to the environment should be reduced to less than 0.006% [Krajewska, 2009a]. In medicine, urea is one of the indicators monitored in the blood to evaluate uremic toxin levels and diagnose kidney diseases [Kumar and Melo, 2006]. As much as 100 to 300 L of dialysate is used in hemodialysis to remove urea and toxic compounds from the blood and the process is expensive. In the food sector, high doses of EC have been shown to have carcinogenic effects in animal tests. For example, the safety level in wines was determined as 15 mg m^{-3} [Andrich et al., 2010]. The removal of urea from aqueous solutions in fields such as chemistry, agriculture, medicine, biomedical, and the food industry is a significant problem [Krajewska, 2009a; Yurekli and Altinkaya, 2011].

Various methods are used for the removal of urea from aqueous solutions, such as catalytic and electrochemical degradation, oxidation, adsorption with strong oxidants, non-enzymatic hydrolysis, and biological conversion of urea nitrogen to dinitrogen. The polar and non-ionic characteristics of urea reduce its affinity for adsorbents. Other methods require high temperature and pressure and are complex and costly [Krajewska, 2009a;

Yurekli and Altinkaya, 2011]. Enzymatic hydrolysis is the most frequently used urea removal method in the literature.

Urease (urea amidohydrolase; E.C. 3.5.1.5.) is a nickel-containing metalloenzyme in the hydrolase family that hydrolyzes urea, ammonia, and carbon dioxide. The hydrolysis reaction in aqueous media can be written as follows [Kumar and Melo, 2016]:

$$NH_2CONH_2 + 3H_2O \xrightarrow{\text{Urease}} 2NH_4^+ + HCO_3^- + OH^- \quad (1)$$

Urease is found in many plants, algae, fungi, bacteria, in numerous organisms including invertebrates, as well as in the soil as a soil enzyme. It hydrolyzes urea 10^{14} times faster than the uncatalyzed reaction and is found in high levels in *Canavalia ensiformis* (jack bean), which is extremely rich in protein content [Akgöl et al., 2002; Yurekli and Altinkaya, 2011; Sujoy and Aparna, 2013; Zeinali and Lenjannezhadian, 2018].

Krajewska comprehensively reviewed important applications of urease, including medical, analytical, environmental, and engineering approaches, as well as its functional and practical characteristics [Krajewska, 2009 a, b]. Urease is used in hypertension treatment as an anticancer agent, as a component of the *Helicobacter Pylori* vaccine, in the wine industry, in nitrogen metabolism, and in the germination of nitrogen-limited *Arabidopsis thaliana* seeds [Sujoy and Aparna, 2013].

The purification of enzymes for use as biological catalysts is difficult and costly, and free enzymes are generally stable only in narrow ranges of temperature and pH range, increasing the importance of enzyme immobilization in enzyme studies. Immobilized urease is used in biosensors for diagnosis, urea determination in biological fluids, artificial kidney devices for the removal of extracorporeal urea from blood for detoxification purposes, urea conversion in fertilizer wastewater, enzyme reactors, removal of urea from food and beverages in the food industry, analysis of heavy metals, determination of creatinine, arginine, IgG, and *Escherichia coli*, and in urease-supported biomineralization processes [Akgöl et al., 2002; Desimone et al., 2008, Krajewska, 2009a].

The immobilization of enzymes on solid carriers can be achieved using chemical (e.g., covalent coupling, entrapment, encapsulation, and cross-linking) and physical (e.g., nonspecific adsorption, ionic binding, hydrophobic adsorption, affinity binding) methods [Kunduru et al., 2017]. Among the numerous methods proposed for enzyme immobilization, adsorption is the most important and useful. Adsorption occurs through physical interactions between the carrier and the enzyme, including van der Waals forces, ionic interactions, and hydrogen bonding. Binding is very weak and the enzyme is not inactivated because its natural structure is only slightly affected [Jesinowski et al., 2014].

The immobilization of urease has been studied comprehensively. Krajewska (2009a) reviewed the urease immobilization studies performed over the last 20 years, including the immobilization techniques and supporting materials used. Some support materials used in immobilization by adsorption include silica/silica gel, poly(2-hydroxyethyl methacrylate-

co-N-methacryloyl-L-histidine-methyl ester) beads, polyaniline membrane, aminated polysulphone membrane, TiO_2, montmorillonite, $Al(OH)_3$, kaolinite, hydroxyapatite, and gelatin beads activated with GA [Krajewska, 2009a].

Bentonite (B), chitosan (Ch), and polyacrylamide (PAAm) are the components of the composites (PAAmB, PAAmCh, and PAAmBCh) investigated in this study. Bentonite (montmorillonite) is a good natural support material owing to its high number of active centers located on the surfaces between its outer layers. It can be used in aqueous environments suitable for enzymes owing to its swelling characteristic and being insoluble in water. Despite its regular structure and adsorbent properties, using aggregated bentonite particles as a support material is disadvantageous in environments with varying temperature and solution electrolytic properties. This disadvantage can be overcome by compositing bentonite with various polymeric materials, which may further improve the immobilization and adsorbent properties depending on the properties of the polymer [Baybaş and Ulusoy, 2011; Jesinowski et al., 2014; An et al., 2015].

Chitosan is a polysaccharide obtained by deacetylation of chitin. It is a non-toxic, biocompatible, gel-forming cationic compound that can be easily prepared in different geometrical forms, such as membranes, beads, nanoparticles, fibers, hollow fibers, or sponges. Because it is an inexpensive, inert, hydrophilic, biocompatible support, it has become an attractive carrier for enzyme immobilization [Jesionowski et al., 2014]. Krojewska comprehensively reviewed chitosan-based materials used in urease immobilization (including chitosan–triphosphate beads, chitosan beads, chitosan membrane, chitosan–PVA capsules, chitosan–PGMA precipitate, chitosan-coated alginate beads, and chitosan–organosilane particles) [Krajewska, 2009a].

As well as natural polymers, synthetic polymers are also used as enzyme carriers. This is because synthetic polymers can be designed according to the desired properties or the polymer structure can be modified after preparing the composite. Poly(vinyl alcohol) (PVA), crosslinked PVA, poly(*N*-methyloxylamide), polypropylene, foam polystyrene, poly(acrylic acid-co-acrylamide), hydrotalcite nanocomposite hydrogels, poly(hydroxybutyrate) nanoparticles, poly(acrylonitrile) (PAN), polystyrene (PS), and polyamide (PA) are examples of enzyme-carrying polymers [Jesionowski et al., 2014].

This article presents the enzyme immobilization properties of PAAmB, PAAmCh, and PAAmChB composites, which have not been investigated in previous studies, as well as their enzymatic properties by providing the urease example. Detailed information on the synthesis, characterization, and metal adsorption of the composites used in the present study can be found in previous MSc and PhD theses [Baybaş, 2009 and Eninanç, 2013]. The PAAmB, PAAmCh, and PAAmChB composites prepared in this study were characterized by FT-IR, XRD, DSC, and SEM analyses. Urease was immobilized to PAAmB, PAAmCh, and PAAmChB carrier materials by adsorption method and the optimum parameters were determined. The effects of temperature, pH, and concentration parameters on the activity of the immobilized enzyme were investigated and the results were compared with those for the free enzyme. The reusability of the immobilized enzyme was also investigated.

2. Methods

2.1. Materials

Na-bentonite, chitosan with > 85% deacylation degree (Alfa Aesar, Germany), and acrylamide monomer (Merck, Germany) were used without any chemical treatment. Other chemicals used were epichlorohydrin (ECH) (Sigma-Aldrich, Germany), *N,N*'-methylenebisacrylamide (Alfa Aesar, Germany), *N,N,N*',*N*'-tetramethylenediamine (TEMED) (Sigma, China), sodium tripolyphosphate (NaTPP) (Merck, Germany), and ammonium peroxodisulfate (Merck, Germany).

In the enzyme studies, Type III 15,000–50,000 units (g solids)$^{-1}$ enzyme (Sigma, USA) with U1500–20KU activity obtained from jack bean (*Canavalia ensiformis*), urea (Merck, Germany), mono- di- and tri-sodium phosphate (NaH_2PO_4, Na_2HPO_4, Na_3PO_4) (Sigma-Aldrich, Germany), glacial acetic acid (Merck, Germany), NH_4OH (Merck, Germany), phenol (Merck, Germany), sodium nitroprusside ($NaNO_2Prs$) (Sigma, USA), sodium hydroxide (Merck, Germany), and sodium hypochlorite (with 15.46% active chlorine content) were used.

All chemical substances used in this study were analytically pure and the water used in the preparation of the solutions was double-distilled.

2.2. Preparation of Composites

First, the chitosan–bentonite (ChB) composite was synthesized for the three-component PAAmChB composite. In order to synthesize 4 g of ChB, 2 g of chitosan (Ch), and 2 g of bentonite (B) were stirred in 80 mL of 5% (v/v) acetic acid with a magnetic stirrer for 2 hours until a homogeneous suspension was achieved and then kept overnight. Next, 80 mL of ECH (0.001M pH = 10) was added to the obtained suspension and stirred for 1 hour with a magnetic stirrer. The resulting viscous suspension was taken into a syringe and added dropwise to a beaker containing 250 mL of 0.05 M NaTPP to form ChB beads and stirred for 3 hours with a magnetic stirrer. Impurities in the beads were removed by washing with water until the wash water conductivity reached that of distilled water. The purified ChB composite was dried, ground, and stored in closed containers [Eninanç, 2013].

For the preparation of the PAAmB, PAAmCh, and PAAmChB composites, 2 g of bentonite, chitosan, or ChB was added to 10 mL of water and mixed with a magnetic stirrer. Next, 10 mL of an aqueous solution of 4 g of acrylamide monomer and 0.4 g of *N,N'*-methylenebisacrylamide (crosslinker) was added to these suspensions. They were stirred until a homogeneous suspension was obtained and then 2 mL of ammonium peroxodisulfate (100 mg mL^{-1}) and 500 μL of TEMED (accelerator) were added to form B, Ch, or ChB immobilized polyacrylamide (PAAm). The resulting composites were washed with pure water until the conductivity of the washing water was equal to that of pure water. The composites were then separated by decantation and dried in the oven at 40°C. The obtained composites were ground and stored in closed containers [Baybaş, 2009; Eninanç, 2013].

2.3. Characterization

The FT-IR spectra of the PAAmB, PAAmCh, and PAAmChB composites and their components (PAAm, B, Ch, ChB) were obtained using an FT-IR instrument (BRUKER TENSOR II) with the KBr pellet technique. XRD spectra were taken using an XRD device (BRUKER AXS D8 ADVANCE) and SEM images were obtained using a high resolution scanning electron microscope (TESCAN MIRA3, XMU). DSC (SHIMADZU-50) analyses were performed under a nitrogen atmosphere.

2.4. Urease Immobilization

Urease was immobilized on the PAAmB, PAAmCh, and PAAmChB composites by adsorption method and the optimum conditions for immobilization were determined by varying the adsorption parameters. All experiments were carried out at 298 K, except for the investigation of adsorption at different temperatures. In all experiments, except for investigation of the adsorption change with pH, solutions of urease in pH 7 phosphate buffer (10 mM) were used. The adsorption change with concentration was investigated between 50 and 2000 mg L^{-1} while the adsorption change with time and temperature was investigated at 2000 mg L^{-1}. The adsorbent doses used in the experiments [V (solution volume, mL)/w (adsorbent mass, g)] were 10/0.1 for temperature and concentration and 20/0.2 for time.

The urease concentrations remaining in the equilibrium solutions at 24 hours or planned time intervals were determined by means of conductivity measurement using an ECOSCAN (Con 5) device. Urease solutions were prepared in the concentration range of 50–3000 mg L^{-1} in phosphate buffer (NaH_2PO_4/Na_2HPO_4). 1 mL of these solutions was diluted to 9 mL with buffer solution (pH=7) and 1 mL of a 10000 mg L^{-1} aqueous solution of urea was added to initiate the urea decomposition reaction. Increased conductivity values owing to the increased concentration of reaction products were recorded and evaluated at 30, 60, 120, 180, 240, and 300 s after the initiation of the reaction.

The reason for using a conductivity method in the urease adsorption experiments rather than the Bradford spectrophotometric protein determination method is that the Bradford method gives incorrect results in Ch-containing media [Su et al., 2010].

2.5. Activity of Immobilized and Free Urease

The high amount of urease adsorbed by the ground composite samples (with high surface area) used revealed both economic properties for the evaluation of urease activity and limiting properties (opalescence solution formation and inhibition of solution fluidity in column applications) for investigator purposes. This negative effect was overcome by the use of millimetric-sized composite particles.

For maximum enzyme adsorption determined by urease adsorption experiments, the change of immobilized enzyme (PAAmB-U, PAAmCh-U and PAAmChB-U; all were denoted as Comp-U hereafter) activity prepared under optimum conditions with pH, substrate (urea) concentration and temperature were compared with the amount of free enzyme (U) activity adsorbed to the composite under the same conditions. In addition, the reusability of the immobilized enzyme was investigated.

The urease adsorption and activity measurements were made using the Berthelot method based on measurement of the NH^{4+} ion concentration. The Berthelot method uses two reagents: Reagent A is a solution of 10 g of phenol and 0.05 g of sodium nitroprusside in 1 L of water and reagent B is a solution of 5 g of sodium hydroxide and 0.42 g of sodium hypochlorite in 1 L of water. The chemical stability of the prepared reagents was maintained by keeping them in a refrigerator in colored bottles. Reagents A and B (5 mL each) were added to 1 mL of immobilized or free enzyme-activated urea solutions, and their absorbance values at 630 nm stabilized after approximately 1 hour at room temperature were measured by UV-Vis spectrophotometer [Chaney and Marbach, 1962]. The immobilized enzyme activity is determined by measuring the NH^{4+} concentration in the sample taken from the solution–composite medium. This requires the use of a reagent to inhibit the free enzyme for instant measurement of free enzyme activity. The suitability of reagent B for enzyme inhibition has been proved in preliminary trials.

2.5.1. Enzyme Activity and Substrate Concentration

Step i) Enzyme Immobilization to Support Materials

0.1 g fractions of milimetric-sized composite particles were weighted, then 10 mL of 200 mg L^{-1} urease solution was added, and their equilibrium solutions were separated after

24 hours of interaction. The composites with the enzyme adsorbed (Comp-U) were washed with 20 mL of phosphate buffer solution. One mL fractions of equilibrium, washing and 200 mg mL^{-1} urease solutions were diluted to 9 mL with phosphate buffer. Then, hydrolysis reaction was initiated by addition of 1 mL of 10000 mg mL^{-1} urea solution to each. After 5 minutes, a known fractions of the solutions were removed from the reaction medium and the reaction in these fractions were terminated by adding Berthelot reagent B at a 1:1 ratio by volume. The NH^{4+} concentration of this solution was determined by the Berthelot method.

Step ii) Determination of the Immobilized Enzyme Activity

9 mL of phosphate buffer was added on the Comp-U samples obtained from *Step i*, then 1 mL of urea solution at varying concentrations (0, 100, 1000, 5000, 10000, or 20000 mg L^{-1}) was added to initiate the enzymatic hydrolysis reaction. The NH^{4+} concentrations of the solution samples taken 5 minutes after the initiation were measured by the Berthelot method.

Step iii) Determination of the Free Enzyme Activity

8 mL of phosphate buffer was added to 1 mL of a solution containing 1 mL of free enzyme in the concentration range of 10–200 mg L^{-1} and the method described at Step ii was repeated.

Calibration curves of the absorbance (A) vs. $[NH_4^+]$ and the free urease (U) vs. $[NH_4^+]$ determined 5 minutes after the initiation were generated, and used for measuring the amount of adsorbed urease and the immobilized enzyme activity from the NH_4^+ concentrations.

For the free and immobilized enzymes, the initial substrate concentration ($[S]_0$) corresponding to the reaction rate (v_0) at 5 minutes was plotted and Michaelis–Menten plots were obtained.

2.5.2. Enzyme Activity and pH

Urease was adsorbed onto 0.1 g of PAAmB, PAAmCh, and PAAmChB under the optimum conditions with a 200 mg L^{-1} urease concentration and the adsorbed amounts were determined by measuring the urease activity remaining in the equilibrium solution using the Berthelot method.

9 mL of 10 mM phosphate buffer at pH 2, 3, 6, 7, and 11 prepared using H_3PO_4, NaH_2PO_4, Na_2HPO_4, or Na_3PO_4 was added to the Comp-U samples and then the reaction was initiated by adding 1 mL of 10000 mg of L^{-1} urea solution. The reaction was terminated after 5 minutes. The amount of ammonia deposited in the medium during that period was measured by the Berthelot method and the % activity values of immobilized urease were calculated. This procedure was repeated for 1 mL of 200 mg L^{-1} of free enzyme.

2.5.3. Enzyme Activity and Temperature

After adding 9 mL of 10 mM pH 7 buffer solution into Comp-U samples prepared by the method described above, they were incubated at 4, 15, 25, and 35°C. After that, 1 mL of 10000 mg L^{-1} urea was added to initiate the reaction. Similar processes were repeated for the free enzyme under identical conditions. The amount of NH^{4+} formed in the medium was measured 5 minutes after the initiation of the reaction and the specific and % activity values were calculated.

2.5.4. Reusability of Immobilized Enzyme

One of the most important advantages of immobilization is the ability to reuse the enzyme. To investigate the reusability of the Comp-U samples, samples prepared as described above were subjected to consecutive urea hydrolysis reactions and the composites were washed with 20 mL of 10 mM phosphate buffer after each use. This procedure was repeated seven times and the activity was calculated for each repeat.

3. Results and Discussion

3.1. Characterization

The results and evaluation of the FT-IR, XRD, SEM, and DSC analyses of the PAAmB, PAAmCh, and PAAmChB composites and their components are presented in the following sections.

3.1.1. FT-IR Analysis

The FT-IR spectra of the composites and their components are shown in Figure 1.

When comparing the peaks observed in the FT-IR spectra of pure B and PAAm with those observed in the PAAmB spectrum, it can be clearly seen that the N-H peak (3200–3500 cm^{-1} wide band), C-H peak (2950 cm^{-1}), amide-end C=O and N-H peaks (1670 and 1610 cm^{-1}), and C-H (1450–1455 cm^{-1}) peaks in the PAAmB composite are attributed to PAAm.

The O-H (3200–3500 cm^{-1} wide band), Si-O (1060-1120 cm^{-1}), and Al-O-Si and Si-O-Si (530 and 460 cm^{-1}) peaks in the spectrum for PAAmB are clearly related to B [Bai et al., 2007; Anirudhan et al., 2008]. As a result, the spectrum of PAAmB proves that this composite is composed of PAAm and B.

When comparing the peaks observed in the FT-IR spectra of pure PAAm and Ch with those observed in the PAAmCh spectrum, the number of distinct peaks representing PAAm or Ch is limited owing to the presence of similar chemical groups in both structures.

The strongest evidence for the presence of PAAm in the composite spectrum is amide-end C=O and N-H (1670 and 1610 cm^{-1}) peaks. Representative changes observed by comparing the bands for PAAmCh between 3200–3500 cm^{-1} and 2800–3000 cm^{-1} with those for Ch and PAAm in the same wavelength range should be considered as additional evidence of the contributions of PAAm and Ch to PAAmCh [Bai et al., 2007; Queiroz et al., 2015].

Similarly, when comparing the peaks observed in the FT-IR spectra of pure PAAm, Ch, and B with those observed in the PAAmChB spectrum, amide-end C=O and N-H (1670 and 1610 cm^{-1}) peaks from PAAm, C=O and C-N (1645 and 1325 cm^{-1}) peaks from Ch and Al-O-Si, and Si-O-Si (530 and 460 cm^{-1}) peaks from B indicate their contribution to the composite spectrum. These data prove that the PAAmChB composite is formed by PAAm, Ch, and B.

3.1.2. XRD Analysis

XRD analysis was used for the structural characterization of PAAm, B, Ch, PAAmB, PAAmCh, and PAAmChB. Bragg's equation (2) was used to calculate the interlayer spacing.

$$d_{hkl} = \frac{n\lambda}{2\sin\theta} \text{ Bragg Equation} \qquad (2)$$

where d is the interlayer spacing (defined with Miller indices hkl), λ is the wavelength of the X-ray beam (Cu Kα) (0.154 nm), and θ is the reflection angle.

The XRD spectra of the composites and their components are shown in Figure 2. The results obtained are in accordance with previous evaluations [Baybaş, 2009; Eninanç, 2013].

In the spectrum in Figure 2a, the peaks of B at $2\theta = 7.1°$ (001) and $2\theta = 19.7°$ (002) are related to the characteristic crystal plane reflections of B [montmorillonite, Na0.3A12(SiAl)4O10(OH)2.2H2O] [Baybaş and Ulusoy, 2011]. In the d001 reflection, considered as the reference in interlayer spacing calculations, the values calculated for B and PAAmB using Bragg's equation (2) were 1.24 nm and 1.42 nm, respectively. This difference proves the increase of the interlayer distance (0.18 nm) owing to the addition of the amorphous component PAAm. The resulting structure is a nanocomposite [Alexandre and Dubois, 2000].

The peak at $2\theta = 19.4°$ belongs to polymorph Ch and is shown in the XRD spectra for the PAAmCh composite (Figure 2b). The spectrum of PAAm and its composites prove that PAAm increases the amorphism of the composites.

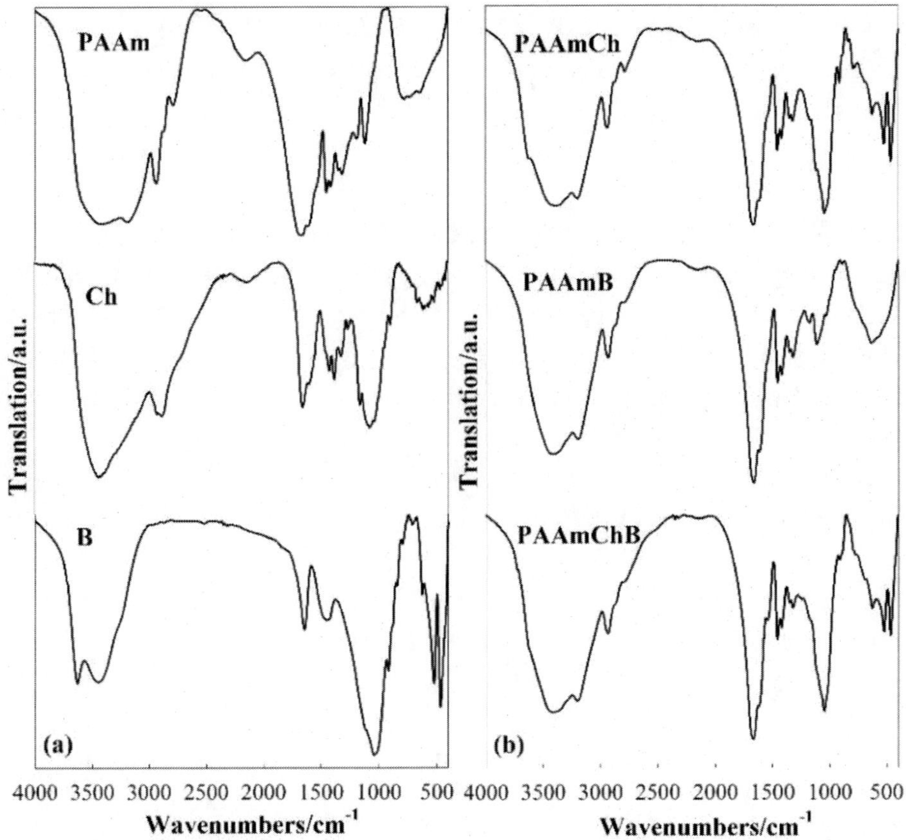

Figure 1. FT-IR spectra of (a) composite components and (b) composites.

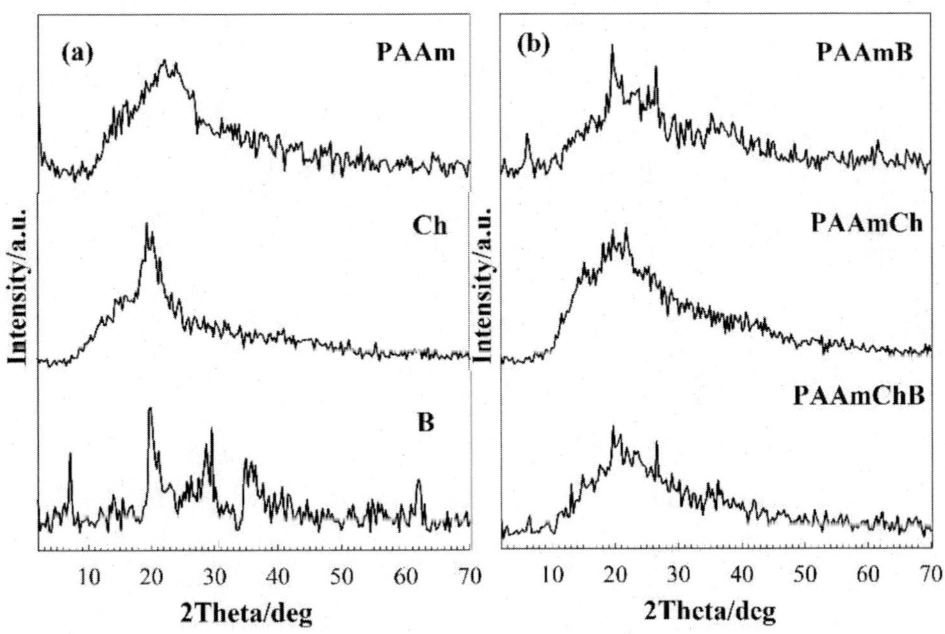

Figure 2. XRD patterns of (a) the composite components and (b) the composites.

The interlayer distance in the PAAmChB composite was calculated as 1.36 nm for the d_{001} reflection. This value indicates that the space increase between bentonite layers in PAAmChB is 0.12 nm.

3.1.3. DSC Analysis

DSC thermograms of composites and their components are shown in Figure 3.

Wide endothermic peaks in the temperature range of 65–100°C are common to all thermograms. These peaks may be related to water loss during the analysis process. In composites with Ch (77°C for PAAmCh, 83°C for PAAmChB), the temperature of this endothermic peak is higher than that observed for pure Ch (73°C). Peak temperatures of 69 and 74°C were observed for PAAm and PAAmB, respectively. These changes are due to differences in the water retention capacities and water–polymer interaction forces of the macromolecules [Kittur et al., 2002].

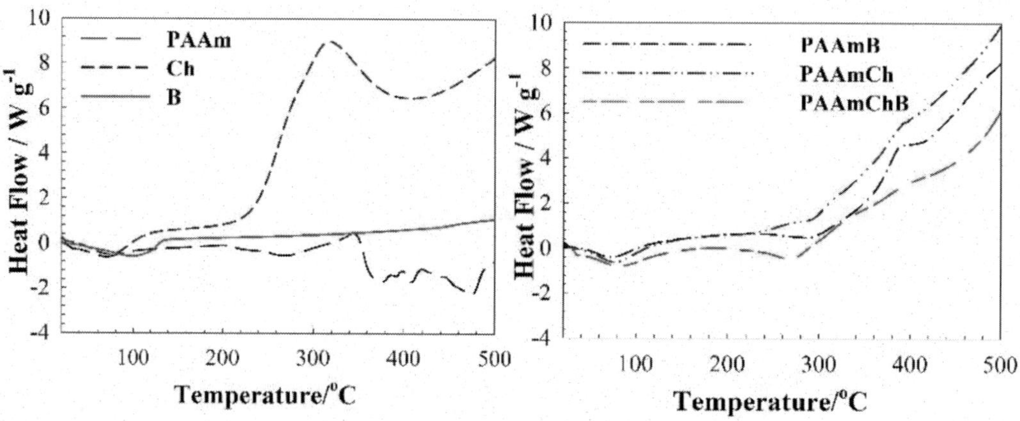

Figure 3. DSC thermograms of (a) the composite components and (b) the composites.

Silva et al. (2000) and Işikver (2017) reported that the T_g of polyacrylamide were 199°C and 190°C, respectively [Silva et al., 2000; Işikver, 2017]. In this study, the T_g of PAAm was 205°C, and the decomposition temperature (T_d) of PAAm was 377°C. The T_g of PAAmB and PAAmChB were reported as 256 and 214°C, respectively. The increase in the T_g values can be explained by the fact that two nitrogen-bound hydrogen atoms in the PAAm structure form hydrogen bonds with other composite components. This effect reduces the mobility of polymer chains and increases the T_g.

No glass transition temperature was observed for Ch and PAAmCh. Similarly, in a study performed by Kittur et al. (2002), it was reported that T_g could not be observed in the second heating run of chitin, chitosan, and their carboxymethyl derivatives.

The exothermic peak observed at 317.5°C is due to monomer dehydration, glycoside bond breakage, and breakage of acetyl and deacetylated units [Dey et al., 2016]. This peak was not observed in polymers containing Ch, which may indicate an increased degree of polymerization and an increase in thermal stability. However, this may be because of the low amount of Ch in the composite and thus the low amount of acetyl groups.

3.1.4. SEM Images

SEM images of composites and their components are shown in Figures 4 and 5.

The morphology of the PAAmB composite, formed by the addition of B with clearly visible mineralogical crystals into the PAAm amorphous structure, reflects the appearance of PAAm-coated mineralogical crystals. PAAmCh, formed by the addition of amorphous Ch with a leaf stacking appearance into the amorphous PAAm structure, appears to be a hybrid of its components. Similarly, the morphology of PAAmChB is completely hybrid in appearance.

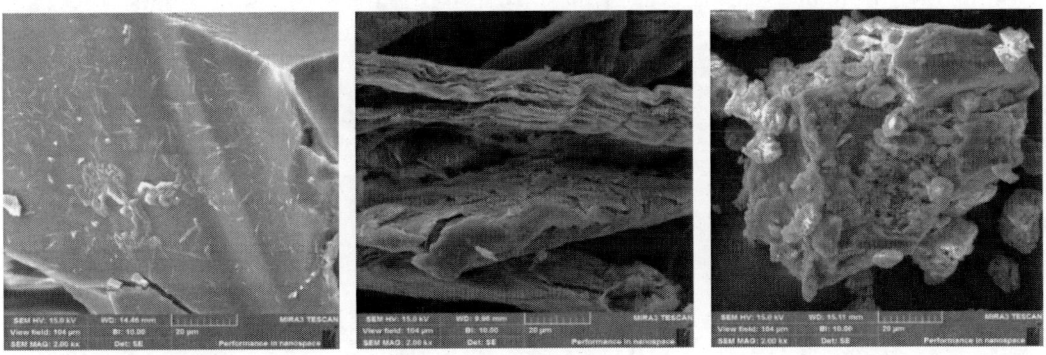

(a) PAAm (b) Ch (c) B

Figure 4. SEM images of composite components.

(a) PAAmB (b) PAAmCh (c) PAAmChB

Figure 5. SEM images of composites.

3.2. Urease Immobilization

In this section, the immobilization of urease on the composites by adsorption is examined and discussed with the effect of concentration, time, and temperature.

3.2.1. Adsorption Determination by Conductivity

A conductivity measurement method was used in all immobilization (adsorption) experiments. In this method, standard enzyme solutions of different concentrations were prepared from the same batch. The slope of 'conductivity vs time' plot for each standard obtained by using the conductivity data measured in the 30–300 s range of the reactions initiated by adding urea at a constant concentration (10000 mg L^{-1}) provided the initial rates for each urease standard (Figure 6). Then, 'the initial rate (v_0) vs standard enzyme concentration (C_{std})' was depicted (Figure 7) to obtain the calibration curve. The enzyme content of the equilibrium solutions obtained after adsorption was calculated by replacing the initial rates found by adding urea to these solutions in the calibration equation.

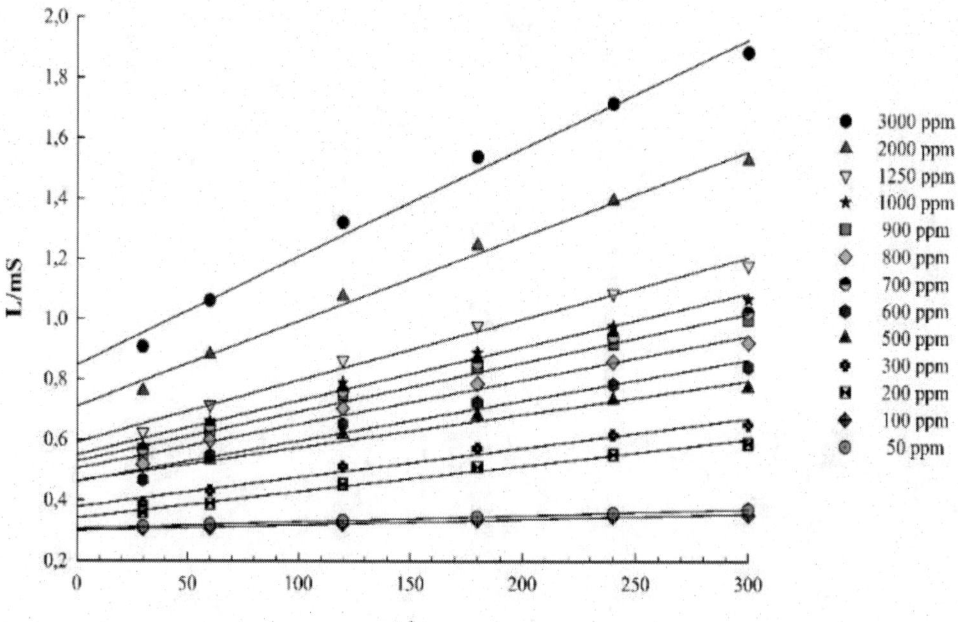

Figure 6. Representative graph used to find the initial rates of standard urease solutions by conductivity measurements.

The amount of enzyme immobilized (Q) in immobilization studies is calculated using Equation (3).

$$Q = \left[\frac{C_i - C_e}{m}\right] V \quad (3)$$

where C_i is the initial concentration (mg L^{-1}) of the enzyme interacted with the adsorbent, C_e is the equilibrium concentration (mg L^{-1}) after adsorption, m is the adsorbent mass (g), and V is the solution volume (L).

3.2.2. Adsorption Dependence on Urease Concentration

The equilibrium urease concentrations obtained from the composites interacted with different concentrations of urease were determined by conductivity method. The amount of immobilized urease Q (mg g^{-1}) was calculated from Equation (3) and Q-C_e isotherms were formed (Figure 8). The obtained isotherms were evaluated according to the Langmuir and Freundlich models.

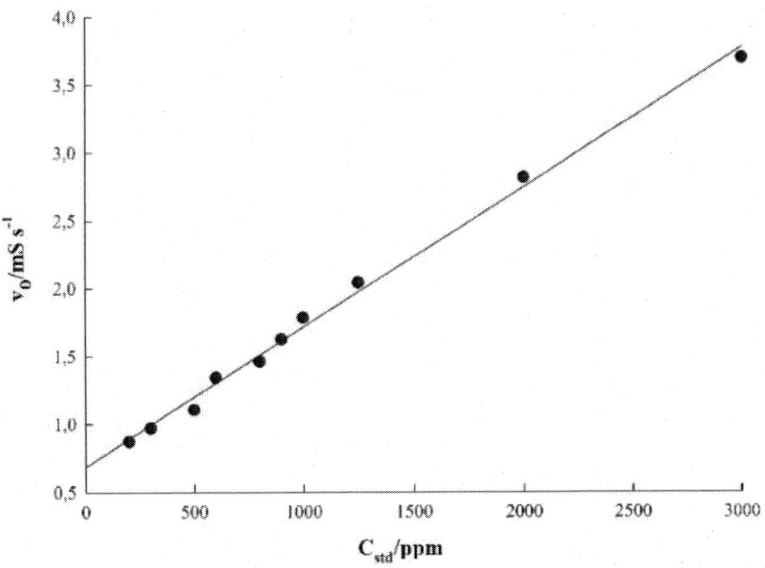

Figure 7. Representative calibration curve used for urease determination in equilibrium solutions after the adsorption on the composites.

In the Langmuir model (4), X_L is the adsorption capacity and K_L is the Langmuir constant (L mg^{-1}) related to adsorption energy.

$$Q = \frac{K_L X_L C_e}{1+K_L C_e} \text{ Langmuir Modeli} \quad (4)$$

In the Freundlich model (5), X_F is the constant related to adsorption intensity and β is the adsorbent surface heterogeneity

$$Q = X_F C_e^{\beta} \text{ Freundlich Modeli} \quad (5)$$

Isotherms for PAAmB, PAAmCh, and PAAmChB as well as the parameters determined from their fit to related models are shown in Figure 8 and Table 1.

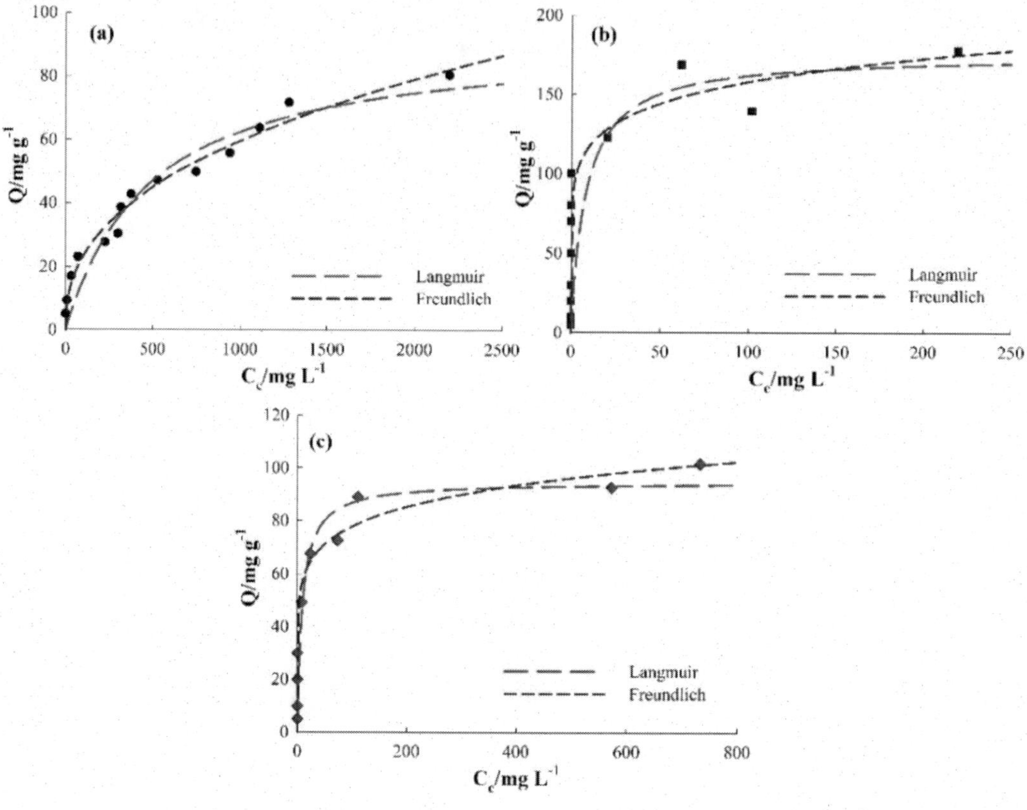

Figure 8. Compatibility of experimentally obtained urease adsorption isotherms to the Langmuir and Freundlich models for (a) PAAmB, (b) PAAmCh, and (c) PAAmChB.

Table 1. Urease adsorption parameters derived from Langmuir and Freundlich models

	PAAmB	PAAmCh	PAAmChB
Langmuir			
X_L/mg g^{-1}	93.0	174.0	94.9
$K_L \times 10^2$/L mg^{-1}	0.2	13.0	10.8
R^2	0.927	0.731	0.858
Freundlich			
X_F	3.6	79.11	42.4
β	0.41	0.15	0.13
R^2	0.976	0.910	0.861

The PAAmCh and PAAmChB isotherms are H-type, which explains the high affinity of urease for these adsorbents according to the Giles classification. The PAAmB isotherm is of the L-type and shows good compatibility with the Langmuir model ($R^2 = 0.927$).

The maximum adsorption capacity of the composites calculated according to the Langmuir model is quite high. The values found in the literature for some adsorbents used in urease adsorption are compared in Table 2.

The adsorption efficiency is evaluated using Equation (6):

$$R_L = \frac{1}{1+K_L C_o} \qquad (6)$$

where R_L is a dimensionless separation factor, K_L is the Langmuir constant (L mg^{-1}), and C_0 is the initial concentration (mg L^{-1}) of the adsorbent. Depending on the magnitude of R_L, adsorption is inefficient (useless) for $1 < R_L$, linear (efficient) for $1 = R_L$, positive (convenient) for $0 < R_L < 1$, and irreversible for R_L=0 [Baybaş and Ulusoy, 2011].

To predict the suitability of the composites in terms of the dimensionless factor, the amount of theoretical adsorbent required to reduce urease from C_0 to C_e was calculated using Equation (7).

Table 2. Comparison of the maximum urease (from jack bean) adsorption capacities (A.C.) for different carriers

Carrier	A.C./mg g^{-1}	Reference
Ni(II)-incorporated microporous polyamide hollow-fiber membrane	78	Akgöl et al. 2002
Poly (HEMA-MAH) poly(HEMA-MAH)-Ni (II)	47.8 66.1	Bayramoğlu et al. 2005
Silica structure PE-MCM-41, MCM-41, SGA	 102 56 21	Hossain et al. 2008
Zn_2 Al – Cl Zn_3 Al – Cl Zn_4 Al – Cl	325 228 215	Vial et al. 2008
Natural halloysite nanotubes	11.7	Zhai et al. 2010
Poly(HEMA-GMA)-IDA-Ni(II) cryogel	11.3	Uygun et al. 2013
Nanoceria (cerium oxıde)	359.6	Kadhem et al. 2016
Epichlorohydrin crosslinked carboxymethyl cellulose (ECH-CMC) graft polyacrylamide (PAm) beads	80-212	Alatawi et al., 2018
PAAmB PAAmCh PAAmChB	93.0 174.0 94.9	This study

$$\frac{W}{V} = \frac{C_o - C_i}{\left(\frac{K_L X_L C_e}{1+K_L C_e}\right)} \qquad (7)$$

where C_0 is the initial concentration (mg L^{-1}), C_e is the equilibrium concentration (mg L^{-1}), W is the mass (g) of adsorbent required, and V is the solution volume (L).

The R_L and absorbent doses (W/V) calculated from Equations (6) and (7) are shown in Table 3. Because all the R_L values are between 0 and 1, all of the adsorbents can be said to be efficient for enzyme immobilization. The fact that the RL values for the adsorbents with Ch are close to 0 at high concentrations can be considered as evidence that PAAmCh and PAAmChB are more efficient than PAAmB.

The absorbent dose (w/v) is important for the industrial use of adsorbents. BATF (Revision and Recodification of Wine Regulations) has allowed the addition of urease to alcoholic and/or fermented beverages to remove carcinogenic ethyl carbamate [Url 1 and 2]. As can be seen from Table 3, the composite mass that needs to be added to a 1 L solution of urease at a concentration of 100 mg L^{-1} to immobilize the enzyme by 90% is less than 2 g for PAAmCh and PAAmChB (0.9 and 1.8 g, respectively). When considering the reusability of the immobilized enzyme and the ease of separation after use, the amount of enzyme used in industry will be greatly reduced and ease of use will be increased. Similarly, enzyme immobilization can be achieved by using low amounts of composites in dialysis machines.

3.1.3. Adsorption Dependence on Time

The change in the urease immobilization on PAAmB, PAAmCh, and PAAmChB at constant concentration and temperature over time was investigated. Pseudo-first-order and pseudo-second-order kinetic models were used in the kinetic evaluation of enzyme immobilization.

The pseudo-first-order (Lagergren) and pseudo-second-order nonlinear rate expressions are given in Equations (8) and (9).

Table 3. Required adsorbent dose and dimensionless factor (R_L) for 50% and 90% adsorption from 1 L enzyme solutions with initial concentrations of 1, 10, 100, and 1000 mg L^{-1}

Composites	C_o/mg L^{-1}	R_L	(w/v)/ g L^{-1}	
			%50	%90
PAAmB	1	1.00	5.24	47.13
	10	0.98	5.29	47.22
	100	0.83	5.77	48.09
	1000	0.33	10.61	56.80
PAAmCh	1	0.88	0.05	0.40
	10	0.43	0.07	0.45
	100	0.07	0.33	0.92
	1000	0.01	2.93	5.59
PAAmChB	1	0.90	0.10	0.89
	10	0.48	0.15	0.97
	100	0.08	0.63	1.83
	1000	0.01	5.38	10.38

$$Q_t = Q_k\left(1 - e^{-kt}\right) \quad (8)$$

$$Q_t \frac{k_2 Q_k^2 t}{1 + k_2 Q_k t} \quad (9)$$

where Q_t (mg g^{-1}) is the experimentally determined amount of enzyme adsorbed on the composite after any time, t, from the initiation of the adsorption and Q_k (mg g^{-1}) is the amount of enzyme adsorbed determined from the fitting of experimental Q_t-t data to this rate model. The constants k_1 (min^{-1}) and k_2 (g mg^{-1} min^{-1}) indicate first- and second-order adsorption rate constants. The Q_k value determined in the first- and second-order rate models defines the theoretical amount adsorbed in the equilibrium adsorption state. The proximity of the Q_e values obtained experimentally in the equilibrium state to the Q_k values is a criterion indicating whether the duration of the adsorption is sufficient. Equations (10)–(13) are used to calculate the initial rate (H: mg g^{-1} min^{-1}) and half-life ($t_{1/2}$: min) for the pseudo-first-order and pseudo-second-order models of adsorption [Baybaş and Ulusoy, 2011].

$$H_1 = k_1 Q_k \quad (10)$$

$$H_2 = k_2 Q_k^2 \quad (11)$$

$$\left(t_{1/2}\right)_1 = \frac{\ln 2}{k_1} \quad (12)$$

$$\left(t_{1/2}\right)_2 = \frac{1}{k_2 Q_k} = \frac{Q_k}{H_2} \quad (13)$$

The Weber–Morris intraparticle diffusion–adsorption model is expressed by Equation (14).

$$Q_t = k_i t^{0.5} \quad (14)$$

where k_i indicates the diffusion rate constant. Q_t-t plotted for pseudo-first-order, t/Q_t-t for pseudo-second-order, and Q_t-$t^{0.5}$ for intraparticle diffusion and their compatibility with the related kinetic models are shown in Figure 9. Table 4 shows the parameters derived from the fit of experimental data to pseudo-first-order, pseudo-second-order, and Weber–Morris models.

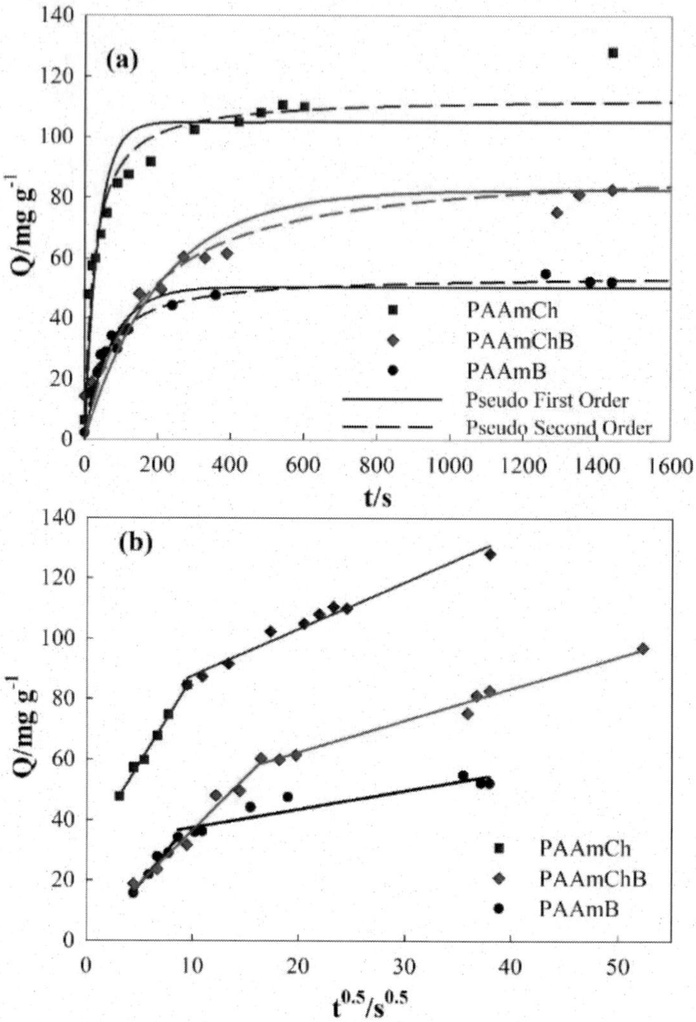

Figure 9. Fitting of urease immobilization kinetics to (a) pseudo-first-order (Lagergren) and pseudo-second-order and (b) Weber–Morris kinetic models.

When examining the regression coefficients given in Table 4, it is observed that the values related to pseudo-second-order kinetics are greater than those for pseudo-first-order and the immobilization kinetics fits the pseudo-second-order model better. The theoretical Q value found from the pseudo-second-order kinetic model is close to the experimental value. The initial adsorption rates derived from this model and half-life values calculated show that immobilization occurs rapidly, especially for PAAmCh.

The intraparticle diffusion rates calculated from the Weber–Morris model were found to be very small for all adsorbents. Two linear components related to the rapid diffusion to the adsorbent surface and intraparticle diffusion obtained from the Weber–Morris model show that the adsorption is chemical and dependent on the concentration.

Table 4. The parameters obtained from the pseudo-first-order (Lagergren), pseudo-second-order, and Weber–Morris kinetic models

Kinetic Model	Adsorbent		
	PAAmB	PAAmCh	PAAmChB
Pseudo-First-Order			
$k_1 \times 10^3$/mg g^{-1} min^{-1}	13.4	26.9	4.8
Q_k/mg g^{-1}	50.4	104.9	82.7
$(t_{1/2})_1$/min	52	26	145
H_1/mg g^{-1} min^{-1}	0.67	2.83	0.39
R^2	0.945	0.858	0.904
Pseudo-Second-Order			
$k_2 \times 10^3$/g mg^{-1} min^{-1}	0.34	0.35	0.07
Q_k/mg g^{-1}	54.7	113.5	92.2
Q_e/mg g^{-1}	51.8	112.4	87.1
$(t_{1/2})_2$/min	54	25	159
H_2/mg g^{-1} min^{-1}	1.02	4.50	0.58
R^2	0.984	0.942	0.942
Weber–Morris			
Diffusion to surface			
k_i/mg g^{-1} min^{-1}	4.31	5.74	3.52
R^2	0.968	0.993	0.973
Intraparticle diffusion			
k_i/mg g^{-1} min^{-1}	0.61	1.55	1.06
R^2	0.883	0.974	0.982

3.1.4. Adsorption Dependence on Temperature

To study the change in urease adsorption with temperature, the composites were each added to solutions containing 2000 mg L^{-1} urease at 4, 15, 25, and 35°C and equilibrium solutions were taken after 24 hours of interaction. The amount of urease adsorbed (Q) was calculated by referencing the measured concentrations (C_0) of the standard 2000 mg L^{-1} urease solutions, which were incubated at optimum temperatures for 24 hours, against the urease content of the equilibrium solutions.

Figure 10 shows the equilibrium urease concentrations derived from the change in the urease immobilization temperature (at 2000 mg L^{-1} concentration) and t-% immobilization curves derived from calculated Q values. The curves show that the percentage immobilization is quite high for Ch-containing composites and does not vary with temperature.

The mean Q values and standard deviations for the four different temperatures are 179.0 ± 20.2 mg g^{-1} for PAAmCh and 176.0 ± 22.4 mg g^{-1} for PAAmChB, with percentage adsorption 98.4 ± 1.9 and 96.6 ± 2.8%, respectively. For PAAmB, the amount and percentage adsorption are lower than those for the Ch-containing composites, but are independent of temperature.

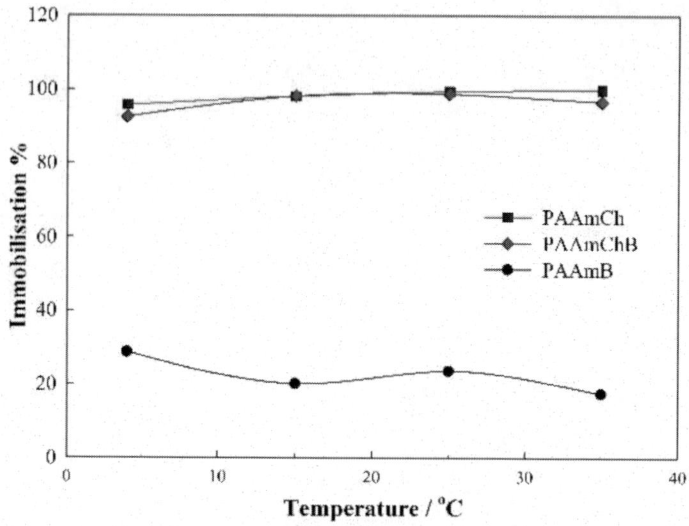

Figure 10. Change of urease immobilization with temperature.

3.4. Urease Activity

The spectrophotometric method and equations used in the activity calculations are explained below.

3.4.1. Measurement of Enzyme Activity by Berthelot Method

The amounts of urease immobilized on the composite samples (Comp-U), and in equilibrium and washing solutions were determined by the Berthelot method (see Sections 2.5 and 2.5.1). Two calibration studies were obtained to measure the amount of urease reciprocal to the [NH^{4+}] concentration. In the first, the absorbance values were measured for standard NH^{4+} solutions and the calibration curve '[NH^{4+}] vs A' was obtained (Figure 11a). In the second, the NH^{4+} concentrations after the hydrolysis reaction in different standard urease-containing solutions were determined using the calibration curve '[NH^{4+}] vs A' and the urease concentrationdependent calibration curve '[C_{urs}] vs [NH^{4+}]' was formed (Figure 11b). This curve was used to find the urease content of the equilibrium and washing solutions depending on their NH^{4+}liberating capacity.

The calibration curve in Figure 11a was also used to investigate the change in the free and immobilized enzyme activity depending on the research variables. Initial rates (v_o; μmol min^{-1} mg^{-1}) were calculated based on the amount of enzyme.

The change of the reaction rate with substrate concentration is expressed by the V_0 - [S] nonlinear Michaelis–Menten equation.

$$v_o = \frac{v_{max}[S]}{K_m+[S]} \quad (15)$$

3.4.2. Changes in Immobilized and Free Enzyme Activity by Substrate Concentration and Determination of Kinetic Parameters

The NH^{4+} concentrations at the end of the 5th minute of the hydrolysis reaction initiated with different urea concentrations and 200 mg L^{-1} free urease was determined using the Berthelot method and the enzyme activity (vo/mM min^{-1}) was calculated for the free enzyme. The values obtained were converted to μmol min^{-1} $mg_{urs}{}^{-1}$.

Figure 12 displays the Michaelis–Menten curves while Table 5 presents the K_m and V_{max} values. Table 6 compares kinetic constants obtained with those reported in previous studies.

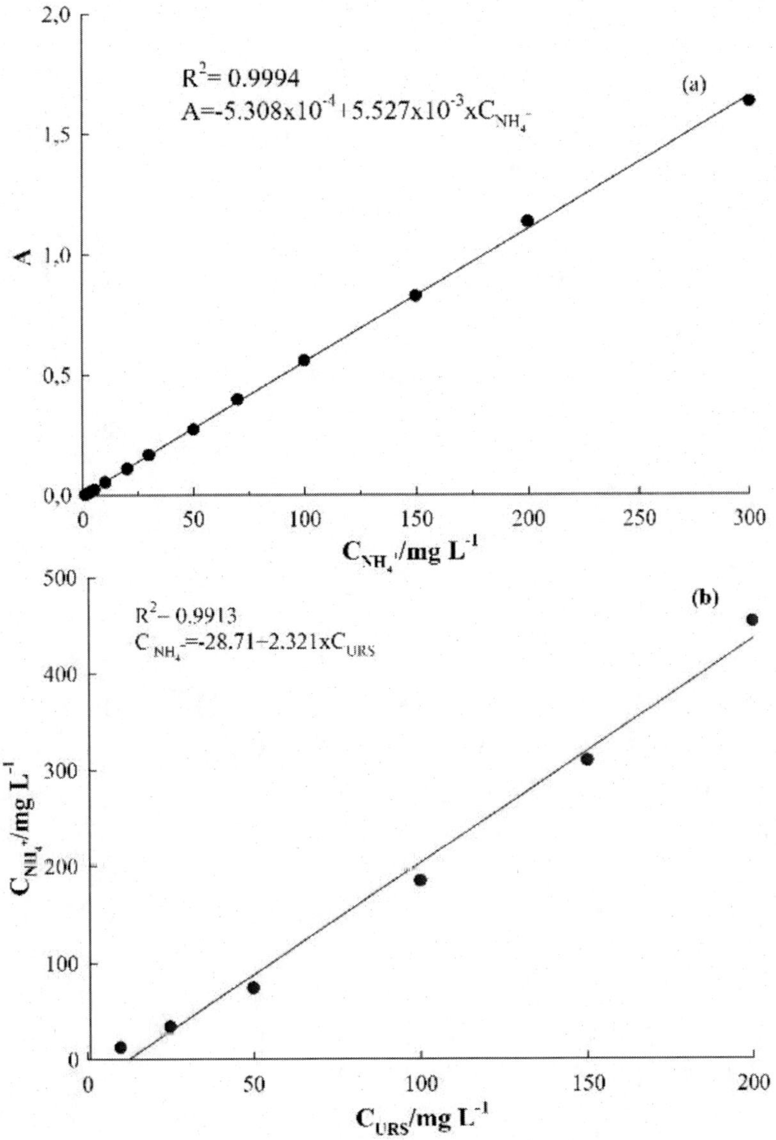

Figure 11. Calibration curves used in enzyme activity calculations (Berthelot method).

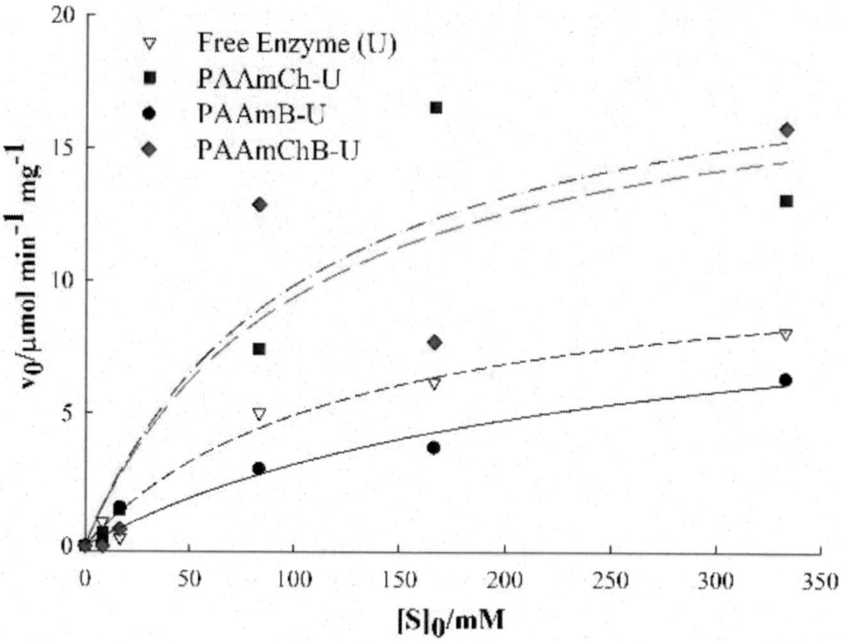

Figure 12. Michaelis–Menten curves obtained for free and immobilized enzymes.

Table 5. Kinetic parameters derived from the compatibility of free and immobilized urease activities to Michaelis–Menten model

Enzyme	R^2	V_{max}/unit mg^{-1}	K_m/mM	(V_{max}/K_m)
Free enzyme (U)	0.988	11.3	128.2	0.0881
PAAmB-U	0.943	10.6	242.6	0.0437
PAAmCh-U	0.890	20.1	103.8	0.1936
PAAmChB-U	0.826	19.0	102.4	0.1855

The K_m value is generally different from the free enzyme value owing to diffusion limitations, steric hindrance, and ionic strength. Because the immobilization process causes structural changes in the enzyme and makes the substrate accessing the active site of the immobilized enzyme more difficult, the affinity of the enzyme for the substrate changes [Bayramoğlu et al., 2003]. It is known that immobilization generally increases the K_m values of enzymes, i.e., the substrate affinity of enzymes immobilized on a solid support decrease. This was repeated in the majority of the studies in Table 6. According to the results of the present study, the K_m values of urease immobilized on the PAAmB composite increased while they decreased in chitosan enzymes. This can be interpreted as indicating that the active regions of the enzyme immobilized to the Ch-containing composites were preserved. A similar situation was observed in the studies by Akgöl and Pithawala et al. (Table 6) [Akgöl et al., 2002; Pithawala et al., 2010].

When comparing the V_{max} values of free and immobilized urease, a decrease is seen for the enzyme immobilized on the composites. This can be explained by the fact that immobilization causes steric-induced restriction of the enzyme–substrate interaction. This decrease is also observed in the studies presented in Table 6, but in one study immobilization increased the V_{max} value [Akgöl et al., 2002].

Table 6. Examples of urease immobilization studies in the literature

Support Material	Experimental conditions			V_{max} (U mg^{-1})	K_m (mM)	Equation	Reference
	C_{URS}[1]	t/°C	pH				
Hydroxyapatite	304	25	8	FE[2]:230.7 İE[3]: 152.9	FE[2]: 7.45 İE[3]: 6.89	M.M.[4]	Marzadori et al., 1998
Polyamide hollow-fiber membrane	2000	25	6	FE[2]:25.9 İE[3]: 59.7	FE[2]: 18 İE[3]: 22	L.B.[5]	Akgöl et al., 2002
Plain poly(HEMA-GMA2) Spacer-arm-attached poly(HEMA–GMA2) film	1000	22	7.5	FE[2]:59.7 İE[3]: 26.6 33.4	FE[2]:18 İE[3]: 32 24	L.B.[5]	Bayramoğlu et al., 2003
Poly(HEMA-MAH) Poly(HEMA-MAH)–Ni	2000	22	6.5	FE[2]:63.4 İE[3]: 24.2 27.2	FE[2]:16 İE[3]: 26 21	L.B.[5]	Bayramoğlu et al., 2005
Ca alginate Paraffin wax Lac	1000		7	FE[2]:154 İE[3]: 154 100 125	FE[2]:0.50 İE[3]: 0.36 0.40 0.26	L.B.[5]	Pithawala et al., 2010
TiO_2 TiO_2-Ch	1500 1000	25	7	FE[2]:1.22U İE[3]:0.899U 0.402 U	FE[2]:0.071 İE[3]: 0.321 0.121	L.B.[5]	Doğaç et al., 2014
ECM-g-Pam	0.3	35	7	FE[2]: 5 ± 0.2 İE[3]: 2 ± 0.2	FE[2]:11 ± 0.5 İE[3]:14 ± 0.7.	L.B.[5]	Alatawi et al., 2018

[1]mg L^{-1}; [2]Free Enzyme; [3]Immobilized Enzyme; [4]Michealis–Menten nonlinear plot; [5]Lineweaver–Burk linear plot

3.4.3. Change in Immobilized and Free Urease Activity with pH

Free and immobilized urease activity was determined using 10000 mg L^{-1} (166.5 mM) urea solutions prepared in 10 mM phosphate buffers at different pH values. Activities were calculated by measuring the amount of ammonium formed after 5 minutes and the results were plotted as pH vs. % activity (Figure 13). Because the active surfaces of the enzymes and composites were considered to be affected differently by the ions of the salts (Na_3PO_4, Na_2HPO_4, NaH_2PO_4) used in pH adjustment, all pH values were adjusted with phosphate salts and H_3PO_4.

The optimum pH for the free enzyme was 7.5; the same value was found for PAAmB, while it was ~6.4 for PAAmChB, and between 3 and 6.4 for PAAmChB. The activity value for PAAmCh remained the same at pHs of 3.1 and 6.4. The optimum pH for PAAmB could not be determined under these conditions.

From Figure 13, it can be said that the immobilized enzymes are active at different pH values. In particular, urease immobilized on PAAmCh is less affected by pH changes.

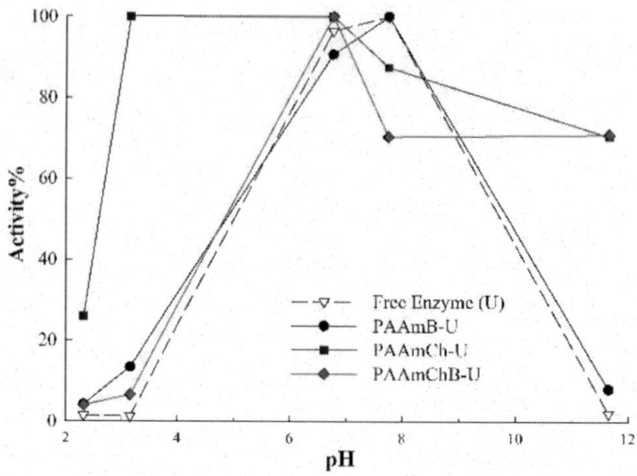

Figure 13. Change in free and immobilized enzyme activities with pH.

3.4.4. Change in Immobilized and Free Urease Activity with Temperature

The effect of temperature on the activity of the free and immobilized enzymes is shown in Figure 14. The temperature at which the enzyme shows 100% activity is called the optimum temperature. It can be seen from Figure 15 that the optimum temperature of the free enzyme is 35°C. The optimum temperature for PAAmCh-U and PAAmChB-U with the immobilized enzyme was outside of the studied temperature range. The increase in the optimum temperature of the Ch-containing composites suggests that immobilized urease is more thermally stable. This proves that the Ch–urease complex structure, which provides additional stability to urease, contains strong molecular bonds. There are cases in the literature where similar shifts in the optimum temperature have been observed for immobilized urease. For example, the optimum pH for dye-Ni-loaded hollow-fiber membranes shifted from 45°C (free enzyme) to 55°C [Akgöl et al., 2002].

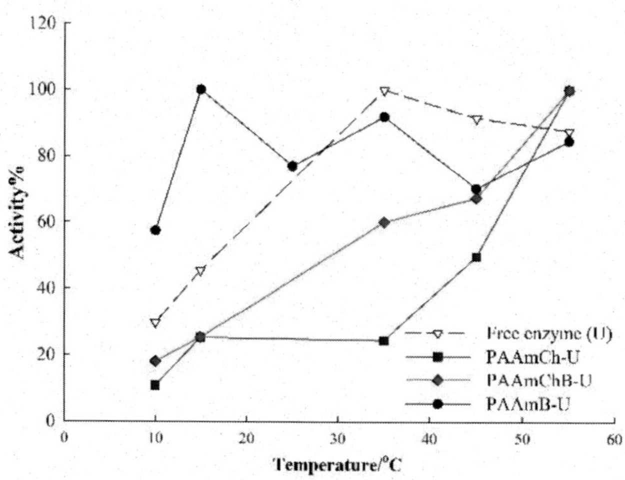

Figure 14. Change of free and immobilized enzyme activities with temperature.

3.4.5. Reusability of Immobilized Enzyme

The composite-immobilized enzymes at 25°C and pH 7 were reused seven times and washed with pH 7 phosphate buffer after each consecutive use. The NH^{4+} concentrations (mM) were calculated after 5 minutes interaction with urea for each use.

The mean activity values and standard deviations were 191.0 ± 30.6 unit g^{-1} for PAAmB, 189.8 ± 23.5 unit g^{-1} for PAAmCh, and 198.5 ± 58.5 unit g^{-1} for PAAmChB. Although the amounts of enzyme immobilized on the different composites used in this study differ significantly, the immobilized enzymes are suitable for seven consecutive uses and remain stable.

4. Conclusion

The results of the FT-IR, XRD, DSC, and SEM analyses prove that the PAAmB, PAAmCh, and PAAmChB composites are hybrid structures formed from the components PAAm, Ch, and B.

The experiments involving urease immobilization on the fine grained composites showed that:

- PAAmCh and PAAmChB isotherms are H-type while the PAAmB isotherm is L-type, according to the Giles classification. This implies that the Ch-containing composites have a higher affinity for urease.
- The amount of composite that needs to be added to a 1 L urease solution to immobilize the enzyme at a concentration of 100 mg L^{-1} is very low, especially for PAAmCh and PAAmChB.
- Initial adsorption rates and half-lives derived from the pseudo-second-order kinetic model show that immobilization occurs rapidly, especially for PAAmCh.
- The immobilization at different temperatures revealed that the percentage immobilization is quite high for Ch-containing composites and does not change with temperature.

The change in the activity with substrate concentration, pH, and temperature was investigated using coarse-grained composites. The results obtained from immobilization and activity experiments show that Ch-containing composites are useful support materials for urease immobilization and reusable.

ACKNOWLEDGMENT

This study was supported by Sivas Cumhuriyet University Scientific Research Projects (CÜBAP) coordination department in project number F-512.

REFERENCES

Akgöl, S., Yalçınkaya, Y., Bayramoğlu, G., Denizli, A. & Arıca, M. Y. (2002). Reversible immobilization of urease onto Procion Brown MX-5BR-Ni(II) attached polyamide hollow-fibre membranes. *Process Biochemistry*, *38*, 675-683.

Alatawi, F. S., Monier, M. & Elsayed, N. H. (2018). Amino functionalization of carboxymethyl cellulose for efficient immobilization of urease. *International Journal of Biological Macromolecules*, *114*, 1018–1025.

Alexandre, M. & Dubois, P. (2000). Polymer-layered silicate nanocomposites: preparation, properties and uses of a new class of materials. *Materials Science and Engineering*, *28*, 1-63.

An, N., Zhou, C. H., Zhuang X. Y., Tong, D. S. & Yu, W. H. (2015). Review article: Immobilization of enzymes on clay minerals for biocatalysts and biosensors. *Applied Clay Science*, *114*, 283–296.

Andrich, L., Esti, M. & Moresi, M. (2010). Urea degradation kinetics in model wine solutions by acid urease immobilised onto chitosan-derivative beads of different sizes. *Enzyme and Microbial Technology*, *46*, 397–405.

Anirudhan, T. S., Suchithra, P. S. & Rijith, S. (2008). Amine–modified polyacrylamide–bentonite composite for the adsorption of humic acid in aqueous solutions. *Colloids and Surfaces A: Physicochemical and Engineering Aspects*, *326*, 147–156.

Bai, J., Li, Y., Dub, J., Wang, S., Zheng, J., Yang, O. & Chenc, X. (2007). One-pot synthesis of polyacrylamide-gold nanocomposite. *Materials Chemistry and Physics*, *106*, 412–415.

Baybaş, Demet. (2009). "The characterization of polyacrylamide-aluminosilicate composites and the investigations for their adsorptive features for thorium" PhD thesis, Cumhuriyet University, 2009. https://tez.yok.gov.tr/UlusalTezMerkezi/giris.jsp (Thesis No: 245935).

Baybaş, D. & Ulusoy, U. (2011). The use of polyacrylamide-aluminosilicate composites for thorium adsorption. *Applied Clay Science*, *51*, 138–146.

Bayramoğlu, G., Altınok, H., Bulut, A., Denizli, A. & Arıca, M. Y. (2003). Preparation and application of spacer-arm-attached poly(hydroxyethyl methacrylate-co-glycidyl methacrylate) films for urease immobilisation. *Reactive and Functional Polymers*, *56*, 111–121.

Bayramoğlu, G., Yalçın, E. & Arıca Y. (2005). Immobilization of urease via adsorption onto L-histidine–Ni(II) complexed poly(HEMA-MAH) microspheres: Preparation and characterization. *Process Biochemistry*, *40*, 3505–3513.

Chaney, A. L. & Marbach E. P. (1962). Modified reagents for determination of urea and ammonia. *Clinical Chemistry*, *8*(2), 130-132. http://clinchem.aaccjnls.org/content/clinchem/8/2/130.full.pdf.

Desimone, M. F., Matiacevich, S. B. Buera, P. & Diaz, L. E. (2008). Effects of relative humidity on enzyme activity immobilized in sol–gel-derived silica nanocomposites. *Enzyme and Microbial Technology*, *42*, 583–588.

Dey, S. C., Al-Amin, M., Rashid, T. U., Sultan, Md. S., Ashaduzzaman, Md., Sarker M. & Shamsuddin, S. Md. (2016). Preparation, characterization and performance evaluation of chitosan as an adsorbent for remazol red. *International Journal of Latest Research in Engineering and Technology*, *2*(2), 52-62.

Doğaç, Y. I., Deveci, I., Teke, M. & Mercimek, B. (2014). TiO_2 beads and TiO_2-chitosan beads for urease immobilization. *Materials Science and Engineering C*, *42*, 429–435.

Eninanç, Onur. "Synthesis and characterization of polyacrylamide-chitosan/zeolite and investigation for its adsorptive features for Th, Ac-228, Pb-212, Bi-212 and Tl-208." MSc thesis, Cumhuriyet University, 2013. https://tez.yok.gov.tr/UlusalTezMerkezi/giris.jsp (Thesis No: 338180).

Hossain, K. Z., Monreal, C. M. & Sayari, A. (2008). Adsorption of urease on PE-MCM-41 and its catalytic effect on hydrolysis of urea. *Colloids and Surfaces B: Biointerfaces*, *62*, 42–50.

Işikver, Y. (2017). Removal of some cationic dyes from aqueous solution by acrylamide- or 2-hydroxyethyl methacrylate-based copolymeric hydrogels. *Fibers and Polymers*, *18*, 2070-2078.

Jesionowski, T., Zdarta, J. & Krajewska, B. (2014). Enzyme immobilization by adsorption: a review. *Journal of the International Adsorption Society*, *20*(5-6), 801-821.

Kadhem, A. H. H., Kadhem, K. M. & Andersonb, G. E. (2016). Immobilization of urease enzyme on nanoceria modifies secondary and tertiary protein structures. *Acta Chimica Slovaca*, *9*, 44-53.

Kittur F. S., Prashanth K. V. H., Sankar K. U. & Tharanathan, R. N. (2002). Characterization of chitin, chitosan and their carboxymethyl derivatives by differential scanning calorimetry. *Carbohydrate polymers*, *49*, 185-193.

Krajewska, B. (2009)a. Ureases. II. Properties and their customizing by enzyme immobilizations: A review. *Journal of Molecular Catalysis B: Enzymatic*, *59*, 22–40.

Krajewska, B. (2009)b. Ureases I. Functional, catalytic and kinetic properties: A review. *Journal of Molecular Catalysis B: Enzymatic*, *59*, 9–21.

Kumar, J. & Melo, J. S. (2016). Jackfruit membrane as new biomaterials for bio-catalytic application in detection of urea. *Journal of Biosensors and Bioelectronics*, *7*(4), 1-5. DOI: 10.4172/2155-6210.1000226.

Kunduru, K. R., Raju Kutcherlapati, S. N., Arunbabu, D. & Jana, T. "Armored Urease: Enzyme-Bioconjugated Poly(acrylamide) Hydrogel as a Storage and Sensing Platform" In *Methods Enzymol*, *590*, 143-167. doi: 10.1016/bs.mie.2017.02.008. Epub 2017.

Marzadori, C., Miletti, S., Gessa, C. & Ciurli, S. (1998). Immobilization of jack bean urease on hydroxyapatite: urease immobilization in alkaline soils. *Soil Biology and Biochemistry*, *30*(12), 1485-1490.

Öner, Serhat. "The preparation and characterization of bentonite/ chitosan/poly(acrylamide) composites: the use of urease immobilization." MSc thesis, Sivas Cumhuriyet University, 2018. https://tez.yok.gov.tr/UlusalTezMerkezi/giris.jsp (Thesis No: 338180).

Pithawala, K., Mishra, N. & Bahadur, A. (2010). Immobilization of urease in alginate, paraffin and lac. *Journal of Serbian Chemical Society*, *75* (2), 175–183.

Queiroz, M. F., Melo, K. R. T., Sabry, D. A., Sassaki, G. L. & Rocha, H. A. O. (2015). Does the use of chitosan contribute to oxalate kidney stone formation, *Marine Drugs*, *13*, 141-158.

Silva, M. E. S. R., Dutra, E. R., Mano, V. & Machado, J. C. (2000). Preparation and thermal study of polymers derived from acrylamide. *Polymer Degradation and Stability*, *67*, 491-495.

Su, Z. Q., Wu, S. H., Zhang, H. L. & Feng, Y. F. (2010). Development and validation of an improved Bradford method for determination of insulin from chitosan nanoparticulate systems. *Pharmaceutical Biology*, *48*, 966-973.

Sujoy, B. & Aparna, A. (2013). Enzymology, Immobilization and applications of urease enzyme. *International Research Journal of Biological Sciences*, *2*(6), 51-56.

Url 1: https://www.fda.gov/food/chemicals/ethyl-carbamate-preventative-action-manual#part5.

Url 2: https://eur-lex.europa.eu/legal-content/EN/TXT/ ?uri=CELEX:32000R1622.

Uygun, M., Akduman, B., Akgöl, S. & Denizli, A. (2013). A new metal-chelated cryogel for reversible ımmobilization of urease. *Applied Biochemistry Biotechnology*, *170*, 1815–1826.

Vial, S., Prevot, V., Leroux, F. & Forano, C. (2008). Immobilization of urease in ZnAl layered double hydroxides by soft chemistry routes. *Microporous and Mesoporous Materials*, *107*, 190–201.

Yurekli, Y. & Altinkaya, S. (2011). Catalytic performances of chemically immobilized urease under static and dynamic conditions: A comparative study. *Journal of Molecular Catalysis B: Enzymatic*, *71*, 36–44.

Zeinali, M. & Lenjannezhadian, H. (2018). Degradation of urea by entrapped cross-linked urease aggregates: a combinatorial approach to urease stabilization for environmental and industrial applications. *International Journal of Environmental Science and Technology.*, *15*(1), 49-56.

Zhai, R., Zhang, B., Liu, L., Xie, Y., Zhang, H. & Liu, J. (2010). Immobilization of enzyme biocatalyst on natural halloysite nanotubes. *Catalysis Communications*, *12*, 259–263.

In: An Introduction to Aluminosilicates
Editor: Nero Regina Blevins
ISBN: 978-1-53617-250-8

Chapter 7

CRYSTALLINE ALUMINOSILICATES FOR RADIONUCLIDE REMEDIATION

Brian J. Riley and Saehwa Chong
Pacific Northwest National Laboratory, Richland, WA, US

ABSTRACT

Crystalline aluminosilicates have a wide range of applications such as media for the capture and/or immobilization of radionuclides. Several different types of metal-exchanged zeolites have been successfully demonstrated as effective sorbents for gaseous, aqueous, or solid-state capture and immobilization of radionuclides including iodine [e.g., $I_{2(g)}$, $CH_3I_{(g)}$], xenon, krypton, radium (e.g., ^{226}Ra and ^{228}Ra), selenium (e.g., $^{79}Se^{6+}$), technetium (e.g., $^{99}TcO_4^-$), strontium (e.g., $^{90}Sr^{2+}$), cesium (e.g., $^{135}Cs^+$, $^{137}Cs^+$), mixed alkali halide streams (i.e., iodide and chloride salts), and actinides (e.g., U^{6+}, Th^{4+}). These zeolites include *sodalite* (SOD), *mordenite* (MOR or Z-type), *faujasite* (FAU), *Linde type-A* (LTA), *cancrinite* (CAN), *gismondine* (GIS), and *heulandite* (or clinoptilolite) (HEU). The structural building blocks of these seven materials vary from cages (e.g., α-cages, β-cages, ε-cages) to channels linking various cage types and these are mostly comprised of aluminosilicates. This chapter will delve into the crystallographic differences between these framework materials and provide a brief overview of literature data where these seven crystalline aluminosilicates have been utilized for the capture and/or immobilization of radionuclides. Metal-exchanged zeolites (e.g., MOR, FAU, LTA) have often been the source of study for gaseous capture of iodine where the more promising compositions include Ag-exchanged sorbents; these types of materials can be made in a variety of robust forms including berl saddles and extrudates. Sodalite can be used as an immobilization strategy for halide waste streams (e.g., Cl, I), including those from electrochemical reprocessing of used nuclear fuel, as well as TcO_4^-. Cancrinite has been demonstrated to enhance sorption capacity of sediments for several problematic ionic radionuclides including Cs^+, SeO_4^{2-}, and TcO_4^-. Actinides such as U^{6+} and Th^{4+} have been successfully captured from solutions using MOR, FAU, LTA, and HEU-type aluminosilicates. Finally, mixtures of aluminosilicates have been included in a set of zeolite materials that show promise as sorbents for ^{226}Ra and ^{228}Ra removal from mine waters.

1. INTRODUCTION

Crystalline aluminosilicates have a lot of documented uses but some of the more prevalent applications include the purification of aqueous and gaseous streams. This is typically demonstrated through one or more of the following processes: ion-exchange, chemisorption, physisorption, and/or molecular sieving. Consequently, aluminosilicates can be used for the remediation of radionuclides. The primary aluminosilicate systems that will be discussed herein include sodalite (SOD), mordenite (MOR), faujasite (FAU), Linde Type-A (LTA), cancrinite (CAN), gismondine (GIS), and heulandite (HEU). While these materials can be comprised of nonaluminosilicates, the focus of the current summary will center on the aluminosilicates.

Figure 1 provides an overview of the general structures for the aluminosilicate networks discussed herein along with schematics of the building blocks and secondary and composite building units.

While different space groups are possible in these systems, even within a given family, example space groups with unit cell parameters are provided in Table 1. Also, the crystal structure can be visualized in several different ways from simple to more complex as shown in Figure 2 for SOD.

The SOD crystal structure is comprised of ***sod*** building blocks and ***6*** secondary building units. The MOR crystal structure is comprised of ***mor*** composite building blocks and ***5-1*** secondary building units. The FAU crystal structure is comprised of ***d6r*** and ***sod*** building blocks along with secondary building units of ***6-6***, ***6-2***, ***6***, ***4-2***, ***1-4-1***, or ***4***. The LTA crystal structure is comprised of ***d4r***, ***sod***, and ***lta*** building blocks as well as secondary building units of ***8***, ***4-4***, ***6-2***, ***6***, ***1-4-1***, or ***4***. The CAN crystal structure is comprised of ***dzc*** and ***can*** building blocks along with ***12***, ***6***, or ***4*** secondary building units. The GIS crystal structure is comprised of ***gis*** and ***dcc*** building blocks along with ***8*** or ***4*** secondary building units. The HEU crystal structure is comprised of ***bre*** composite building blocks and ***4-4=1*** secondary building units.

In most cases, these types of aluminosilicates can be altered chemically where the alkali and alkaline earths can be replaced for different species using ion exchange techniques. Additional processes can be used to ion exchange such as soaking the crystalline materials in molten salts; this process was used for sodalites by Stein et al. [2].

Table 1. Summary of crystal structure information for the crystalline aluminosilicates discussed herein. Note that SG = space group (# = SG number); *a*, *b*, and *c* denote unit cell parameters (if only one value is present, then $a = b = c$); α, β, and γ denote unit cell angles; V denotes unit cell volume; and ρ_c denotes the calculated density

Type	Example chemistry	SG (#)	*Unit cell parameter(s) a, b, c* (Å)	α, β, γ (°)	V (Å^3)	ρ_c (Mg m^{-3})	Reference
SOD	$Na_8(AlSiO_4)_6Cl_2$	$P\bar{4}3n$ (218)	8.8767	90	699.5	2.30	[3]
MOR	$Na_8(Al_8Si_{40}O_{96})\cdot 24H_2O$	*CmCm* (63)	18.13, 20.49, 7.52	90	2793.6	2.10	[4]
FAU	$Na_{56}(Al_{56}Si_{136}O_{384})\cdot nH_2O$	$Fd\bar{3}m$ (227)	24.7499	90	15160.7	1.40	[5]
LTA	$Na_{12}(Al_{12}Si_{12}O_{48})\cdot nH_2O$	$Pm\bar{3}$m (221)	12.32	90	1869.9	1.99	[6]
CAN	$Na_6Ca(AlSiO_4)_6(CO_3)\cdot 2H_2O$	$P6_3$ (173)	12.75, 12.75, 5.14	90, 90, 120	723.6	2.27	[7]
GIS	$Ca_2(Al_4Si_4O_{16})\cdot 8H_2O$	$P2_1/c$ (14)	10.02, 10.62, 9.84	90, 92.42, 90	1046.2	2.22	[8]
HEU	$Ca_{2.9}$ $Na_{1.1}(Si_{28.8}Al_{7.2})O_{72}\cdot 20.5H_2O$	*C*2/m (12)	17.672, 17.930, 7.443	90, 116.5, 90	2110.6	2.066	[9]

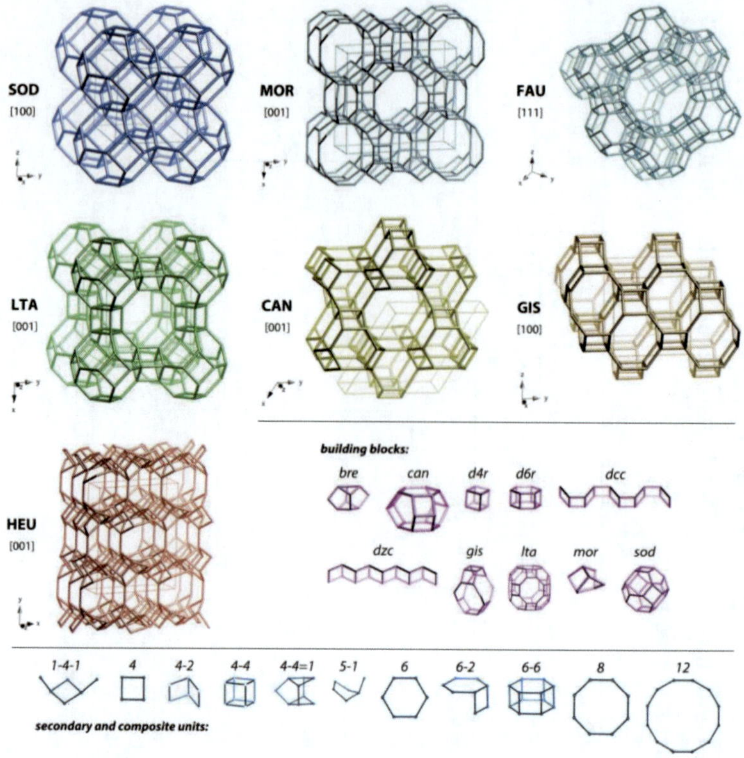

Figure 1. Summary of crystal structures and morphologies for sodalite (SOD), mordenite (MOR), faujasite (FAU), Linde type-A (LTA), cancrinite (CAN), gismondine (GIS), and heulandite (HEU) with the structural building blocks as well as secondary and composite units (see text for further description). These images were modified from Baerlocher, McCusker, and Olson [1] (© Elsevier 2007).

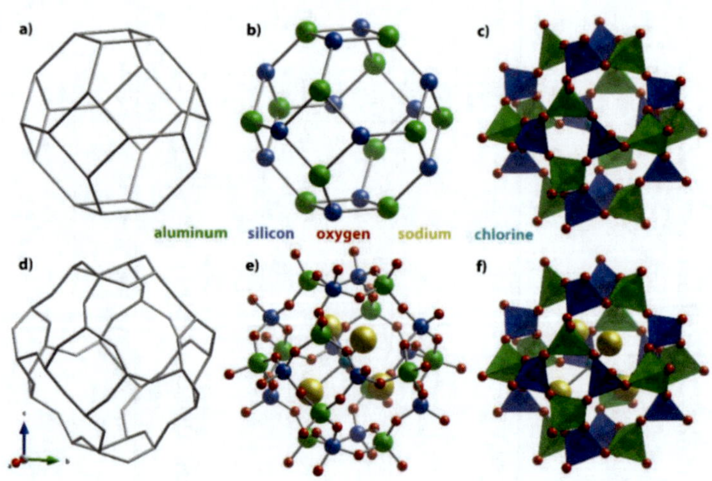

Figure 2. Different visualizations of the SOD (sodium chlorosodalite) unit cell including (a) the framework without oxygen, (b) as ball-and-stick without oxygen, (c) the polyhedra without the $ClNa_4$ tetrahedron, (d) the framework with oxygen, (e) as ball-and-stick with all atoms, and (f) the polyhedra with all atoms. The reader is referred to the online version of this figure for color references to the elements.

2. STRUCTURAL COMMENTARY

2.1. SOD Structure

Sodalite has a general chemical formula of $M_8(AlSiO_4)_6X_2$, where *M* and *X* represent +1 cations (e.g., Li^+ [10], Na^+ [3, 10, 11], K^+ [10], Ag^+ [12, 13], Tl^+ [13]) and -1 anions (e.g., OH^- [11], Cl^- [3, 14, 15], Br^- [14, 15], I^- [14, 15], IO_3^- [16]), respectively. Sodalites with mixed +1 cations, +2 cations (e.g., Fe^{2+} [17], Mg^{2+} [18], Ca^{2+} [19]), and -2 anions (e.g., CO_3^{2-} [20], MoO_4^{2-} [21], WO_4^{2-} [19]) are reported in the literature, but are less common. The framework of the aluminosilicate β-cage contains 6-membered rings and 4-membered rings formed by alternating AlO_4 and SiO_4 tetrahedra connected by shared oxygen atoms (Figure 3). The *X* anion is at the center of the β-cage, tetrahedrally coordinated by four *M* cations, and the cubic unit cell parameters increase with larger ions in the cage. Sodalites with +2 cations or without an anion [e.g., $Na_8(AlSiO_4)_6 \cdot 8H_2O$] inside the β-cage have two less cations per unit cell for charge balance. Sodalite generally has a cubic space group of $P4\bar{3}n$ [3, 10-13], but other symmetries including $P\bar{4}3m$ [22], $I\bar{4}3m$ [21], $Pm\bar{3}n$ [18], $P23$ [20], $P6_3$ [23], $Aba2$ [19], $Abm2$ [24], and $P1$ [25] have been reported. Sodalite is one of the very limited aluminosilicate structures that can contain a halide anion, and thus has several niche applications that do not apply for the other species discussed herein.

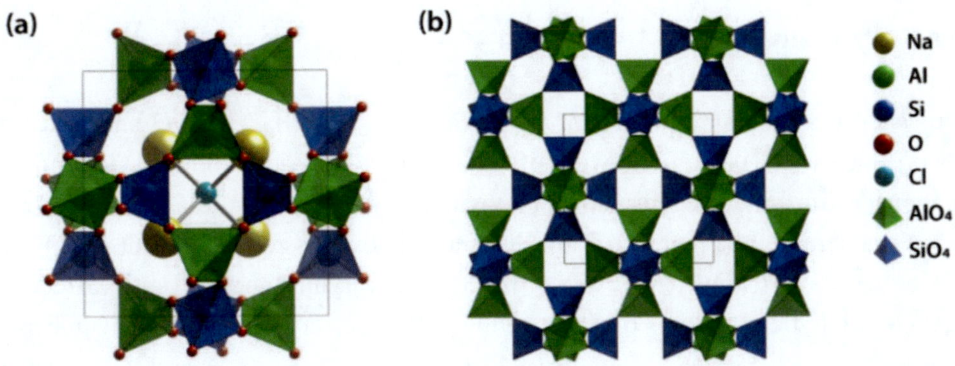

Figure 3. (a) Structure of the β-cage formed by AlO_4 and SiO_4, showing 6-membered and 4-membered rings and a Cl^- anion tetrahedrally coordinated by four Na^+ cations. (b) Stacking of the β cages in the sodalite structure viewed along the [100] direction. The squares represent the unit cell.

2.2. MOR Structure

Mordenite has a general formula of $M_8(Al_8Si_{40}O_{96}) \cdot n\, H_2O$, where *M* can be +1, +2, or mixed cations (e.g., Na^+ [4], Rb^+ [26], Ca^{2+} [27], Ba^{2+} [28]; the moles of *M* are halved for +2 cations). The framework of mordenite contains channels with 12-membered and 8-

membered rings interconnected by smaller channels along the [001] direction, and *M* cations are positioned in the channel of 8-membered rings (Figure 4). The Si and Al atoms are positioned at the same sites with different occupancies in the structure, and varying the Si/Al ratio can shift the positions of *M* cations in the structure [29]. Mordenite has an orthorhombic space group of *CmCm* [4] or less commonly, $Cmc2_1$ [30] or *Pbcn* [28].

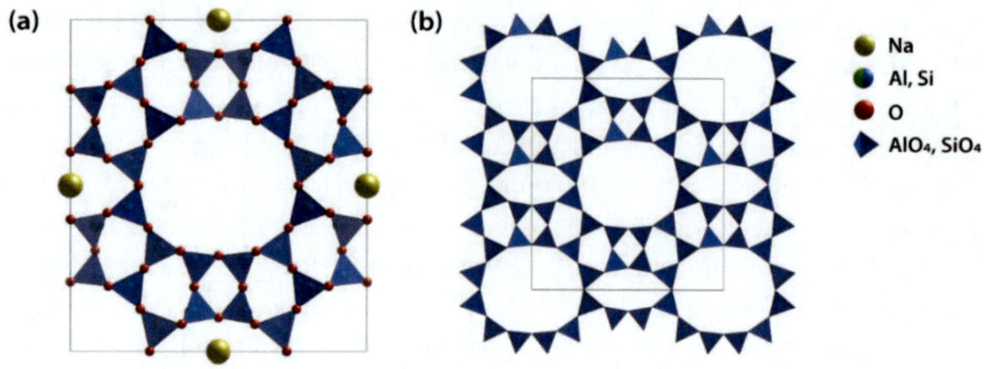

Figure 4. (a) Framework of MOR showing 12-membered rings and Na containing 8-membered ring channels. (b) Interconnection of large channels by smaller channels viewed along the [001] direction. The squares represent the unit cell.

2.3. FAU Structure

Faujasite has a general formula of $M_{56}(Al_{56}Si_{136}O_{384}) \cdot n\ H_2O$ with +1 and +2 cations (e.g., Na^+ [5], Ca^{2+} [31]). The framework of faujasite is composed of the β-cages connected by hexagonal prism units, creating supercages with ~12 Å diameter and 12-membered rings with ~8 Å diameter [32] (Figure 5). Faujasite is also known as zeolite X (SiO_2/Al_2O_3 ratio of 2–3) or zeolite Y (SiO_2/Al_2O_3 ratio >3) [33], and varying the Si/Al ratio affects the physical and catalytic properties of faujasite [32]. Faujasite crystallizes in the cubic space group of $Fd\overline{3}m$.

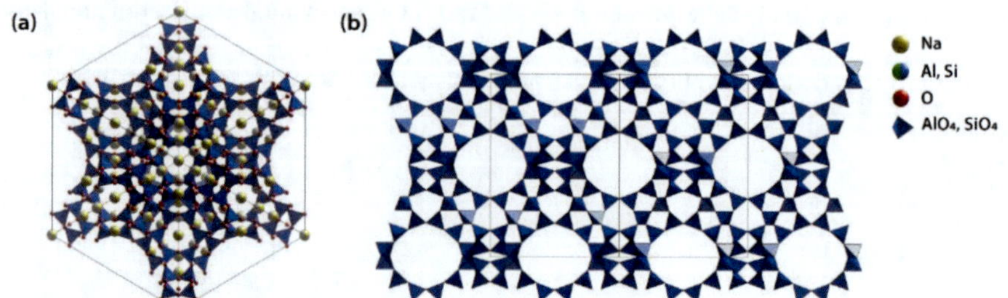

Figure 5. (a) Structure of FAU framework viewed from the [111] direction.
(b) Stacking of supercages showing 12-membered rings viewed along the [110] direction. The black lines represent the unit cell.

2.4. LTA Structure

Linde Type-A (LTA), also known as zeolite A, has a general formula of $M^{+}_{12}(Al_{12}Si_{12}O_{48}) \cdot n\, H_2O$ where *M* represents +1 cations (e.g., Li^+, Na^+, Ag^+, Tl^+), and *n* varies from 20 to 30 [6]. LTA with mixed +1 cations (e.g., $Cs_7Na_5(Al_{12}Si_{12}O_{48}) \cdot x\, H_2O$) [34], mixed +1 and +2 cations [e.g., $Mn_{4.5}Na_3(Al_{12}Si_{12}O_{48}) \cdot x\, H_2O$] [35], and +2 cations [e.g., $Ca_6(Al_{12}Si_{12}O_{48}) \cdot x\, H_2O$] [6] are reported in the literature. The aluminosilicate framework is composed of α-cages and β-cages formed by AlO_4 and SiO_4 tetrahedra. The α-cage contains 8-membered rings, 6-membered rings, and 4-membered rings with the cavity diameter of 11.4 Å, and the β-cage contains 6-membered rings and 4-membered rings with the cavity diameter of 6.6 Å in the case of Na-containing LTA (Figure 6) [6]. Zeolite A generally has a cubic space group of $Pm\bar{3}m$ [6] in which Al and Si equally occupy the same Wyckoff position of 24*k* but also has a less common space group of $Fm\bar{3}c$ [36].

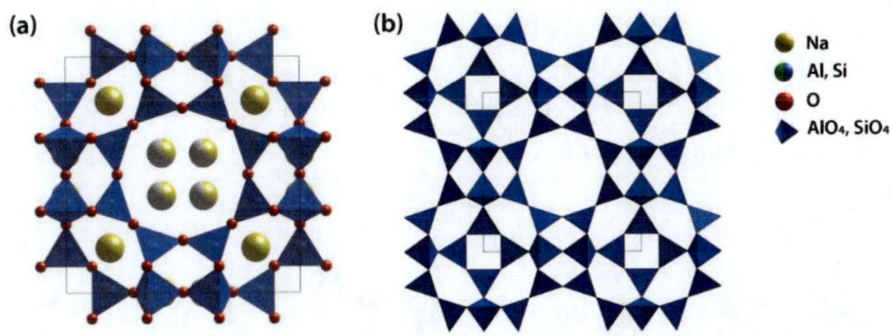

Figure 6. (a) Structure of the α-cage formed by AlO_4 and SiO_4 tetrahedra and distribution of cations within the cage. (b) Framework of LTA showing the α-cage at the center of the unit cell and β-cages at the corners. The structures are viewed along the [100] direction, the white area in the atoms shows the unoccupied percentage, and the black squares represent the unit cell.

2.5. CAN Structure

Cancrinite has a general formula of $M_0(AlSiO_4)_6(CO_3) \cdot 2H_2O$ where *M* denotes +1, +2, or mixed cations (e.g., Na^+, Ca^{2+}), and CO_3^{2-} can be replaced by two -1 anions such as OH^-. The framework of cancrinite contains 12-membered and 6-membered ring channels formed by alternating AlO_4 and SiO_4 tetrahedra along the [001] direction, and these rings are interconnected by small ε-cages (Figure 7). The cations are positioned in the channels and cages, and the mixed cations of +1 and +2 can occupy the same sites with partial occupancies, whereas the CO_3^{2-} anion is positioned at the center of the 12-membered ring channel. Cancrinite generally crystallizes in the hexagonal space group of $P6_3$.

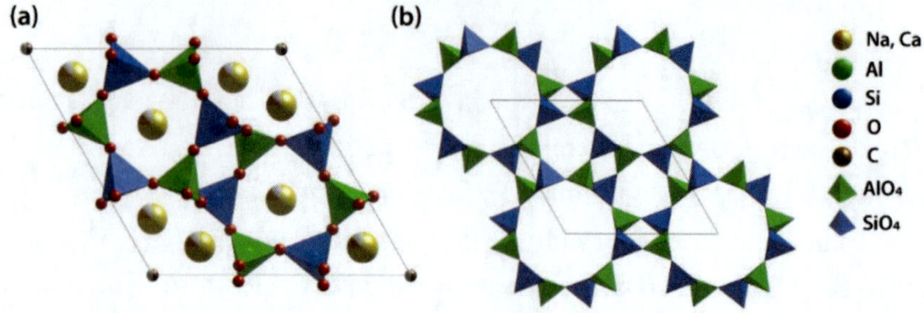

Figure 7. (a) Structure of CAN channels by AlO_4 and SiO_4 tetrahedra and distribution of cations and anions within the unit cell. (b) Stacking of channel and cage structures viewed from the [001] direction.

2.6. GIS Structure

Gismondine has a general formula of $Ca_2(Al_4Si_4O_{16}) \cdot 8H_2O$, but an Na^+-containing structure with a similar topology has been reported [37]. The framework of gismondine is composed of 8-membered and 4-membered rings (Figure 8). The ribbons formed by AlO_4 and SiO_4 tetrahedra run along *a*- and *c*-axes and are interconnected to form 8-membered rings parallel to (100) and (001) planes. Gismondine generally crystallizes in the monoclinic space group of $P2_1/c$.

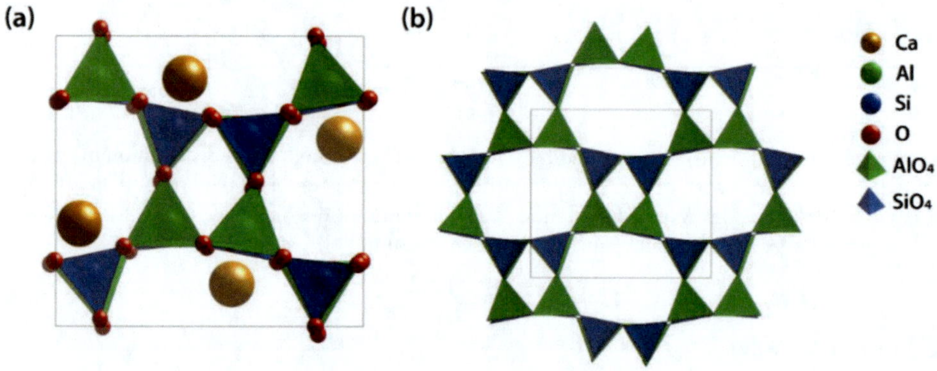

Figure 8. (a) Structure of GIS unit cell showing distribution of atoms.
(b) The 4-membered and 8-membered rings formed by the AlO_4 and SiO_4 ribbons.

2.7. HEU Structure

Clinoptilolite and heulandite have a general formula of $M_8[Al_6(Al,Si)_2Si_{28}O_{72}] \cdot n\,H_2O$ where *M* denotes +1, +2, or mixed cations (e.g., Na^+, K^+, Cs^+, Ca^{2+}). Clinoptilolite is included under the heulandite group due to the isostructural characteristics, but generally contains higher Si/Al ratios and higher alkali content [38]. The framework contains channels of 10-membered rings and 8-membered rings formed by AlO_4 and SiO_4 tetrahedra, and they are interconnected by small cages while the cations are positioned in these channel (Figure 9). Clinoptilolite and heulandite generally crystalize in the monoclinic space group of *C*2/*m*.

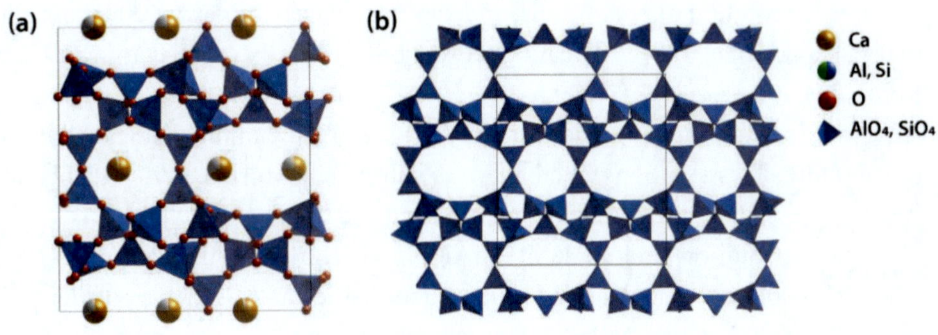

Figure 9. (a) Structure of heulandite unit cell showing 10-membered rings and 8-membered rings forming channels along the [001] direction with cations in the channel. (b) Interconnection of large channels by smaller cage units shown from the [001] direction.

3. Remediation of Radiohalides

Isotopes ^{129}I and ^{36}Cl are long-lived halide radioisotopes that can be present in used nuclear fuel (depending on the composition of fuel). The half-lives ($t_{1/2}$) of ^{129}I and ^{36}Cl are 1.57×10^7 y (ß decay) and 3.01×10^5 y (ß decay), respectively. Additional halide isotopes exist including ^{131}I ($t_{1/2} = 8.04$ d, ß decay). These species are typically present in ionic form (e.g., in caustic scrubber solutions) but can also be present in a vapor state [e.g., $ICl_{(g)}$, $I_{2(g)}$, $Cl_{2(g)}$]. If present in the soil, due to high water solubilities, they can transport easily into groundwater. Also, radioiodine can be readily absorbed into the human metabolic process through the thyroid making its effective capture a requirement for human health. This section discusses methods that have been demonstrated to capture and/or immobilize these species.

3.1. Radioiodine Remediation with SOD, FAU, LTA, and MOR

Sodalite has been demonstrated to be an affective host for radioiodine due to the incorporation of the large iodide anion in the center of the β-cage, where iodide is tetrahedrally coordinated by four cations inside the β-cage, resulting in tightly immobilized iodide in the mineral. The framework of the β-cage expands with a larger unit cell when incorporating a large anion. For an example, the unit cell of iodosodalite is 9.009 Å compared 8.8767 Å of natural chlorosodalite due to the incorporation of the larger iodide anion compared to chloride anion.

In the study by Strachan and Babad [39, 40], examples for iodosodalite synthesis included:

1. An aqueous solution of NaI, NaOH, and kaolin clay were mixed together and allowed to dry in an oven at 100°C. The product was hot uniaxially pressed at 1080°C at 13.8 MPa.
2. An aqueous solution was made from $NaAlO_2$ and NaI. Colloidal silica was added to this mixture. This solution gelled and was subsequently dried at 70°C. This powder was hot-pressed at 900°C at 13.8 MPa to form sodalite.
3. An aqueous solution was made of $NaAlO_2$ and KIO_3. To this was added colloidal silica. This solution gelled and was subsequently dried at 70°C. This powder was hot-pressed at 900°C at 13.8 MPa to form iodate sodalite.

In a study by Nam et al. [41], the effects of drying temperature, NaOH concentration, and precursor variation on the aqueous synthesis of iodosodalite were investigated to find the optimal aqueous synthesis conditions. Different Al and Si precursors including $NaAlO_2$, colloidal SiO_2, kaolinite [$Al_2Si_2O_5(OH)_4$], and metakaolin ($Al_2Si_2O_7$) were used with NaI to synthesize iodosodalite. In addition, the sintering conditions with different sodium borosilicate glass binders were studied. The iodosodalite powder was mixed with different amounts of glass binders, pressed into pellets, and then heat-treated at 650 to 850°C to convert into waste forms. From this study, Nam et al. drew the following conclusions:

1. Kaolinite was the best precursor for the aqueous synthesis of iodosodalite under the given synthesis conditions.
2. Mixing 10 mass% of glass binder and heat-treatment at 750°C yielded the highest fraction of iodosodalite.
3. A significant fraction of iodine was removed during heat-treatment due to decomposition of sodalite, transition of sodalite to nepheline, and volatilization of iodine from the amorphous fraction.
4. Chemical durability testing showed that the iodine release rate was low after several days, and lower than other elements that were in the samples including Al, Si, and Na.

In studies by Chong et al. [42, 43], hydrothermally grown iodosodalite and sintered glass-bonded sodalite were characterized. The effects of process variables of hydrothermal synthesis including pH, aging time, temperature, Si/Al molar ratio, and precursors were investigated. Iodosodalite synthesized with metakaolin were mixed with different amounts of sodium borosilicate glass binders and pressed into pellets, followed by heat-treatment at 650 to 850°C to synthesize the waste forms. From this study, Chong et al. drew the following conclusions:

1. Iodosodalite produced hydrothermally showed the highest iodosodalite yield with process conditions of pH 14 (controlled by NaOH concentration), 20 d under hydrothermal conditions at 150-180°C and a Si/Al molar ratio of 1.
2. Different precursors used for hydrothermal processing resulted in similar iodosodalite yields.
3. Having smaller amorphous fractions in the initial iodosodalite powder resulted in a higher mass fraction of iodosodalite after firing at 650–850°C.
4. At higher sintering temperatures, more nepheline was present from the decomposition of sodalite.
5. The iodine release rate was similar to sintered pellets synthesized by sodalite produced with the aqueous method described by Nam et al. [41].

Chemical durability studies on iodosodalite synthesized by hot isostatic pressing (HIP) at (900 and 1100°C, 200 MPa, 2 h) [44], hydrothermally followed by heat-treatment at 750°C [43], aqueously followed by heat-treatment at 750°C [41], and by a solid state method at 900°C [45] showed inconsistencies in iodine release rates. These differences are likely due to differences in sample preparation and experimental conditions. Maddrell et al. [44] reported the iodine release rate of iodosodalite by HIP as 0.005–0.01 g m^{-2} d^{-1} for the 7–14 d period where as Chong et al. [43] and Nam et al. [41] reported the iodine release rates ~20 times higher. These discrepancies are likely due to the inherent porosities in the ceramic products produced without a hot-pressing method.

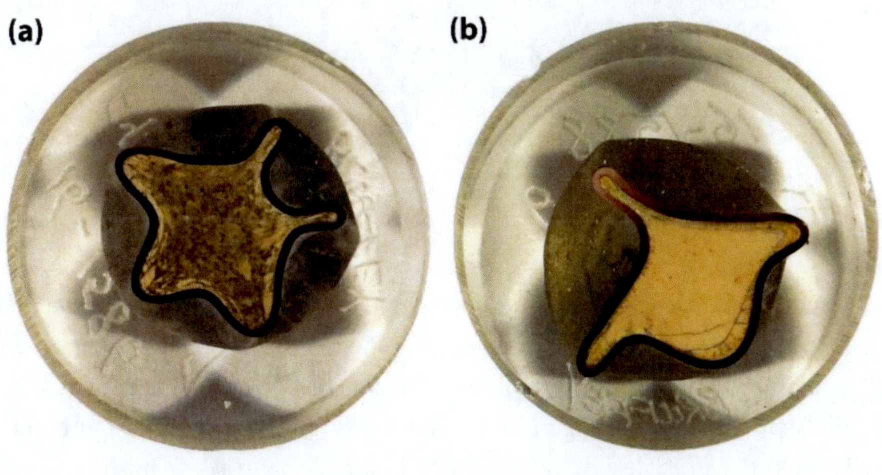

Figure 10. Pictures of (a) HIP-7 (I-AgX) and (b) HIP-9 (I-AgA) from Bruffey et al. [47]. HIP-7 contained both sodalite and nepheline while HIP-9 only contained sodalite based on X-ray diffraction (XRD) analysis.

For capturing gaseous radioiodine, various silver-based zeolite sorbents including Ag-exchanged faujasite (AgX), Ag-exchanged mordenite (AgZ), and Ag-exchanged zeolite A (AgA) have been studied extensively. Once radioiodine is captured in the zeolites, it can be processed directly into waste forms by hot-pressing methods such as HIP. During HIP processing (900–1100°C, 175–300 MPa, 2–3 h), iodide-loaded AgA and AgX zeolite phases generally transformed into iodosodalite [i.e., $Ag_8(AlSiO_4)_6I_2$ – see Figure 10] whereas iodide-loaded AgZ did not transform into sodalite, but crystallized into AgI phases [46-48]. No differences were observed visually or detected with XRD in samples based on the pressures used for the HIP process. The framework of SOD is only composed of the β-cages whereas zeolites A and X have additional bridging units between the β-cages; during HIP processing, the bridging units of zeolites likely collapse, and the β-cages are rearranged to form sodalite. The mordenite framework does not contain the β-cages, and perhaps it helps explain the unlikeliness of mordenite to transform into sodalite under the given HIP conditions. Another likelihood is that the Al:Si stoichiometry in the base mordenite is not optimum for forming sodalite. HIPed iodine-loaded Ag-containing zeolites usually contain secondary crystalline phases including nepheline [e.g., $K_{1.2}Na_{2.8}Fe_{0.1}Al_{3.9}Si_4O_{16}$], quartz, and Ag metal phases as well as an amorphous phase [46, 47].

Figure 11. Picture of Ionex® Type Ag900 Ag-mordenite (courtesy of Molecular Products Group Ltd.) [49].

The AgZ sorbent is ideal for capturing $I_2(g)$ due to the strong chemisorption that occurs between $I_2(g)$ an Ag^+ to form AgI, a very stable chemical compound. Several versions of AgZ are available commercially in engineered forms containing binders (e.g., clays) such cylindrically shaped extudates (e.g., Ionex Type Ag900) [49] (shown in Figure 11) and berl saddles. These types of materials are available in large quantities to meet the demands of large facilities, such as a nuclear fuel reprocessing facility where these types of off-gases would be prevalent.

Gaseous iodine capture has also been demonstrated by Maddrell et al. [50]. In this study, a similar approach was taken to that of Bruffey et al. [47] discussed previously, but with some additional experiments. While they performed five separate experiments, only the consolidation of the iodine-loaded commercial silver-exchanged zeolite (i.e., $Ag_{84}Na_2Al_{86}Si_{106}O_{384}$) will be discussed here. Iodine loading was done by adding 130 mg g^{-1} for sodalite conversion. This material showed partial conversion to the $Ag_8(AlSiO_4)_6I_2$ sodalite structure following HIP processing at 5°C min^{-1} to 900°C, a 2-h dwell at 100 MPa of applied pressure, and cooling at 5°C min^{-1}. These sodalite regions also contained some AgI crystals.

Additional metal-exchanged FAU zeolites have been a commonly explored topic for the capture of iodine in various forms, e.g., $I_{2(g)}$, $CH_3I_{(g)}$. In a study by Chebbi et al. [51], periodic dispersion-corrected density functional theory modeling was performed to evaluate inhibiting species, bonding characteristics, and probability of formation of iodine-based FAU zeolites exchanged with monovalent cations including Ag^+, Cs^+, Cu^+, H^+, K^+, Li^+, Na^+, and Rb^+. It was discovered that CO and H_2O were found to be the most inhibiting of the secondary species evaluated and that Ag and Cu showed the most promise for strongly absorbing $I_{2(g)}$ and $CH_3I_{(g)}$. The assessment included the evaluation of atomic displacement from initial positions. Chebbi et al. observed that the alkalis, once exchanged, remained in the 6MR (hexagonal) plane while transition metals tended to move to the side of the window near Al atoms following adsorption. It was found that alkali-exchanged FAU could be affected by the presence of $CH_3Cl_{(g)}$ and $Cl_{2(g)}$, both of which showed similar adsorption energies to the iodine-based analogs and could inhibit adsorption through binding competition. The alkali-exchanged zeolites remain sensitive to water, which is likely an issue for most industrial applications, whereas this was not the case for Ag- and Cu-exchanged FAU. Issues were identified for Ag- and Cu-exchanged FAU, however, where CO and NO could interfere with complexation.

An experimental study was performed by Chebbi et al. [52] evaluating the capture of $CH_3I_{(g)}$ on Ag-FAU using spectroscopic techniques including diffuse-reflectance ultraviolet-visible spectroscopy and diffuse-reflectance infrared Fourier transform spectroscopy. The base Ag-FAU used in this study had a Si:Al ratio of 2.5 and was loaded with 23 mass% Ag, which was distributed as Ag^+ isolated cations or clusters of charged or metallic Ag. The sorbents showed a loading capacity of 0.209 g g^{-1} and, once captured, 79 mass% was converted to AgI precipitates within the FAU structure.

3.2. Halide Salt Waste Immobilization in SOD

The accepted "baseline" waste form option for used electrochemical reprocessing salt wastes is the glass-bonded SOD (GBS) ceramic waste form (CWF) [53, 54]. After processing several batches of fuel, the electrorefiner contains a wide range of fission products released from the fuel. To make this waste form, zeolite 4A (LTA-type) is reacted with used salt at 500°C to occlude the salt in the cages of the zeolite. Then, a glass binder is added to the salt-occluded zeolite and the mixture is heated at 925°C to produce the GBS CWF. In this waste form, the alkali halides incorporate into the sodalite phase and the other fission products released from the fuel enter into the glass phase.

3.3. Molten Salt Reactor Waste Immobilization in SOD

In a recent study by Riley et al. [55], SOD was identified as a potential host for salt waste streams coming out of molten salt reactors (MSRs). The MSR concept is one of the Generation IV nuclear reactor designs that utilizes a molten salt either as the coolant (with solid fuel) or as both the coolant and the fuel. The traditional salt types used in these concepts include fluorides (e.g., $^{7}LiF\text{-}BeF_2$ or FLiBe) and chlorides, although other salts have been considered. These types of reactors have some advantages over traditional light water reactor but also come with some complications such as the complex waste streams generated. While an option likely exists to dispose of the used MSR salts in a salt repository, it is also possible that these streams can be immobilized in a glass-bonded sodalite, e.g., $M_8^+(AlSiO_4)_6F_2$ or $M_8^+(AlSiO_4)_6Cl_2$ for fluoride or chloride salts, respectively, where M^+ denotes a +1 cation.

4. Remediation of Technetium

Technetium is a radioactive element that is generated during fission of actinide fuels. Technetium can be found within nuclear wastes in different forms with different oxidation states including charged species (e.g., Tc^{4+}, Tc^{7+}) [56] or as metallic species (i.e., Tc^0) within nuclear fuel as the ε-metal phase (i.e., Tc-Pd-Rh-Ru-Mo) [57], but it is most commonly present as ^{99}Tc (ß decay) with a $t_{1/2}$ of 2.13×10^5 y. The most common oxidation state is Tc^{7+} followed by Tc^{4+} in both solution and in solid salt forms [56]. This section describes methods for the capture and/or immobilization of technetium in crystalline aluminosilicates.

4.1. Immobilizing Technetium in SOD

The TcO_4^- anion is highly mobile within the environment and requires an effective method of immobilization for disposal. One such option is the SOD structure in the form of $Na_8(AlSiO_4)_6(^{99}TcO_4)_2$. In previous studies by Mattigod et al. [58], Dickson et al. [59,

60], and Luksic et al. [61], potential TcO_4^- incorporation into SOD was assessed using solution-based processes with the nonradioactive surrogate anion perrhenate (i.e., ReO_4^-).

In the study by Mattigod et al. [58], $Na_8(AlSiO_4)_6(ReO_4)_2$ was synthesized using a hydrothermal reaction with $NaReO_4$, NaOH, $NaAlO_2$, and $Na_2SiO_3 \cdot 9H_2O$ in water in a 125-mL Teflon®-lined Parr autoclave at 175°C for 24 h. The resulting product was washed and XRD showed a pure phase with cubic $P\bar{4}3n$ space group (a = 9.1528 Å).

For the work by Dickson et al. [59], a similar hydrothermal approach was conducted but with different competing ions (i.e., Cl^-, CO_3^{2-}, SO_4^{2-}, MnO_4^-, and WO_4^{2-}) using zeolite and NaOH in autoclaves at 90°C for 96 h. This study showed that ionic competition was apparent when trying to incorporate ReO_4^- into the SOD structure and showed insignificant ReO_4^- incorporation into SOD when the difference in ionic radii between the competing ions was >12%.

In separate work by Dickson et al. [60], competing ions of OH^-, NO_2^-, NO_3^-, and Cl^- were evaluated for ReO_4^- incorporation into SOD. Samples were synthesized using NaOH, $NaAlO_2$, Na_2SiO_3, and various salts including NaCl, $NaNO_2$, $NaNO_3$, and $NaReO_4$. Despite changes in pH, SOD was observed to form in the presence of Cl^- and NO_2^-.

Luksic et al. [61] evaluated the retention of perrhenate SOD [i.e., $Na_8(AlSiO_4)_6(ReO_4)_2$] during simulated glass melting tests – see Figure 12 for a picture of the SOD used in this study. Since the common form of technetium is pertechnetate in Hanford waste glass feeds and that is volatile during the melting process (it leaves into the off-gas stream), the goal of this study was to evaluate if adding the Re^{7+} to the feed as SOD instead of Re_2O_7 prior to melting would improve retention during the melting process. These tests showed that adding the Re as SOD0 improved retention during melting by 21% and 85% for high-level waste and low-activity waste glasses, respectively.

Figure 12. Picture of $Na_8(AlSiO_4)_6(ReO_4)_2$ SOD from Luksic et al. [61]. This was reprinted with permission from Elsevier ©2016.

In a study by Pierce et al. [62], TcO_4^- was incorporated into SOD along with ReO_4^- to create mixed-guest $Na_8(AlSiO_4)_6(ReO_4)_{2-x}(TcO_4)_x$ sodalite. This material was synthesized hydrothermally with zeolite 4A, NaOH, $NaReO_4$, and $NaTcO_4$ at 225°C (400 PSI) for 7 days. The incorporation of TcO_4^- into the SOD structure was confirmed with X-ray absorption spectroscopy, XRD, and scanning electron microscopy with energy dispersive spectroscopy. This is the first fully documented characterization and confirmation of SOD containing TcO_4^-.

4.2. Immobilizing Technetium in CAN

Dickson et al. [60] also demonstrated ReO_4^- incorporation into CAN. Cancrinite formation was favored with NO_3^- ions present in 1 m or 16 m OH^- (molal) solutions. The ReO_4^- that was incorporated into nitrate-based CAN/SOD and nitrate-based CAN remained stable and difficult to remove under ion exchange conditions evaluated in this study, suggesting that it would be a good host for sequestration of TcO_4^-.

5. Remediation of Radioactive Alkalis and Alkaline Earths

5.1. Remediation of Cesium and Strontium

Isotopes ^{135}Cs, ^{137}Cs, and ^{90}Sr are radionuclides generated from fission of nuclear fuels. The high activities of ^{137}Cs and ^{90}Sr are of utmost concern with $t_{1/2}$s of 30.17 y (ß-decay) and 28.8 y (ß-decay). Also, ^{135}Cs has a $t_{1/2}$ of 2.3×10^6 y (ß-decay). Cesium isotopes are commonly present as salts and can easily transport into the environment if improperly contained while strontium can be incorporated into bones [63].

5.1.1. Immobilizing Cesium in SOD and CAN

In a study by Mon et al. [64], Cs^+ incorporation into SOD was evaluated for looking at remediation of $^{137}Cs^+$ ions leaking out of the Hanford waste storage tanks into the nearby soil. The main concern here is that it will reach the groundwater and be transported into nearby waterways. In this study, SOD was one of four different aluminosilicates evaluated, and showed one of the tightest bindings observed across the samples tested including SOD, CAN, LTA, and allophane [i.e., $(Al_2O_3)(SiO_2)_{1.3–2}\cdot(H_2O)_{2.5–3}$]. For this study, Cs-exchanged materials were exchanged three times with solutions of 0.5 N Na^+, K^+, or Ca^{2+} at 80°C and the exchange was quantified after acid digestions of the samples. Following the exchanges, <20% of the Cs^+ was exchanged from the SOD, showing evidence of a tight bond. This is in comparison of 94–99% of the Cs being exchanged in the LTA and allophane being readily exchangeable, with SOD being the most resistant to ion exchange (i.e., allophane < LTA < CAN < SOD). For the ion exchange, the cations most readily exchangeable for Cs^+ were $K^+ > Ca^{2+} > Na^+$. While CAN was not as effective as SOD in retaining Cs^+ ions in the chemical structure, it did perform well at retaining 43–78% of the Cs from exchanges with Na^+, K^+, or Ca^{2+} which ranked in performance as $Na^+ > Ca^{2+} > K^+$.

5.1.2. Remediation of Cesium and Strontium with GIS

In a study by Vereshchagina et al. [65], (Cs,Na)P1, (Sr,Na)P1, and (Cs,Sr,Na)P1 GIS-like zeolite microspheres called cenospheres were evaluated as potential precursor hosts for immobilizing ^{135}Cs, ^{137}Cs, and ^{90}Sr. The process utilized involved three separate steps of (1) hydrothermal treatment of the base cenospheres (referred to as *zeolitization*), (2) sorption of the Cs^+/Sr^{2+} ions (referred to as *precursor formation*), and annealing (referred to as *recrystallization*). The base materials were coal fly ash cenospheres with a Si:Al molar ratio of 2.7. The Cs and Sr loadings were performed using an ion exchange process with nitrates and, following Cs and Sr exchanges, these materials were transformed into glass-ceramics at temperatures of 900–1000°C. During the heat-treatment process, the microspheres converted into a variety of different aluminosilicate minerals including composites of feldspar, nepheline, and pollucite with the resulting chemistry forming a glass around these crystalline materials. The crystalline content in these composites was around 60 mass% of pollucite and Sr-feldspar with approximately 10–20 mass% glass phase. Overall, this process shows promise as a potential technique to capture and immobilize these high-activity radionuclides into stable glass-ceramic composites using this cost-effective approach.

5.2. Remediation of Radium

Radium is a radioactive element that can be found in minerals uncovered during mining activites. Two isotopes discussed below include ^{226}Ra ($t_{1/2}$ = 1599 y, α decay) and ^{228}Ra ($t_{1/2}$ = 5.76 y, ß decay).

Synthetic Na-P1, a GIS-like framework, was used in a group with other synthetic zeolites including NaX (Na-FAU) and SOD as well as mixtures of Na-P1/Na-X/LTA and Na-P1/LTA zeolites to assess the removal of ^{226}Ra and ^{228}Ra from mine water [66]. Here, a 30-cm diameter column filled with 26 kg of sand and 2 kg of zeolite mixtures were used. The mine water contained 4.216±0.534 kBq m^{-3} of ^{226}Ra and 7.81±1.46 kBq m^{-3} of ^{228}Ra with a thermal conductivity of 0.119 S cm^{-1} and a salinity of 110 g dm^{-3}. Then, 120 dm^3 of salt water was passed through the column and radium was measured using liquid scintillation. The results indicated that 98% of both ^{226}Ra and ^{228}Ra were removed during the experiment. The presence of these zeolites in solution did not affect the levels of Ca or

Mg in solution or cause significant changes in the levels of monovalent metals in solution (e.g., Na^+, K^+).

6. Remediation of Radioactive Noble Gases (Xenon, Krypton)

A wide range of noble gas fission products are generated during fission of actinides, including those in a molten salt reactor, which include short lived radioisotopes such as ^{90}Kr ($t_{1/2}$ = 32.3 s), ^{139}Xe ($t_{1/2}$ = 39.5 s, ß decay), ^{89}Kr ($t_{1/2}$ = 3.18 min, ß decay), ^{137}Xe ($t_{1/2}$ =3.83 min, ß decay), ^{135m}Xe ($t_{1/2}$ = 15.3 min, isometric transition or IT and ß decay), ^{88}Kr ($t_{1/2}$ = 2.84 h, ß decay), ^{135}Xe ($t_{1/2}$ = 9.1 h, ß decay), ^{133m}Xe ($t_{1/2}$ = 2.19 d, IT), and ^{133}Xe ($t_{1/2}$ = 5.25 d, ß decay); a longer-lived isotope includes ^{85}Kr ($t_{1/2}$ = 10.7 y, ß decay) [55, 67].

Filled valence electron shells found in noble gases make them highly unreactive, although some compounds have been discovered (summarized previously [68]). However, some porous materials can be used to capture these species through weak bonding interactions or molecular sieve processes.

A polymer composite of AgZ MOR was used to capture noble gases from used nuclear fuel (i.e., Xe and Kr). In a study by Garn, Greenhalgh, and Law [69], sodium mordenite (NaZ) was imbedded in a polymer matrix of polyacrylonitrile (PAN), a passive and porous scaffold, and then exchanged to the AgZ by soaking these beads in a solution of $AgNO_3$ followed by rinsing with nanopore water and drying in a vacuum oven at 50°C. (The process of making PAN-sorbent composites along with additional applications is explained in more detail elsewhere [70].) The AgZ-PAN composites showed Xe capacities of ~28–32 mmol kg^{-1} at ambient temperatures and 360–460 mmol kg^{-1} at reduced temperatures (220 K). The Kr capacities at 220 K ranged from ~ 2.4–3 mmol kg^{-1}. In addition to these loading measurements, selectivity measurements were also performed between Xe and Kr using Equation (1)

$$S_{\mathrm{Xe,Kr}} = \frac{x_{\mathrm{Xe}} y_{\mathrm{Xe}}}{x_{\mathrm{Kr}} y_{\mathrm{Kr}}} \quad (1)$$

where X_{Xe} and X_{Kr} denote the mol fractions of Xe and Kr in the adsorbed phase (capacity), respectively, and Y_{Xe} and Y_{Kr} denote the mole fractions of Xe and Kr in the feed gas (feed concentration), respectively. At ambient temperature, $S_{\mathrm{Xe,Kr}}$ values were 6.9 and 7.8 in air and He, respectively, while values at 220 K were 22.4 and 22.3 in air and He, respectively.

7. REMEDIATION OF ACTINIDES (URANIUM AND THORIUM)

Uranium is a naturally occurring radioactive actinide that is present in the environment as ^{238}U (>99.2%, $t_{1/2} = 4.5\times10^9$ y, α decay), ^{235}U (< 0.8%, $t_{1/2} = 7\times10^8$ y, α decay), and ^{234}U (< 0.01%, $t_{1/2} = 2.5\times10^5$ y, α decay). The oceans are the largest source of uranium on the planet, where it is present as UO_2^{2+} (uranyl ion), but its low concentration and vast array of competing ions (e.g., more concentrated heavy metals) and binding anions (e.g., CO_3^{2-}) make capture/removal difficult [71]. Thorium is another radioactive actinide with the most stable isotope being ^{232}Th (>99.9%, $t_{1/2} = 1.4\times10^{10}$ y, α decay) with a ~3-fold higher abundance in the earth's crust over uranium; the other primary isotope found in nature is ^{230}Th ($t_{1/2} = 7.5\times10^4$ y, α decay). One of the primary applications for both actinides is as fuel for nuclear reactors. The speciation of U^{6+} and Th^{4+} can change drastically in solution as a function of pH as shown in Figure 13, which makes capture even more complicated.

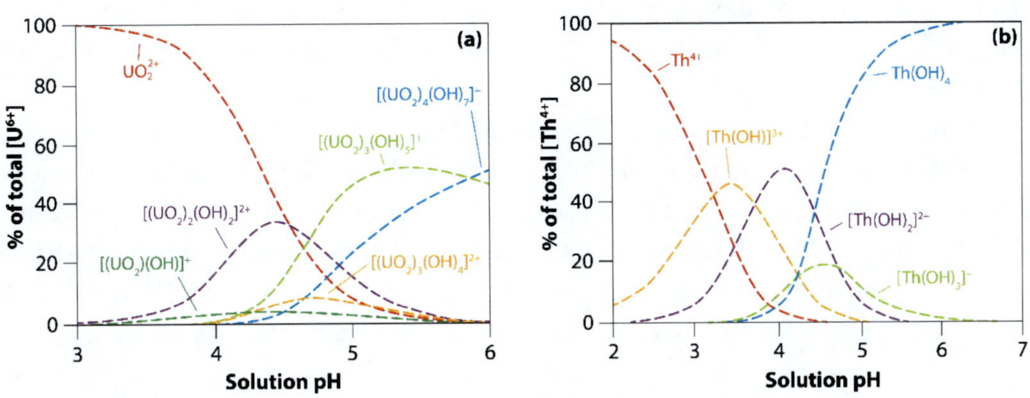

Figure 13. Summary of (a) U(VI) and (b) Th(IV) speciation as a function of pH [72, 73].

A wealth of literature exists on uranium capture with HEU-type zeolites. For example, in a study by Misaelides et al. [74], uranium and thorium uptake were evaluated for a material consisting primarily of HEU-type zeolites, but also containing additional phases of SiO_2, feldspars, and phyllosilicates. Solutions were prepared using nitrate salts [i.e., $Th(NO_3)_4$, $UO_2(NO_3)_2$] between 50 and 20,000 mg L^{-1}. The maximum Th and U uptake values ranged between 1.57–12.41 mg g^{-1} and 0.98–8.70 mg g^{-1}, respectively. The uptake of these metal ions is dependent on concentration and pH and can range from ion-exchange to adsorption or surface precipitation. Based on pH, the speciation of uranium and thorium can change drastically, which must be accounted for when doing these types of experiments.

In a paper by Olguín et al. [75], zeolite X (FAU) was demonstrated to effectively capture uranium present in solutions as U(IV) (added as UCl_4) and UO_2^{2+} ions [added as $UO_2(NO_3)_2$], where the captured uranium ions incorporated into cation sites within the FAU crystal structures. If solution concentrations >0.01 N were utilized, following uranium uptake, the crystal structure was affected (i.e., lost). Maximum uptake values were higher for U(IV) ions (2.86 meq g^{-1} FAU) than for UO_2^{2+} (1.86 meq g^{-1} FAU) and no desorption was observed for either species after 200 h of contact time. It was noted that less uranium entered the zeolite than Na^+ that was removed from the zeolite during the ion exchange process.

Brindley and Bastovanov evaluated uranium adsorption with uranyl acetate using Na-type [i.e., $Na_{12.12}(Si_{12.07}Al_{11.93}O_{48.10})\cdot 23.8H_2O$], K-rich [i.e., $K_{7.80}Na_{1.63}Ca_{0.04}(Si_{12.21}Al_{11.79}O_{48.15})\cdot 20.4H_2O$], and Ca-rich [i.e., $Ca_{4.53}Na_{2.98}K_{0.13}(Si_{12.26}Al_{11.74}O_{48.22})\cdot 22.2H_2O$] LTA zeolites using solutions ranging 100–3,700 ppm UO_2^{2+} [76]. The crystal structures were modified during UO_2^{2+} uptake to compreignacite-like for the Na-LTA and K-LTA zeolites and to becquerelite-like structure for the Ca-LTA zeolite. The initial pH values of the solutions were 3.4–3.6 (i.e., for 1,000 ppm – 0.1 N concentrations), but these pHs increased during the experiment; this pH change was attributed to the removal of H^+ from solution as it exchanged with the cations in the zeolites (i.e., Na^+, K^+, Ca^{2+}).

Andreeva and Chernyavskaya [77] evaluated the sorption of UO_2^{2+} ions on synthetic Na-MOR (26–34 μm) and natural HEU (clinoptilolite, 0.1–0.5 mm) in pH 2–4; hydrogen-forms of these zeolites were also evaluated. They determined that the capture mechanism of UO_2^{2+} ions is where these ions enter the zeolite channels. As the pH was increased, the sorption capacities of UO_2^{2+} ions increased. As UO_2^{2+} ions were captured, H^+ ions from the H-forms of the zeolites were released into solution lowering pHs during the tests.

8. Remediation of Radioactive Chalcogens (i.e., Se^{6+})

Radioactive chalcogens are present in nuclear waste streams including selenium and tellurium isotopes. Some examples of radioactive chalcogens released during the Fukushima Daiichi nuclear power plant accident include selenium-79 (^{79}Se, $t_{1/2} = 6.5\times10^4$ y, ß decay), tellurium-127m (^{127m}Te, $t_{1/2} = 109$ d, IT/ß decay), and tellurium-129m (^{129m}Te, $t_{1/2} = 33.6$ d, IT/ß decay) [78, 79].

In a study by Pahlevi et al. [79], selenate (i.e., SeO_4^{2-}) was immobilized in CAN-SOD-GIS composites with a hydrothermal approach by reacting a solution of $NaAlO_2$, NaOH, $Na_2SiO_3\cdot 9H_2O$, and Na_2SeO_4 at 90°C for 6–720 h in a Teflon®-lined autoclave. During the hydrothermal reaction, cancrinite coprecipitated within the crystalline phases. The concentration of SeO_4^{2-} in the SOD increased as a function of reaction time in the autoclave as did the phase distribution between CAN, SOD, and Na-zeolite P (GIS). This approach can be used to pull SeO_4^{2-} ions out of solutions for immobilization in crystalline aluminosilicate matrices.

CONCLUSION

Several different types of crystalline aluminosilicates were discussed in this chapter for the capture and/or remediation of radionuclides ranging from solid to aqueous to gaseous species. These aluminosilicates included sodalite (SOD), mordenite (MOR), faujasite (FAU), Linde Type-A (LTA), cancrinite (CAN), gismondine (GIS), and clinoptilolite/heulandite (HEU). The range of radionuclide remediation applications for these materials is based off the diverse set of properties coupled to the chemistry. Remediation and immobilization options are all linked back to the specific structures that comprise the aluminosilicate networks. The channels and cages within the structures provide locations for ions to interact with the aluminosilicate network through ion-exchange, chemisorption, physisorption, and/or molecular sieving. The data presented herein are a brief summary of the vast amount of literature data on this topic and new applications are being discovered every year.

REFERENCES

[1] Baerlocher, C.; McCusker, L. B.; Olson, D. H. *Atlas of Zeolite Framework Types*; Elsevier; Oxford, 2007;pp. Pages|.

[2] Stein, A.; Ozin, G. A.; Macdonald, P. M.; Stucky, G. D.; Jelinek, R., *J. Am. Chem. Soc.* **1992**, *114,* 5171-5186.

[3] Siqueira, G. O.; Gravina, É. G.; Resende, J. A. L. C.; Fernandes, N. G., *Powder Diffrac.* **2009**, *24,* 41-43.

[4] Meier, W. M., *Z. Kristallogr.* **1961**, *115,* 439-450.

[5] Yonkeu, A.; Buschmann, V.; Miehe, G.; Fuess, H.; Goossens, A. M.; Martens, J. A., *Cryst. Eng.* **2001**, *4,* 253-267.

[6] Reed, T. B.; Breck, D. W., *J. Am. Chem. Soc.* **1956**, *78,* 5972-5977.

[7] Jarchow, O., *Z. Kristallogr.* **1965**, *122,* 407-422.

[8] Fischer, K. F., *Am. Mineral.* **1963**, *48,* 664-672.

[9] Bresciani-Pahor, N.; Calligaris, M.; Nardin, G.; Randaccio, L.; Russo, E.; Comin-Chiaramonti, P., *J. Chem. Soc. Dalton Trans.* **1980**, *9,* 1511-1514.

[10] Brenchley, M. E.; Weller, M. T., *Zeolites* **1994**, *14,* 682-686.

[11] Kendrick, E.; Dann, S., *J. Solid State Chem.* **2004**, *177,* 1513-1519.

[12] Eiden-Aßmann, S.; Schneider, A. M.; Behrens, P.; Engelhardt, G.; Mändar, H.; Felsche, J., *Eur. J. Inorg. Chem.* **2001**, *2001,* 1527-1534.

[13] Latturner, S. E.; Sachleben, J.; Iversen, B. B.; Hanson, J.; Stucky, G. D., *J. Phys. Chem. B* **1999**, *103,* 7135-7144.

[14] Nielsen, N. C.; Bildsøe, H.; Jakobsen, H. J.; Norby, P., *Zeolites* **1991**, *11,* 622-632.

[15] Gesing, T. M., *Z. Kristallogr.* **2007**, *222,* 289-296.

[16] Buhl, J.-C., *Thermochim. Acta* **1996**, *286,* 251-262.

[17] Armstrong, J. A.; Dann, S. E.; Neumann, K.; Marco, J. F., *J. Mater. Chem.* **2003**, *13,* 1229-1233.

[18] Han, S.; Smith, J. V.; Pluth, J. J.; Richardson, J. W., *Eur. J. Mineral.* **1990**, *2,* 787-798.

[19] Depmeier, W., *Acta Cryst. C* **1984**, *40,* 226-231.

[20] Fechtelkord, M., *Micropor. Mesopor. Mater.* **1999**, *28,* 335-351.

[21] van Smaalen, S.; Dinnebier, R.; Katzke, H.; Depmeier, W., *J. Solid State Chem.* **1997**, *129,* 130-143.

[22] Barth, T. F., *Z. Kristallogr.* **1932**, *83,* 405-414.

[23] Gesing, T. M., *Z. Kristallogr.* **2000**, *215,* 510-517.

[24] Többens, D.; Depmeier, W., *Z. Kristallogr.* **1998**, *213,* 522-531.

[25] Gualtieri, A. F.; Aprea, P., *Micropor. Mesopor. Mater.* **2006**, *96,* 276-286.

[26] Schlenker, J. L.; Pluth, J. J.; Smith, J. V., *Mater. Res. Bull.* **1978**, *13,* 77-82.

[27] Mortier, W. J.; Pluth, J. J.; Smith, J. V., *Mater. Res. Bull.* **1976**, *11,* 15-21.

[28] Schlenker, J. L.; Pluth, J. J.; Smith, J. V., **1978**, *13,* 169-174.

[29] Shiokawa, K.; Ito, M.; Itabashi, K., *Zeolites* **1989**, *9,* 170-176.

[30] Simoncic, P.; Armbruster, T., *Am. Mineral.* **2004**, 421-431.

[31] Pluth, J. J.; Smith, J. V., *Mater. Res. Bull.* **1972**, *7,* 1311-1322.

[32] Kaduk, J. A.; Faber, J., *Rigaku J.* **1995**, *12,* 14-34.

[33] Garcia, G.; Cardenas, E.; Cabrera, S.; Hedlund, J.; Mouzon, J., *Micropor. Mesopor. Mater.* **2016**, *219,* 29-37.

[34] Vance Jr, T. B.; Seff, K., *J. Phys. Chem.* **1975**, *79,* 2163-2167.

[35] Yanagida, R. Y.; Vance, T. B.; Seff, K., *Inorg. Chem.* **1974**, *13,* 723-727.

[36] Turner, S.; Sieber, J. R.; Vetter, T. W.; Zeisler, R.; Marlow, A. F.; Moreno-Ramirez, M. G.; Davis, M. E.; Kennedy, G. J.; Borghard, W. G.; Yang, S., *Micropor. Mesopor. Mater.* **2008**, *107,* 252-267.

[37] Armstrong, J. A.; Weller, M. T., *J. Am. Chem. Soc.* **2010**, *132,* 15679-15686.

[38] Alberti, A., **1975**, *22,* 25-37.

[39] Babad, H.; Strachan, D. M., "Method for immobilizing radioactive iodine," patent application 1980.

[40] Strachan, D. M.; Babad, H., *Am. Ceram. Soc. Bull.* **1979**, *58,* 327.

[41] Nam, J.; Chong, S.; Riley, B. J.; McCloy, J. S., *MRS Adv.* **2018**, *3,* 1093-1103.

[42] Chong, S.; Peterson, J.; Nam, J.; Riley, B.; McCloy, J., *J. Am. Ceram. Soc.* **2017**, *100,* 2273-2284.

[43] Chong, S.; Peterson, J. A.; Riley, B. J.; Tabada, D.; Wall, D.; Corkhill, C. L.; McCloy, J. S., *J. Nucl. Mater.* **2018**, *504,* 109-121.

[44] Maddrell, E.; Gandy, A.; Stennett, M., *J. Nucl. Mater.* **2014**, *449,* 168-172.

[45] Nakazawa, T.; Kato, H.; Okada, K.; Ueta, S.; Mihara, M., pp. 51–57, *Proc. of Mater. Res. Soc.*, 663. Edited by Hart, K. and Lumpkin, G., 2000.

[46] Vance, E. R.; Gregg, D. J.; Grant, C.; Stopic, A.; Maddrell, E. R., *J. Nucl. Mater.* **2016**, *480,* 177-181.

[47] Bruffey, S. H.; Jubin, R. T.; Jordan, J. A. 2016. Fundamental Aspects of Zeolite Waste Form Production by Hot Isostatic Pressing. FCRD-MRWFD-2016-000267, ORNL/SR-2016/759. Oak Ridge National Laboratory, Oak Ridge, TN.

[48] Sheppard, G. P.; Hriljac, J. A.; Maddrell, E. R.; Hyatt, N. C., pp. 36.31, *Proc. of Mater. Res. Soc.*, 932. Edited by Van Iseghem, P., 2006.

[49] IONEX type AG900 (2019) https://www.molecularproducts.com/products/ionex-type-ag900.

[50] Maddrell, E. R.; Vance, E. R.; Gregg, D. J., *J. Nucl. Mater.* **2015**, *467,* 271-279.

[51] Chebbi, M.; Chibani, S.; Paul, J.-F.; Cantrel, L.; Badawi, M., *Micropor. Mesopor. Mater.* **2017**, *239,* 111-122.

[52] Chebbi, M.; Azambre, B.; Cantrel, L.; Koch, A., *J. Phys. Chem. C* **2016**, *120,* 18694-18706.

[53] Ebert, W. L. 2005. Testing to Evaluate the Suitability of Waste Forms Developed for Electrometallurgically-Treated Spend Sodium-Bonded Nuclear Fuel for Disposal in the Yucca Mountain Repository. ANL-05/43. Argonne National Laboratory-East, Argonne, IL.

[54] Riley, B. J.; Vienna, J. D.; Frank, S. M.; Kroll, J. O.; Zhu, Z.; Canfield, N. L.; Kruska, K.; Schreiber, D. K.; Peterson, J. A.; Crum, J. V., *J. Nucl. Mater.* **2017**, *489,* 42-63.

[55] Riley, B. J.; McFarlane, J.; DelCul, G. D.; Vienna, J. D.; Contescu, C. I.; Forsberg, C. W., *Nucl. Eng. Des.* **2019**, *345,* 94-109.

[56] Darab, J. G.; Smith, P. A., *Chem. Mater.* **1996**, *8,* 1004-1021.

[57] Crum, J. V.; Strachan, D.; Rohatgi, A.; Zumhoff, M., *J. Nucl. Mater.* **2013**, *441,* 103-112.

[58] Mattigod, S. V.; Peter McGrail, B.; McCready, D. E.; Wang, L.-Q.; Parker, K. E.; Young, J. S., *Micropor. Mesopor. Mater.* **2006**, *91,* 139-144.

[59] Dickson, J. O.; Harsh, J. B.; Lukens, W. W.; Pierce, E. M., *Chem. Geol.* **2015**, *395,* 138-143.

[60] Dickson, J. O.; Harsh, J. B.; Flury, M.; Pierce, E. M., *Micropor. Mesopor. Mater.* **2015**, *214,* 115-120.

[61] Luksic, S. A.; Riley, B. J.; Parker, K. E.; Hrma, P., *J. Nucl. Mater.* **2016**, *479,* 331-337.

[62] Pierce, E. M.; Lilova, K.; Missimer, D. M.; Lukens, W. W.; Wu, L.; Fitts, J.; Rawn, C.; Huq, A.; Leonard, D. N.; Eskelsen, J. R.; Woodfield, B. F.; Jantzen, C. M.; Navrotsky, A., *Environ. Sci. Technol.* **2017**, *51,* 997-1006.

[63] Bryant, F. J.; Loutit, J. F., pp. 449-465, *Proc. of Royal Soc. Lond.*, 159B, 1963.

[64] Mon, J.; Deng, Y.; Flury, M.; Harsh, J. B., *Micropor. Mesopor. Mater.* **2005**, *86,* 277-286.

[65] Vereshchagina, T. A.; Vereshchagin, S. N.; Shishkina, N. N.; Vasilieva, N. G.; Solovyov, L. A.; Anshits, A. G., *J. Nucl. Mater.* **2013**, *437,* 11-18.

[66] Franus, M.; Wdowin, M.; Bandura, L.; Franus, W., *Fresenius Environ. Bull.* **2015**, *24,* 854-866.

[67] Riley, B. J.; McFarlane, J.; DelCul, G. D.; Vienna, J. D.; Contescu, C. I.; Hay, L. M.; Savino, A. V.; Adkins, H. E. 2018. Identification of Potential Waste Processing and Waste Form Options for Molten Salt Reactors. NTRD-MSR-2018-000379, PNNL-27723. Pacific Northwest National Laboratory, Richland, WA.

[68] Subrahmanyam, K. S.; Spanopoulos, I.; Chun, J.; Riley, B. J.; Thallapally, P. K.; Trikalitis, P. N.; Kanatzidis, M. G., *ACS Appl. Mater. Interfac.* **2017**, *9,* 33389-33394.

[69] Garn, T. G.; Greenhalgh, M.; Law, J. D., *J. Nucl. Sci. Technol.* **2015**, *53,* 1484-1488.

[70] Riley, B. J.; Garn, T. in *Purification of Gas and Liquid Streams Using Composite Sorbents Embedded in a Polyacrylonitrile Matrix*; Wythers, M.; Ed.: Nova Science Publishers, Inc.; 2019; Vol. 35 pp. 119-147.

[71] Lu, Y., *Nature Chemistry* **2014**, *6,* 175-177.

[72] Sylva, R. N.; Davidson, M. R., *J. Chem. Soc. Dalton Trans.* **1979**, 465-471.

[73] Langmuir, D.; Herman, J. S., *Geochim. Cosmochim. Acta* **1980**, *44,* 1753-1766.

[74] Misaelides, P.; Godelitsas, A.; Filippidis, A.; Charistos, D.; Anousis, I., *Sci. Tot. Environ.* **1995**, *173-174,* 237-246.

[75] Olguín, M. T.; Solache-Ríos, M.; Acosta, D.; Bosch, P.; Bulbulian, S., *Micropor. Mesopor. Mater.* **1999**, *28,* 377-385.

[76] Brindley, G. W.; Bastovanov, M., *Clays Clay Minerals* **1982**, *30,* 135-142.

[77] Andreeva, N. R.; Chernyavskaya, N. B., *Radiokhim.* **1982**, *24,* 9-13.

[78] Takahashi, S.; Kawashima, S.; Hidaka, A.; Tanaka, S.; Takahashi, T., *Nucl. Technol.* **2018**, *205,* 646-654.

[79] Pahlevi, N. D.; Guo, B.; Sasaki, K., *Ceram. Int.* **2018**, *44,* 8635-8642.

In: An Introduction to Aluminosilicates
Editor: Nero Regina Blevins
ISBN: 978-1-53617-250-8

Chapter 8

SYNTHESIS OF LAYERED ALUMINOSILICATE BY A DIRECT METHOD

T. T. Sundaram* and N. Prasanna
[1]Department of Chemistry, Adhiyamaan College of Engineering, Hosur, Tamil Nadu, India
[2]Department of Chemistry, Government Arts and Science College, Hosur, Tamil Nadu, India

ABSTRACT

Layered aluminosilicates are well known in the name of zeolite, which are generated by the basic crystallographic building unit of TO_4 (T = Si, Al) tetrahedral, with shared corners of their unit among itself with in a two-dimensional manner. The dimensional array of layered aluminosilicate is defined by channels and a porous structure. Si^{4+} created a neutral framework, whereas Al^{3+} created a negative charge in the framework and is neutralized by alkali metal cations.

Layered aluminosilicates are synthesized in the source of the tetrahedral atom of 'Si' and 'Al', surfactant as the template or structure-directing agents (SDAs), use mineralizers as a method of maintaining pH, the solvent, etc. The mixture makes gel and is heated at a temperature in the range of 140-150°C for a period of time. The factors influencing the formation of layered aluminosilicates were the time period of synthesis, the pH, the total amount of solvent, SDA, gel composition, and temperature. The formation mechanism followed the liquid-crystal templating/ADOR (assembly-disassembly-organization-reassembly) mechanism. However, the partially inadequate experimental details of the formation mechanism were a blockade of custom design and need complete exploration of aluminosilicate in an experimental approach. Increasing the pressure on a scientist to

* Corresponding Author's Email: sundaramtt@gmail.com.

synthesize the layered aluminosilicate in the green is a rational pathway to decrease hazardous waste.

The Physiochemical properties of as-synthesized layered aluminosilicate have been Characterized by Small angle PXRD, FTIR, Porosity studies, Thermal studies, ^{27}Al MAS NMR, and Morphological studies.

Layered aluminosilicate acting as Bronsted and Lewis acid, well-defined porous and its channels, undergo adsorption and desorption phenomenon. Most of its parameters, like porosity, size, and morphology, were tunable.

It has used as a Heterogeneous catalyst, adsorption-desorption, separation, membrane, petrochemical industry, drug delivery, tissue engineering, and sensor. In the forthcoming article, it is explored in the field of advanced modern construction material, electronic, and agricultural engineering.

Keywords: layered aluminosilicate, sol-gel synthesis, influencing parameter, characterization

ABBREVIATIONS

ADOR	Assembly-Disassembly-Organization-Reassembly
BJH	Barrett-Joyner-Halenda
CMC	Critical Micelles Concentration
HLB	Hydrophile-Lipophile Balance
MAS NMR	Magic Angle Spinning Nuclear Magnetic Resonance
NP	Nonylphenol
PXRD	Powdered X-Ray Diffraction
SEAD	Selected Area Electron Diffraction pattern
SDA	Structural Directing Agents
TEM	Tunneling Electron Microscope
TEOS	Tetraethyl-Orthosilicate
TMOS	Tetramethyl Orthosilicate
TPD	Temperature-Programmed Desorption
WAXD	Wide-Angle X-Ray Diffraction

1. INTRODUCTION

A scientist looks easily tunable in Physico-chemical properties, flexibility, and feasibility in synthesis to get a targeted product. The adequate experimental information and computational works might help to synthesize the desire-targeted product in modern inorganic chemistry. The environmental production law enforced to follow the green pathway within an industrial chemical procedure and not a very harmful industrial chemical process.

Scientists and industrial chemists were continuously searching for a novel new material that has to fulfill the mitigation of environmental impacts of industrial activity and reduction on all aspects of structure to composition to physiochemical property. The need for new material creates voids in industrial needs. The multi-functionality of material is promised to be highly tunable according to the industrial needs. It offers tunable size, Physio-chemical properties, diverse morphology, high thermal stability, chemical stability, mechanical stability, and no toxicity, which would provide more support to a wide range of applications [1, 2]. This tunable character mainly depends on the variable stoichiometric composition of the initial reaction mixture [3]. Sufficient knowledge in the field of layered structures and its properties can comprehend the feasibility in the novel layered structure synthesis. Moreover, less dimension of the material is an opportunity to apply in the areas of electrical, optical, etc. Table 1 cleared out important over silicate framework, and it is one of the material, might fill the voids of industrial requirements in most of the aspects. Layered aluminosilicate materials have excellent chemical, thermal, electrical, and other properties. The scientist looks at these properties for applying or by adopting the material to the required application in industry. Polymorphs of silicate tetrahedra are the neutral framework. The substitution of Al^{3+} in the silicate framework unit develops a net negative charge due to the formation of $[AlO_4]^{5-}$ and is neutralized by Na^+, K^+, Ca^{2+}, H^+, NH_4^+, water molecule, etc. Tetrahedra of $[SiO_4]^{4-}$, $[AlO_4]^{5-}$, simple primary building units sharing their corner apical oxygen (T-O-T) in their convenient spatial orientation to form desired secondary building units [4]. However, the O-T-O bond angle closes to the 109°28' for the rigid tetrahedron [5]. The T-O bond length depends on the respective central atom for Si-O ~ 1.64°A and Al ~ 1.73°A [6, 7]. Isomorphous substitution of aluminum atoms in the silicate framework makes it a significant material in modern material science and gives it wide industrial applications. The followings were feasible with layered aluminosilicate: 1. the flexibility of frame with starting substance, 2. Tunable physiochemical properties 3. Tunable porous dimensions and surface character 4. Tunable electrical fields 5. Feasible to get specific structure and properties. Using a layered aluminosilicate would make a unique three-dimensional zeolite for various potential applications, membranes with desirable properties, employing unique chromatographic methods, innovative building materials, sensing microfilms, and agricultural engineering, etc.

2. Synthesis of Layered Aluminosilicate

The making of a clear solution or homogenous colloidal gel is the first step of synthesis for getting targeted layered aluminosilicate. 'Si,' and 'Al' sources undergo hydrolysis in an acidic medium to form tetrahedral of $[SiO_4]^{-4}$ and $[AlO_4]^{-5}$ units and is followed by the condensation of these units to amorphous oligomeric that are prenucleation species of the layered zeolite precursor. Structural directing agents form different shapes like lamellar, cubic, square, and hexagonal, which have more active surfaces, adsorb amorphous oligomeric species, and make themselves as ordered, well defined mesoporous to the desired orientation is dependent upon reaction condition.

Water is used as a solvent to make a homogenous gel. The mineralizer regulates the reaction and its condition. 'Si' sources, 'Al' sources, SDA, water, and mineralizers are mixed thoroughly, to make a homogenous gel and started the primary nucleation process. The homogenous gel was converted into layered aluminosilicate material by either hydrothermal method or sol-gel method, which is the second step of the synthesis. The sol-gel process provides a variety of different products, such as molded gels, spun fibers, thin films, molecular cages, aerogels, xerogels, and mesoporous materials for a variety of applications such as gas and liquid separations, optical coatings, protective films, membranes, and catalysis [8]. This method is more effective and convenient for the synthesis of layered aluminosilicate.

Table 1. Difference between Silicate and Aluminosilicate

Framework	Silicate	Aluminosilicate
Structural units	Only Polymorphs of $[SiO_4]^{4-}$ units	Polymorphs of $[SiO_4]^{4-}$, $[AlO_4]^{5-}$
Physical nature of the framework	Neutral framework	Net negative charge raises due to Al^{3+} ion.
Chemical Nature of framework	Hydrophobic nature	Tunable hydrophilic nature
Applicability	Limited, mostly inactive	Wide range of application
Active sites	Mostly inactive sites	Tunable to sites in the range of acidic to the basic
Influence of framework 'atom'	Different composition of 'Si' cause homogeneity and does not much alter the physiochemical properties	Stoichiometry ratio of 'Al' atom in framework influence the most of physiochemical properties and develops heterogeneity

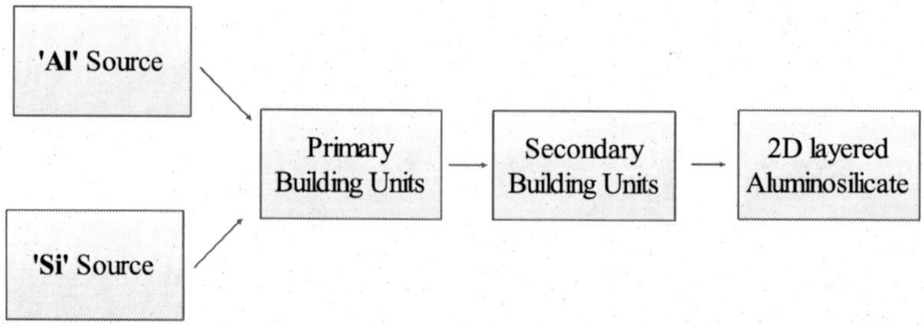

Figure 1. Schematic Diagram of Layered Aluminosilicate Formation.

The Sol-gel method allows the solvents to escape, homogeneous gel reaches supersaturation, and finally gets targeted zeolitic-layered crystalline substance. Secondary nucleation procedure accompanied by regeneration of small nuclei of secondary construction unit to larger one via a layer by layer mechanism. The shape of the layer depends on micelles of SDA that vary from lamellar, cubic, square, and hexagonal, which depends on the nature of SDA, pH, temperature, ionic strength of the solution, etc. The primary building units are adsorbed by micelles shape, which is more surface-active and acts as a precast of aluminosilicate framework unit and oriented mutual structure organization with the help of a structural directing agent to form secondary building units. These secondary or pre-building units very slowly condense and increasing electrostatic interaction between each other to an organized two-dimensional spatial arrangement with defined internal cavities and external pore openings.

The final solid product is filtered and washed with deionized water and dried at 100°C in a hot air oven for 1 hour to remove the adsorbed water and is then heated at 300 to 600°C in a muffle furnace for 2 to 6 hours to remove any organic impurities. A schematic shown in Figure 1.

The layered aluminosilicate synthesis carried out by the hydrolysis of Silicon and aluminum precursors by the sol-gel method and condensed the hydrolyzed precursors upon thermal technique. In a typical synthesis, Sodium Metasilicate acts as a 'Si' source, anhydrous aluminum chloride acts as an 'Al' source, deionized water acts as a solvent, and concentrated hydrochloric acid acts as a mineralizer. The Si/Al ratio is 1. Slowly mix both solution and stirrer until turbid homogenous obtained. To make them clear, a mineralizer such as con hydrochloric acid is added. Silicate precursor and aluminum precursor, joined to form gel, it is dissolved in mineral acid pH of the solution below 1. After getting a homogenous solution, Nonylphenol added to the homogenous solution. NP, non-ionic surfactant act as structural directing agent, behave as a space filler and help to form stabilizing inorganic structure thermodynamically favorable by successive transformation towards (Ostwald rule of successive transformations) a stable system. During heating Nonylphenol, precursors of aluminum and silicon undergo a collision, then from the condensed product. NP interacts inorganic structure with an electron-rich phenoxy group perpendicular to the inorganic structural surface by electrostatic Coulomb interaction. It follows the $S^0H^+X^-I^+$ synthetic route, where S^0 - non-ionic surfactants; I^+ - silicate or aluminate species; X^- - Cl^-, Br^-, I^-, SO_4^{2-}, NO_3^- under $pH < 2$ [9, 10]. Continue the stirrer with 140-150°C for 3 hrs, evaporation of water lead to condense the solution become gel consistency, continuous heating it slowly converted into white powder mass. Finally, the driven force of the complex system, the aluminosilicate structure transforms toward a platy shape in the presence of NP. So Potential transformation enabled in open space air heating and thermodynamically favorable thermally stable white platy aluminosilicate staying in the vessel.

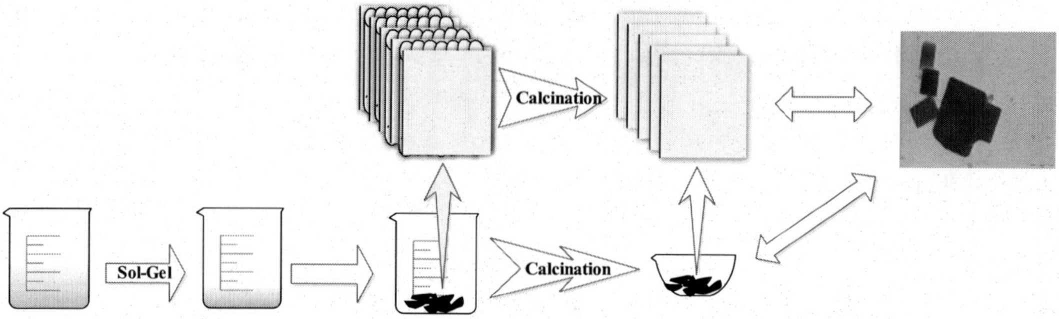

Figure 2. Schematic diagram for synthesis of layered aluminosilicate.

After getting white platy aluminosilicate powder mass, washed with deionized water, then dried in a hot air oven at 110°C for 1hr for removal of adsorbed water molecules then dried white powder calcined in the open air for removal of a template and adsorbed water molecules at 300°C for 4 hrs. The following parameter influenced the nature, type, crystallinity, and morphology of the final product [11, 12].

2.1. Nature of Reactants and Their Pretreatments

The source of 'Si,' 'Al' influence morphology, size, particle distribution, structure, and type of framework. The common sources of 'Si' are fumed silica, sodium water-glass, colloidal silica sol, tetramethylorthosilicate (TMOS), and tetraethyl orthosilicate (TEOS). Aluminum isopropoxide, sodium aluminate, pseudo-boehmite, aluminum sulfate, aluminum hydroxide, aluminum nitrate, and Aluminum metal are utilized as an Al source in zeolite synthesis.

Inorganic precursor yielded a more hydroxylated product. Organic precursors are easily inserted in the metal into the framework. If the solubility of resources is more, that makes high surface area with regular, ordered structure and tiny crystals due to greater interaction with structural directing agents and reacting resources. However, reduced solubility of resources and SDA leads to a decrease in the surface area, and thermodynamic equilibrium favors the larger the size of crystals [13]. Depending upon the medium, the precursor of aluminum atoms changes its nature. The Al^{3+} concentration in acidic medium is influenced by temperature and pH. Increase the temperature, the formation of $AlH_3SiO_4^{2+}$ complex become an increase in pH above 3 [14]. In basic solution, it mostly exits as an Al $(OH)_4^-$ and higher the pH value the Al $(OH)_4^-$ will be transferred to AlO_2^-.

2.2. Structural Directing Agents

Generally, SDA acts as a precast mold and shaping the zeolite and influence its nature, structure, pH, ionic strength in solution, and temperature. Higher concentrations of it result in mostly layered structure. Non-ionic surfactants have reduced cost, non-toxicity, biodegradability, low critical micelles temperature, and be comfortable from the synthesis of mesoporous solids. The regularity of the mesoporous increase with low critical micelles concentration (CMC) value [15]. Larger the value of the CMC trend to form a lamellar structure. The hydrophile-lipophile balance (HLB) of non-surfactants is which range from 4 to 12 favors the synthesis of 3 D mesoporous structure. Lyophilic surfactants with HLB values lower than 4 always direct the formation of lamellar mesostructure [9].

2.3. Si/Al Ratio

Determine the nature of the framework, thermal stability, reactivity, retention of water, morphology, and porous. High Si/Al ratio gels prepared with weak alkalinity, whereas low Si/Al ratio needs strong alkaline condition for preparing starting gel. Beneath the isoelectronic point, both 'Al' and 'Si' sources are soluble in acidic medium. The amount of aluminum plays a very important role in modern molecular science and influences most of the properties of zeolite. The higher aluminum content in the framework resulted in, increase in acidity, active sites, ion exchange capacity, hydrophilicity, reactivity in organic reaction, and adsorption of specific gases.

2.4. Temperature

Temperature of reaction plays a significant role in the synthesis of most of all types of zeolites. It increases the rate of formation and enhances the contraction of building units. In particular, temperature range can form the desired type zeolite phase. Higher the temperature range of the reaction medium, increase both the nucleation rate and the crystal growth rate, leads to larger crystals have been obtained.

2.5. Reaction Time

Homogenous gel achieved rapidly; the conversion of zeolite takes few hours to few days. The crystallinity increases with time.

2.6. Mineralizer

It regularizes the homogeneity from the solution and maintains the pH of the reaction medium. To get the hydrothermal condition, alkalinity as a mineralizer (pH > 10). Alkalinity is just the OH-/Si ratio. Crystallization raised the solubility of 'Si,' and 'Al' resources. It reduced the amount of polymerization of anion, decrease the induction period, and accelerates by a growing rate of polymerization of anions that are aluminate. Also, it reveals in smaller the particle size and narrow distribution. For the sol-gel process, acidity as a mineralizer (pH < 2) and involves soluble or hydrolysis of a precursor, which depends on the pH of the reaction medium, consequently condensation of it to form small sub-units, results in these sub-units linked together form an inorganic network. The synthesis route plays a significant role in determining the properties of the product. The isoelectronic point of precursors determines the molecular-scale morphology of the product. Below the isoelectronic point, the reactants are in charged intermediate in the form of a highly branched structure that was oligomeric through duster-cluster aggregation processes. Above the isoelectronic point, the formation of negatively charged species are joined in a denser cluster. In most of the evaporation-induced synthesis, reaction carried out in high acidic medium, the rate of hydrolysis are fast in the medium, and crystallization depends on temperature [8].

2.7. Inorganic Cations

Alkali and alkaline earth metal cations, which are usually introduced to the synthesis mixture in the form of basic compounds to adjust the pH, can have significant effects on nucleation, crystal growth, and morphology. It can control the $OH^-/F^-/H_3O^+$ in the solution, which affects the dissolution and polymerization-depolymerization reactions of silicates and aluminosilicates. It also orders the water molecule around itself by successive replacement of silicate and aluminate species partially and organize the nucleation as a center, creating a cage like structure [16]. Even in the preparation of pure-silica solid, small alkali-metal cations are required for practical kinetics. If a large amount of it is added to the reaction medium, the results are layered solid as a product.

2.8. Solvents

The role of solvents is the ability to interact with reacting species. A balanced interaction with reacting species is dependent on the right choice of solvent. It contributes to a homogenous reaction medium. The water utilized as a solvent is used for the preparation of layered aluminosilicate. Greater amounts of water lower supersaturation that favors the formation of a large crystal of zeolites [17]. It also significantly influences the architecture of pore size and pore volume. Different amounts of solvent give different pore architecture for the same structural directing agent.

2.9. Ageing

The ageing process is started by the mixing of reagents into a homogenous solution to the formation of a product that is solid and depends upon the reaction temperature. It increased the nucleation rate and the crystal population. It decreased the crystal size, induction period, and duration of crystallization [18, 19]. Usually, the ageing period is between 22 hr to 30 days. More extended ageing periods result in minimized induction, crystallization period, decrease the crystal size, and increase the number of crystals in the reaction medium. Further, the ageing process also enhances the condensation of unreacted species and makes a cross-link during the drying process under acidic condition.

2.10. Stirring

Modified crystallization kinetics enhance the small crystals, rapid supersaturation, and selectivity in zeolite formation. The unstirred synthetic mixture tends to form intergrown structures. Fast stirring the synthetic mixture leads to layered solid [19].

3. CHARACTERIZATION

3.1. Small-Angle Powdered X-Ray Diffraction

Small-angle powder XRD was used to assess the structural ordering of the synthesized material via the sol-gel process. An intense high peak at low angle region is associated with layered aluminosilicate [20]. The synthesized and calcined layered aluminosilicate have shown a high intense peak below $2\theta < 2^{\circ}$ reflection (Figure 3) due to 2D ordered mesoporous structure present in it. The calcined layered aluminosilicate always showed less intense peak with a shift of the interplanar distance was observed because of the lattice contraction of interlayers during calcination process. Generally, PXRD gives a badly resolved diffraction pattern (well defines neither the nuclear scale nor mesoscopic scale) with inadequate information to choose any crystal symmetry for some ordered mesoporous material [21]. Unit cell values not refined that clearly showed, crystal despite their amorphous nature may be the presence of high symmetry ordered orientation of mesoscopic level [8, 22]. But, both small angle powder XRD Patterns show a shallow background sign and little sharp reflections in $2\theta = 0.5\text{Å}$ indicating enclosure of mesophase in the aluminosilicate framework.

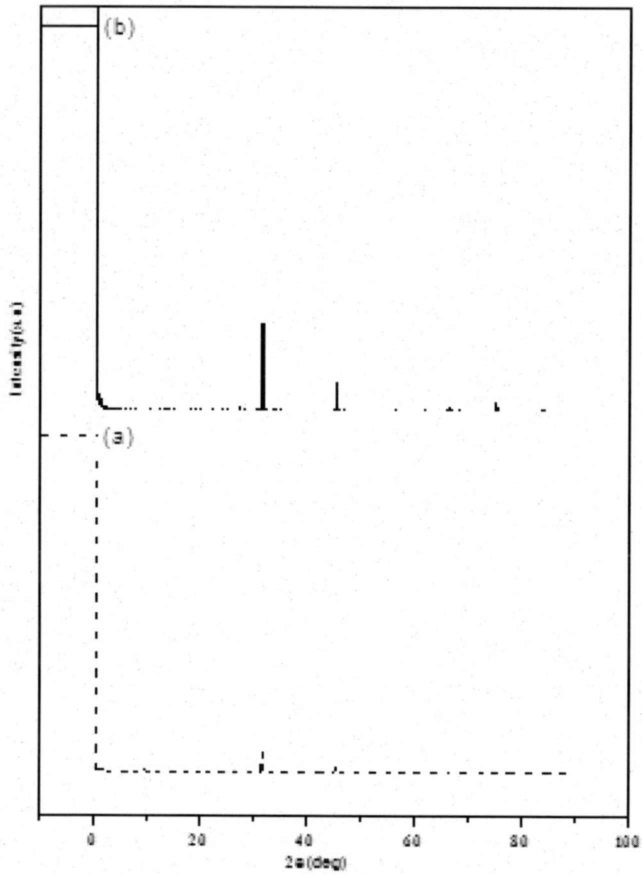

Figure 3. (a) Small angle XRD of as - synthesized aluminosilicate. (b) small angle XRD of calcined Layered aluminosilicate.

The absence of Bragg diffraction peaks in the wide-angle X-ray diffraction regime clearly shows the absence of a structural registry in the layered aluminosilicate framework. The WAXD regime increases the intensity of some peaks shows that the calcination process framework contraction and thickening of the pore wall.

3.2. FTIR Spectroscopy

The absorption of the water molecule is a unique property of zeolite, mainly aluminosilicate framework due to heterogeneity and hydrophilic character.

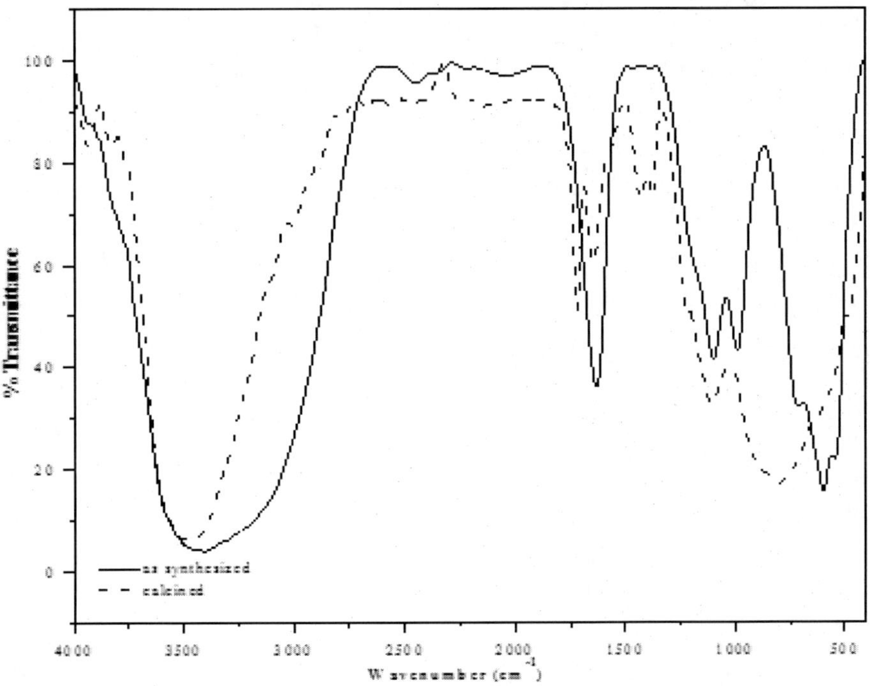

Figure 4. FTIR of as - synthesized and calcined layered aluminosilicate.

The fundamental hydroxyl stretching vibrations ν_{OH} is unique, identification of aluminosilicate framework, and ν_{OH} of calcined zeolite in the range of 3200-3800 cm^{-1}. This range inclusion of terminal M—OH and Si—OH on the external surface, the hydroxyl group at extra-framework metal species, bridging hydroxyl groups in different size of cages and channels, internal Si—OH and hydroxyl proton interacts with framework oxygen [23-31]. Figure 4 showed the FTIR spectra of as-synthesized and calcined aluminosilicate.

The hydroxyl is characterized by arising of broadband centered at 3406 cm^{-1} for as-synthesized aluminosilicate framework and attribute to overtones of physically adsorbed surface molecular water. It was hydrogen-bonded to each OH group of Si and Al and extended from 2650 cm^{-1} to 4000 cm^{-1} with a less distinct shoulder at 3952 cm^{-1} [32-34]. The hydroxyl stretching region of open-air calcined layered aluminosilicate solid was centered at 3497 cm^{-1} and shifted to higher region to the as- synthesis counterpart and the range extending from 2709 cm^{-1} to 3880 cm^{-1} with appearance of less intense peaks at 3824 cm^{-1} and 3936 cm^{-1}, which may attribute to adsorption or formation of molecular hydrogen in framework [35, 36]. The broadness of the stretching region in as-synthesized wasn't just Al-OH, Si-OH, and also moieties that bounded to organic and water molecules in pores which are open by Vander walls attraction, and voids of silicon and aluminum building unit which are tetrahedral. These moieties were merged, producing vibration, which shows broadband in this region. After calcining the layered aluminosilicate framework, some of bounded physisorbed water and organic molecules (structural-directing agents) were eliminated which causes reduced broadness in the lower region of the FTIR spectra due to absence of water and organic molecules, which escape through the calcination, maybe creating a pore opening and template (SDA) removal diminishing Hydrogen bonding with the framework. The as-synthesis material shows a strong adsorption peak on 1631 cm^{-1} due to physisorbed molecular water that attributed to the bending mode of the water molecule [37, 38]. Whereas calcined layered aluminosilicate solid shift this peak to the higher wavelength region and shows strong band in 1717 cm^{-1} due the molecular water become ions of a dimer $(H_2O)_2$ ---H^+. The bending mode of molecular water peak shifted into 1642 cm^{-1}, which could depend on the surface morphology, the charge density of the surface, and the pore size of the aluminosilicate framework. Calcination creates a porous structure that reabsorbs atmospheric water and carbon dioxide. Bands arise in the 1000-1700 cm^{-1} region, causing physisorption of atmospheric water and Carbon dioxide in calcined layered aluminosilicate material, which rehydrated in the open air. The bands 1642, 1427, and 1372 cm^{-1} are attributed to physisorption atmospheric water and carbon dioxide, which were absent or shifted in as-synthesized layered aluminosilicate material [39]. The most important and intense band should appear in the 950-1200 cm^{-1} region for stretching vibration of Si (Al)-O.

The band near 1100 cm^{-1} attributes asymmetric stretching mode framework or ring structure of the aluminosilicate framework in as-synthesized material. This band shifted in the calcined material for 1108 cm^{-1} due to the contraction of the aluminosilicate framework. Generally, the amount of aluminum atoms (Al) increases in the aluminosilicate framework that vibrates at the high-frequency region and vice versa (Lighter atoms show absorption bands at larger wavenumbers).

T-O-T symmetric Stretching band for as-synthesized material shows at 600 cm^{-1}, whereas calcined material appears at 803 cm^{-1} [40]. Si-O-Si deformation frequency raised in less shoulder peak of T-O-T symmetric stretching frequency, which centered around 803 cm^{-1}. This huge broadband in the framework region clearly shows that different framework building units and species interacted with each other. The peak boarding and shifting in 800-400 cm^{-1} region cleared out that the Al-O-Al bond in as-synthesized framework interact with structural directing agents and packed regular layers in order. Calcination process leads to exfoliate the layer morphology, resultant stacks the layer causes more Al-O-Al interaction with other building species, results in wide boarding in the calcined layered aluminosilicate framework [41].

The broadness and intensity of the hydroxyl band decreases with the increase of 'Si' atoms in the framework. The Si/Al ratio 1, the hydroxyl band broadness due to extra framework Al atoms bonded with OH group, which makes strong hydrogen bonding between the interlayer surface. If a sufficient increase in the amount of 'Si' in the reaction medium or increase Si/Al ratio that decreases the Al in the tetrahedral sites which interconnect the frame network, decreasing the Al (VI) state, results decrease the extra framework Al in external surface.

There was a sharp band at 1634 cm^{-1} due to the bending vibration of the hydroxyl group attributed to an absorption water molecule in the porous lattice, which does not depend on the Si/Al ratio of the framework. Due to additional lattice created with a calcined sample, Si/Al = 1 raised to shoulder peak around 3936 cm^{-1} [42]. The framework region was 400-1300 cm^{-1} [32]. The Si/Al = 1 had board band compared with other ratios, results the framework region was merged and shown board band due to high aluminum content and influenced by the framework oxygen, neighboring group, Al-O-Al, Si-O-Al and Si-O-Si network bands.

The broadness of the band reduced due to an increase in the Si/Al ratio of the framework, and it gradually reduces the own characteristic peaks of individual groups with the decrease of Si/Al ratio in the framework.

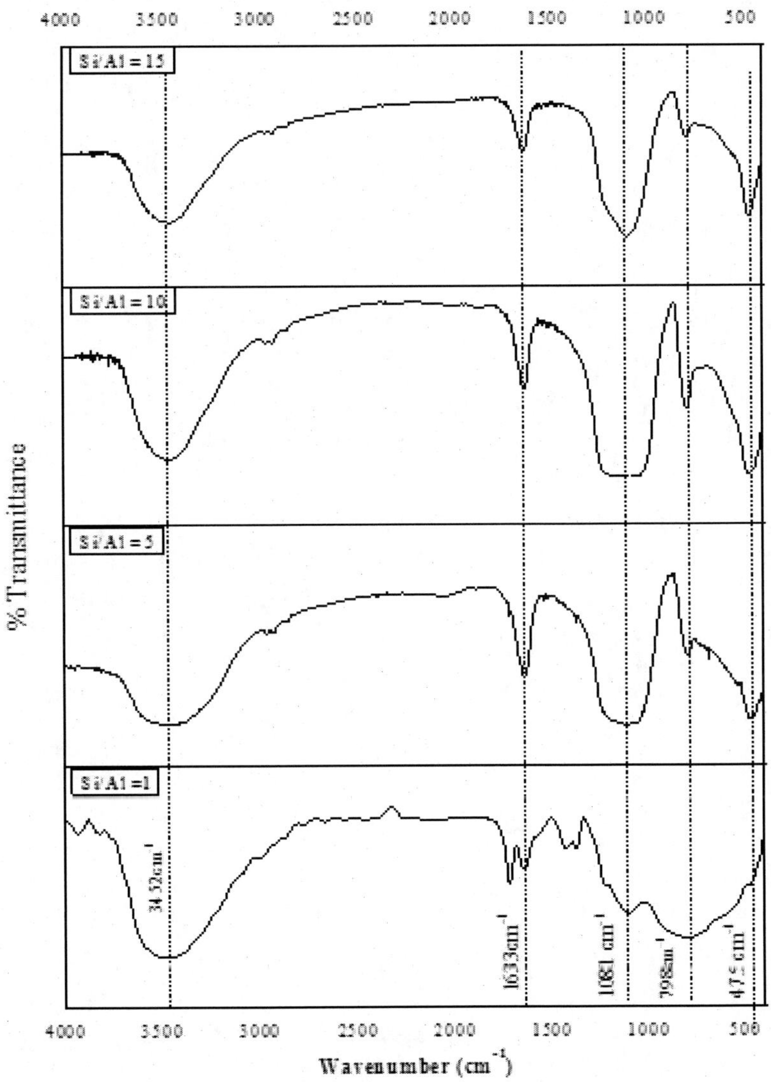

Figure 5. FTIR spectroscopy for Different Si/Al ratio calcined Layered aluminosilicate.

The Si-O-Si bond depends on the amount of aluminum atoms in the framework. It influences the initial reaction mixture, and the formation of Si-O-Si linkages decreases for a low Si/Al ratio. The board band in the range in 400-700 cm^{-1} is due to the Al-O-Al network [41].

The framework vibrational frequencies mainly influenced by cation-oxygen coordinated groups and interactions with neighboring groups [43]. These influencing factors applied to the nature of Al-O stretching frequencies in different Si/Al ratio and study framework changes in it. Si/Al ratio 1 shows the characteristic vibration frequency of a sharp peak near around 800 cm^{-1} that corresponds to the isolated AlO_4 tetrahedra, which have characteristic range 650-800 cm^{-1}.

It also noted that AlO_4 tetrahedral stretching frequencies of vibrational modes must remain in 700-800 cm^{-1} region. This region attributes to T-O-T stretching vibration of the framework.

3.3. Acidity

Acidity is an important unique property of layered aluminosilicate. The uniqueness of layered aluminosilicate could make a promising application to fine chemical synthesis as a solid acid catalyst, particularly in petrochemical industries, to the utilization of commercially important hydrocarbon transfer reactions. The acidic property of aluminosilicate mainly depends on the amount aluminum atoms in their framework. As the Si/Al ratio increases, the acidity of aluminosilicate decreases, and thermal stability increases. The layered aluminosilicate consists of Bronsted and Lewis acidic sites in the framework, and it could be arises by high-temperature activation.

The chemical method, temperature-programmed desorption (TPD), uses the NH_3, pyridine as a base, allow bind to the acidic sites of aluminosilicate, and find stronger and weaker sites based on desorption using thermal energy.

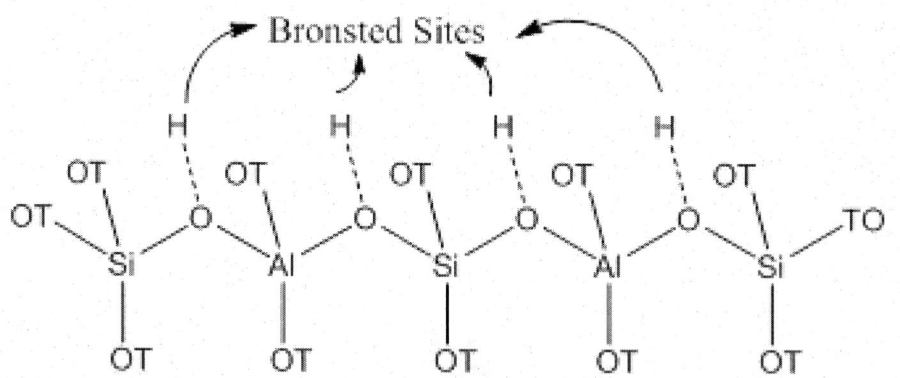

Figure 6. Bronsted Sites in Layered Aluminosilicate framework.

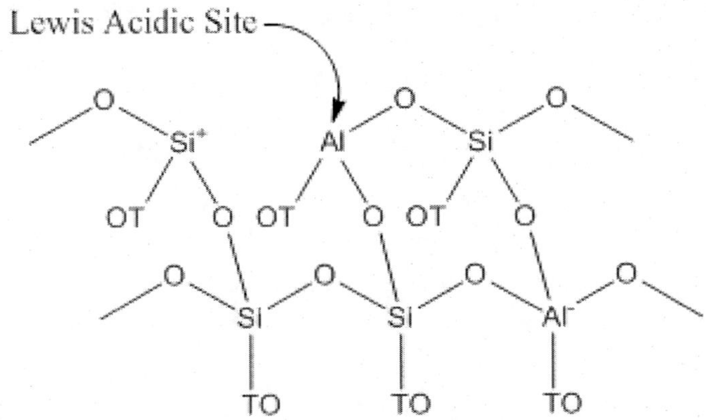

Figure 7. Lewis Acidic Site in Layered aluminosilicate framework.

Instead of this method, FTIR used to differentiate between Bronsted site and Lewis site by the adsorption of NH_3 or Pyridine. The adsorption of weak bases can also be used for the determination of the distribution of strength of the acidic site. The strength of acidic sites in aluminosilicate also determined with weak bases that adsorbed at the acidic site of it. The pyridinium adsorbed on Bronsted acidic sites and showed the band at 1520 and 1620 cm^{-1}. It also coordinated to Lewis sites and arose the band at 1450 and 1600 cm^{-1} [44].

3.3.1. Bronsted Acid

The isomorphous substitution of aluminum atoms in the silicate framework generate negative charge located in oxygen atoms (Al-OH-Si), neutralized by proton, these proton is called Bronsted acidic site. Also observed in hydrated aluminosilicate, the strong electrostatic field interaction of adsorbed water molecule with the exchange cations and the AlO_4^- unit causes the polarization of water molecules and has strong acidic Bronsted properties [30]. The number of aluminum atom in the framework are higher in adsorbed water molecules, resulting in higher acidic character.

3.3.2. Lewis Acid

The center is capable of accepting the pair of electrons from adsorbed species. They are different aluminum in the form of extra framework Al^{3+} species, cations, framework trigonal Al atoms, formed by thermal dehydroxylation of the Bronsted acidic sites, and located in defeat centers [45]. Lewis acidity is much weaker than Bronsted acidity in aluminosilicate.

3.4. Porosity Studies

The calcined and degassed porous samples are used to measure the porosity. The nitrogen sorption isotherms of porous samples are measured at 77K, the boiling point of liquid nitrogen. The quantity of adsorbed/desorbed nitrogen is measured as a function of the applied pressure, giving an increase to the adsorption/desorption isotherm. The linear isotherm plot corresponds to nonporous, horizontal deviation goes to microporous, and vertical deviation exposes to mesoporous. A typical example of a layered aluminosilicate profile of porosity shown in Figure 8.

In the P/P_0-0.2-0.45 range, the surface of layered has a microporous structure filled by nitrogen. The relative pressure range (0.45-0.9) is filling with mesoporous/capillary condensation in the interparticle voids. According to the classification of IUPAC, the layered aluminosilicate could be type IV with a hysteresis loop that revels to multilayer physisorption associated with capillary condensation in mesoporous solids.

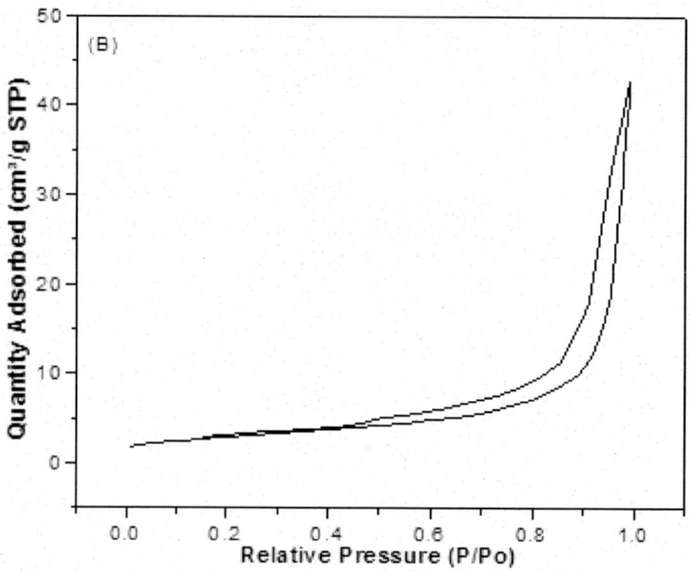

Figure 8. N_2 adsorption-desorption isotherms of calcined Layered Aluminosilicate.

Table 2. Types of Hysteresis

TYPE	Description
Type1-H1	Narrow distribution and uniform mesoporous solids.
Type2-H2	Difficult to identify, complex distribution pores with narrow necks and wider shape.
Type3-H3	Non-rigid aggregates of plate-like particles with macro pores network.
Type4-H4	Microporous associated with mesoporous in narrow slit pores.

Hysteresis stretches give information regarding pore shapes like inkbottle, narrow capillary, and wide. The layered aluminosilicate agreed with type IV with hysteresis type4-H4 attributes to narrow microporous network with large mesoporous [46]. The aluminum content was increased in the framework to decrease the porosity of the framework decreased with an increase of aluminum content. The pore-size distribution and pore volume were determined by the BJH (Barrett-Joyner-Halenda) method.

3.5. Thermal Studies

The thermal studies are very important to focus on the stability of the heterogeneous catalyst. It provides the analyst with a quantitative measurement of any weight change associated with a transition, like higher temperature calcination of the material, collapse of the framework by contraction, reduced surface area, and pore volume. It finds applications in the determination of drying range, calcination temperature, phase composition, percentage weight loss, and stability limits of a catalyst.

The two endothermic peaks cause weak interaction with surrounding atmospheric species. The physisorbed water on the surface of layered aluminosilicate framework is completely desorbed at 81.38°C due to weak Vander wall climb the surrounding atmospheric moisture [47], the porous framework with filled in interlayer adsorption and absorbed water desorbed at 165.93°C.

At the temperature range of 150-300°C, the removal of tightly bounded template molecules on the interlayer surface of as-synthesised layered aluminosilicate framework and desorbed, shown strong endothermic reaction at 231.06°C [48, 49]. The hydrophilic character of layered aluminosilicate is generated by the isomorphous substitution of aluminum in the framework.

The removal of a physisorbed water molecule from layered aluminosilicate depends on the amount of aluminum content in the layered framework. The higher the aluminum content, the more heat that is required to remove the adsorbed water. The amount of heat required decreases with decreasing aluminum content. The temperature range was above 740°C, and the aluminosilicate framework undergoes phase change or collapse of layer structure at 760°C (Figure 9).

Generally, aluminosilicate materials are hygroscopic in nature. During heating, the material above 100°C that supplies thermal energy to the framework and exceeds the binding energy of the water molecule, which detaches from the holes and escapes through open pores or windows. Later return to room temperature. Rehydration takes place. The tetrahedral unit occludes the atoms and ions. So the framework retains its connectivity.

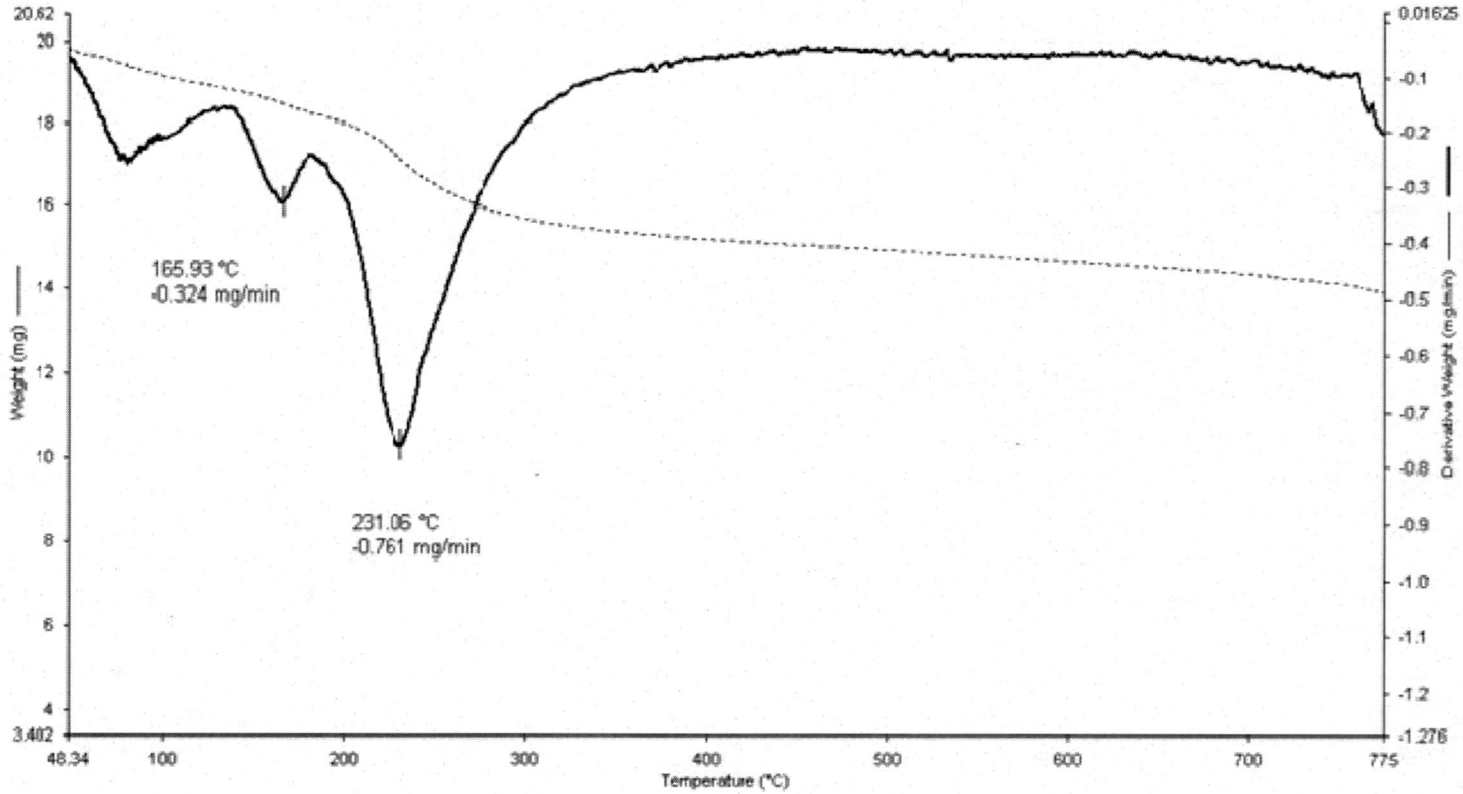

Figure 9. Thermal studies of as-synthesized aluminosilicate.

Table 3. Water desorbed temperature of different Si/Al ratio aluminosilicate

Si/Al ratio	Physisorbed Water removal T°C	Chemisorbed Water removal T°C
1	81	165
5	50	--
10	53	135
15	52	105

3.6. ^{27}Al MAS NMR

It helps to find the nature of aluminum atom coordination and incorporation in framework, and further identify the location, mainly the extra-framework aluminum in pore opening areas. Any adsorbed molecules that form a complex are also identified. The framework and extra-framework aluminum is characterized by the ^{27}Al MAS NMR spectra. The formation of the Al-O-Al bond is forbidden, based on Lowenstein's rule. Based on the rule, the framework contained only Al $(OSi)_4$ units.

The AlO_4 units, usually the chemical shift of about 60 ppm, the hydrated zeolite, and these lines range from about 55 to 68 ppm. However, the dehydrated zeolite could raise much broader resonance due to the distortion of a framework and causes extensive quadrupole interactions.

The non-framework usually raises the signals at 0 ppm due to AlO_6. In most of the cases, these octahedral aluminum undergone distortion heavily due to hydration of it and cause strong quadrupole line broadening observed.

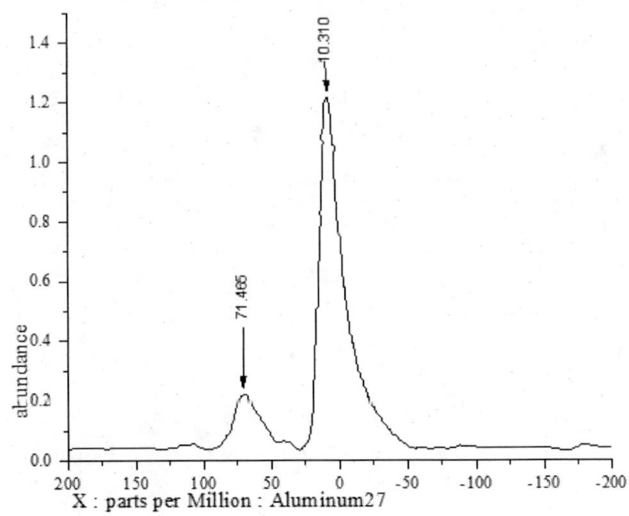

Figure 10. ^{27}Al MAS NMR of Calcined Layered aluminosilicate.

The ^{27}Al resonance at 71.465 ppm suggests a framework environment, aluminum in the form of AlO_4 tetrahedra linked with Q^3 SiO_4 [AlO_4 (Q^3 SiO_4)]. From literature reports, aluminosilicate material to have these units shows ^{27}Al chemical shift range 70-80 ppm [50] and confirms the existence of structurally deformed aluminum sites. The broad and strong peak raised in 10.3 ppm due to presence of Aluminum in VI state. The aluminum content is higher in frameworks, causing higher octahedral aluminum. These aluminum species undergo reversible coordination in aluminosilicate frameworks [51]. The below equation shows the reversible coordination in aluminosilicate frameworks.

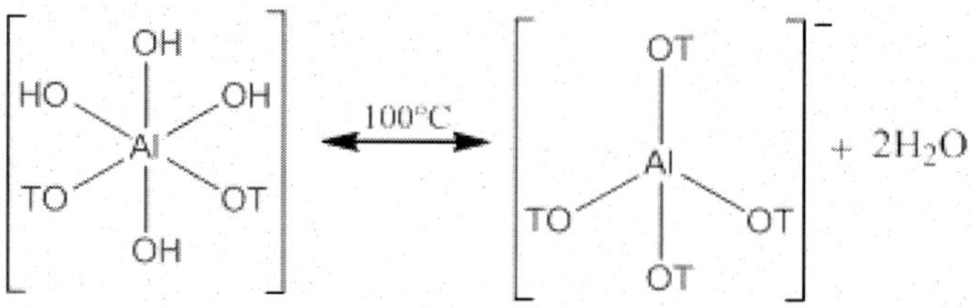

3.7. Morphological Studies

The surface of as-synthesized aluminosilicate is highly powdery and heterogeneous due to the coexistence of different zeolite phases together with other crystalline layered and amorphous mass. Crystals of various shapes and sizes, together with amorphous masses, incorporate into friable grains. Overall, it appears as a hardy tight mass with a rough surface. Open-air calcined aluminosilicate material had the platy silt type of material with a smooth surface adsorbed as a dirty mass. During heating, aluminosilicate framework undergoes multiple species gathered to regular order structure.

Figure 11. SEM image of as-synthesized aluminosilicate sample.

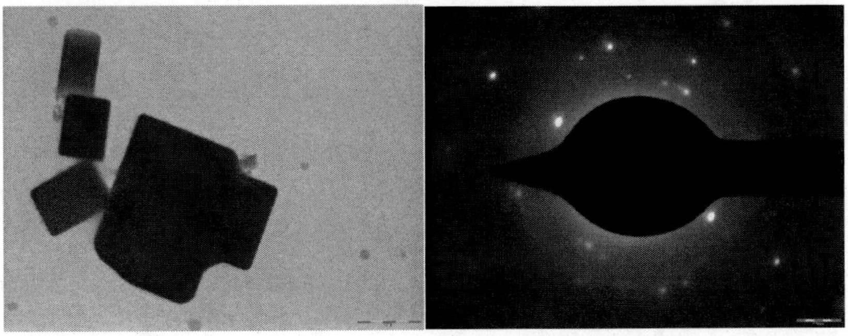

Figure 12. TEM studies and SAED Pattern of Calcined Layered aluminosilicate.

TEM images of calcined aluminosilicate samples clearly show the layered crystalline structure, but amorphous to Powder XRD. The structure is further confirmed with a Selected Area Electron Diffraction pattern of the sample. SEAD showed the regular arrangement of atomic units. The average size of the particle length 0.8-1.8 μm and breath 0.5 μm.

CONCLUSION

The sol-gel method provides the mesoporous layered aluminosilicate by a direct method. Its mechanism was complex to understand and followed the liquid-crystal template mechanism. The presence of aluminum atoms in silicate has offered numerous possibilities in chemistry as well as increases the opportunity in molecular science and industrial application. This layered aluminosilicate has the following advantage over their silicate counterpart

- Increased reactivity of the framework by increasing acidic sites
- Increased ion exchange capacity
- The tunability of the framework mainly depends on aluminum content
- Able to create different sites Bronsted acidic, Lewis acidic and Lewis basic sites
- Retention of water content in the framework by absorbing the atmospheric moisture
- Dealumination, creation of new aluminum species

Disadvantages: The aluminum atom could induce the adsorption of a water molecule from the surrounding atmosphere into the framework, and it creates a hygroscopic nature. This hygroscopic tendency increases by increasing the number of framework aluminum atom and the Si/Al ratio reaches to one, become deliquescing nature. Pre-heat treatment required before the starting of any experiment and is very sensitive to the surrounding atmosphere. So, precautions are required in the handling and storage of the material. The following points are observed if the Si/Al ratio decreases compared with their silicate counterpart,

- Decreased hydrothermal stability
- Decreased thermal stability
- Decreased stability of the framework
- Decreased e surface area
- Rising aluminum content creates more stress and strain in the framework

The rational methods and computational techniques can eliminate these kinds of problems in the framework.

REFERENCES

[1] Keane, M. A. (2003). Ceramics for Catalysis. *Journal of Materials Science,* 8 (38): 4661-75.

[2] Perez, O. J. O., Borade, R. B. and Clearfield, A. (1998). Synthesis of a Mesoporous Aluminophosphate. *Journal of Molecular Structure,* 470 (1-2): 221-28.

[3] Cundy, C. S. and Cox, P. A. (2003). The Hydrothermal Synthesis of Zeolites: History and Development from the Earliest Days to the Present Time. *Chemical Reviews,* 103 (3): 663-701.

[4] Stuckenschmidt, E., Joswig, W. and Baur, W. H. (1996). Flexibility and Distortion of the Collapsible Framework of NAT Topology: The Crystal Structure of H_3O-Natrolite. *European Journal of Mineralogy*, 85-92.

[5] Hammonds, K. D., Heine, V. and Dove, M. T. (1998). Rigid-Unit Modes and the Quantitative Determination of the Flexibility Possessed by Zeolite Frameworks. *The Journal of Physical Chemistry, B* 102 (10):1759-67.

[6] Baur, W. H. (1978). Variation of Mean Si-O Bond Lengths in Silicon-Oxygen Tetrahedra. *Acta Crystallographica Section B: Structural Crystallography and Crystal Chemistry,* 34 (6): 1751-56.

[7] Downs, R. T., Gibbs, G. V, Bartelmehs, K. L. and Boisen, M. B. (1992). Variation of Bond Lengths and Volumes of Silicate Tetrahedra with Temperature. *American Mineralogist*, 77 (7-8): 751-57.

[8] Alothman, Z. A. (2012). A Review: Fundamental Aspects of Silicate Mesoporous Materials. *Materials*, 5 (12): 2874-2902.

[9] Zhao, D., Huo, Q., Feng, J., Chmelka, B. F. and Stucky, G. D. (1998). Nonionic Triblock and Star Diblock Copolymer and Oligomeric Surfactant Syntheses of

Highly Ordered, Hydrothermally Stable, Mesoporous Silica Structures. *Journal of the American Chemical Society,* 120 (24): 6024-36.

[10] Zhao, D., Feng, J., Huo, Q., Melosh, N., Fredrickson, G. H., Chmelka, B . F., and Stucky, G . D. (1998). Triblock Copolymer Syntheses of Mesoporous Silica with Periodic 50 to 300 Angstrom Pores. *Science,* 279 (5350): 548-52.

[11] Georgiev, D., Bogdanov, B., Hristov, Y. and Markovska, I. (2011). Synthesis of NaA Zeolite from Natural Kaolinite. *Oxidation Communications,* 34 (4): 812-19.

[12] Cejka, J., Van Bekkum, H., Corma, A. and Schueth, F. (2007). *Introduction to Zeolite Molecular Sieves*. Vol. 168. Elsevier.

[13] Breck, D. W. (1974). Structure, Chemistry and Use of Zeolite Molecular Sieves. Wiley, New York.

[14] Pokrovski, G. S., Schott, J., Harrichoury, J. C. and Sergeyev, A. S. (1996). The Stability of Aluminum Silicate Complexes in Acidic Solutions from 25 to 150°C. *Geochimica et Cosmochimica Acta* 60 (14): 2495-2501.

[15] Huo, Q., Margolese, D . I., Ciesla, U., Demuth, D. G., Feng, P., Gier, T. E., and Sieger, P. (1994). Organization of Organic Molecules with Inorganic Molecular Species into Nanocomposite Biphase Arrays. *Chemistry of Materials,* 6 (8): 1176-91.

[16] Feijen, E. J. P., Martens, J. A. and Jacobs, P. A. (1994). Zeolites and Related Microporous Materials: State of the Art. *Studies in Surface Science and Catalysis,* 84 (Part A): 3-21.

[17] Morris, R. E. and Weigel, S. J. (1997). The Synthesis of Molecular Sieves from Non-Aqueous Solvents. *Chemical Society Reviews,* 26 (4): 309-17.

[18] Bronić, J., Subotić, B., Smit, I. and Despotović, L. J. A. (1988). Influence of Gel Ageing on Zeolite Nucleation Processes. In: *Studies in Surface Science and Catalysis*, 37:107-14. Elsevier.

[19] Twomey, T. A. M., Mackay, M., Kuipers, H. and Thompson, R. W. (1994). *In Situ* Observation of Silicalite Nucleation and Growth: A Light-Scattering Study. *Zeolites,* 14 (3): 162-68.

[20] Jackowski, A., Zones, S. I., Chaudhuri, K., Lacheen, H. S. and Yeh, S. W. A (2014). A Hydrothermal Synthesis of a 2-Dimensional Layered Silicate Followed by a Transition to a 3-Dimensional Aluminosilicate Zeolite Dedicated to the Memory of Our Wonderful Zeolite Research Colleague, Lun-Teh Yuen. *Microporous and Mesoporous Materials,* 197: 33-39.

[21] Diaz, I. and Mayoral, A. (2011). TEM Studies of Zeolites and Ordered Mesoporous Materials. *Micron,* 42 (5): 512-27.

[22] Solovyov, L. A. (2013). Diffraction Analysis of Mesostructured Mesoporous Materials. *Chemical Society Reviews,* 42: 3708-20.

[23] Ward, J. W. (1976). Zeolite Chemistry and Catalysis. *ACS Monograph,* 171 (80): 10.

[24] Karge, H. G., Hunger, M. and Beyer, H. K. (1999). Characterization of Zeolites - Infrared and Nuclear Magnetic Resonance Spectroscopy and X-Ray Diffraction. In: *Catalysis and Zeolites*, 198-326. Springer.

[25] Hunger, M., Ernst, S., Steuernagel, S. and Weitkamp, J. (1996). High-Field 1H MAS NMR Investigations of Acidic and Non-Acidic Hydroxyl Groups in Zeolites H-Beta, H-ZSM-5, H-ZSM-58 and H-MCM-22. *Microporous Materials*, 6 (5-6): 349-53.

[26] Datka, J. and Tużnik, E. (1985). Hydroxyl Groups and Acid Sites in Na ZSM-5 Zeolites Studied by IR Spectroscopy. *Zeolites*, 5 (4): 230-32.

[27] Jentys, A., Mirth, G., Schwank, J. and Lercher, J. A. (1989). Interaction of Hydrocarbons and Water with ZSM-5. In: *Studies in Surface Science and Catalysis*, 49:847-56. Elsevier.

[28] Uytterhoeven, J. B., Schoonheydt, R., Liengme, B. V and Hall, W. K. (1969). Studies of the Hydrogen Held by Solids: XVI. Infrared Spectroscopy of X- and Y-Type Zeolites Containing Univalent and Divalent Cations. *Journal of Catalysis*, 13 (4): 425-34.

[29] Vijayasankar, A. V., Mahadevaiah, N., Bhat, Y. S., and Nagaraju, N. (2011). Mesoporous Aluminophosphate Materials: Influence of Method of Preparation and Iron Loading on Textural Properties and Catalytic Activity. *Journal of Porous Materials* 18 (3): 369-78.

[30] Ward, J. W. (1968). Spectroscopic Study of the Surface of Zeolite Y. II. Infrared Spectra of Structural Hydroxyl Groups and Adsorbed Water on Alkali, Alkaline Earth, and Rare Earth Ion-Exchanged Zeolites. *The Journal of Physical Chemistry*, 72 (12): 4211-23.

[31] Zecchina, A., Geobaldo, F., Spoto, G., Bordiga, S., Ricchiardi, G., Buzzoni, R., and Petrini, G. (1992). Low-Temperature Fourier-Transform Infrared Investigation of the Interaction of CO with Nanosized ZSM5 and Silicalite. *Journal of the Chemical Society, Faraday Transactions,* 88 (19): 2959-69.

[32] Luan, Z. and Fournier, J. A. (2005). *In Situ* FTIR Spectroscopic Investigation of Active Sites and Adsorbate Interactions in Mesoporous Aluminosilicate SBA-15 Molecular Sieves. *Microporous and Mesoporous Materials,* 79 (1-3): 235-40.

[33] Rege, S. U. and Yang, R. T. (2001). A Novel FTIR Method for Studying Mixed Gas Adsorption at Low Concentrations: H_2O and CO_2 on NaX Zeolite and γ-Alumina. *Chemical Engineering Science,* 56 (12): 3781-96.

[34] Madejová, J. (2003). FTIR Techniques in Clay Mineral Studies. *Vibrational Spectroscopy,* 31 (1): 1-10.

[35] Huang, Y., Jiang, Z. and Schwieger, W. (1999). Vibrational Spectroscopic Studies of Layered Silicates. *Chemistry of Materials,* 11 (17): 1210-17.

[36] Kazansky, V. B., Serykh, A. I., Anderson, B. G. and Santen, R. A. V. (2003). The Sites of Molecular and Dissociative Hydrogen Adsorption in High-Silica Zeolites Modified with Zinc Ions. III DRIFT Study of H_2 Adsorption by the Zeolites with

Different Zinc Content and Si/Al Ratios in the Framework. *Catalysis Letters*, 88 (3-4): 211-17.

[37] Agmon, N. (2016). The Acid Test for Water Structure. *Nature Chemistry*, 8 (3): 206-7.

[38] Thämer, M., De Marco, L., Ramasesha, K., Mandal, A. and Tokmakoff, A. (2015). Ultrafast 2D IR Spectroscopy of the Excess Proton in Liquid Water. *Science,* 350 (6256): 78-82.

[39] Liu, J., Yu, Y., Mu, Y. and He, H. (2006). Mechanism of Heterogeneous Oxidation of Carbonyl Sulfide on Al_2O_3: An *in Situ* Diffuse Reflectance Infrared Fourier Transform Spectroscopy Investigation. *The Journal of Physical Chemistry. B,* 110 (7): 3225-30.

[40] Singha, M. and Singh, L. (2016). Vibrational Spectroscopic Study of Muscovite and Biotite Layered Phyllosilicates. *Indian Journal of Pure and Applied Physics,* 54 (February): 116-22.

[41] Babonneau, F., Coury, L. and Livage, J. (1990). Aluminium Sec-Butoxide Modified with Ethylacetoacetate: An Attractive Precursor for the Sol-Gel Synthesis of Ceramics. *Journal of Non-Crystalline Solids,* 121 (1-3): 153-57.

[42] Selvaraj, M., Sinha, P. K. and Pandurangan, A. (2004). Synthesis of Dypnone using SO_4^{2-}/Al-MCM-41 Mesoporous Molecular Sieves. *Microporous and Mesoporous Materials,* 70 (1-3): 81-91.

[43] Tarte, P. (1967). Infra-red Spectra of Inorganic Aluminates and Characteristic Vibrational Frequencies of AlO Tetrahedra and AlOs Octahedra. *Spectrochimica Acta,* 23A: 2127-43.

[44] Payra, P. and Dutta, P. K. (2003). Zeolites: A Primer. In: *Handbook of Zeolite Science and Technology*, 13-36. CRC press.

[45] Ravishankar, R., Kirschhock, C. E. A., Gerrits, P. P. K., Feijen, E. J. P., Grobet, P. J., Vanoppen, P., and Schryver, F. C. D. (1999). Characterization of Nanosized Material Extracted from Clear Suspensions for MFI Zeolite Synthesis. *The Journal of Physical Chemistry B,* 103 (24): 4960-64.

[46] Sing, K. S. W. (1982). Reporting Physisorption Data for Gas/Solid Systems with Special Reference to The Determination of Surface Area and Porosity. *Pure and Applied Chemistry,* 54 (11): 2201-18.

[47] Fernandez, R., Martirena, F. and Scrivener, K. L. (2011). The Origin of the Pozzolanic Activity of Calcined Clay Minerals: A Comparison between Kaolinite, Illite and Montmorillonite. *Cement and Concrete Research,* 41 (1): 113-22.

[48] Drits, V. A., Lindgreen, H., Salyn, A. L., Ylagan, R. and McCarty, D. K. (1998). Semiquantitative Determination of Trans-Vacant and Cis-Vacant 2:1 Layers in Illites and Illite-Smectites by Thermal Analysis and X-Ray Diffraction. *American Mineralogist,* 83 (11-12 Part 1): 1188-98.

[49] Somani, O. G., Choudhari, A. L., Rao, B. S. and Mirajkar, S. P. (2003). Enhancement of Crystallization Rate by Microwave Radiation: Synthesis of ZSM-5. *Materials Chemistry and Physics,* 82 (3): 538-45.

[50] Singh, P. S., Bastow, T. and Trigg, M. (2005). Structural Studies of Geopolymers by ^{29}Si and ^{27}Al MAS-NMR. *Journal of Materials Science*, 40 (15): 3951-61.

[51] Omegna, A., van Bokhoven, J. A. and Prins, R. (2003). Flexible Aluminium Coordination in Alumino-Silicates. Structure of Zeolite H-USY and Amorphous Silica-Alumina. *The Journal of Physical Chemistry B*, 107 (34): 8854-60.

Dr. T. T. Sundaram

Affiliation: Life Time Member, ISTE.

Education: M. Sc., M. Phil., Ph. D.

Business Address:
Department of Chemistry, Adhiyamaan College of Engineering,
Dr. M. G. R Nagar, Hosur, Tamilnadu, 635109, India.

Research and Professional Experience: Totally 15years

Professional Appointments:
Assistant Professor, Adhiyamaan College of Engineering, hosur - July 2008 - tilldate
Senior Lecturer, Er. Perumal Manimekalai College of Engineering, Hosur - August 2007 - May 2008
Lecturer, Pee Gee College of Arts and Science, Dharmapuri, -June 2003 - July 2005.

Honors: Nil

Publications from the Last 3 Years:

1. Sundaram, T. T., N. Prasanna, and P. Jayamurugan. (2016). "Facile Synthesis of Aluminosilicate." *Oriental Journal of Chemistry* 32 (3): 1711-15. https://doi.org/10.13005/ojc/320351. ISSN: 0970-020X/2231-5039.
2. Srinivasan, Venkatesan, Subramani Kullagounder, Sibi Srinivasan, and Sundaram Thiraviyam. (2015). "Synthesis of Mesoporous Silicon with Uniform Pore Size Using the Ashes of Husks from Panicum Miliare Millet: A Novel Recyclable Bio-Waste." *Silicon*, 1-7. ISSN: 1876-990X/1876-9918.

3. Srinivasan, Venkatesan, Sundaram Thiraviam, Kullagounder Subramani, Sibi Srinivasan, Pragathiswaran Chelliah, and Stanley Anthuvan Babu. (2014). "*In Situ* Disassembling Behavior of Composite Hydrogels for the Efficient Removal of Crystal Violet Dye from Aqueous Solution." *ISRN Polymer Science* 2014: 1-10. https://doi.org/10.1155/2014/632348.
4. Kannan, Chellapandian, Thiravium Sundaram, and Thayumanavan Palvannan. (2008). "Environmentally Stable Adsorbent of Tetrahedral Silica and Non-Tetrahedral Alumina for Removal and Recovery of Malachite Green Dye from Aqueous Solution." *Journal of Hazardous Materials* 157 (1): 137-45. https://doi.org/ 10.1016/j.jhazmat.2007.12.116. 0304-3894/1873-3336.
5. Kannan, C., A. Selvaraj, T. T. Sundaram, and T. Palvannan. (2007). "Soil Pollution by Malachite Green Dye and Dye Recovery from Effluent." *Environmental Science* 2 (2): 10-16.

In: An Introduction to Aluminosilicates
Editor: Nero Regina Blevins
ISBN: 978-1-53617-250-8

Chapter 9

CLINOPTILOLITE-RICH TUFF – A NATIVE ALUMINOSILICATE NANOCOMPOSITE FOR THE REMOVAL OF TRACE QUANTITIES OF POLLUTANTS

Eva Chmielewská[1,*], Renáta Górová[1], Włodzimierz Tylus[2], Mária Kovaľaková[3] and Jozef Kravčák[3]

[1]Faculty of Natural Sciences, Comenius University, Bratislava, Slovakia
[2]Faculty of Chemistry, Wroclaw University of Science and Technology, Wroclaw, Poland
[3]Faculty of Electrical Engineering and Informatics, Technical University of Košice, Košice, Slovakia

ABSTRACT

Thanks to recent biosynthesis achievements, the development of versatile protocols in architectural design and control has advanced the research in adsorbent chemistry beyond the conventional one, which has not processed multifunctional heterostructures before. It should also be pointed out that usually novel and state-of-the art biosynthesis protocols allow for the simplification of experimental design and apply renewable and ubiquitous biological materials, thereby minimizing energy requirements and waste generation. In

[*] Corresponding Author Email: chmielewska@fns.uniba.sk.

recent decades, magnetic nanoparticles have proven to be frequently occuring building blocks for hybrid structure construction and moreover, have shown innovative prospects as multifunctional adsorbents. Such magnetic heterostructures are commonly prepared by using aluminosilicates (zeolites), silicates, carbon or biopolymer shells as linkers for further functionalization or protective coatings. Recently, some studies have illustrated that nanoparticles have the ability to enter organisms during ingestion or inhalation and can translocate within the body to various organs and tissues to exert toxicity effect. In order to depress the toxic effects of nanoparticles, various surface coatings have been employed. This can greatly reduce the contact of nanoparticles with living bodies and thus their toxicity impact. Therefore, immobilized metal oxides nanoparticles have been videly used in environmental applications and have shown promising performance in pollutant removal and toxicity mitigation. Among the most widely-used nanoparticles, magnetic nanoparticles of iron and manganese oxides have sparked an immense interest in research of engineering applications for the treatment of contaminated water or subsurface environments. These nanoparticles not only have a large removal capacity, fast kinetics and high reactivity for specific pollutants removal due to their extremely small particle size and thus high surface area to volume ratio, but also one more important aspect: they are economically feasible. Their cleanup process of pollutants removal is mostly controlled by complexation. Zeolites occuring in nature are highly porous aluminosilicates of volcanic origin. They have been applied for the immobilization of Fe – and Mn-oxide nanoparticles in current research. Zeolite rock is essentially composed of an active component, i.e., industrially most frequently used clinoptilolite as well as some other geological minerals, potentially nanodispersed in rock matrix, such as volcanic glass, feldspar, quartz and clays, which may reach up to several tens % of the rock. Pharmaceuticals, which are known today as emerging contaminants (ECs), can enter the environment through a number of pathways and can be further distributed to various environmental media. One prominent pathway could be the use of wastewater sludge or wastewater for field fertilization and irrigation. In aquatic environments, a large variety of these compounds and their metabolites have been detected. For this reason, the objective of the work presented was to examine several natural, commercial or even waste products in selected pharmaceutical (diclofenac DCF) removal, by using synthetic water solutions in laboratory measurements. Several surface treated zeolite samples, such as FeO(OH)-zeolite, MnOx - zeolite, carbonized and surfactant coated zeolite, were also evaluated and compared for DCF removal efficiency. In contrast to a voluminous organic and dissociated DCF molecule, another model pollutant Zn(II) cation was chosen for the evaluation of those zeolite samples, which did not indicate any or only low DCF uptake ability. It was chosen particularly to demonstrate the samples removal efficiency to a different type of pollutant. Subsequently, some complementary analytical methods (XPS, XRD, SEM, SEM-EDS, BET, MAS NMR) were used for the characterization of the material studied.

Keywords: clinoptilolite-rich tuff, aluminosilicate nanocomposites, removal of pharmaceuticals and Zn(II), adsorption, XPS, HPLC, MAS NMR.

INTRODUCTION

Adsorption is recognized as an efficient and cost-effective separation technique for water decontamination, especially when local accessible, low cost, natural and environmental friendly materials are available [1, 2]. The worldwide interest in renewable resources, reduced greenhouse emissions and more effective treatment of solid waste have supported renewed interest in the design of environmental adsorbents. These follow along the lines of commodities for adsorption use to be occasionally applied: chitosan, zeolites, clay, certain industrial or agricultural waste products like fly ash, low rank coal, lignite, natural metal oxides, peat, sawdust, waste slurry, lignine, shungite, coconut shell charcoal, waste biomass and cork (Quercus suber L.) [3, 4].

Zeolite aluminosilicate frameworks consist of AlO_4 and SiO_4 tetrahedras, set in a repeating pattern to form a microporous three-dimensional structure (micropores are defined by IUPAC as being less or equal to 2 nm in diameter). Oxygen atoms form the link between adjacent silicon and aluminum atoms. The negative charge of their frameworks is balanced by exchangeable cations that reside as guests within the pores. The general formula for the composition of a zeolite is as follows:

$$A^{m+}_{y/m} [(SiO_2)_x . (AlO_2^-)_y] . zG$$

where A is the cation (of charge *m*) and G is the neutral guest. Today, over 80 various natural zeolite species are known, while mostly clinoptilolite, which is found in Slovakia too, most often plays the most important role [5, 6].

Wet chemicals, mostly sol-gel methods, are used extensively for the preparation of nanoparticles and nanostructures of various dimensions, such as silver clusters, FeO(OH) – oxihydroxides or TiO_2 semiconductor, on the surface of porous materials including zeolite in order to upgrade their adsorption potential. They may interact with each other by hydrogen, covalent, van der Vaals, vacancy or donor reaction centres, adhesion, complementary and other specific and non-specific forces [7].

To enhance the adsorption ability of natural zeolite (clinoptilolite-rich tuff), Chmielewská et al. [8] crosslinked flexible biogenic alginate with zeolite powder using Fe(III) and Ca(II) salts and obtained promising results. Another excellent example how to functionalize zeolite surface and thus improve their adsorption properties is the use of expandable biogenic surfactants. Various aspects regarding surfactant coated zeolites have been examined in recent decades, based on zeolite's superior hydrodynamic properties and sufficiently high capacity toward a broad range of pollutants [9]. Cationic octadecylammonium (ODA) surfactant, which consists of long, 18C straight hydrocarbon chains, was chosen for zeolite hydrophobization as an inland economically accessible product, applied with a long tradition for domestic bentonite beneficiation.

For several reasons, we chose to study diclofenac (DCF) removal on natural and surface treated zeolites and then compare them to other products, i.e., plenty of adsorption materials that were imported to the domestic market, such as granulated ferric hydroxide (GEH), Reo Amos, Absodan, Nanofer, Bayoxide and others. Some of them are extremely expensive, despite consisting of natural or waste materials [10].

In contrast to a voluminous organic and dissociated DCF molecule, the pollutant Zn(II) cation was chosen for the evaluation of those zeolite samples, which did not indicate any or only low DCF uptake ability. It was chosen particularly to demonstrate the removal efficiency of those zeolite samples to a different type of pollutant. Subsequently, some complementary analytical methods (XPS, XRD, SEM, SEM-EDS, BET, MAS NMR) were used for the characterization of the material studied.

METHODS

Adsorbents Examined

The following adsorbents for DCF or Zn(II) uptake from synthetic solutions were examined: The natural zeolite clinoptilolite-rich tuff (quarry nearby Nižný Hrabovec in Eastern Slovakia). The zeolite was surface-treated and covered with MnO_X or oxi(hydr)oxides (denoted as FeO(OH) – zeolite) as well as by mixing with powdered activated carbon Norit GL 50 and offered to the market by the company Zeocem, under the commercial name KlinoCarb for 700 Euros per ton. A detailed description of FeO(OH) – zeolite sol-gel synthesis was published in paper [7]. The FeO(OH)–MnOx – zeolite was synthesized analogically as FeO(OH) – zeolite, however, instead of the natural material, the commercial Mn-Klinopur was used as the starting material.

The commercial Mn-Klinopur, obtained from Zeocem (Slovakia), was prepared by simple and inexpensive large-scale redox-synthesis at ambient temperature using a 5% $KMnO_4$ solution. Imported adsorption or waste products, such as Absodan DN2, Reo Amos EcoDry Plus, German activated carbon Silcarbon, Czech industrial waste Chezacarb, Reo Amos Peatsorb CB18, cationic surfactant octadecylammonium ODA-zeolite, carbonized zeolite (synthesized in our laboratory), German granulated ferric hydroxide GEH, Chitosan, Spilkleen SK1 and domestic geocomposite Slovakite, were also examined for DCF uptake.

In recent decades, chitin and its deacetylated derivative chitosan have appeared as potential biosorbents. Chitosan possesses appropriate mechanical strength, rigidity and settling characteristics, which are a perfect alternative to other biomass-adsorbents. Chitin is a non-toxic, odourless, moisture-retaining and healing polymer, found in animals, in exoskeleton shells of fungi, yeasts, arthropods and marine invertebrates. In fact, it is one of the most abundant and renewable natural biopolymers (polysaccharide).

One product that is useful for acids and bases neutralization and petroleum compounds removal is Spilkleen SK1, manufactured from recycled cellulose, calcium carbonate and clay. Slovakite, which is a commercial domestic geocomposite, is manufactured by IPRES engineering, Ltd. Bratislava from inland-rich aluminosilicate resources, such as dolomite, bentonite, diatomic clays, alginite and zeolite. It is justified only with clinker and finally pressurized. The price of Slovakite is 700 Euros per ton. The foreign adsorbent of the trademark Spilkleen SK2 is similar to the SK1 product and consists of a mixture of kaolin, calcium carbonate and cellulose, applied mostly for petroleum emulsions removal. The commercial product Absodan DN2 was manufactured from natural aluminosilicates resources as well and resembles volcanic glass. Granulated ferric hydroxide (GEH) is an approved commercial adsorbent manufactured by GEH Wasserchemie GmbH & Co. KG Osnabrück (Germany) selling for the price of 3750 Euros per ton. Powdered peat (here denoted as Peatsorb CB 18) contains lignin, cellulose, fulvic and humic acids as its major constituents, which contribute in chemical bonding with polar functional groups, such as alcohols, aldehydes, ketones and ethers. German provenience activated carbon Silcarbon was produced by the company Silcarbon Aktivkohle GmbH (Kirchhundem) from coconut shells, charcoal (wood) or alternatively coal at the temperature 700 – 900°C. Granulated Silcarbon SC-40 is offered on the market for 10, 310 Euros per ton. Industrial ashes Chezacarb is essentially amorphous carbon as a waste product from Unipetrol RPA Litvínov, Czech Republic.

Material Characterization and Analytical Equipment

Scanning Electron Micrographs SEM of the samples studied, as well as their elemental analysis, were done on a JEOL JSM 6390LV Electron Microscope after being stuck to a carbon band and metallized with gold using the coating apparatus JEOL JFC-1200. Tables 2 and 4 represent the average weight percentage of the elements in individual samples calculated from 5 to 10 measurements, depending on sample homogeneity.

Solid-state NMR measurements were performed on a Varian 400 MHz solid-state NMR spectrometer (Palo Alto, CA, USA) at a resonance frequency of approximately 100 MHz with the use of 4 mm ZrO_2 rotors and magic angle spinning (MAS) frequency of 10 kHz at room temperature. The spectra of pure diclofenac sodium and the sample of clinoptilolite with adsorbed diclofenac sodium were recorded using the cross-polarization (CP) technique with ^1H 90° pulse of 2.6 μs, contact time of 2 ms, acquisition time of 20 ms, high power proton decoupling of 88 kHz, recycle delay of 6 s and averaging over 14 000 scans. The high-resolution single-pulse ^{13}C NMR spectrum of KlinoCarb with adsorbed diclofenac was recorded with ^{13}C 90° pulse width of 2.5 μs, acquisition time of 20 ms, recycle delay of 8 s and averaging over 8672 scans.

External surface area and porosity of selected samples were determined at liquid nitrogen temperature (76 K) using Micromeritics ASAP 2400 Apparatus with enhanced gas analysis manifold, a lower pressure transducer, by applying the static volumetric technique and t-plot methods including BHJ pore diameter computation. Prior to the measurements, the samples were vacuum-degassed overnight at 2 Pa by T = 623.15 K.

X-ray photoelectron spectroscopy (XPS) studies were carried out using a SPECS PHOIBOS-100 hemispherical spectrometer with a Mg source (1253.6 eV) operating at 250W for high resolution spectra. The spectrometer energy scale was calibrated using the Au 4f7/2 and Cu 2p 3/2 lines at 84.2 and 932.4 eV, respectively. The powdered samples (with a diameter under 100 μm) were pressed into a molybdenum sample holder. To better understand the nanomorphology of modified zeolites, grain-sized (0.2 – 0.5 mm) samples, denoted as coarse, were analysed in the "as received" form or after Ar^+ sputtering. The C1s peak of the carbon at 284.8 eV was taken as reference in calculating BEs and accounting effects. The spectra were collected and processed by SpecsLab II and CasaXPS v.2.3.19 software.

The X-ray powder diffraction (XRPD) patterns were accomplished at room temperature on a BRUKER D8 Advance Apparatus with Cu antikathode, Ni K_β filters and LynxEye detector at 40 kV and 40 mA. The resulted patterns were compared with the JCPDS Cataloque (Joint Commettee on Powder Diffraction Standards No. 22-1236, 13-0304 for clinoptilolite verification).

Laboratory Setup

Conventional batch adsorption experiments were performed for isotherms calculation and plotting. For the most part, the 30 mL aliquots of synthetic solutions with variable concentration of DCF (or Zn^{2+} - solution) and 0.3 g of adsorbents (weight with analytical precision) were mixed intensively (speed of 180 rpm using the Biosan SIA Multi-Rotator, Latvia) for 2 hours in 50 mL polyethylene tubes (sufficiently long to reach an equilibrium of the system). Afterwards, the supernatant was separated through a 0.45 mm membrane filter and the residue was analysed. All measurements were done in triplicate. The equilibrium uptake capacity a_{eq} (in mg/g) for each sample was calculated according to the following mass balance equation:

$$a_{eq} = \left(\frac{C_i - C_{eq}}{m} \right) V$$

where c (i) and c (eq) were initial and equilibrium concentrations of the studied pollutant (in mg/L), m was the mass of the adsorbent examined (in g) and V was the volume of the solution in liters (L).

HPLC analyses of diclofenac in aqueous solutions were performed on Agilent Technologies 1200 Series Liquid Chromatograph, equipped with an automatic sampler and DAD detector. The Zorbax SB-C18 chromatographic column 3.5 μm, 150 x 2.1 mm was utilized for the analyses, the analyte was monitored at 278 nm and quantified based on a calibration curve. Unified initial concentration of DCF in D.I. water Co = 300 mg/L was initially used for all adsorbents examined to find the best one among them all. The stock solution was prepared from diclofenac sodium salt, p.a. (obtained by Sigma-Aldrich, Belgium).

The chemicals necessary for the $ZnSO_4$ stock solution preparation were purchased from Lachema Brno (made in Czech Republic) with analytical grade quality. Aqueous solutions of the pollutant as Zn(II) were determined by means of Isotachophoresis on the Analyser ZKI 02 (Villa Labeco, Slovakia). The measured data were processed by ITP PRO 32 program (Kas Comp, Ltd. Bratislava).

Results and Discussion

Removal of Diclofenac from Aqueous Solution with Natural and Geocomposed Materials

Diclofenac [sodium 2-(2,6-dichlorofenylamino) fenyl acetate] medication works by reducing hormones that cause inflammation and pain in the body [11]. Today, the amount of these drugs used in various countries is very alarming.

Natural zeolites (mostly species clinoptilolite), as versatile materials also for pharmaceutical applications, are known to be unharmful to animals and humans, moreover, with appropriate inertness. Several authors have reported that decreasing the particle size of clinoptilolite and modifying its surface significantly increased the adsorbed drug amount [12].

Clay minerals, which are abundant in nature, have a unique role in water decontamination, however, conventional and expensive activated carbon is usually applied to organic pollutants including pharmaceuticals removal. Because of the enormous and high-quality deposit of clinoptilolite-rich tuff in Slovakia, as well as increasing concentrations of emerging contaminants in waters, the objective of the work presented was to examine and compare a broad range of various adsorption materials including surface modified clinoptilolite-rich tuff for DCF uptake in order to assist in solving the above mentioned acute problem.

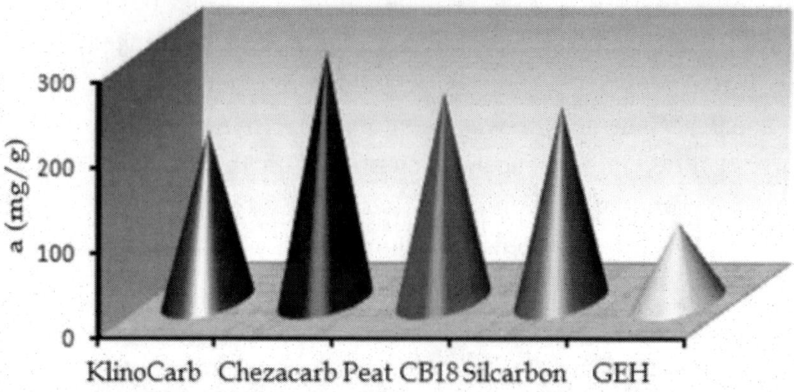

Figure 1. (Continued)

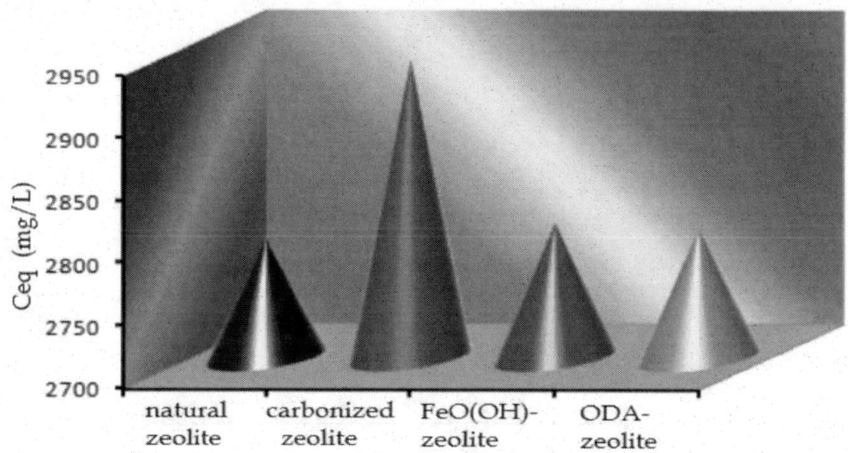

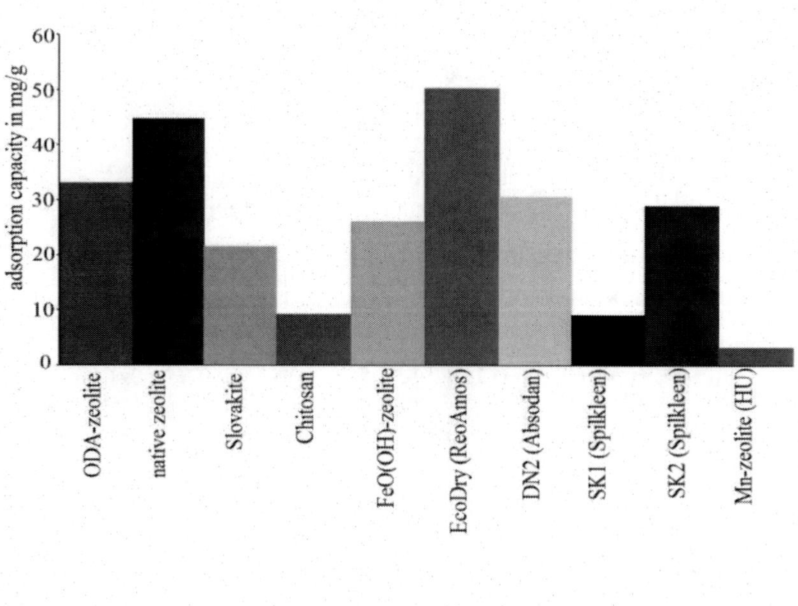

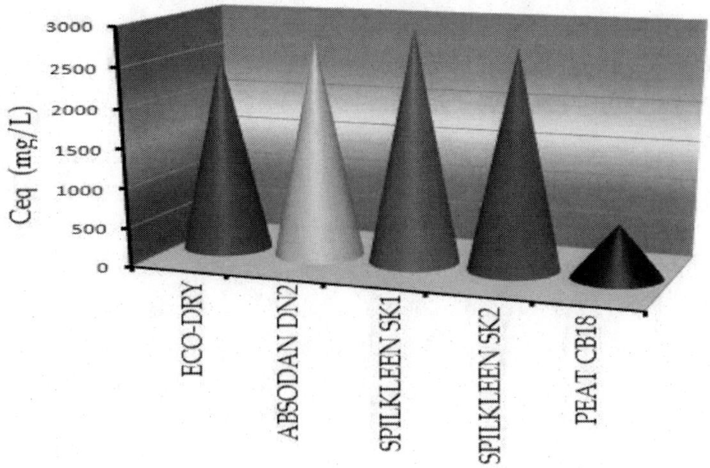

Figure 1. (Continued)

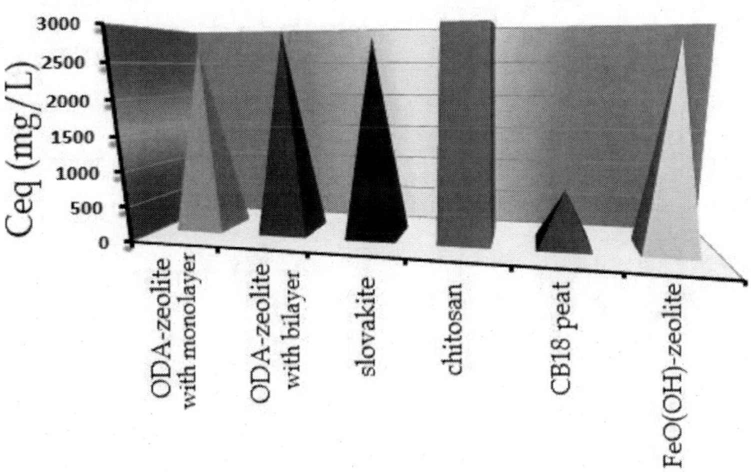

Figure 1. Efficiency of various natural or commercial adsorbents to remove diclofenac (with the same initial concentration of diclofenac in solution 300 mg/L)

As Figure 1 displays, clinoptilolite-rich tuff and its surface-treated forms, such as ODA-zeolite, carbonized and FeO(OH)-zeolite, did not reach an adsorption capacity toward DCF higher than 40 mg/g. Among them, only a native form showed an adsorption capacity toward DCF somewhat higher than 40 mg/g. The imported commodities, Absodan DN2 and the trademark Spilkleen SK2, reached the capacity to DCF a little less than 40 mg/g. Clinoptilolite-rich tuff, which was surface-coated with manganese polyoxides and denoted here as Mn-zeolite (HU), proved to be of lowest adsorption potential to DCF. From the capacity data plotting in Figure 1, it can be further estimated that the commercial products Chitosan and Spilkleen SK1 exhibit approximately 3-times higher capacity values to DCF than the capacity of Mn-zeolite.

Slovakian geocomposites of the trademark Slovakite appeared in comparison to Chitosan twice as effective in DCF uptake. Among the imported series of adsorbents (Absodan, Reo Amos, Spilkleen), the absolute best and most effective for DCF uptake, as well as simultaneously totally biodegradable, was Absodan CB18 Peatsorb; the second best was the product Reo Amos EcoDry with the capacity toward DCF nearly 50 mg/g, than the highly porous Absodan DN2 and the last position belongs to the totally biodegradable and combustible Spilkleen SK2 adsorbent. Nevertheless, the powdered peat, which is offered to the market as CB18 Peatsorb, reached the capacity value for DCF more or less comparable even with the most efficient industrial ashes (250 mg/g). The activated carbon Silcarbon of German provenience and the Slovak commercial product KlinoCarb showed somewhat lower adsorption performance in DCF uptake.

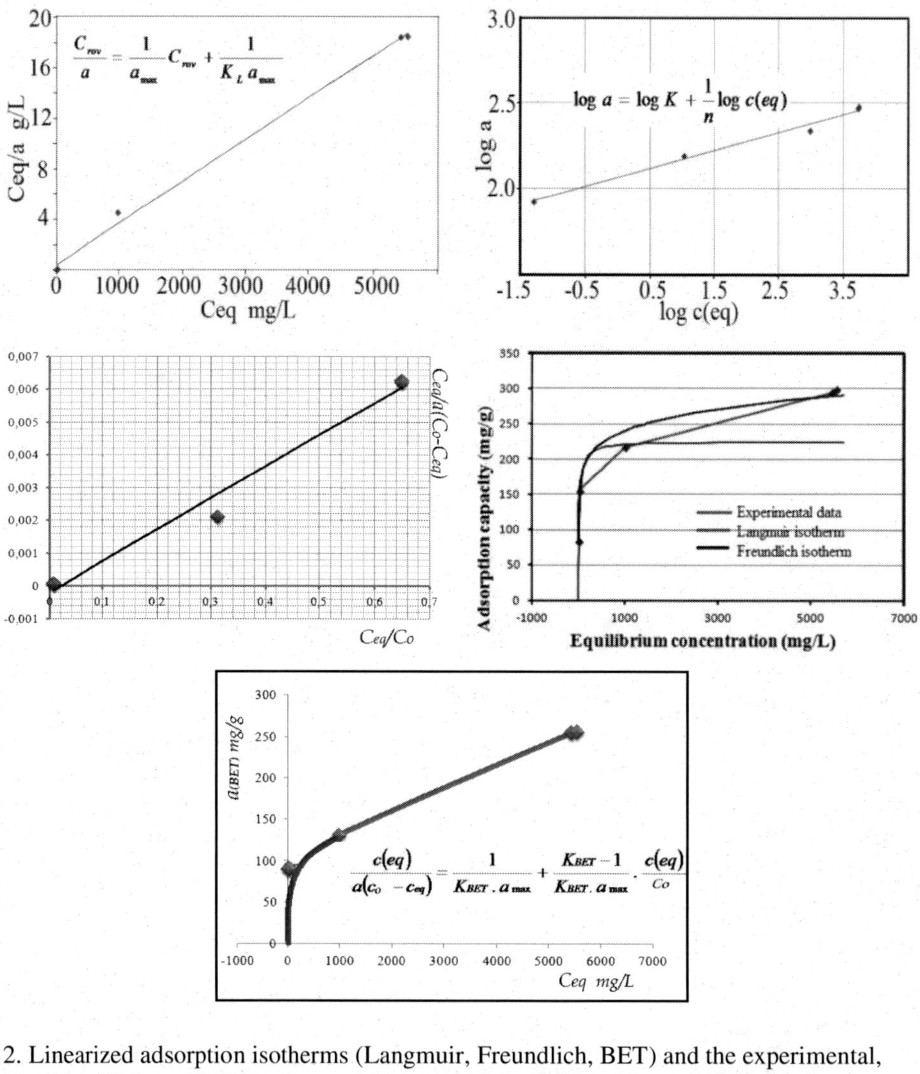

Figure 2. Linearized adsorption isotherms (Langmuir, Freundlich, BET) and the experimental, Langmuir, Freundlich and BET isotherms (downwards) for DCF uptake onto KlinoCarb.

In order to conclude and summarize the results discussed above, the following observations need be pointed out. In our research, the most effective adsorbents examined in DCF uptake from aqueous solutions were as follows: Chezacarb with a capacity of nearly 300 mg/g, Peatsorb CB18 with a capacity close to 250 mg/g, the German Silcarbon and the Slovakian commercial product KlinoCarb from the company Zeocem with a capacity around 200 mg/g. Because the peat, as a low cost natural material with confirmed high adsorption potential toward DCF, does not possess acceptable hydraulic and settling properties, further attention was focused on the characterization of the inland economically accessible and most effective KlinoCarb. Two basic monomolecular adsorption isotherms, Langmuir and Freundlich, and one polymolecular BET isotherm were calculated and drawn for DCF uptake onto KlinoCarb thereafter. As Table 1 and Figure 2 present, Langmuir and BET isotherms are better suited for the mathematical description of DCF adsorption onto KlinoCarb ($K_L = 0.0095$, $R^2 = 0.9969$ and $K_{BET} = -49.3942$, $R^2 = 0.9916$) than the Freundlich isotherm ($K_F = 113.07$, $n = 9.1575$, $R^2 = 0.9792$).

Table 1. Experimental and calculated isotherm data for DCF adsorption onto KlinoCarb

Experimental data			Freundlich isotherm			Langmuir isotherm		BET isotherm		
C_0 (mg/L)	C_{eq} (mg/L)	a (mg/g)	log C_{eq}	log a	a_F (mg/g)	C_{eq}/a	a_L (mg/g)	C_{eq}/C_0	$C_{eq}/a_{BET}(C_0-C_{eq})$	a_{BET}
7,157	0,050	83,830	-1,301	1,923	81,521	0.0006	0,559	0.0069	0.000084	90.57
1559,772	10,897	154,771	1,037	2,190	146,776	0.0704	79,175	0.0069	0.000045	90.49
3158,942	985,972	217,080	2,994	2,337	240,073	4.5419	220,600	0.3121	0.00209	130.62
8383,649	5434,972	294,570	3,735	2,469	289,274	18.4505	224,264	0.6483	0.006257	255.48
8538,745	5544,622	298,964	3,744	2,476	289,905	18.5461	224,281	0.6493	0.006194	256.24
Isotherm constants:			K_F = 113.074 1/n = 0.1092 Determination coefficient: R^2 = 0.9792			K_L = 0.0095 a_{max} = 225.093 Determination coefficient: R^2 = 0.9969		K_{BET} = - 49.3942 Determination coefficient: R^2 = 0.9916		

SEM - EDS, ^{1}H - ^{13}C CP MAS NMR and BET Analyses

Table 2 presents SEM-EDS data of the most effective organic carbon-rich adsorbents studied in this research for DCF uptake including natural zeolite (clinoptilolite-rich tuff). A comparable C – contents in Chezacarb, Silcarbon and KlinoCarb products could be observed. According to the information obtained by a personal discussion with a responsible manager from the company Eveco Brno, who provided KlinoCarb manufacturing, the product consists of 80% finely powdered zeolite and 20% powdered activated carbon Norit GL 50 (both powdered fractions under 100 μm). Opposite or reverse ratio, i.e., 80% of activated carbon and 20% of zeolite was identified in the above product composition by SEM – EDS (Table 2). Thus, a potential technology failure was argued for the delivered inhomogeneity of the product, likely caused during regular aeration, sedimentation as well as transportation of the commodity. As the manager explained, the reason might be that the light fraction, i.e., activated carbon of the mixture with zeolite moved much easier to the border of a discharge hopper [13, 14].

The XRD diffractogram of KlinoCarb (not recorded in the Chapter) did not confirm any crystal structure from the mixture produced, nevertheless, the diffractogram resembled activated carbon, which means a substance of amorphous nature.

Table 2. SEM-EDS composition of natural zeolite (clinoptilolite-rich tuff), Chezacarb, Silcarbon and KlinoCarb

Element Wt.% ± Error	Zeolite		Chezacarb		Silcarbon		KlinoCarb	
Oxygen	57.880	1.320	7.792	0.345	8.986	0.343	10.434	0.829
Magnesium	0.417	0.052	-		0.125	0.025	0.302	0.066
Aluminum	5.163	0.153	1.486	0.045	0.757	0.032	1.115	0.083
Silicon	27.320	0.647	0.128	0.028	0.306	0.027	0.763	0.076
Potassium	2.052	0.094	-		0.815	0.040	-	
Calcium	1.435	0.078	-		-		0.245	0.063
Iron	1.013	0.098	-		-		-	
Carbon	-		89.912	0.344	89.012	0.344	86.823	0.821
Sulfur	-		0.683	0.040	-		0.318	0.067

NMR spectroscopy was also used to confirm or disprove the predominant content of activated carbon in the obtained KlinoCarb product. The NMR spectrum of DCF (Figure 3) outlines signals at chemical shifts of 182.5 ppm (carbon 8), 141.2 – 115.7 ppm (carbons 1-6 and 1'-6'), and 44.5 and 43.2 ppm (carbon 7), characteristic for carbons in DCF structure according to [15, 16]. The NMR spectrum of clinoptilolite-rich tuff with adsorbed DCF confirmed the presence of ionized DCF. In general, the DCF signals are broadened, which is also reflected in less resolved spinning sidebands. The signals from carbon 7 (CH_2) are shifted to lower values, the signal from carbon 8 is split into two lines, which might indicate some changes in the environment of this carbon, probably due to the interaction of DCF with clinoptilolite surface or other ions, since DCF ions are too large to enter clinoptilolite channels. The NMR spectrum of DCF adsorbed onto KlinoCarb displays a broad signal expressly characteristic for activated carbon. These signals cannot be resolved due to a large broadening caused likely by a strong interaction of DCF with activated carbon. In addition, by using this complementary NMR analysis, a dominant presence of activated carbon in the KlinoCarb sample was ultimately confirmed (Figure 3).

Figure 4 illustrates the most interesting and typical morphologies of the adsorbents examined under Scanning Electron Microscopy (SEM) investigations. Photomicrographs of activated carbon, Peatsorb, carbonized zeolite, Eco Dry Plus, ODA-zeolite and Chezacarb (Figure 4b-h) indicate the presence of many pores, cracks, holes, plates, fibres and oval-shaped objects on the surface and nearly rough, more or less various-shaped, aggregate-like surface morphology. Clinoptilolite shows typical sheets, well-defined and tabular-shaped crystals with excellent crystal edges. Absodan DN2 presents typical needle morphology. When the surface is covered by FeO(OH) species or ODA-surfactant, smaller, more agglomarated crystals and poorly defined crystal edges were observed, in some spots with a profound change of typical lattice structure. The euhedral crystals of clinoptilolite and their apparent sharpness decreased with ODA surfactant and iron oxides coverage (Figure 4i). Some crystal habits of iron oxide are observed as round or disc-shaped hematite (light spots of images), coordinating dark rod-shaped goethite(Figure 4ij) [17].

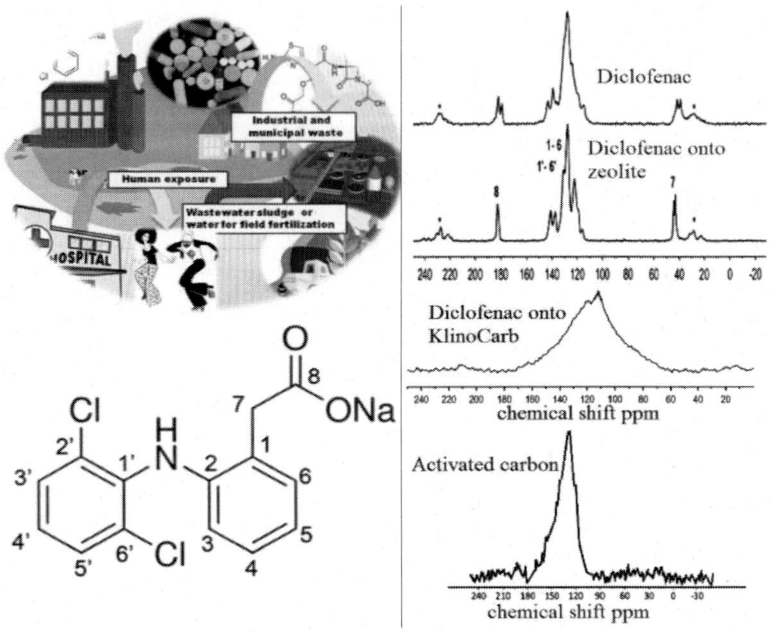

Chemical structure of diclofenac (lower) with pollutiom motive (upper picture).

Figure 3. ^{1}H-^{13}C CP MAS NMR spectra of diclofenac, diclofenac adsorbed onto zeolite, KlinoCarb and NMR spectra of Silcarbon (spinning sidebands are denoted by asterisks).

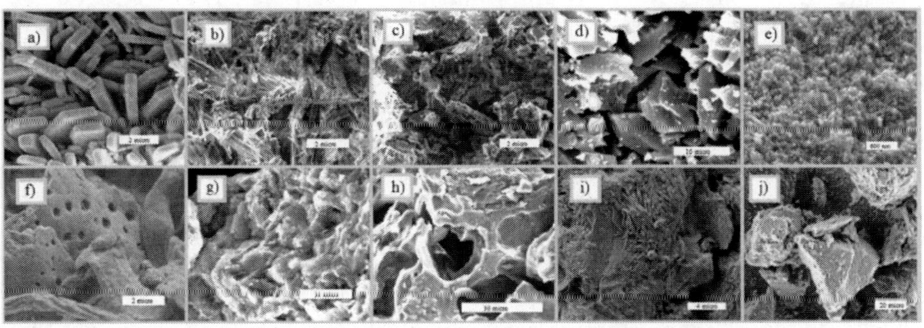

Figure 4. Scanning Electron Micrographs of a) natural zeolite (clinoptilolite tuff), b) Absodan DN2, c)Rep Amos EcoDry plus, d) activated carbon, e) Chezacarb, f) Reo Amos Peatsorb CB18, g) ODA-zeolite, h) carbonized zeolite, i) FeO(OH)-zeolite, j) GEH.

Table 3. Specific surface areas and porosity values of some materials studied (grain-size fr.under 100 μm except the carbonized zeolite)

Sample	S_{BET} (m^2/g)	S_t (m^2/g)	V_{micro} (cm^3/g)	*BJH* (*nm*)
Slovakian zeolite (clinoptilolite-rich tuff)	59.2	21.4	0.0045	14.5
Carbonized zeolite (0.2-0.5 mm)	23.7	17.6	0.0026	11.8
Fe-alginate-zeolitic pellet (1:2)	27.5	19.3	0.001	13.9
GEH (German granulated ferric hydroxide)	298	299	0	3.7
German Activated carbon Silcarbon	942	55.6	0.459	2.6
Commercial product Slovakite	61.5	47.7	0.007	8.9
Chezacarb (Czech industrial ashes)	1078	1041	0.010	9.0
KlinoCarb (Zeocem commercial product)	715	385	0.174	4.2

For material characterization and structural investigation of the adsorbents studied, the external surface area S_t with a specific surface area S(BET) and porosity were determined at liquid nitrogen temperature. Table 3 summarizes all the data measured. As can be observed, mainly the carbon-rich adsorption materials (Chezacarb, Silcarbon, KlinoCarb) distinguish high S(BET) values, which corresponded well with their efficiency in DCF uptake.

Mono- or Bimetallic Fe and Mn Oxides Supported Clinoptilolite-Rich Tuff for Alternative Zn(II) Removal

Since clinoptilolite-rich tuff and its surface treated forms, such as FeO(OH)-zeolite and Mn-zeolite did not extend toward DCF some competitive removal efficiency, as an alternative pollutant Zn(II) cation was chosen to demonstrate better and more effective adsorption ability of the zeolitic aluminosilicate toward this different type of pollutant. According to the experimental results and calculated adsorption isotherms, the following order with decreasing adsorption capacities of individual adsorbents to Zn(II) was found: MnOx-zeolite (35.1 mg/g) > FeO(OH)-MnOx-zeolite (23.5 mg/g) > FeO(OH)-zeolite (12.9 mg/g) > natural zeolite (8 mg/g). As it can be seen, MnOx-zeolite reached in regard to natural zeolite a 4-times higher capacity toward Zn(II) ions. The more effective adsorption of FeO(OH)- and FeO(OH)-MnOx-zeolites toward Zn(II) might be related to the presence of various finely dispersed Fe(III) hydrogels immobilized onto zeolitic surface during additional modification with alkalic Fe(III) – solution, as well as abundant surface hydroxyl groups, which probably bound this pollutant using complexation reactions.

In addition to the results above, as well as the mathematical isotherm calculations for the system studied, X-ray Photoelectron Spectroscopy (XPS), as a surface-sensitive quantitative spectroscopic technique, was used to measure the iron and manganese composition and their chemical states in order to help clarify the surface binding interactions between the adsorbents synthesized and the Zn(II) pollutant. Based on X-ray

Photoelectron Spectroscopy measurements and deconvolution of the Mn2p envelope of MnOx-zeolite, it was concluded that the only form of manganese compound on the surface of the modified zeolite is MnO_2 (Figure 5).

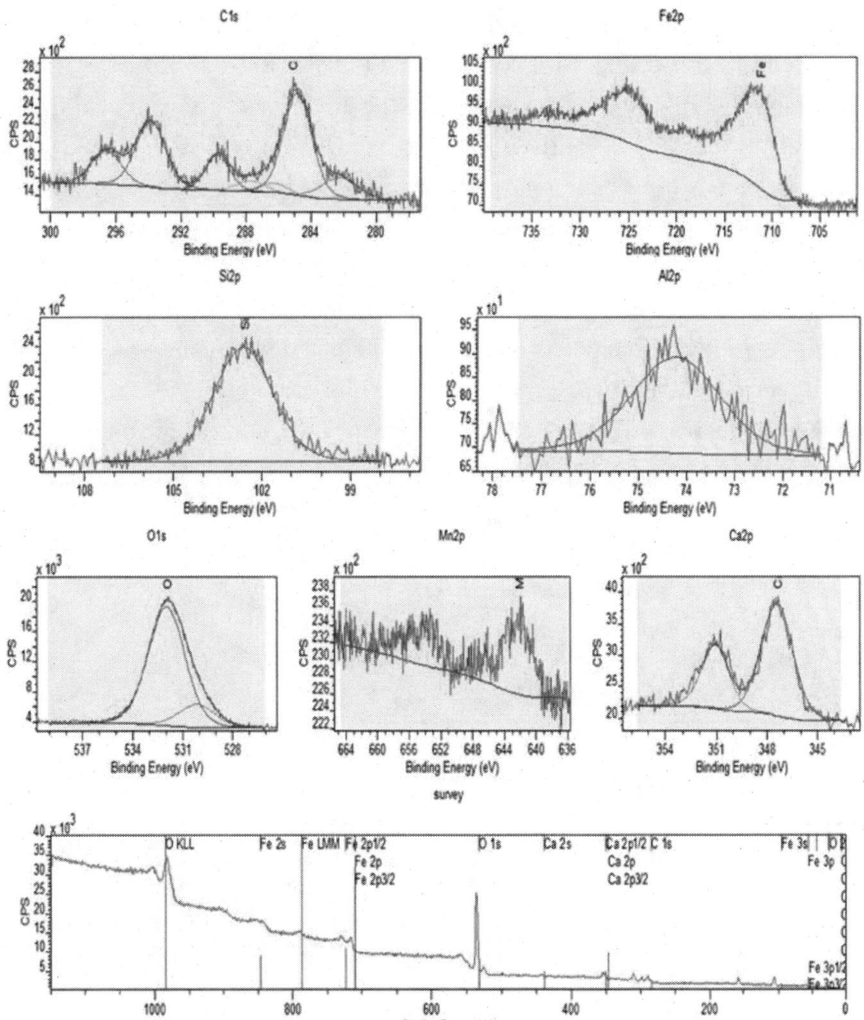

Figure 5. The set of XPS spectra for FeO(OH)-MnOx-zeolite (coarse 0.2-0.5 mm) which showed a close similarity with the spectra of the FeO(OH)-zeolite.

It was supposed that Mn(IV), which occurred as MnO_2, was in deep profiles of a zeolitic structure as crystalline or chemically bound. Furthermore, as the XPS results confirmed, the immobilization of FeO(OH) onto MnOx-zeolite caused significant drop of Mn content, which means that the surface oxidation with $KMnO_4$ solution was less invasive on the zeolite structure than the alkalic FeO(OH) precipitation, by which even the structure Al(III) ions might undergo some partial replacement with Fe(III) ions. This fact corresponds well with the Si/Al ratio, arising from the value of 4.2 for natural zeolite to 4.5 for MnOx-zeolite, to 5.3 for FeO(OH)-MnOx-zeolite and to 6.2 for FeO(OH)-zeolite (Table 4). At the same time, the occurrence of akaganeite β-FeOOH and lepidocrocite γ-FeO(OH) as well as pyrolusite MnO_2 was supposed to be identified on the surface of zeolites modified with manganese or iron oxihydroxides. Some carbonate groups at ca. 290 eV on the surface of both coarse MnOx – and FeO(OH)-MnOx –zeolite as $CaCO_3$ were observed, respectively (Figure 5). In conclusion, the XRD measurements (in this Chapter not recorded) for all the recorded diffractometer traces were typical for well-crystallized clinoptilolite, i.e., consisting of the characteristic Braggs reflections at 10, 22 and in the range of 25 - 30 of 2 theta. Even so, it was supposed that the new immobilized FeO(OH) hydrogels on the natural, as well as MnOx-zeolite in the form of amorphous phase with the content below 10% wt., might result in an undetectable change of both X-ray diffractograms. Only the visible peaks of primary deposited manganese dioxide (MnO_2), a high probability of pyrolusite on the surface of natural zeolite or the commercially sold product (Mn-Klinopur) was determined, where the Mn % wt. content in some of the darkest surface spots reached even 38 % wt. according to SEM-EDS.

Table 4. Surface composition (% at.) using the XPS method of all the samples studied of grain-size 0.2-0.5 mm

Sample in % at.	CO_3^-	C-O	O 1s	K 2p	Ca 2p	Si 2p	Al 2p	Fe 2p	Mn 2p	Mn/Fe	Mn/Al.	Mn/Si	Fe/Al.	Fe/Si	Si/Al
Natural zeolite	0.39	0.62	52.5	1.61	1.2	30.04	7.08	0.78	0	0	0	0	0.110	0.026	4.2
MnOx -zeolite	4.14	1.73	49.97	0.64	5.48	19.42	4.29	0.36	1.04	2.89	0.242	0.054	0.084	0.019	4.5
FeO(OH)-MnOx-zeolite	2.83	0.71	51.28	2.42	4.41	19.72	3.71	2.1	0.37	0.18	0.100	0.019	0.566	0.106	5.3
FeO(OH)-zeolite	0	1.34	50.98	2.91	<0.1	28.94	4.65	1.28	0	0	0	0	0.275	0.044	6.2

Table 5. An average SEM-EDS composition of natural zeolite (clinoptilolite-rich tuff) and 3 zeolite modified samples

Element % wt. ± Error	Natural zeolite		FeO(OH)-zeolite		MnOx-zeolite		FeO(OH)-MnOx-zeolite	
Oxygen	57.187	2.020	57.187	2.020	41.621	0.860	55.426	3.372
Magnesium	0.417	0.052	0.451	0.132	0.094	0.154	0.497	0.133
Aluminum	5.163	0.153	4.343	0.269	2.114	0.190	4.260	0.330
Silicon	27.320	0.647	20.972	0.832	6.860	0.280	25.168	1.570
Potassium	2.052	0.094	3.669	0.252	0.89	0.165	5.143	0.386
Calcium	1.435	0.078	3.365	0.981	9.149	0.337	2.742	0.258
Manganese	-		0.09	0.311	8.424	0.799	1.106	0.264
Iron	1.013	0.098	3.751	1.005	0.843	0.524	4.485	0.473

The Braggs reflection, at around 27-28 of 2 theta, confirmed the new crystallized pyrolusite mineral. While the results obtained by SEM-EDS in % wt. (Table 5) represent a surface profile of nanocomposites studied approximatelly up to 50 nm deep, the results summarized in Table 4 and provided by the XPS method are in % and mean analyses up to 5 nm under the outer surface. In general, when comparing both analytical techniques, an inhomogeneous coating and dispersion of Fe(III) and Mn(IV) metallic species on the zeolitic surface was observed.

CONCLUSION

To date, more than 200 different pharmaceuticals alone have been reported in river waters globally, with concentrations up to a maximum of 6.5 mg /L for the antibiotic ciprofloxacin [18]. An increase of drugs in surface waters may be anticipated during music festivals, public holidays, major sporting events and by students during exam periods (Figure 3, upper left). The presence of these chemicals in the environment is more serious considering that they do not appear individually, but as a complex mixture, which could lead to unwanted synergistic effects. According to European water quality monitoring, accomplished by the University of York (Great Britain), the highest water pollution with pharmaceuticals was detected in the river Danube. It must be said that detailed monitoring campaigns have been started only recently, in spite of substantial evidence of aquatic contamination by pharmaceuticals over the past 20 years. Nonsteroidal anti-inflammatory drugs (NSAID) are a class of drugs that provides analgesic, antipyretic, and at higher doses, anti-inflammatory effects without addictive reactions. From this group, diclofenac was found in concentrations of about 2000 ng/L, but its concentration often varied in the range of 3000–4000 ng/L in Slovak wastewaters [11].

The European legislation concerning the protection of aquatic environments and related organisms is mainly represented by the Water Framework Directive 2000/60/EC (WFD), according to which all the European water bodies have to reach a "good" ecological and chemical status by 2015, through the adoption of measures aimed at reducing, limiting and preventing pollution. Moreover, the EU Directive 2013/39/EU has already recommended to monitor in waters 45 priority compounds including pharmaceuticals [19, 20].

Wastewater treatment using the activated sludge process, depending on its age and other parameters, may partially decrease the concentration of pharmaceuticals in water, but not thoroughly. Additionally, during the anaerobic digestion, biosolids (or treated sludge) are generated and often applied to agricultural land as a fertiliser in many countries. Despite lengthy digestion (in avarage 4 weeks) and outdoor storage for up to six months following treatment, some emerging contaminants have shown to persist. Membrane processes, based on high solubility of pharmaceuticals in waters, are not effective for their removal and therefore some German plants have begun to use ozonization with activated carbon adsorption. Among the traditional geomaterials used in biomedicine such as pharmaceutical excipients or nanotransporters (nanobots), clinoptilolite-rich tuff with hydrophobized external surface has also recently occurred as a potential representant of aluminosilicates to be able to effectively remove pharmaceuticals from waters. This chapter attempts to refer to the recent research results on the suitability of using several aluminosilicates including natural zeolite for currently emergent pharmaceuticals (diclofenac), as well as Zn(II) removal from aqueous media. While some superior efficiency for DCF uptake showed only the KlinoCarb among the zeolites, and in a much lower extent the native form of zeolite, the manganese or iron and manganese oxides coated zeolites were more effective for alternative Zn(II) removal.

Acknowledgments

XPS analyses were financed by a statutory grant allocation from the Polish Ministry of Science and Higher Education for the Faculty of Chemistry of Wrocław University of Science and Technology. The authors express their thanks for XPS measurements and results evaluation to Dr. W. Tylus and to the Polish Ministry of Science and Higher Education for their financial support. The authors also thank to reviewers Dr. J.Lesný and Dr. I. Horváth (former senior scientists of University in Trnava and Slovak Academy of Sciences) for their valuable refinements and improvement of the whole Chapter.

References

[1] Králik, M. (2014). Adsorption, chemisorptions, and catalysis. *Chemical Papers – Chemické Zvesti* 68: 1625-1638, doi: 10.2478/s 11696-014-0624-9.

[2] Dabrowski, A. (2001). Adsorption – from theory to practice. *Advances in Colloid and Interface Science* 93: 135 – 224.

[3] Gupta, V., K. (2009). Application of low-cost adsorbents for dye removal-a review. *Journal Environmental Management* 90: 2313-42. doi: 10.1016/j.jenvman.2008.

[4] Crini, G. (2005). Recent development in polysaccharide-based materials used as adsorbents in wastewater treatment. *Progress in Polymer Science* 30: 38-70.

[5] Galamboš, M., Suchánek, P., Rosskopfová, O. (2012). Sorption of anthropogenic radionuclides on natural and synthetic inorganic sorbents. *Journal of Radioanalytical and Nuclear Chemistry* 293: 613-633.

[6] Pandová, I. (2011). *Zeolites and their application for environmental protection,* Technical University in Košice, Prešov, 93 p. (in Slovak) ISBN 978-80-553-0634-6.

[7] Chmielewská, E., Tylus, W., Drábik, M., Majzlan, J., Kravčak, J., Williams, C., Čaplovičová, M., Čaplovič, Ľ. (2017). Structure investigation of nano-FeO(OH) modified clinoptilolite tuff for antimony removal. *Microporous and Mesoporous Materials* 248: 222 – 233.

[8] Chmielewská, E., Sabová, L., Peterlik, H., Wu, A. (2011). Batch-wise adsorption, SAXS and microscopic studies of zeolite pelletized with biopolymeric alginate. *Brazilian Journal of Chemical Engineering* 28: 63-71.

[9] Haggerty, G. M. and Bowman, R. S. (1994). Sorption of chromate and other inorganic anions by organo-zeolite. *Environmental Science & Technology* 28: 452.

[10] De Gennaro, B., Mercurio, M., Cappelletti, P., Catalanotti, L., Dakovič, A., De Bonis, A., Grifa, C., Izzo, F., Krakovic, M., Monetti, V., Langella, A. (2016). Use of surface modified natural zeolite (SMNZ) in pharmaceutical preparations. Part 2. A new approach for a fast functionalization of zeolite-rich carriers. *Microporous and Mesoporous Materials* 235: 42-49.

[11] Mackuľak, T., Birošová, L., Grabic, R., Škubák, J., Bodík, I. (2014). National study of illicit drug use in Slovakia based on wastewater analysis. *Science of theTotal Environment* 494: 158-165.

[12] Krajišnik, D., Dakovič, A., Milojevič, M., Malenovič, A., Kragovič, M., Bajuk, D., Bogdanovič, D., Dondur, V., Milič, J. (2011). Properties of diclofenac sodium sorption onto natural zeolite modified with cetylpyridinium chloride. *Colloids and Surfaces B: Biointerfaces* 83: 165–172.

[13] https://www.zeocem.com/en/products/enviro/zeowater (personal communication with Project Manager Enviro&Industry A. Cseteová, March 2018).

[14] http://www.evecobrno.cz/en/aktivity/cleaning-exhaust-and-flue-gases (personal communication with R. Puchýř, March 2018).

[15] Hruby, M., Agrawal, K., Policianova, O., Brus, J., Skopal, J., Svec, P., Otmar, M., Dzubak, P., Stepanek, P., Hajduch, M. (2016). Article in Biomedical papers of the Medical Faculty of the University Palacky, Olomouc, Czechoslovakia, DOI: 10.5507/bp.2016.013.

[16] Courtier-Murias, D., Farooqa, H., Longstaffe, J. G., Kelleher, B. P., Hart, K. M., Simpson, M. J., Simpson, A. J. (2014). *Geoderma* 226–227: 405–414. https://doi.org/10.1016/j.geoderma.2014.03.006.

[17] Krauklis, A., Ozola, R., Burlakovs, J., Rugele, K., Kirillov, K., Trubaca-Boginska, A., Rubenis, K., Stepanova, V., Klavins, M. (2017). FeOOH and $Mn_8O_{10}Cl_3$ modified zeolites for As(V) removal in aqueous medium. *Journal of Chemical Technology and Biotechnology* 92: 1948–1960.

[18] Bottoni, P., Caroli, S., Caracciolo, A. B. (2010). Pharmaceuticals as priority water contaminants. *Toxicological & Environmental Chemistry* 92, 3, 549-565.

[19] European Chemicals Bureau. Technical Guidance Document on Risk Assessment. Dublin, UK. Institute for Health and Consumer Protection, European Commission

2003 http://echa.europa.eu/documents/10162/16960216 /tgdpart2_2ed_en.pdf. Accessed July 24, 2014.

[20] European Protection Agency. Office of Research. Nanomaterial case study: nanoscale silver in disinfectant spray. EPA/600/R-10/081F, 2012. http://cfpub.epa.gov/ncea/cfm/recordisplay.cfm?deid=24166. Accessed July 24, 2014.

In: An Introduction to Aluminosilicates
Editor: Nero Regina Blevins
ISBN: 978-1-53617-250-8

Chapter 10

MODERN SYNTHESES AND THE USE OF MOR TYPE ZEOLITE IN ADSORPTION AND CATALYSIS

Travkina Olga[1,*], Ahmed K. Ramadan[2] and Kutepov Boris[1]

[1]Institute of Petrochemistry and Catalysis Russian Academy of Sciences, Ufa, Bashkortostan, Russia

[2]Department of Chemistry, Faculty of Sciences, University of Zakho, International Road Zakho, Duhok, Kurdistan Region - Iraq

ABSTRACT

This review discusses modern methods for the synthesis of highly dispersed micro- and micromesoporous mordenite with a high degree of crystallinity, as well as high crystallinity mordenite with a hierarchical porous structure, the granules of which represent single crystal intergrowths of various sizes, methods for modifying the mentioned crystalline aluminosilicates as a result of postsynthetic treatments (ion exchange, dealumination), and modern adsorption and catalytic systems.

Keywords: zeolite type of MOR, hierarchical porous structure, granules, ion exchange, dealumination

[*] Corresponding Author's E-mail: simchanka@mail.ru.

INTRODUCTION

Currently, catalysts containing zeolites of various structural types are increasingly used in heterogeneous catalytic processes of petrochemistry and basic organic synthesis [1-5]. One of these zeolites is MOR (mordenite) type zeolite, that belongs to the class of medium silicon zeolites (ratio Si/Al = 4,5-13).

The chemical composition and structure of the crystal lattice of mordenite provide high thermal stability and specific catalytic properties in the transformation of various molecules. The outcomes of the research studies on the synthesis and use of powdered mordenite carried out before 1990 are summarized in [6–8]. Since then, updated information on the synthesis, modification by postsynthetic treatment, application of highly dispersed, and granular mordenite-containing material in sorption and catalysis, including the research articles of our research group, have been published in academic papers.

The present review addresses modern methods for synthesis of highly dispersed micro- and micro-mesoporous mordenite of high crystallinity degree, as well as high crystallinity mordenite of hierarchical porous structure with the granules representing single crystal intergrowths of various sizes, and methods of modifying the mentioned crystalline aluminosilicates by postsynthetic treatment (ion exchange, dealumination), modern catalytic systems based on them.

1. COMPOSITION, STRUCTURE AND METHODS FOR SYNTHESIS OF HIGHLY DISPERSED OF MOR TYPE ZEOLITE

1.1. Composition and Structure of MOR Type Zeolite

Synthetic MOR type zeolite, that is mordenite, belongs to the class of medium-silica zeolites [6, 7]. The considerably high modulus of mordenite results in higher thermal stability and acid resistance, as compared to low-silicon of FAU type zeolites, and facilitates modification of the sodium form due to ion exchange or dealumination.

Idealized composition of the unit cell of mordenite is as follows:

$$\mathbf{Na_8[Al_8Si_{40}O_{96}] \cdot 24H_2O.}$$

The unit cell has a rhombic form with the following parameters: a = 18,13Å, в = 20,49Å, c = 7,52Å. MOR type zeolites are combined into spatial crystallographic C_{mcm}-group (Figure 1). The internal volume of spatial blocks makes the porous structure of zeolite.

Figure 2 shows a diagram of the cross section of the structure of MOR type zeolite in the plane perpendicular to *C* axis.

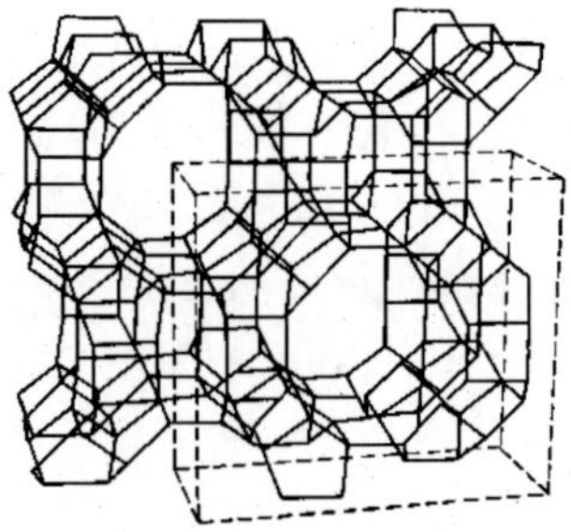

Figure 1. Stereodiagramma of framework of MOR type zeolite.

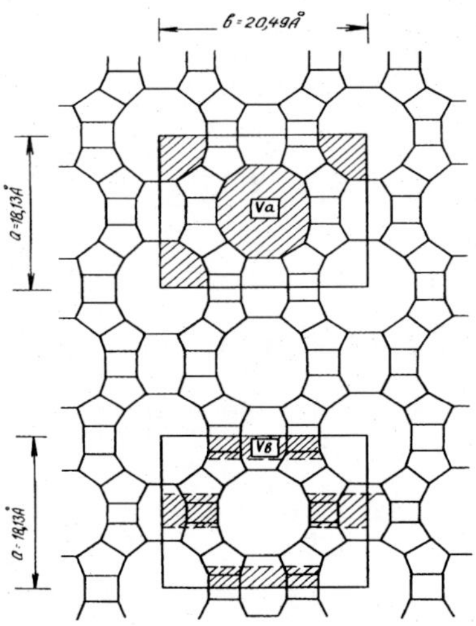

Figure 2. Schematic representation of a cross-section of the structure of the MOR type zeolite. Shaded large channels (Va) and side niches (Vв).

The unit cell includes two large channels with additional cavities (niches) with windows along the walls, making it 3.87 × 4.72 Å in size. The cavities of adjacent channels are displaced relative to each other, while molecules larger than 2.8 Å cannot move from one channel to another.

In a narrowly porous modification of mordenite, windows into the channel system are partially blocked by alkali metal cations. The projection of the structure of the MOR type zeolite (Figure 3) in the direction of large channels and the localization site of cations shows that only in one of the positions (VI) cations are localized in large channels and prevent the adsorption of large molecules.

The total number of localization sites significantly exceeds the number of cations in the unit cell. This creates an opportunity for different distribution of cations at localization sites. Depending on the degree of occupation of Na^+ cations of position (V), narrow or large-pore of MOR type zeolites can be formed. The large-pore modification appears to be less stable, with gradual redistribution of cations taking place at prolonged crystallization. The instability of large-pore MOR type zeolite in Na-form is further illustrated by the fact that only narrow-pore modification of MOR type zeolite is found in natural zeolites. In turn, the localization of Na^+ cations may depend on the arrangement of Al atoms. The change in the temperature and time of crystallization obviously has a definite influence on the distribution of Al atoms and, therefore, on the localization of Na^+ cations.

Large-pore mordenite adsorbs molecules with critical diameter of about 6 Å (benzene, n-heptane, cyclohexane). The narrow-pore modification adsorbs molecules that are no more than 5 Å in diameter (nitrogen, oxygen, argon, etc.). Narrow-pore modification can be transformed into large-porous by chemical modification (for example, decationization). Consequently, regardless of the type of the original sodium form, typical size of molecules adsorbed by decationized form of mordenite makes ~ 6 Å.

1.2. Modern Methods for Synthesis of Highly Dispersed Mordenite

The nature of chemical composition and structure of the crystal lattice impart high thermal stability, specific adsorption, and catalytic properties to MOR-type zeolite.

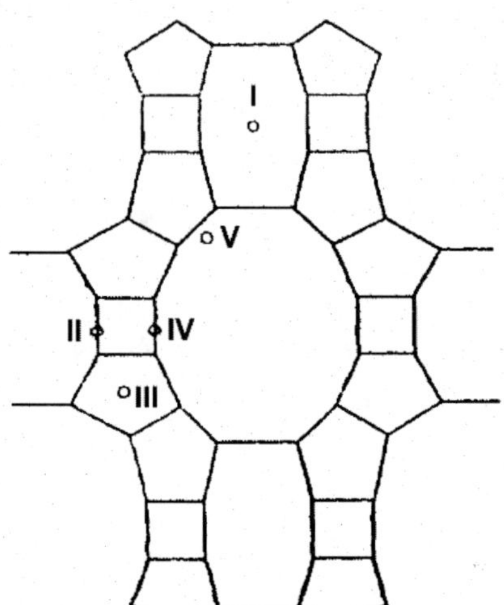

Figure 3. Projection of the structure of the MOR type zeolite in the direction of large channels and five sites of cation localization.

High phase purity, crystallinity degree, and dispersion of crystals in adsorption and catalysis ensure effective application of any zeolite, including synthetic mordenite.

Crystal-forming elements of the sodium form structure of zeolite are Si, Al, Na, and H_2O. They offer various raw materials, both synthetic and natural. For example, aluminum salts, aluminum hydroxide, and clay minerals are used as a source of aluminum, while silica-containing raw materials are silicas, silica gels, sodium silicate, silicic acid hydrogels, X-ray amorphous silica, and natural aluminosilicates [9].

The results of studies on the synthesis of powdered mordenite carried out before 1990 are summarized in [7-9]. The following outlines the latest data on synthesis of mordenites.

A variant of mordenite synthesis available for practical use with a molar ratio Si/Al of 9.0 ÷ 10.0 of high phase purity and crystallinity degree was detailed in [10-12]. Reaction mixture of composition $(2.2\text{-}2.4)Na_2O{\cdot}Al_2O_3{\cdot}(12\text{-}13)SiO_2{\cdot}(280\text{-}300)H_2O$ is used for synthesis of mordenite with a molar ratio Si/Al of at least 9.0. The preparation of the latter using only aqueous solutions of sodium silicate and sodium aluminate is impossible, as an additional source of silicon is necessary. In [13, 14], they applied silicon oxide with the properties listed in Table 1.

Silica-aluminohydrogel of sodium was prepared by mixing aqueous solutions of silicate (with silica concentration of 110–200 $g{\cdot}l^{-1}$ and silicate module of 2.6–3.0), sodium aluminate (with concentration of aluminum oxide of 250-270 $g{\cdot}l^{-1}$), and silicon oxide (see Table 1). In the resulting reaction, the mixture of the composition $2.2Na_2O{\cdot}Al_2O_3{\cdot}12SiO_2{\cdot}300H_2O$, 10% wt. powdered mordenite (seeds) was added and crystallized at 150-170°C under stirring. Once the synthesis was over, the solid phase was separated from the mother liquor, washed with distilled water (up to pH = 8–9 of the washing water), and dried at 120–140°C.

The silicon oxide used must completely go into solution before crystallization starts [14]. Otherwise, formed gel composition is incompatible with the above reaction mixture composition. As a result, crystallization of several zeolite phases, such as ANA (analcime), GIS (P zeolite), and PHI (phillipsite), is practicable. Preliminary transfer of silicon oxide to a solution of sodium silicate increases the dispersion of the crystals formed.

It is noteworthy that the method for crystallization of powdered zeolite with addition of crystalline seed is not rather technological.

In order to increase the dispersion of crystals, two variants of hydrothermal synthesis using seeds of mordenite crystals without an organic template were carried out in [15]. In the first case, gel of the following composition was used: $1.21Na_2O\ 1.24K_2O\ Al_2O_3\ 18SiO_2\ xH_2O$, which had been previously subjected to aging for 2 to 7 days (x = 600, 280 and 22.5) before crystallization. In the second case, colloidal seed of the above composition was applied. The crystallization kinetics of mordenite with different seed content (1, 2 and 5% wt.) was examined as well. The use of the seeding agent enabled the researchers to synthesize highly dispersed mordenite crystals. The experiments showed that the crystals size distribution strongly depends on the water content in the initial gel and the amount of the seed added. They proved that the content of seeding agent and concentration of water in the initial gel are the key factors influencing the physicochemical properties of crystalline products and the duration of crystallization.

In order for industrial processes for preparing zeolites to be effective, a reduction in their duration is required. For this purpose, it was proposed to use microwave radiation for the synthesis of pure mordenite [16]. The influence of molar ratio Si/Al, the duration of aging and crystallization on the phase purity, and the crystallinity degree and morphology of the products formed were examined in detail. As a result, synthesis conditions were determined that allow crystallization of high crystallinity mordenite within 12 hours without the use of seed crystals of mordenite and the use of an organic structuring agent (OSDA). Adding two different OSDAs, namely o-phenylenediamine (OPDA) and tetramethylammonium hydroxide (TEAOH), caused the crystallization period to decreases to 6 hours; while it takes 24 hours with OSDA and 48 hours without it in the traditional heating way. Under these conditions, rectangular and spherical mordenite crystals form.

The synthesis of zeolites (including mordenites) using seeds was also described in [17].

Table 1. The chemical composition of kaolin and silicon oxide [13]

Raw materials	Chemical composition, % wt.		
	Na_2O	Al_2O_3	SiO_2
Silicon oxide	0,2	0	99,8
Kaolin	0,07	42,6	57,3

High-performance mordenite membranes were successfully synthesized from fluoride-containing dilute solution without organic template when heated with microwave radiation in a short synthesis period [18]. The influence of crystallization temperature, the concentration of fluoride ion and source of aluminum on the morphology of the membrane, and effectiveness of its use for the dehydration of acetic acid were systematically studied in this work. The experiments illustrated that high-performance mordenite membranes about 5 μm thick can be prepared from dilute solution of the composition SiO_2: 0,06Al_2O_3: 0,24Na_2O: 0,3NaF: 120H_2O using various sources of aluminum at 165°C for 4 hours.

Ultrafast synthesis of mordenite without organic structurant (OSDA) in a rapidly heated tubular reactor is presented in [19]. Through the typical $Na_2O/SiO_2/Al_2O_3$ system, they reduced the synthesis period to 10 minutes, which is much less than the duration of conventional syntheses. Ground crystals of mordenite about 200 nm in size used as crystalline seed proved to be more effective than crystals 2 microns in size.

In [20], the researchers suggested to carry out synthesis of nanoscale crystals of mordenite in three stages. At the first stage, gel is prepared, then crystallized, then washed with distilled water and dried.

The patent project [21] details a method for preparing mordenite with an average size of primary crystals less than 80 nm in size and specific surface area of mesopores more than 30 $m^2 \cdot g^{-1}$.

In this study, zeolite mordenite (MOR-ST) was directly synthesized by hydrothermal method using starch as a template successfully. The morphological characterization of the samples was fulfilled by XRD, SEM, and BET. SEM showed that the as-synthesized MOR-ST displayed a flammulina-like morphology and smaller crystalline size in comparison with the conventional mordenite (MOR-TEA) synthesized by TEAOH template. Additionally, MOR-ST presented a bigger BET surface area, smaller bulk density, and higher hydrothermal stability [22].

During crystallization of mordenite, kaolin is one of the starting components [13]. Kaolin is a natural crystalline aluminosilicate. The crystal structure of kaolin is stable in alkaline solutions, therefore, kaolin is preliminarily calcined at 600-650°C for 4-6 hours to obtain zeolite (Scheme 1).

During calcination, kaolin is dehydrated to form X-ray amorphous metakaolin ($Al_2Si_2O_7$), which can be applied as aluminosilicate raw material in the synthesis of powdered MOR type zeolite. The crystal-forming components missing in the reaction mixture were introduced with sodium silicate solution, in which the crystallization was carried out [13]. The reaction mixture composition (2,2Na_2O Al_2O_3 12SiO_2 300H_2O) was crystallized at 150°C with constant stirring. They previously introduced 10% wt. seed crystals (powdered MOR type zeolite) in the reaction mixture and it was soaked for 24 hours at 30°C before crystallization.

X-ray powder diffraction and infrared spectroscopy data indicate that 14 hours is sufficient time for the synthesis of mordenite with a degree of crystallinity close to 100%. The required adsorption capacity values for H_2O, C_6H_6, and C_7H_{16} vapors of the obtained mordenite are reached after 14 hours (Table 2). These values coincide with the values of the adsorption capacities of a powdered zeolite of the MOR type, obtained from synthetic raw materials.

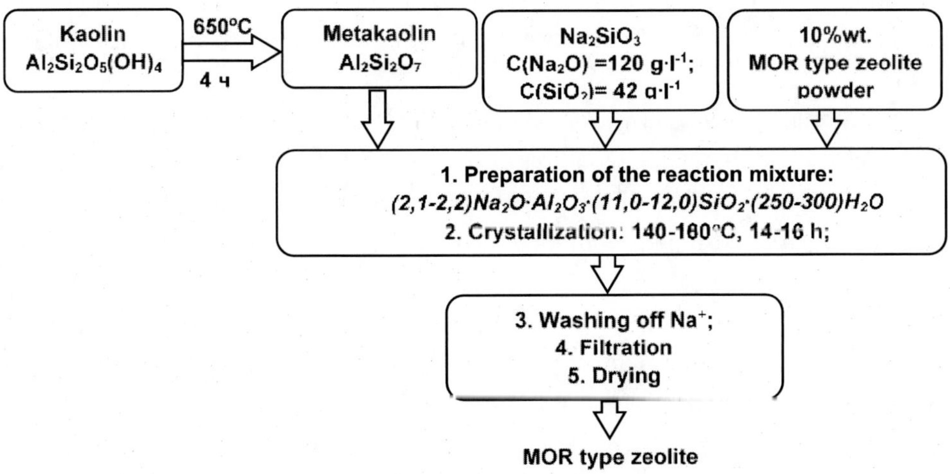

Scheme 1. Preparation of highly dispersed mordenite using kaolin [13].

Table 2. Effect of crystallization time of metakaolin in sodium silicate solution on the equilibrium adsorption capacities for water vapor, benzene and heptane of the obtained product [13]

Time, hour	Equilibrium adsorption capacities ($cm^3 \cdot g^{-1}$) for vapor:		
	H_2O	C_6H_6	C_7H_{16}
4	0,04	0,03	0,04
7	0,07	0,06	0,08
10	0,15	0,12	0,12
12	0,16	0,12	0,13
14	0,16	0,12	0,14
16	0,16	0,12	0,14

The research results show that zeolite with a lower modulus (Si/A = 4,0) crystallizes from metakaolin than the zeolite obtained from synthetic raw materials. The decrease in modulus is presumably caused by higher alkalinity (the molar ratio of sodium oxide to aluminum oxide) in the reaction mixture during crystallization from metakaolin.

The patent project [23] also describes a method for preparing mordenite by hydrothermal crystallization involving natural aluminosilicate minerals as common sources of silicon and aluminum. The method for preparation of mordenite envisaged in the present invention broadens the raw material base, which is characterized by simple technology and low cost of molecular sieve synthesis.

In [24], mordenite was synthesized from diatomite without templates. During crystallization, the main 4-membered rings turned to emerge at a later stage than the 5-membered rings. It was also demonstrated that diatomite consists of 5-membered rings and various rings longer than the 5-membered ring. The authors suggest that during the formation of mordenite from diatomite, 5-membered rings emerge, which interact with 4-membered rings located around them, finally forming the porous crystalline structure of mordenite.

In [25], mordenite was synthesized involving various diatomites for sorption of radioactive cesium. A high value of cation-exchange capacity was reached when diatomite had a high Al/Si atomic ratio. The degree of extraction of Cs^+ cations by synthesized mordenite (1.0 g) from cesium-containing solutions prepared using 100 ml of distilled and sea water, made about 100% and 83% in 100 ml, respectively. The composite powder material consisting of mordenite and nanoscale magnetite was synthesized from mixed suspension of diatomite and nanoscale magnetite. The magnetic collection rate for composite material using neodymium magnet made more than 90% for 20 and 30% wt. of magnetite-containing composite materials. The total degrees of extraction of Cs^+ cations using magnetic collection after adsorption of Cs^+ in water were 92.7% and 97.2% for 20 and 30% wt. of the magnetite-containing composite material, respectively.

Hydrothermal synthesis of mordenite of high degree of crystallinity in sodium form from natural materials such as kaolin and rice husk ash is outlined in [26-28].

Hierarchical natural mordenite has been synthesized [29] from low-cost natural zeolites through a high-energy ball milling and a combined ball milling-recrystallization method. In this study, the particle size of natural mordenite (MOR) was reduced into micro- and nanosize by high-energy ball milling with variations in milling time. Furthermore, the selected milled MOR was recrystallized by using a silicate solution that favored the growth of the mordenite phase to recover the demolished mordenite structure. The mesopore volume and external surface area were increased significantly after milling treatments and remained large even after recrystallization.

1.3. Synthesis of High-Modulus Mordenite

The research work [30] details an experiment to synthesize high-modulus mordenite in the presence of monoethanoamine (MEA) or isobutyl alcohol (IBA). As a result, mordenite of high phase purity, modulus of 13-22 and crystallinity degree close to 100% wt was synthesized. It is based on hydrothermal crystallization of silica-aluminum hydrogels of the following composition: $2.5MEA\cdot(2.2\text{-}7.3)Na_2O\cdot Al_2O_3\cdot(12\text{-}40)SiO_2\cdot(300\text{-}1200)H_2O$, prepared by mixing fumed silica or waste of crystalline silicon production (content of $SiO_2 \geq 98{,}0\%$) with silicate and aluminate sodium solutions in the presence of MEA or IBA. Crystallization is carried out at 130-160°C for 72-96 h. The prepared zeolites have adsorption capacity of 0.10 - 0.11 $cm^3\cdot g^{-1}$ for water vapour and 0.08-0.09 $cm^3\cdot g^{-1}$for n-heptane vapour.

The study outcomes in [31] present crystallization of mordenite in the presence and absence of organic template. It was demonstrated that an increase in the SiO_2/Al_2O_3 ratio in the initial gel of up to 20 in the absence of organic template produces a pure mordenite phase with the maximum silicate modulus of 12. Mordenite with 95-96% crystalline phase is formed from aluminosilicate gel with $2Na_2O\cdot 15SiO_2\cdot Al_2O_3\cdot 450H_2O$ molar composition without a structure-forming component at 135-140°C within 60-72 h. In the presence of tertethylammonium bromide (TEABr) applied as template, mordenite with a high crystallinc phasc content crystallizes from gel of $3{,}0SiO_2\cdot Al_2O_3\cdot 8TEABr\ 9{,}6Na_2O\cdot 640H_2O$ molar composition at the same temperature within 120-144 hours. Molar ratio Na_2O/SiO_2 in the initial aluminosilicate gels was found to have a significant impact on the crystallization of mordenite: the minimum ratio of Na_2O/SiO_2 in the gels for crystallization of mordenite with silicate modulus of 24-25 makes 0.32.

Hydrothermal synthesis of high-modulus mordenite samples was studied in [32], involving two different types of seed crystals (mordenite and ferrierite) at 170°C, SiO_2/Al_2O_3 ratios of 20, 60, and 100 and OH/SiO_2 ratio of 0.2, 0.3, and 0.4, as well as various durations of crystallization. High crystallinity MOR samples turned out only at OH/SiO_2 ratio of 0.4.

The ratio SiO_2/Al_2O_3 is one of the most important parameters for zeolite materials. For mordenites, it usually varies in the range of 4 to 32. High crystallinity mordenite samples with the ratio $SiO_2/Al_2O_3 = 11 - 263$ were prepared in [33]. The synthesis is based on the preparation of gel by mixing silica, previously obtained by acid hydrolysis, sodium aluminate and ionic liquid (1-butyl-3-methylimidazole bromide), followed by crystallization of the dried gel at 140°C for 4 days.

In the experiment [34], an attempt was made to synthesize mordenite with a Si/Al ratio in the range of 5 to 25, using silica gel and sodium aluminate as sources of Si and Al, which was specified by the composition of the initial gel solution. Mordenite of high crystallinity and phase purity was successfully synthesized from gels of a ratio in the range of 10 to 15. Amorphous materials were obtained from the initial solution with a Si/Al ratio of 25. The crystalline mordenite samples have prismatic, hexagonal shape, while the other samples have fibrous or needle structure.

1.4. Synthesis of Mordenite with a Hierarchical Porous Structure

Conventional microporous molecular sieves are not sufficiently effective in catalytic conversions with the molecules of starting materials or products being more than 10 Å due to the spatial limitations of the porous structure. In addition, diffusion limitations can lead to an accelerated loss of catalyst activity. The production of materials with a micro-meso-macroporous structure (hierarchy), ensuring effective diffusion of the reactants to the active sites and reverse diffusion of the formed reaction products, could be a practical solution of the aforementioned problems.

The synthetic methods of obtaining various types of micro/mesoporous materials can be divided into the following four basic groups: (I) direct synthesis using various template synthesis approaches, (II) generating mesopores in zeolite crystals by means of post-synthetic treatment, (III) zeolitization of mesoporous materials, and (IV) zeolite recrystallization [3, 4, 34 and 35].

Facile method to synthesize H-mordenite nanosheet assemblies without using any template agent are reported in the literature [36]. This synthesis methodology avoids the disadvantages of high cost and pollution emissions. Briefly, the silica gel was slowly dropped into the sodium metaaluminate alkaline solution under continuous stirring. The obtained gel was aged, crystallized, and then calcined to get the HMOR zeolite samples.

The hierarchical mordenite was successfully one-pot synthesized by using the diquaternary ammonium-type surfactant $C_{18}H_{37}–N^+(CH_3)_2–C_6H_{12}–N^+(CH_3)_2–C_6H_{13}(Br^-)_2$, designated as $C_{18\text{-}6\text{-}6}Br_2$ hereafter, as a meso-template [37]. The synthesis was carried out by mixing sodium hydroxide, sodium aluminate (41.3% wt. Al_2O_3), and distilled water under stirring condition to obtain a clear solution. Then, fumed silica (92% wt. SiO_2) was added under vigorous stirring. The mixture was aged at 100°C for 15 h. Then, $C_{18\text{-}6\text{-}6}Br_2$ was added to the obtained precursor and stirred for another 5 h at 60°C. The resulted suspension with the molar composition of $11.4Na_2O:2Al_2O_3:40SiO_2:2C_{18\text{-}6\text{-}6}Br_2:2500H_2O$ was kept at 160°C for 120 h. The final product was filtered off, washed with distilled water, dried and calcined in air at 550°C for 8 h to remove the organic template. The hierarchical mordenite has loose aggregation which consists of rod-like nanocrystals with widths of about 30-40 nm.

Yangyang Yuan et al. [38] developed a facile synthetic strategy to fabricate hierarchical MOR type zeolites that results in nanocrystal assembled morphology composed of 20–50 nm crystallites, dramatically reducing mass transport limitation. A typical synthesis procedure was as follows: NaOH and surfactant $C_{12}H_{25}\text{-}N^+(CH_3)_2\text{-}C_2H_4\text{-}N^+(CH_3)_2\text{-}C_{12}H_{25}Br_2$ ($C_{12\text{-}2\text{-}12}$) were dissolved in H_2O to prepare a clear solution. Thereafter, tetraethylammonium hydroxide (TEAOH) was added to the solution dropwise, followed by the addition of $NaAlO_2$. Then, colloidal silica was added to the mixture under vigorous stirring. The final mixture with a composition of $SiO_2:0.033Al_2O_3:0.25Na_2O:0.10TEAOH:0.05C_{12\text{-}2\text{-}12}:40H_2O$ was stirred for an additional 3 h to obtain a homogenous gel prior to being transferred into a stainless steel autoclave. The sol-gel was subjected to hydrothermal crystallization at 130°C for 5–7 days and rotated at a speed of 60 rpm. The products were recovered by centrifugation, washed with deionized water, dried overnight at 120°C, and finally calcined at 550°C for 5 h to remove the organic species. Detailed investigation shows that the use of TEAOH itself at lower crystallization temperatures could decrease the crystal size and generate hierarchical structures constructed by the interconnection of nanocrystals. The introduction of the cationic $C_{12\text{-}2\text{-}12}$ surfactant in the TEAOH-containing sol-gel further reduces the crystal size and improves the porosity of the products.

Shiyao Wang et al. [39] synthesized the hierarchical MOR type zeolites via a one-step dual-functional templating strategy utilizing gemini organic surfactant $[C_{18}H_{37}(CH_3)_2N^+(CH_2)_2N^+(CH_3)_2C_8H_{17}]\,[Br^-]_2$ ($C_{18\text{-}2\text{-}8}$) through hydrothermal process. The synthesis was performed by mixing sodium hydroxide, sodium aluminate, distilled water, and subsequently adding $C_{18\text{-}2\text{-}8}$ to the mixture vigorously at the room temperature for 60 min using a magnetic stirrer. Finally, fumed silica was added to the gel at room temperature to produce a white gel with the following composition: $NaAlO_2:50SiO_2:1.14C_{18\text{-}2\text{-}8}:850H_2O:3.8NaOH$. The gel was set to crystallize in a preheated oven at 130°C for 96 hours. After crystallization, the zeolite product was filtered, washed and dried at 100°C overnight.

Synthesized hierarchical MOR type zeolites has a larger BET (412.0 $m^2\cdot g^{-1}$) and abundant intracrystalline mesopores (average mesopore size distribution of 4.55 nm). The XRD study revealed a long range structural ordering of mesoporous and a good crystallinity of microporous structure. The results indicated that the surfactant acted as a dual-functional template for generating both micropores and mesopores simultaneously.

DAI Guang et al. [40] developed a synthesis method of the hierarchical MOR zeolite nanocrystallines with mesoporous architecture composed of 2~4 nm pores. The synthesis was performed by adding surfactant cetyltrimethylammonium bromide (CTAB) as mesoporous template in the absence of an organic co-solvent and zeolite seeds, in which the mesoporous volume and external surface area could be adjusted by changing the amount of CTAB. The largest external surface area and mesoporous volume were 191 $m^2 \cdot g^{-1}$ and 0.17 $cm^3 \cdot g^{-1}$. Adsorption of mesitylene on the MOR type zeolites samples showed that only a small amount of adsorption could be found in the microporous mordenite, but a large amount of adsorption in hierarchical mordenite, which presented the characteristics of typical IV isotherm indicating a mesoporous structure.

The synthesis of mesoporous mordenite samples [41] was carried out by mixing a given amount of a soft template, sodium hydroxide, sodium aluminate, and amorphous pyrolyzed silica under continuous stirring for 30 min. The resulting suspension was dried in an oven at 100°C for 24 hours. The dry solid was macerated and placed in special autoclaves in the presence of 30 cm^3 of deionized water per gram of solid. In these autoclaves, the deionized liquid water did not enter into physical contact with the solid. The autoclaves were heated at 175°C for 4 days. After this time, the solid was suspended in deionized water, filtered, washed, and oven dried overnight at 100°C.

Bhupendra Kumar Singh et al. [42] developed a facile and template free hydrothermal synthesis of mesoporous mordenite (MOR) by controlling synthesis conditions. The synthesis of mesoporous mordenite samples was carried out by mixing sodium hydroxide, sodium aluminate, distilled water, and colloidal silica. The resultant mixture with a molar ratio Si/Al of 12.5 was stirred for 2 h at ambient temperature. The crystallization of obtained gel was carried out by heating in the autoclave at 180°C for 24 h rotated at a speed of 25 rpm. After a predefined time, the product was filtered, washed with hot distilled water (80°C), and dried at 110°C overnight. The synthesized material exhibited significantly higher surface area than conventional mordenite.

Dealuminated mordenite zeolites are well-known and widely used hierarchical porous catalysts. In addition to the intrinsic unidimensional framework microporosity of the mordenite zeolite structure, postsynthetic dealumination can further introduce extraframework meso- and macroporosity. It is commonly assumed that such extraframework porosity provides more efficient access to the inner micropore volume. Alexey V. Kubarev et al. [43] showed that the area of enhanced reactivity in hierarchical zeolite, examined with super-resolution fluorescence microscopy, is spatially restricted to narrow zones around meso and macropores, as observed with focused ion-beam-assisted scanning electron microscopy. This comparison indicates that reagent molecules efficiently reach catalytic active sites only in the micropores surrounding extraframework porosity and that extensive macroporosity does not warrant optimal reactivity distribution throughout a hierarchical porous zeolite.

Base-treatment on mordenite [44] for selective removal of silica from zeolitic framework could effectively improve its catalytic stability in heavy alkylbenzene transalkylation. It has been demonstrated that the degree of stabilization and pore structure of mordenite could be tuned by the choice of different base-treating protocols for manipulation of silica dissolution rate. Mordenite subjected to single-cycle base-treatment for extending base-treating time maintained its original activity with marginally improved catalytic stability. On the other hand, multiple-cycle base-treatment protocol enhanced both the catalytic activity and stability of mordenite. The better enhancement in the catalytic performance from the multiple-cycle basetreatment protocol than the single-cycle protocol is attributed to the improved diffusivity associated with interconnected larger mesopores. Samples prepared from repeating base-treatment for 15 min for two cycles as an optimum condition possessed a maximum mesopore size of 14 nm, exhibiting the highest activity and stability for heavy alkylbenzene transalkylation reaction.

Desilication of mordenite crystals previously enriched in silica by dealumination has been investigated [45] in order to better understand the process of mesopore formation. The alkaline treatments were carried out using sodium hydroxide (NaOH) as the desilicating agent. The experimental procedure was the following: mordenite were stirred in an aqueous solution of NaOH. The NaOH concentration range was 0.2 – 0.4 M and the treatment was carried out at 80°C during 30 min. The generation of mesoporosity is due to the hydrolysis and dissolution of the surface of the crystals, probably initiated by the formation of etch pits at low coordination atoms, followed by surface roughening, formation and growth of internal mesopores, fragmentation, and, ultimately, by a complete dissolution of the crystal. Such a process, which is similar to the one involved in the chemical weathering of rocks, results in a severe loss of crystalline material and of the strong acidity associated to it.

Recent advances in the development of new micro-mesoporous zeolites, including mordenites, obtained by the method of their recrystallization in an alkaline medium in the presence of a template are summarized in [4].

According to data [4], the initial mordenite entirely retains its crystal structure upon recrystallization. The crystallinity of obtained material (RMOR-1) is slightly higher than that of the parent zeolite, moreover the mesoporous phase in the RMOR-1 type materials is not structured. Low temperature nitrogen adsorption isotherms and transmission electron microscopy images indicate the formation of 3–4 nm and >30 nm mesopores. The mesopores whose size is over 30 nm form inside the crystal as a result of alkaline desilication. The 3–4 nm mesopores appear as a result of the self organization of silicate fragments that were transfered into the solution during desilication in the presence of the surfactant. The resulting mesoporous material with disordered wormlike porosity coats the outer surface of crystallites and fills type I pores. In the preparation of the RMOR-1 type materials, the zeolite crystals retain their size and shape, but the outer surface of the particles becomes nonuniform. This is due to the deposition of mesoporous phase and of dissolved zeolitic fragments onto the crystal surface.

Investigation of the acidic properties of RMOR-1 type materials demonstrated that the recrystallization under mild conditions causes no significant changes in the number and strength of acid sites.

The results of a study of the catalytic properties of RMOR-1 samples in petrochemical synthesis reactions are described in Section 4.

2. Regulation of Physicochemical Properties of Mordenite by Postsynthetic Treatment

The catalytic properties of mordenite, as well as zeolites of other structural types, depend on the concentration and strength of acid sites, as well as their availability. The strength and concentration of acid sites of mordenite are regulated by postsynthetic treatment, involving ion exchange of Na^+ to H^+ cations or cations of divalent and trivalent metals, as well as dealumination of the zeolite crystal lattice under various treatments [3, 6]. In this case, a change in the parameters of the pore structure is also possible.

2.1. Regulation of the Acidic and Adsorption Properties of Mordenite with a Hierarchical Porous Structure (MOR_{mmm}) by Decationation

There are no strong acid sites in the Na-form of zeolite of any type. They appear after the replacement of Na^+ cations, primarily, by H^+ cations. Generally, the H-form of mordenite is synthesized by 2-3 exchange treatments in aqueous solutions of ammonium salts and subsequent heat treatment at 450-600°C for 4-6 h in air [13, 50].

Exchange proceeds more slowly in MOR_{mmm} zeolite than in powdered zeolite, and the limiting exchange degree makes 0.93 [13, 50]. The low value of the limiting degree of Na^+ cations exchange in MOR_{mmm} zeolite accounts for the form of the granules representing intergrowths of crystals with a part of intracrystalline space inaccessible for ion exchange.

The values of equilibrium adsorption capacities for benzene vapour and the concentration of strong acid sites of MOR_{mmm} zeolite are presented in Figure 4.

It is demonstrated that with an increase in the exchange degree of Na^+ cations for H^+ cations, the values of equilibrium adsorption capacities for benzene vapours increase by 10-12% (Figure 4a), caused by increased available micropore volume. The concentration of strong acid sites increases as well (Figure 4b).

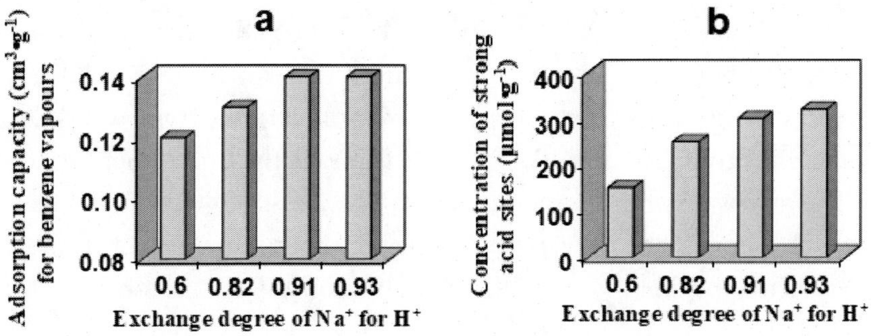

Figure 4. Dependence of the values of equilibrium adsorption capacity for benzene vapours (a) and concentration of strong acid sites (b) of MOR_{mmm} zeolite on the exchange degree of Na^+ cations for H^+ cations [50].

2.2. The Influence of Dealumination of Mordenite on Its Physicochemical Properties

The Si/Al ratio of the frame atoms (modulus) affects the acidic and catalytic properties, as well as thermal stability, texture, and adsorption properties of zeolite of any structural type, including mordenite [3, 51]. The scientists in [6, 19 and 20] have developed the following techniques to increase the mordenite modulus: thermal and hydrothermal exposure, chemical exposure, combination of hydrothermal, and chemical modification.

Calcination of the ammonium form of mordenite in the presence of water vapour results in hydrolysis of Si–O–Al bonds and the release of aluminum from the zeolite lattice. Silica migrating from collapsed portions of the zeolite crystal fills the vacancies [23-25, 51]. Aluminum releasing from the lattice is present in zeolite in the form of extra-lattice compounds, located within existing micropores or in the form of deposits on the outer surface of the zeolite crystals [25]. As a result of dealumination, the Si/Al ratio of the atoms increases and the unit cell parameters decrease, since the Si–O bond is shorter than the Al–O bond. Reduction of the average interatomic distance T–O increases the stability of the zeolite crystal lattice [26]. The process of mordenite dealumination can involve partial amorphization of the crystalline framework and formation of secondary pore structure with pore radius of 10–20 Å [24, 27].

The deeply substituted ammonium forms of mordenite (Na_2O = 0,03%) were found to transform into high-silica zeolites with slight amorphization irrespective of the treatment temperature and the initial modulus [23].

The behavior of mordenites (exchange degree not more than 0.78) during steam treatment differs significantly from the behavior of deeply substituted samples. Regardless of Si/Al ratio varying over a wide range, steam treatment at 700°C results in amorphization of the crystal lattice of mordenites of exchange degree not more than 0.78 [21, 28]. The collapse of the framework appears to be caused by the reaction of residual NaOH with the lattice. The dealumination degree of the lattice is determined by decationation level of the initial sample, while an increase in the exchange degree of Na^+ cations to NH_4^+ cations in the mordenite provides a smooth control of the composition of the framework [21, 28]. The dealumination of the zeolite framework proceeds mainly in the first 4 hours after supplying water vapor. The extension of the duration of treatment decreases the removal rate of metal atoms. The equilibrium is not reached even after 32 hours of steam treatment, though, stemming from different stability of individual regions of the crystal structure [32, 52 and 53].

Steam treatment should be conducted in the temperature range from 400 to 600°C [33]. If the temperature is more than 700°C, the effective concentration of water on the zeolite surface becomes too low to form a sufficient amount of silica acid to fill the lattice, finally resulting in rapid thermal destruction. The chemical effect of water at a relatively high concentration at temperatures below 300°C prevails over the thermal effect of the lattice rearrangement, resulting in destruction of the zeolite structure similar to dissolution, not dealumination.

Normally, hydrothermal dealumination of zeolites produces 5–50 nm mesopores representing either cavities inside zeolite crystals or cylindrical pores connecting the outer and inner surface of the crystal. The amount of mesopores of this kind can make up more than 70% of the total pore volume [34-36].

Dealumination during the chemical treatment of mordenite, depending on the reagents applied, can involve enrichment of the dealuminized zeolite (treatment of mordenite by ammonium fluorosilicate solutions [54]), as well as without it (treatment of mordenite by ethylene diamine-tetraacetic acid solutions [28]). In the first case, silicon atoms in the reagent fill the vacancies formed in the crystal lattice caused by dealumination. In the second case, injection of silicon of that kind does not occur [24].

The acid treatment usually comes as the final stage of zeolite modification to remove amorphous fragments from zeolite pores. Still, more concentrated aqueous solutions of the acid are capable of efficiently hydrolyzing the -Si-O-Al- bond without preliminary steam treatment, extracting aluminum from the framework and forming mesopores in the pore structure of zeolites using both mineral (nitric, hydrochloric, sulfuric) and organic acids (oxalic, acetic, tartaric) [37-39]. Obviously, the nature of the acid and the structure of the zeolite are the key factors contributing to the depth of dealumination. It was demonstrated in the research paper [52] that the dealuminized sample treated with oxalic acid exhibits a larger Si/Al ratio and a smaller amount of mesopores than the sample treated with nitric acid

Hydrophilicity of H-mordenites with a Si/Al molar ratio varying in the range from 5 up to 103 was investigated by IR spectroscopy, TPD of ammonia, and thermogravimetry methods. It was found that dealuminated mordenite retains affinity for water in the entire molar ratio range, in contrast to published results for synthetic pentasil zeolites. In H-mordenites extra-framework aluminum exists in the form of dimers of aluminum hydroxide, those presumably are stabilized in the channels of mordenite. An inversely proportional correlation between the water content in pores and the concentration of channel Lewis sites was observed [52].

The combination of hydrothermal and chemical treatment includes initial conversion of the ammonium form of zeolite of any structural type into an ultra-stable form during hydrothermal treatment. Subsequent chemical treatment removes the extra-lattice aluminum.

Quantitative relationship between the temperature of heat treatment and composition of the framework for the hydrogen forms of mordenites with a different initial Si/Al ratio was calculated in [51]. Based on this relationship, the principles of thermal dealumination of zeolites with different amount of aluminum atoms in the framework were formulated. It is impossible to prepare mordenites with low aluminum content of high crystallinity degree under the conditions of traditional acid treatment [51]. In the same work, a method for preparation of mordenites with the desired ratio of Si/Al combining thermal and acid dealumination was developed. Once aluminum is removed from the mordenite framework, the total number of acid sites decreases and strong Brønsted acid sites bound to isolated aluminum atoms emerge [51]. These sites exhibit increased heats of NH_3 adsorption, 150 to 170 $kJ \cdot mol^{-1}$; their amount increases at up to 50% removal of A1 and decreases under deeper dealumination [51]. Decrease in the aluminum content in the framework reduces the intensity of coke formation; consequently increasing the selectivity of hydrocarbon conversion processes with respect to the target products [51].

Table 3 details dealumination degrees, relative crystallinity degrees, and Si/Al ratios of the crystal lattice modulus of mordenite with a hierarchical porous structure (MOR_{mmm}), MOR_{mmm} sample after steam treatment ($HMOR_{mmm}$-US), and after treatment with citric acid solutions [49, 50].

Table 3. Dealumination degrees, relative crystallinity degrees, crystal lattice modulus of the samples after treatment of $HMOR_{mmm}$ and $HMOR_{mmm}$-US zeolites with citric acid solution [50]

Conditional designation of the sample	Preparation conditions [a]	Dealumina-tion degree [b], %	Relative crystallinity degree, %	Crystal lattice modulus
NH_4MOR_{mmm}	MOR_{mmm} after 4 ion-exchange treatments	-	92	5,0
$HMOR_{mmm}$-0,1N	NH_4MOR_{mmm} after treatment with 0.1N citric acid	0	92	5,0
$HMOR_{mmm}$-0,3N	NH_4MOR_{mmm} after treatment with 0.3N citric acid	4,8	92	5,2
$HMOR_{mmm}$-0,5N	NH_4MOR_{mmm} after treatment with 0.5N citric acid	10.8	92	5,6
$HMOR_{mmm}$-0,9N	NH_4MOR_{mmm} after treatment with 0.9N citric acid	22,8	85	6,5
$HMOR_{mmm}$-US	MOR_{mmm} after steam treatment	-	92	7,7

Table 3. (Continued)

Conditional designation of the sample	Preparation conditions [a]	Dealumina-tion degree [b], %	Relative crystallinity degree, %	Crystal lattice modulus
$HMOR_{mmm}$-US-0,1N	MOR_{mmm}-US after treatment with 0.1N citric acid	5,90	92	7,7
$HMOR_{mmm}$-US-0,3N	MOR_{mmm}-US after treatment with 0.3N citric acid	13,9	92	7,7
$HMOR_{mmm}$-US-0,5N	MOR_{mmm}-US after treatment with 0.5N citric acid	25.1	92	8,0
$HMOR_{mmm}$-US-0,9N	MOR_{mmm}-US after treatment with 0.9N citric acid	40.1	92	8,5

Conditions:

[a] - all samples after treatment with citric acid were calcined at 540 ^{0}C for 4 h in air;

[b] - the dealumination degree of zeolite was calculated relative to the amount of Al_2O_3 in the $HMOR_{mmm}$ sample.

It is apparent from the information supplied that treatment with 0.3N and higher concentration of citric acid solution promotes marked dealumination of the zeolite sample.

The dealumination degree of zeolite under treatment with 0.3N concentration solution makes 4.8% and reaches 22.8% with an increase in the acid concentration to 0.9N as shown in Table 3. Subsequent to the treatment of the $HMOR_{mmm}$-US sample with citric acid solution of the same normality, the dealumination degree appeared to be higher than that of the NH_4MOR_{mmm} sample (Table 3). Its maximum value under the studied conditions made 40.1% at 0.9N concentration of citric acid.

The increased dealumination degree of the $HMOR_{mmm}$-US sample arises from the process of steam treatment when the zeolite sample experienced partial dealumination of the crystalline framework and release of part of the aluminum into the channels. Therefore, this part of aluminum was removed from the zeolite faster than from the crystal lattice at subsequent acid treatment.

Treatment with 0.5N solution of citric acid does not change the crystallinity degree of zeolite (Table 3). An increase in acid concentration to 0.9N reduces the crystallinity degree to 85%. Partial amorphization of the NH_4MOR_{mmm} sample during acid treatment occurs as a result of removal of aluminum directly from the crystal lattice.

When the dealuminating of the MOR_{mmm}-US sample is carried out using citric acid solution, the crystallinity degree remains the same (sample $HMOR_{mmm}$-US-0,9N). The research results are explained by the fact that, when processing the $HMOR_{mmm}$-US sample, extra-lattice aluminum is the first to be removed, and the crystal lattice finally becomes more resistant to treatment with citric acid solution.

Acid treatment of $HMOR_{mmm}$ brings about a slight decrease in the concentration of weak and strong acid sites (Table 4), since aluminum atoms are removed from the zeolite lattice with simultaneous destruction of acid sites (sample $NMOR_{mmm}$-0.9N) [49, 50].

Steam treatment of the $HMOR_{mmm}$-US sample decreases the concentration of strong acid sites from 312 to 250 $\mu mol{\cdot}g^{-1}$ This significant decline cannot be explained only by the removal of a large number of aluminum atoms from the zeolite lattice compared to acid treatment. Apparently, the aluminum oxide clusters formed during steam treatment block the remaining acid sites. The assumption is confirmed by the fact that concentration of strong acid sites during subsequent acid treatment increases and makes 250 μmol·g^{-1} (Table 4).

The HMORmmm sample is characterized by BET specific surface area of 456 m^{2}·g^{-1}, micro- and mesopore volumes of 0.15 and 0.20 cm^{3}·g^{-1}, respectively, as presented in Table 5.

Table 4. The amount of weak and strong acid sites (μmol·g^{-1}) in the samples according to the TPD data of pre-adsorbed ammonia [49]

Sample	Weak[a] acid sites	Strong[b] acid sites	Σ	T_{max}[c] of strong acid sites
$HMOR_{mmm}$	408	312	720	420
$HMOR_{mmm}$-0,5N	254	185	439	422
$HMOR_{mmm}$-0,9N	203	121	324	428
$HMOR_{mmm}$-US	358	204	562	420
$HMOR_{mmm}$-US-0,5N	315	250	565	423
$HMOR_{mmm}$-US-0,9N	246	185	431	440

Conditions:

[a] - the amount of NH_3 desorbed in the temperature range 100-350°C;

[b] - the amount of NH_3 desorbed in the temperature range 350-500°C;

[c] - maximum temperature on the part of the desorbogram characterizing the desorption of ammonia of strong acid sites

Table 5. The pore structure characteristics of the studied samples provided by the data of nitrogen adsorption-desorption [49]

Sample	S_{BET} [a], m^{2}·g^{-1}	V_{micro} [b], cm^{3}·g^{-1}	V_{meso} [c], cm^{3}·g^{-1}
$HMOR_{mmm}$	456	0,15	0,20
$HMOR_{mmm}$-0,5N	398	0,14	0,21
$HMOR_{mmm}$-0,9N	378	0,13	0,21
$HMOR_{mmm}$-US	387	0,12	0,21
$HMOR_{mmm}$-US-0,5N	372	0,13	0.22
$HMOR_{mmm}$-US-0,9N	362	0,12	0,23

Conditions: [a] surface area by method BET

[b] micropore volume

[c] mesopore volume

During acid treatment with 0.9N citric acid solution (sample $HMOR_{mmm}$-0.9N), the micropore volume decreases to 0.13 cm^{3}·g^{-1} due to the partial destruction, while the mesopore volume slightly increases (0.21 cm^{3}·g^{-1}). The maximum mesopore volume equal to 0.23 $cm^3 \cdot g^{-1}$ is reached at combined treatment (sample $HMOR_{mmm}$-US-0.9N). As already noted, the crystalline lattice of zeolite is not destroyed in this case.

3. The Use of Mordenite in Catalysis

At present, catalysts containing zeolites of various structural types are increasingly being used in heterogeneous catalytic processes of oil refining, as well as oil and gas chemistry [1-5, and 55]. Catalytic systems based on dispersed micro- and micro-mesoporous mordenite, composite materials obtained by recrystallization of mordenite, and also granular mordenite of high crystallinity with a hierarchical porous structure are described below.

3.1. Catalytic Systems Based on Highly Dispersed Microporous in MOR Type Zeolite

For the hydroisomerization of n-pentane and n-hexane, introduced into industrial practice, bifunctional catalysts are used, including those based on mordenite in the H-form, containing 0.4-0.6% wt. platinum [56, 57].

Proposed method [58] includes mixing of components, zeolite component-mordenite with binder-aluminum hydroxide, plastification by means of peptizing by acid solution, granulation, platinum deposition, and reduction of catalyst. Components are mixed at mass ratio from 1:9 to 2:3 in terms of calcined mordenite and aluminum hydroxide; after platinum deposition, heat treatment is carried out in two stages at a temperature of 100-110°C at first stage and not above 250-300°C at second stage; reduction of catalyst is performed at temperature not below 500°C. As a result, the selectivity of the catalyst is increased and significantly reduced power requirements.

In [59], it was concluded that the most effective mordenite-containing catalyst in the hydroisomerization of cyclohexane, n-heptane, and benzene is a bifunctional catalyst, in which 0.3% wt. Pt dispersed on the surface of the support is present. The latter contains 30% wt. mordenite (MOR) manufactured by Zeolyst International (with a Si/Al ratio of 10) and 70 wt.% Al_2O_3. The maximum yield of the target isomerization products on the aforementioned catalyst is achieved in the temperature range of 280-310°C, thermodynamically favorable for the formation of methylcyclopentane from benzene.

Alkaline earth metal modification of H-mordenite (HMOR) was examined in the isopropylation of biphenyl (BP) [60]. At low loadings of 0.2–0.5 $mmol \cdot g^{-1}$ of the metals, the isomerization of 4,4′-diisopropybiphenyl (4,4′-DIPB) was effectively prevented at temperatures as high as 300°C, resulting in high selectivities for 4,4′-DIPB. Metals reduce the acid amounts by low loadings, and deactivate external acid sites. At the loading of 2 mmol/g, only BaMOR was active with high selectivities for 4,4′-DIPB; however, other metals lost catalytic activity with the increase in the loading. These differences are ascribed to changes of acidity and pore size. The modification of barium with high loading formed moderate acidities, and kept pores open for the access of bulky molecules, resulting in the selective formation of 4,4′-DIPB; however, other metals declined the activities by choking pores for magnesium and calcium or by losing acidity for strontium.

The authors in [61] propose a method of producing a linear monoalkylaromatic compound with controlled concentration of 2-phenyl isomers and very low colour index after sulphonation, involving use of a catalyst system based on highly stable solid catalysts. The alkylation process is carried out with a catalyst which enables obtaining a linear alkylaromatic compound with minimum content of 2-phenyl isomers, which constitutes 20% wt., contains MOR type zeolite, from 0.01 to 0.20% wt. of at least one metal selected from a group consisting of Li, Na, K, Mg or Ca, with a maximum Na content of 0.01%, and from 0 to 0.5% wt. of at least one metal selected from a group consisting of Ti, Zr and Hf, wherein the catalyst, which enables obtaining at most 20% wt. 2-phenyl isomers, also contains FAU type zeolite, from 0.5 to 2 wt% of at least one metal selected from a group consisting of Li, Na, K, Mg or Ca, and rare-earth metals in amount of: from 4.5 to 10 wt% La, from 1.2 to 4 wt% Ce, from 0.5 to 1.5 wt% Pr and from 2 to 3 wt% Nd. As a result, increased activity and selectivity of the catalyst with respect to linear monoalkylated compounds is observed.

The authors in [62] suggested methods for producing xylene isomers. Application of a specified process ensures a high conversion of aromatic compounds C_9 and methylethylbenzol, and also to high ratios of xylene isomers to ethylbenzol, xylene isomers to aromatic compounds C_9, xylene isomers to aromatic compounds C_{10}, trimethylbenzol to methylethylbenzol, and benzol to ethylbenzol in conversion products.

Zeolite catalyzed Friedel-Crafts reactions were examined using acetic anhydride as an acetylating agent and an acetic acid as a solvent [63]. It revealed that the reaction of anisole smoothly occurred quantitatively for 3 hours using mordenite zeolite with SiO_2/Al_2O_3 = 200, and with SiO_2/Al_2O_3 = 110, and the increasing of Brønsted acidity allowed to completely react within 2 hours. Furthermore, the selectivity of 4-methoxyacetophenone (4-MA) among the isomers was found to be quantitative; no by-products and/or isomers were not detectable. With the excellent recyclability and reusability, the mordenite zeolite exhibited at least 30 times quantitatively both conversion of anisole and selectivity of 4-MA. The mordenite catalysts of fresh and the used after 30 times were characterized. This opportunity obviously indicates the sufficient shape selective catalyst of mordenite zeolite and gives a green synthetic tool for heterogeneous acylation reaction.

The catalytic transalkylation chemistry encountered with mesitylene interacting with benzene has been revealed to be working with the acid zeolites, H-mordenite and HY [64]. Preferential adsorption on the active site of the alkyl group containing-aromatic molecule was very advantageous for transalkylation reactions since otherwise the smaller benzene would be the dominant adsorbed molecule; it was also noted that when the alkyl acceptor was a small molecule, formation of the transition state was favourable. The study has shown that transalkylation was favoured over disproportionation at higher temperatures and this was in line with the earlier conclusions that disproportionation would only be effective at low temperatures. Reaction temperature influences the selectivity of alkyl-transfer reactions. The feed composition seemed to play a significant role in transalkylation and its manipulation may result in effective removal of alkyl groups from the alkyl-containing molecule. The window of operation is reduced somewhat by the finding that working at high temperatures is associated with undesirable higher deactivation rates.

The disproportionation of ethylbenzene and toluene have been compared on two zeolites, H-mordenite and HY [65]. The online examination of products as a function of time on stream reflect the different processes occurring on the two zeolites. The relative activities expressed on a unit aluminium center basis, are in line with their relative cracking (n-heptane) activities reported by others, showing that no extraneous factors affect activity. The data set provides a baseline for the activity of these two zeolites for future use in comparing other more complex alkylation reactions involving mononuclear and binuclear aromatic molecules.

The authors [66] suggest a method of producing lower alkyl ether of lower aliphatic alcohol having formula R_1-COO-R_2, involving reaction of a pre-dried lower alkyl ether having formula R_1-O-R_2, in which R_1 and R_2 independently denote C_1-C_6alkyl groups, provided that the total number of carbon atoms in groups R_1 and R_2 ranges from 2 to 12, or R_1 and R_2 together form a C_2-C_6 alkenyl group, with material which contains carbon monoxide, in the presence of a catalyst which contains mordenite and/or ferrierites in anhydrous conditions. The application of the proposed method allows to increase the yield and selectivity of end product.

In the presence of a catalyst [67] containing mordenite and/or ferrierite under anhydrous conditions, it is proposed to obtain a lower aliphatic carboxylic acid alkyl ester having the formula R_1-COO-R_2 by reacting a pre-dried lower alkyl ester whose composition corresponds to the formula R_1-O-R_2 where R_1 and R_2 are C_1-C_6 alkyl groups, provided that the total number of carbon atoms in the groups R_1 and R_2 is from 2 to 12, or R_1 and R_2 together form a C_2-C_6 alkylene group with a feed containing carbon monoxide.

The present study [68] demonstrates a clean and green approach for the production of bio-fuel additives from renewable glycerol acetalization using microwave irradiation as a heating method. Various transition metal ions (M = Fe-, Co-, Ni-, Cu- & Zn-) promoted mordenite solid acid catalysts were prepared by wet impregnation method and examined for the acetalization of glycerol with acetone. Investigation of the structural and chemical properties of the catalysts was achieved using XRD, FT-IR, TGA-TPD, SEM/EDX, and XPS techniques. Comparative studies using conventional heating were performed in order to study the efficiency of microwave-assisted acetalization reactions. The effect of various operating parameters on the catalytic activity studies was examined. Amongst the metal-promoted mordenite catalysts tested, Cu-MOR had the best performance due to the presence of a large amount of acidic sites and the synergetic effects of metal particles interacting with mordenite. Microwave-assisted acetalization of glycerol into Solketal was found to be an energy efficient route for glycerol valorization, achieving 98% selectivity to solketal at 95% glycerol conversion during a reaction time of 15 min over Cu-Mor catalyst and acetone/glycerol at a molar ratio of 3:1. Reuse of spent catalyst showed good repeatability for up to four reaction cycles with a marginal decrease in conversion.

Modified mordenites have been used [70] in the direct conversion of glucose into 5-hydroxymethylfurfural (5-HMF) in two different media: the ionic liquid 1-butyl-3-methylimidazolium bromide and a biphasic system composed of water–acetone and ethyl acetate. The results revealed that there are crucial differences in the final catalytic performances. Modified mordenite treated with 1 M NH_4Cl only, with a moderate Si/Al ratio of 11.2 and a high content of strong Brønsted acid sites (TPD_{NH3} acidity of 1.39 mmol g^{-1}), possessed the highest catalytic activity resulting in 64% 5-HMF yield with 97% glucose conversion in the ionic liquid system. On the other hand, the mordenite treated with 1 M NH_4Cl and 2.4 M NH_4F, with a total acidity of 1.51 mmol g^{-1}, mainly due to Lewis acid sites, gave 50% 5-HMF yield with 98% conversion in the biphasic system. The presence of strong Lewis acid sites has a major effect on the catalytic activity in aqueous–organic biphasic medium, in agreement with earlier works that showed that the reaction proceeds via Lewis acid-catalyzed isomerization followed by Brønsted acid-catalyzed dehydration. In contrast, the quick adsorption and strong interaction of the ionic liquid with the zeolite, which hinders the diffusion of the products, enhances the role of zeolite mesoporosity in the catalytic activity when the reaction is carried out in ionic liquid solution.

Hydrocracking of α-cellulose has been conducted in a semi-batch reactor at 400, 450, and 500°C with hydrogen flow (30 $mL{\cdot}min^{-1}$.) for 4 h [71]. Mordenite (MOR) and Co, Ni, and Pd metal supported on the MOR were used as solid catalysts. Thermal hydrocracking was also conducted at 450°C with the amount of liquid product being 37.86 wt.%. The highest liquid conversion obtained by mordenite catalyst was 94.66 wt.% at 450°C and the highest liquid conversion (98.08 wt.%) was reached by Pd/MOR catalyst at 400°C.

Ti-modified hierarchical mordenites (Ti/MM) were prepared using tetrabutyl orthotitanate (TBOT) as the precursor of Ti species by incipient wetness impregnation technique [72]. The introduced Ti species were located outside the micropores in mordenite due to the large molecular size of TBOT. As a result, the active Ti species over hierarchical mordenites are easily accessible. The catalytic results indicated that Ti/MM were highly active catalysts in the oxidative desulfurization (ODS) of dibenzothiophene due to rich mesoporosity. The sulfur removal content over a Ti-modified hierarchical mordenite (Ti/MM-0.5) with relatively large mesopore volume reached 98.6% after a reaction time of 120 min at 60°C by using H_2O_2 as oxidant, which is nearly four times as that over Ti-modified conventional mordenite. Accordingly, the sulfur content in model fuel was reduced from 1000 ppmw to 14 ppmw. These results indicated that Ti-modified hierarchical mordenite with rich accessible Ti sites and low cost has potential applications for ODS of transportation fuels.

N-Methyl amines play a major role in the production of medicines, pesticides, surfactants, and dyes. *N*-Methylation of primary or second amines with methanol is considered to be a green path for the synthesis of *N*-methyl amines and the catalyst is key [73]. In this article, the combined Al_2O_3–mordenite catalyst (mass fraction of alumina is 40%) with good activity, selectivity, lifetime and stability was prepared for *N*-methylation of various amines with methanol in a hydrogen free system in a fixed bed reactor.

[74] described a method for preparing a catalyst for the decomposition of nitrous oxide by mixing a zeolite selected from the following structural series: MFI, MEL, BEA, FER, or MOR, with a modified iron binder. Alumina is proposed as a binder with additives selected from a number of the following oxides: silicon oxide, titanium oxide, zirconium oxide, iron oxide, cobalt oxide, copper oxide, lanthanum oxide, and phosphorus oxide. The mass ratio of alumina/any other of the listed oxides varies from 0.01: 100 to 10: 100, and the mass ratio of zeolite/ binder from 1: 9 to 9: 1. The modification is carried out by introducing a dry or hydrated salt and/or solution of an iron salt into the binder, followed by molding and activation to obtain a catalyst in which at least 0.5 to 50% of the iron and not less than $5 \cdot 10^{17}$ iron atoms per gram of catalyst are in reduced form in the form of special complexes of α-centers.

3.2. Catalytic Systems Based on Highly Dispersed Micro-Mesoporous of MOR Type Zeolite

Due to its structure and high Si/Al ratio, mordenite zeolite has high thermal and acidic stability. Mordenite zeolites has been used as catalysts in many industrially important reactions such as hydrocracking, hydroisomerization, alkylation, acid-catalyzed isomerization of alkanes and aromatics, and reforming. In order to overcome the problem of the limited access to the active sites, OSDA-free synthesized mordenite undergoes fluoride etching as a post-synthetic treatment. The post-synthetic treatment is performed with hydrofluoric acid in combination with ammonium fluoride [75]. Thus, the porosity is enhanced additionally without changing considerably the Si/Al ratio of the zeolite framework. All samples have been characterized by X-ray diffraction analysis, nitrogen adsorption, scanning electron microscopy, high-resolution transmission electron microscopy and solid-state nuclear magnetic resonance spectroscopy. The catalytic activity of the samples obtained has been investigated in the reaction of *m*-xylene transformation. All mordenite samples having undergone post-synthetic treatment exhibit catalytic activity higher than that of the parent sample.

An effective and rapid procedure to synthesize hierarchical MOR (Hi-MOR) nanosheets by self-assembly between subnanocrystals and organic templates was reported [76]. The hierarchical pore systems provided a large quantity of Brønsted acid sites. Compared with conventional mordenite, Hi-MOR showed higher activity, xylene selectivity, and longer catalytic life in the reaction of toluene disproportionation thanks to its excellent diffusion from hierarchical structure.

The effect of base-treatment by single-cycle or multiple-cycle protocol on the textural and catalytic properties of mordenite was studied [44]. According to BET (Brunauer, Emmett and Teller), isotherm and temperature programmed desorption of hexane isomers, mordenite after single-cycle base-treatment generated disordered mesopores in the size of 8 nm regardless of base-treating time, and multiple-cycle treatment generated 8–14 nm mesopores interconnected with silica-blocked 12-MR micropore. The latter treatment was more effective for enhancing both of the catalytic activity and stability of mordenite than the former. Study on optimum base-treatment condition by varying cycle number and base-treating time in each cycle was conducted. The improved catalytic performance is attributed to the mesopore formation with enhanced diffusivity.

Hard coke blockage of micropores of acidic zeolites generally causes serious catalytic deactivation for many chemical processes. The H-mordenite nanosheet assemblies exhibit the high catalytic activity for carbonylation of dimethyl ether because of their large quantity of framework Brønsted acids [36]. The specific morphology of the nanosheet unites improves mass diffusion for both reactants and products. Consequently, the coke precursor species can readily migrate from the micropores to the external surface of the assemblies, inducing the improved catalytic stability through inhibiting hard coke formation in frameworks.

Catalytic properties of the MOR zeolitic samples were investigated by benzylation reaction of mesitylene and benzyl chloride [40]. Compared with the traditional microporous mordenite, the reaction conversion on the hierarchical mordenite increased ~ 7 times and the reaction apparent rate constant increased ~ 19 times. This property is due to the expansive external surface areas and mesopores volume in the hierarchical mordenite zeolite, which can effectively improve accessibility of the bulky reactants to acid sites and mass transfer rate of big products, and greatly promote the catalytic efficiency of bulky molecules.

Catalytic efficiencies of samples of mesoporous MOR were tested [42] in two highly applicable Freidel–Crafts catalytic reactions, viz; (i) Friedel–Crafts acylation of 1-methoxynaphthalene (MN) with benzoic anhydride (BAn) and (ii) Friedel–Crafts alkylation of benzene with benzyl alcohol. In order to evaluate the catalytic application of the mesoporous MOR, similar reactions were also performed with two commercial mordenites: CBV 10A (Si/Al = 6.5) and CBV 21 (Si/Al = 10). Synthesized mesoporous MOR demonstrated significantly improved catalytic efficiencies (≈ 100%), 3–5 times higher than its conventional counter-parts.

Hierarchical mordenite catalysts were prepared by HF and NH_4F etching. Materials' zirconia modification was performed by incipient wetness impregnation in order to vary surface acidity type and strength [77]. Catalytic performance was studied for the glycerol esterification with acetic acid. Zr increased reaction turnover activity compared to raw zeolite substrate due to abundant Brönsted and Lewis moiety presence. The demonstrated biomass valorization processes to bio-based platform chemicals are of interest, as they employ biodiesel production waste.

Desilication of mordenite crystals previously enriched in silica by dealumination has been investigated in order to better understand the process of mesopore formation [45]. Pentene oligomerization in the liquid phase in the presence of an alkane solvent has been used to evaluate the catalytic activity of the desilicated zeolite. Relevant physico-chemical characteristics such as texture, porosity, acidity, and composition of zeolite samples desilicated at 80°C in NaOH solutions at different base concentrations were analyzed quantitatively. The generation of mesoporosity is due to the hydrolysis and dissolution of the surface of the crystals, probably initiated by the formation of etch pits at low coordination atoms, followed by surface roughening, formation and growth of internal mesopores, fragmentation, and, ultimately, by a complete dissolution of the crystal. Such a process, which is similar to the one involved in the chemical weathering of rocks, results in a severe loss of crystalline material and of the strong acidity associated to it. The oligomerization catalyst prepared from desilicated mesoporous mordenite proved highly stable and selective for the production of C_{15}–C_{30} oligomers from pentene.

The intrinsic activity of acid sites in the hierarchical MOR zeolites synthesized by the approach introduced in the paper was well investigated [39]. Ethanol dehydration is predominantly a unimolecular reaction path at high temperature. The present study demonstrated that hierarchical MOR zeolites not only have higher activity and stability, but also can avoid side-reaction taking place in ethanol dehydration reactions.

Fabrication of cost-effective catalysts is one of the key strategies to monetize natural gas derivatives such as dimethyl ether to olefins. In this study, the particle size of natural mordenite (MOR) was reduced into micro- and nanosize by high-energy ball milling with variation in milling time [29]. Furthermore, the selected milled MOR was recrystallized by using a silicate solution that favored the growth of the mordenite phase to recover the demolished mordenite structure. The mesopore volume and external surface area were increased significantly after milling treatments and remained large even after recrystallization. A high conversion of dimethyl ether was obtained over the recrystallized MOR (99.7%) and milled natural MOR (54.1%) as compared to the parent (1.2%). Moreover, the milling only and the milling–recrystallization processes improved selectivity toward olefins and prolonged catalyst lifetime. The reduced particle size combined with the hierarchical porous and acidity effectively enhanced catalysts activity and selectivity to olefins.

In this study, zeolite mordenite (MOR-ST) was directly synthesized by a hydrothermal method using starch as a template successfully. The morphological characterization of the samples was fulfilled by XRD, SEM, and BET [22]. SEM showed that the as-synthesized MOR-ST displayed a flammulina-like morphology and smaller crystalline size in comparison with the conventional mordenite (MOR-TEA) synthesized by TEAOH template. Additionally, MOR-ST presented bigger BET surface area, smaller bulk density, and higher hydrothermal stability.

The catalytic performance of MOR-ST and MOR-TEA were studied using Daqing wax oil in a quartz micro-reactor at 500°C. The results revealed that MOR-ST provided remarkably enhanced activity and higher gasoline yield in crack of wax oil in sharp contrast to MOR-TEA (conversion 85.70 wt.% vs. 69.00 wt.%).

In the catalytic performance evaluation of samples of mesoporous mordenite in the n-heptane cracking reaction, a rise in catalytic activity and a significant reduction in sample deactivation rate were observed, thus indicating the prospect of using mordenite for application in reactions involving petroleum cracking of heavy fractions [41].

The benzylation of benzene with benzyl alcohol, which involves large product molecules and requires strong acidity, is used as a probe reaction to test the catalytic performance of the hierarchical mordenite (HMOR) [37]. For comparison, the conventional mordenite (CMOR) and mesoporous AlMCM-41 material with similar Si/Al ratios are also tested in the reaction. A remarkably enhanced catalytic activity is detected on the HMOR compared to the CMOR and AlMCM-41. The apparent reaction rate constant for HMOR is 23 times that for CMOR and 8 times that for AlMCM-41. The combination of superior mass transfer ability and strong acidity of HMOR not only improves the benzylalcohol conversion, but also promotes the selectivity of the conversion of benzylalcohol into biphenylmethane.

The hierarchical MOR show much higher catalytic activity in benzylation of benzene with benzyl alcohol [38]. The results demonstrate that the decreased crystal size of the MOR zeolite, especially in the *c*-direction, could effectively prompt mass transport in the 12-MR channels, leading to superior catalytic efficiency. In addition, the material also displays enhanced activity and anti-deactivation behavior in the carbonylation of dimethyl ether to methyl acetate, which is a novel route to produce ethanol from syngas.

The authors [78 80] describe the process of alkylation of aromatic hydrocarbons on a catalyst containing type Y zeolite and type MOR zeolite having a controlled macroporous structure. Catalytic compositions exhibit reduced deactivation rates during the alkylation process, thereby increasing the life of the catalysts.

3.3. Catalytic Compositions Based on Composite Materials Obtained by Recrystallization of Mordenite

Hydroisomerization of n-hexane, n-octane, and n-hexadecane has been studied over Pt-containing micro/mesoporous materials obtained by mordenite (MOR) recrystallization in alkaline solution in the presence of cethyltrimethylammonium bromide [81-83]. Micro/mesoporous nanocomposites MOR/MCM-41 with intermediate degree of recrystallization showed remarkably high activity and selectivity in hydroisomerization of n-octane and high selectivity in hydroisomerization of n-hexadecane. The effect is due to high accessibility of the active sites, improved transport of bulky molecules provided by mesopores, and optimal size of Pt particles located in mesopores.

The catalytic properties of compositions based on materials obtained by the recrystallization of mordenite have also been studied in the alkylation of aromatic compounds, dehydration of alcohols and oxygenates, skeletal isomerization of alkenes, hydroisomerization and hydrocracking of hydrocarbons, cracking of alkanes and alkylaromatic compounds, and disproportionation and transalkylation of 4-aromatic hydrocarbons. These materials have demonstrated increased activity and selectivity in reactions requiring strong acidity, but having some limitations on the diffusion of reagents or products in zeolite micropores. Using compositions based on materials obtained by recrystallization of mordenite as catalysts, it was possible to significantly improve the processes of alkylation of benzene with dodecene [4, 84 and 85], naphthalene with cyclohexene [86], transalkylation of biphenyl with para-diisopropylbenzene [4, 87], disproportionation of toluene and cumene [4, 88], and hydrocracking of alkanes [4, 88]. In addition, in reactions such as transalkylation of biphenyl with para-diisopropylbenzene [4, 89], isomerization of n-butene [4], and disproportionation of cumene [88], the use of recrystallized zeolites led to a significant increase in the time of stable operation of the catalyst.

A catalyst for the synthesis of isobutylene from acetone based on micro-mesoporous mordenite with a deep degree of recrystallization was developed [90], which ensures 47% acetone conversion and 67% isobutylene selectivity. In addition, a technique was proposed that allows an increase the mordenite stable operation time during butene oligomerization based on stepwise recrystallization in a slightly alkaline solution in the presence of cetyltrimethylammonium bromide. It was shown that the resistance of the catalyst obtained by this method of deactivation is associated with the buildup of a thin mesoporous layer on the surface of the zeolite crystal.

The liquid-phase alkylation of benzene with 1-tetradecene was realized over dealuminated mordenites with different mesoporous surface areas and three different levels of acidity [91]. The number of Brønsted acid sites decreases; however, it represents a higher proportion of the total acidity. Alkylation results demonstrated that the mordenite activity in alkylation accelerated with increasing mesoporous surface area in spite of the lower total acidity. This fact indicates that the accessibility of acid sites and the internal mass transport in mordenite channels play a significant role in liquid-phase alkylation of aromatics with 1-alkenes.

In the isopropylation of naphthalene for the production of 2,6-naphthalenedicarboxylate, the monomer of polyethylene naphthalate plastic, a shape selective mordenite (MOR) zeolite catalyst, provides the best selectivity for the desired 2,6-diisopropylnaphthalene [92]. However, the small pore size of the zeolite limits the naphthalene conversion and lowers the stability because of pore mouth blocking by coke. The researchers discovered that these problems could be mitigated by synthesizing a micro-meso hierarchical pore structure in the MOR zeolite by the recrystallization of MOR with cetyltrimethylammonium bromide as a mesopore-forming surfactant. The recrystallized catalysts allow the facile diffusion of bulky molecules through connected meso- and micropores to and from active catalytic sites located in the small MOR pores. Relative to microporous MOR, the hierarchical MOR catalyst demonstrated a greatly enhanced activity, stability, and coke tolerance, and the intrinsic high shape selectivity of MOR for 2,6-diisopropylnaphthalene was maintained. Mild desilication enlarged the pore volume and formed additional acid sites to increase the activity further.

3.4. Catalytic Systems Based on Granular Mordenite of High Crystallinity with a Hierarchical Porous Structure

Granular mordenite of high crystallinity with a hierarchical porous structure (MOR_{mmm}) zeolite was used to prepare new effective catalyst for hydroisomerization of n-hexane, representing the H-form of the indicated zeolite, promoted with 0.5% wt. Pt ($0.5Pt/HMOR_{mmm}$) [13, 48]. Its catalytic properties in the conversion of n-hexane were compared to the catalytic properties of granular catalyst prepared by mixing 70.0% wt. mordenite in the H-form (Si/Al = 10) with 30.0% wt. AlO(OH), in terms of Al_2O_3, followed by heat-forming at 600°C AlO(OH) and applying 0.5% wt. Pt to the prepared granules (hereinafter 0.5Pt/HMOR with binder). The comparison of the experimental data on both samples revealed that the $0.5Pt/HMOR_{mmm}$ catalyst exhibited higher activity, while the selectivities of the formation of isohexanes are close. For catalyst 0.5Pt/HMOR binder free (conversion of n-hexane not lower than 80% wt), the maximum selectivity of the formation of isohexanes at the level of 84.0% is achieved at 300°C and volumetric feed rate of 1.5 $hours^{-1}$. For the 0.5Pt/H MOR_{mmm} sample, the conversion of n-hexane practically does not change and remains at the level of 80.0 83.0% wt. at 300°C and vol. feed rate of 1.5 $hours^{-1}$ for 50 hours. The selectivity of the formation of isohexanes makes 84-86.0%.

Table 6 shows the characteristics of the hydroisomerization catalysts of the model mixture (20% wt. benzene and 80% wt. n-heptane) [49, 50]. The catalysts were synthesized by impregnation of mordenite samples binder free, H-MORmmm and $HMOR_{mmm}$-US-0.9N

(NH_4MOR_{mmm} sample subjected to post-synthetic thermochemical treatments in water vapor medium and 0.9 N aqueous solutions of citric acid) with H_2PtCl_6 solution for moisture capacity, followed by drying at 120°C and calcination at 500°C for 3 hours. The platinum content in all the prepared samples was the same and made 0.3% wt.

The research studies showed that complete conversion of benzene in the model mixture with heptane is observed over all the studied catalysts in the temperature range 280-300°C, pressure of 1.5 MPa and volumetric feed rate of 2 hours^{-1}. The comparison of the paraffin conversion values over the Cat-1 ÷ Cat-3 catalysts at constant temperature illustrates that they can be arranged in the following row according to their activity in paraffin conversion: Cat-3 > Cat-2 > Cat-1. As a result, it was concluded that the difference in the catalytic properties in the hydroisomerization of the model mixture of benzene-n-heptane of the catalytic systems accounts for both the difference in the strength and concentration of acid sites and the properties of the pore structure. Thus, granular mordenites in the H-form of high crystallinity degree and hierarchical porous structure promoted with 0.3%wt. Pt offer powerful catalytic systems for hydroisomerization of benzene/n-heptane mixture.

Table 6. Physical and chemical properties of mordenite-containing catalysts [49, 50]

Conv. names	Composition of the catalyst	V_{micro} [a], $cm^3 \cdot g^{-1}$	V_{meso} [b], $cm^3 \cdot g^{-1}$	Concentration of acid sites, μmol·g^{-1}		T_{max} of strong acid sites, °C
				weak	strong	
Cat-1	0,3%Pt/(70%HMOR+30% wt γ-Al_2O_3)	0,10	0,15	288	246	420
Cat-2	0,3%Pt/$HMOR_{mmm}$	0,15	0,20	408	312	420
Cat-3	0,3%Pt/$HMOR_{mmm}$-US-0,9N	0,12	0,23	246	185	440

Conditions: [a] micropore volume
[b] mesopore volume

CONCLUSION

Over the past 15–20 years, the crystallization principles of powdered MOR type zeolite in the Na-form have been studied in detail, and various variants for the synthesis with the ratio Si/Al = 4.5 – 13 of high phase purity, crystallinity degree, and dispersion have been developed. Catalysts for transformation of various molecules based on them have been presented.

Whereas microporous mordenites are ineffective in catalytic transformations of bulk molecules of organic compounds due to diffusion restrictions, the following methods have been worked out for producing micro-mesoporous mordenites of hierarchical porous structure: synthesis applying various approaches of template synthesis, formation of mesopores in zeolite crystals by postsynthetic treatment, zeolitization of mesoporous materials, and recrystallization of zeolites. Based on the prepared materials, new catalytic systems combining the advantages of zeolites and mesoporous molecular sieves have been synthesized. These catalysts significantly improved the performance of alkylation of benzene with dodecene, transalkylation of biphenyl with para-diisopropylbenzene, disproportionation of toluene and cumene, vacuum gasoil cracking, and alkane hydrocracking. Moreover, the use of catalysts based on micro-mesoporous mordenites in the transalkylation of biphenyl with para-diisopropylbenzene and disproportionation of cumene significantly increase the time of stable operation of the catalyst.

The methods for production of mordenite-containing catalysts detailed in the scientific literature most often involve synthesis of zeolite in the form of 1-5 μm crystals in the required cation-decationized forms and subsequent molding into granules (1.6-2.0 mm in diameter and 3-6 mm long) in the mixture with binder material. Aluminum hydroxide of pseudoboehmite structure, converted to γ-Al_2O_3 under heat treatment, is commonly employed as the second component. The catalyst contains 20 to 30% wt binder material. Platinum, palladium, and other promoters are most often to be introduced into the catalyst in the mixing of zeolite and binder. The introduction of the latter into the catalyst reduces the concentration of catalytic sites by almost the same amount. Currently, methods for the synthesis of granular mordenite in Na-form of high crystallinity degree and phase purity with hierarchical porous structure (Na-MOR_{mmm}) have been developed. The experiments on the synthesis revealed that the granules should contain at least 50% wt powdered zeolite before crystallization. During crystallization of mordenite-containing granules (the content of the crystalline phase higher than that of the amorphous material), strong supersaturation occurs in the crystallization nuclei and nanocrystals from 20 to 300 nm in size are formed. Then, the formed nanocrystals grow together with each other and with the initial crystals from 2 to 5 microns in size. As a result, a hierarchical porous structure is formed with a meso and macropore volume of ~ 0.20 $cm^3 \cdot g^{-1}$ and ~ 0.18 $cm^3 \cdot g^{-1}$, respectively.

Postsynthetic treatment of MOR type zeolite in the H-form in water vapour at 400 to 600°C is a convenient way for modifying the acid properties and structure in the production of industrial catalysts. By changing the temperature and time of steam treatment, it is possible to produce MOR type zeolites in the H-form with different modulus and ensured formation of optimal mesopore transport system. Subsequent chemical treatment with acid solutions may promote further modification of the zeolites mentioned above. The postsynthetic treatment conditions in water vapour and aqueous solution of citric acid for MOR_{mmm} in the H form, providing an increase in the modulus from 9.0 to 17.0 with constant crystallinity degree caused by dealumination, were specified. The interval indicating the pore size distribution is expanding, and there is an increase in the mesopore volume from 0.20 to 0.23 $cm^3 \cdot g^{-1}$ through the contraction of micropore volume.

Bifunctional catalysts for hydroisomerization of n-hexane and mixture of benzene with n-heptane based on MOR_{mmm} zeolite in decationed form were developed.

In the coming years, the researchers will focus on the following issues:

- development of promising methods for industrial implementation of the preparation of micro-mesoporous mordenites with hierarchical porous structure of high phase purity, crystallinity and dispersion;
- study of the principles of specific catalytic processes in the presence of new mordenite-containing catalysts;
- determination of the reasons for deactivation of these catalytic systems and development of techniques for subsequent effective regeneration;
- development and implementation of effective methods for preparing catalysts based on granular MOR type zeolite of the molar ratio Si/Al = 4.5 – 13 of high crystallinity degree and phase purity with hierarchical porous structure.

REFERENCES

[1] Degnan, T. F. Jr. (2007). Recent progress in the development of zeolitic catalysts for the petroleum refining and petrochemical manufacturing industries. *Studies in Surface Science and Catalysis*, *170* (54), 54-65.

[2] Vermeiren, W. & Gilson, J. P. (2009). Impact of zeolites on the petroleum and petrochemical industry. *Topics in Catalysis*, *52* (9), 1131–1161.

[3] Cejka, J., Corma, A. & Zones, S. (2010). *Zeolites and catalysis: synthesis, reactions and applications*. Weinheim: Wiley-VCH.

[4] Ivanova, I. I., Knyazeva, E. E., Maerle, A. A. & Kasyanov, I. A. (2015). Design of micro/mesoporous zeolite-based catalysts for petrochemical and organic syntheses. *Kinetics and Catalysis*, *56* (4), 549-561.

[5] Shamzhy, M., Opanasenko, M., Concepción, P. & Martínez, A. (2019). New trends in tailoring active sites in zeolite-based catalysts. *Chemical Society Reviews*, *48*, 1095-1149.

[6] Breck, Donald W. (1974). *Zeolite molecular sieves. Structure, chemistry, and use*. New York: John Wiley & Sons.

[7] Barrer, Richard M. (1982). *Hydrothermal chemistry of zeolites*. New York: Academic Press in London.

[8] Rabo, J. (1980). *Chemistry of Zeolites and Catalysis on Zeolites*. Translated by Minachev, Kh.M. Moscow: World.

[9] Clifton, Robert A. (1987). *Natural and synthesis zeolites*. Inf. Circ. U.S. Bur. Mines.

[10] Makhamatkhanov, Rustam. A. (2001). "Development of new synthesis methods, study of the physicochemical and catalytic properties of mordenite-type zeolite." PhD in Chemistry, Institute of Petrochemistry and Catalysis.

[11] Makhamatkhanov, R. A., Pavlov, M. L., Kutepov, B. I. & Patrikeev, D. V. (2006). New methods for the synthesis of mordenite. *Chemical technology*, *7* (1), 6-10.

[12] Pavlov, M. L., Makhamatkhanov, R. A., Kutepov, B. I. & Dzhemilev, U. M. (2001). Synthesis of mordenite-like zeolite from secondary silicon-containing raw materials. *Russian Journal of Applied Chemistry*, *74* (2), 333-334.

[13] Ahmed, Kanaan R. (2013). "Isomerization of n-hexane in the presence of catalysts based on granulated mordenite-type zeolite without binders." PhD in Chemistry, Ufa State Petroleum Technological University.

[14] Gorshunova, K. K., Ahmed, Kanaan R., Travkina, O. S., Pavlova, I. N., Grigor'eva, N. G., Pavlov, M. L. & Kutepov, B. I. (2014). Influence of synthesis conditions on the adsorption and catalytic properties of mordenite zeolite. *Petroleum Chemistry*, *54* (2), 132-136.

[15] Todorova, T. D. & Kalvachev, Yu. A. (2015). Seed-mediated approach to size-controlled synthesis of a mordenite type zeolite from organic template free initial gel. *Bulgarian Chemical Communications*, *47* (3), 409–416.

[16] Khalil, U. & Muraza, O. (2016). Microwave-assisted hydrothermal synthesis of mordenite zeolite: Optimization of synthesis parameters. *Microporous and Mesoporous Materials*, *232*, 211-217.

[17] Zhang, H., Zhang, H., Wang, P., Zhao, Y., Shi, Z., Zhanga, Y. & Tang, Y. (2016). Organic template-free synthesis of zeolite mordenite nanocrystals through exotic seed-assisted conversion. *RSC Advances*, *53* (6), 47623-47631.

[18] Li, L., Yang, J., Li, J., Han, P., Wang, J., Zhao, Y., Wang, J., Lu, J., Yin, D. & Zhang, Y. (2016). Synthesis of high performance mordenite membranes from fluoride-containing dilute solution under microwave-assisted heating. *Journal of Membrane Science*, *512*, 83-92.

[19] Zhu, J., Liu, Z., Endo, A., Yanaba, Y., Yoshikawa, T., Wakihara, T. & Okubo, T. (2017). Ultrafast, OSDA-free synthesis of mordenite zeolite. *Cryst Eng Comm*, *19* (4), 632-640.

[20] Mohammad Hassnain. An Experimental Enquiry into The Growth of Mordenite Nanocrystals Sans Seed Addition. Material Science and Nanotechnology 3/1/17, 21 p.

[21] Pat. US 10017394 B2 (2018). Process for preparing a molecular sieve.

[22] Ren, S., Gong, C., Zeng, P., Guo, Q. & Shen, B. (2016). Synthesis of flammulina-like mordenite using starch as template and high catalytic performance in crack of wax oil. *Fuel*, *166*, 347-351.

[23] Pat. US 9845249 B2 (2017). Method for preparation of mordenite.

[24] Cui, M., Wang, L., Zhang, Y., Wang, Y. & Meng, C. (2015). Changes of medium-range structure in the course of crystallization of mordenite from diatomite. *Microporous and Mesoporous Materials*, *206*, 52-57.

[25] Yamada, K., Johan, E., Matsue, N., Itagaki, Y. & Aono, H. (2015). Preparation of mordenite and its composite material with nano-sized magnetite from diatomites for

radioactive Cs decontamination. *Journal of the Ceramic Society of Japan*, *123* (1435), 129-135.

[26] Rakmae, S. & Wittayakun, J. (2015). The effect of gel volume in autoclave on the synthesis of mordenite from rice husk silica by hydrothermal method. *Suranaree Journal Science and. Technology*, *22* (1), 83-91.

[27] Merdekawati, R., Ediati, R. & Prasetyoko, D. (2015). "Effect of silica to alumina molar ratio on crystallinity of mordenite synthesized using rice husk ash and kaolin fron Bangka Belitung." Paper presented at the International Conference on Science and Science Education, Indonesia.

[28] Rahmawati, S., Ediati, R. & Prasetyoko, D. (2015). "Direct synthesis of mordenite from kaolin and rice husk ash." Paper presented at the International Conference on Science and Science Education, Indonesia.

[29] Kurniawan, T., Muraza, O., Miyake, K., Hakeem, A. S., Hirota, Y., Al-Amer, A. M. & Nishiyama, N. (2017). Conversion of dimethyl ether to olefins over nanosized mordenite fabricated by a combined high-energy ball milling with recrystallization. *Industrial & Engineering Chemistry Research*, *56* (15), 4258-4266.

[30] Pavlov, M. L., Makhamatkhanov, R. A., Kutepov, B. I., Kudasheva, F. Kh., Apkarimova, G. I. & Buyanova, O. S. (2004). Synthesis of high-ratio mordenite. *Russian Journal of Applied Chemistry*, *77* (4), 683-684.

[31] Gizetdinova, Anastasiya F. (2015). "Synthesis and study of zeolites of structural types Beta and Mordenite for the process of medium temperature isomerization of pentane-hexane fractions." PhD in Chemistry, Irkutsk State Technical University.

[32] Ma, Z., Xie, J., Zhang, J., Zhang, W., Zhou, Y. & Wang, J. (2016). Mordenite zeolite with ultrahigh SiO_2/Al_2O_3 ratio directly synthesized from ionic liquid-assisted dry-gel-conversion. *Microporous and Mesoporous Materials*, *224*, 17-25.

[33] Gili, M. B. Z. & Conato, M. T. (2018). Synthesis and characterization of mordenite-type zeolites with varying Si/Al ratio. *Materials Research Express*, *6* (3), 015515.

[34] Tao, Y., Kanoh, H., Abrams, L. & Kaneko, K. (2006). Mesopore-modified zeolites: preparation, characterization, and applications. *Chemical Reviews*, *106*, 896-910.

[35] Pe´rez-Ramı´rez, J., Christensen, C. H., Egeblad, K., Christensen, C. H. & Groen, J. C. (2008). Hierarchical zeolites: enhanced utilisation of microporous crystals in catalysis by advances in materials design. *Chemical Society Reviews*, *37*, 2530–2542.

[36] Liu, Y., Zhao, N., Xian, H., Cheng, Q., Tan, Y., Tsubaki, N. & Li, X. (2015). Facilely synthesized H-mordenite nanosheet assembly for carbonylation of dimethyl ether. *ACS Applied Materials & Interfaces*, *7* (16), 8398–8403.

[37] Li, Y., Sun, C., Fan, W., Wang, Y., Lan, A., Han, P., Li, X. & Dou, T. (2015). One-pot synthesis of hierarchical mordenite and its performance in the benzylation of benzene with benzyl alcohol. *Journal of Materials Science*, *50* (14), 5059–5067.

[38] Yuan, Y., Wang, L., Liu, H., Tian, P., Yang, M., Xu, S. & Liu, Z. (2015). Facile preparation of nanocrystal-assembled hierarchical mordenite zeolites with remarkable catalytic performance. *Chinese Journal of Catalysis*, *36* (11), 1910-1919.

[39] Wang, S., He, B., Tian, R., Sun, C., Dai, R., Li, X., Wu, X., An, X. & Xie, X. (2018). Synthesis and catalytic performance of hierarchically structured MOR zeolites by a dual-functional templating approach. *Journal of Colloid and Interface Science*, *527*, 339-345.

[40] Guang, D., Huan, X., Wen-Ming H., Jing-Hong, M. & Rui-Feng, L. (2018). Synthesis and characterization of hierarchical mordenite zeolite. *Journal of Inorganic Materials*, *33* (4), 397-402.

[41] Santos, P. S., Almeida, N. C., Santos, I. S. Marques, M. de F. V. & Fernandes, L. D. (2018). Synthesis and characterization of mesoporous mordenite zeolite using soft templates. *Catalysis Letters*, *148* (7), 1870–1878.

[42] Singh, B. K., Kim, Y., Baek, S. B., Meena, A., Sultan, S., Kwak, J. H. & Kim, K. S. (2018). Template free facile synthesis of mesoporous mordenite for bulky molecular catalytic reactions. *Journal of Industrial and Engineering Chemistry*, *57*, 363-369.

[43] Kubarev, A. V., Janssen, K. P. F. & Roeffaers, M. B. J. (2015). Noninvasive nanoscopy uncovers the impact of the hierarchical porous structure on the catalytic activity of single dealuminated mordenite crystals. *Chem Cat Chem*, *7*, 3646 – 3650.

[44] Tsai, S. T., Chao, P. Y., Chao, P. H., Du, K. J., Fang, M. J., Liu, S. B. & Tsai, T. C. (2016). Study on optimum base-treatment of mordenite for catalyticalkylbenzene transalkylation. *Catalysis Today*, *259*, 423-429.

[45] Bertrand-Drira, C., Cheng, X.-w., Cacciaguerra, T., Trens, P., Melinte, G., Ersen, O., Minoux, D., Finiels, A., Fajula, F. & Gerardin, C. (2015). Mesoporous mordenites obtained by desilication: mechanistic considerations and evaluation in catalytic oligomerization of pentene. *Microporous and Mesoporous Materials*, *213*, 142-149.

[46] Travkina, O. S., Kuvatova, R. Z. & Kutepov, B. I. (2017). "Mordenite based catalysts. Synthesis and properties." In *Obtaining, structure and application of petrochemical products and organic synthesis*, edited by Talipov, R.F., 104-124. Ufa: Publishing House of Ural State Technical University.

[47] Giudici, Reinaldo. (1999). "Synthesis, dealumination and adsorption behavior of mordenite extrudates." Doctor of Technical Sciences, Zurich.

[48] Gorshunova, K. K., Travkina, O. S., Pavlov, M. L., Kutepov, B. I., Kuvatova, R. Z. & Amineva, N. A. (2013). Synthesis of Binder-Free Granulated MOR-Type Zeolite with Hierarchic Pore Structure. *Russian Journal of Applied Chemistry*, *86*, 1805–1812.

[49] Travkina, O. S., Khazipova, A. N., Pavlova, I. N., Akhmetov, A. F., Kutepov, B. I. & Galyautdinova, L. S. (2018). Physicochemical and catalytic properties of granular mordenites of high crystallinity degree with a hierarchical porous structure: influence

of post-synthetic treatment. *International Research Journal of Pure & Applied Chemistry*, *17* (1), 1-9.

[50] Kuvatova, Rezeda. Z. (2018). "Physico-chemical principles of crystallization of granular zeolites of the structural type MOR with a hierarchical porous structure." PhD in Chemistry, Bashkir State University.

[51] Mishin, Igor' V. (1995). "Regulation of the structural, adsorption and catalytic properties of zeolites by changing the composition of the framework." Doctor of Chemical Sciences, Moscow.

[52] Petranovskii, V., Pestryakov, A., Hernández, M. Á., Rivas, F. C., Claverie, A. L. & Fuentes, S. (2015). Hydrophilicity of mordenites with different SiO_2/Al_2O_3 molar ratio. *Procedia Chemistry*, *15*, 72 – 78.

[53] Cysneiros, O. M. S., Silva, B. J. B., Silva, A. O. S., Alencar, S. L., Santos, R. B., Soares, P. F. M., Mendonca, T. R. D., Sousa Junior, L. V. & Santos, J. R. (2016). Crystallization of mordenite zeolite with seed addition in reaction mixtures with different SiO_2/Al_2O_3 ratios. *Scientia Plena*, *12* (4), 1-11.

[54] Zhang, J., Mao, Y., Li, J., Wang, X., Xie, J., Zhou, Y. & Wang, J. Ultrahigh mechanically stable hierarchical mordenite zeolite monolith: Direct binder-/template-free hydrothermal synthesis, *Chemical Engineering Science*, *138*, 473-481.

[55] Xie, Z., Liu, Z., Wang, Y., Yang, Q., Xu, L. & Ding, W. (2010). An overview of recent development in composite catalysts from porous materials for various reactions and processes. *International Journal of Molecular Sciences*, *11*, 2152-2187.

[56] Stocker, M. (2005). Gas phase catalysis by zeolites. *Microporous and Mesoporous Materials*, *82* (3), 257-292.

[57] Agabekov, V. E. & Senkov, G. M. (2006). Catalytic isomerization of light paraffin hydrocarbons. *Catalysis in industry*, *5*, 31-41.

[58] Pat. RU 2287369 C1. (2006). Method of preparation of catalyst for benzene hydroisomerization process.

[59] Belopukhov, E. A., Beliy, A. S., Smolikav, M. D., Kiryanov, D. I. & Gulyaeva, T. I. (2012). Hydroisomerization of benzene on the catalysts $Pt/MOR/Al_2O_3$. *Catalysis in industry*, *3*, 37-43.

[60] Sugi, Y., Tamada, H., Kuriki, A., Komura, K., Kubota, Y., Joseph, S., Chokkalingam, A., El. Newehy, M., Al-Deyab, S. S., Jang, H. G., Kim, J. H., Seo, & Vinu, A. (2015). Alkaline earth metal modified H-Mordenites. Their catalytic properties in the isopropylation of biphenyl. *Industrial & Engineering Chemistry Research*, *54* (49), 12283–12292.

[61] Pat. RU 2460715 C2. (2012). Method of producing monoalkylated aromatic compounds with high linearity and controlled isomery.

[62] Pat. RU 2354640 C2. (2009). Xylene isomerisation process for mixtures containing aromatic compounds C_9.

[63] Makihara, M. & Komura, K. (2017). A novel Friedel-Crafts acylation reaction of anisole for production of 4-methoxyacetophenone with high selectivity and sufficient reusability of mordenite zeolite catalyst. *Green and Sustainable Chemistry*, *7* (3), 185-192.

[64] Mokoena, K. & Scurrell, M. S. (2017). Alkyl transfer reactions on solid acids. The transalkylation of benzene with mesitylene on H-mordenite and HY zeolites. *Microporous and Mesoporous Materials*, *241*, 28-35.

[65] Mokoena, K. & Scurrell, M. S. (2018). Alkyl transfer reactions on solid acids. The disproportionation of ethylbenzene and toluene on H-mordenite and HY zeolites. *Petroleum Science and Technology*, *36* (16), 1208-1215.

[66] Pat. RU 2411232 C2. (2011). Method for carbonylation of alkyl ethers.

[67] Pat. RU 2457030 C2. (2012). Method for carbonylation of dimethyl ether.

[68] Priya, S. S., Selvakannan, P. R., Chary, K. V. R., Kantam, M. L. & Bhargava, S. K. (2017). Solvent-free microwave-assisted synthesis of solketal from glycerol using transition metal ions promoted mordenite solid acid catalysts. *Molecular Catalysis*, *434*, 184-193.

[69] Grigor'eva, N. G., Kutepov, B. I., Makhamatkhanov, R. A., Bikbaev, R. T., Khazipova, A. N. & Kovtunenko, I. A. (2002). Oligomerization of α-methylstyrene in the presence of mordenite type zeolite. *Russian Journal of Applied Chemistry*, *75* (9), 1539-1540.

[70] Mamo, W., Chebude, Y., Márquez-Álvarez, C., Díaz, I. & Sastre, E. (2016). Comparison of glucose conversion to 5-HMF using different modified mordenites in ionic liquid and biphasic media. *Catalysis Science & Technology*, *6*, 2766-2774.

[71] Trisunaryanti, W., Triyono, T., Armunanto, R., Hastuti, L. P., Ristiana, D. D. & Ginting, R. V. (2018). Hydrocracking of α-cellulose using Co, Ni, and Pd supported on mordenite catalysts. *Indonesian Journal of Chemistry*, *18* (1), 166-172.

[72] Leng, K., Sun, Y., Zhang, X., Yu, M. & Xu, W. (2016). Ti-modified hierarchical mordenite as highly active catalyst for oxidative desulfurization of dibenzothiophene. *Fuel*, *174*, 9-16.

[73] Su, J., Li, X., Chen, Y., Cui, Y., Xu, J., Qian, C. & Chen, X. (2016). N-Methylation of amines with methanol in a hydrogen free system on a combined Al_2O_3–mordenite catalyst. *RSC Advances*, *6*, 55643-55649.

[74] Pat. RU 2477177 C2. (2013). Method of preparing catalyst for decomposing nitrous oxide and process of decontaminating gas emissions containing nitrous oxide.

[75] Kalvachev, Y., Todorova, T., Nihtianova, D., Lazarova, H. & Popova, M. (2017). Fluoride etching of mordenite and its influence on catalytic activity. *Journal of Materials Science*, *52* (9), 5297–5308.

[76] Liu, M., Jia, W., Li, J., Wang, Y., Ma, S., Chen, H. & Zhu, Z. (2016). Catalytic properties of hierarchical mordenite nanosheets synthesized by self-assembly between subnanocrystals and organic templates. *Catalysis Letters*, *146* (1), 249–254.

[77] Popova, M., Lazarova, H., Kalvachev, Yu., Todorova, T., Szegedi, Á., Shestakova, P., Mali, G., Dasireddy, V. D. B. C. & Likozar, B. (2017). Zr-modified hierarchical mordenite as heterogeneous catalyst for glycerol esterification. *Catalysis Communications*, *100*, 10-14.

[78] Pat. US 6977319. (2005). Alkylated aromatic compositions, zeolite catalyst compositions and processes for making the same.

[79] Pat. US 7109141. (2006). Alkylated aromatic compositions, zeolite catalyst compositions and process for making the same.

[80] Pat. US 7183452. (2007). Alkylated aromatic compositions, zeolite catalyst compositions and processes for making the same.

[81] Konnov, S. V., Sushkevich, V. L., Monakhova, Yu. V., Yuschenko, V. V., Ponomareva, O. A. & Ivanova, I. I. (2008). Hydroisomerization of n-octane over Pt-containing micro/ mesoporous molecular sieves. *Studies in Surface Science and Catalysis*, *174*, 1167-1170.

[82] Konnov, S. V., Monakhova, Yu. V., Knyazeva, E. E., Yushchenko, V. V., Ponomareva, O. A. & Ivanova, I. I. (2009). Hydroisomerization of n-octane on platinum-containing micro-mesoporous molecular sieves. *Petroleum Chemistry*, *49* (1), 79-85.

[83] Konnov, S. V., Ivanova, I. I., Ponomareva, O. A. & Zaikovskii, V. I. (2012). Hydroisomerization of n-alkanes over Pt-modified micro/mesoporous materials obtained by mordenite recrystallization. *Microporous and Mesoporous Materials*, *164*, 222-231.

[84] Ponomareva, O. A., Timoshin, S. E., Monakhova, Yu. V., Knyazeva, E. E., Yuschenko, V. V. & Ivanova, I. I. (2010). Benzene alkylation with dodecene-1 over micro-mesoporous molecular sieves. *Petroleum Chemistry*, *50* (6), 427-436.

[85] Ponomareva, O. A., Timoshin, S. E., Knyazeva, E. E., Ordomskii, V. V., Yushchenko, V. V., Kulikov, N. S., Ivanova, I. I. & Zaikovskii, V. I. (2011). Physicochemical and catalytic properties of micro-mesoporous zeolite materials. *Russian Journal of Physical Chemistry A*, *85* (12), 2103-2108.

[86] Ivanova, I. I., Kuznetsov, A. S., Knyazeva, E. E., Fajula, F., Thibault-Starzyk, F., Fernandez, C. & Gilson, J. P. (2011). Design of hierarchically structured catalysts by mordenites recrystallization: Application in naphthalene alkylation. *Catalysis Today*, *168*, 133-139.

[87] Ivanova, I. I., Kuznetsov, A. S., Yuschenko, V. V. & Knyazeva, E. E. (2004). Design of composite micro/mesoporous molecular sieve catalysts. *Pure and Applied Chemistry*, *76* (9), 1647-1657.

[88] Ordomsky, V. V., Ivanova, I. I., Knyazeva, E. E., Yuschenko, V. V. & Zaikovskii, V. I. (2012). Cumene disproportionation over micro/mesoporous catalysts obtained by recrystallization of mordenite. *Journal of Catalysis*, *295*, 207-2016.

[89] Ivanova, I. I., Kuznetsov, A. S., Ponomareva, O. A., Yuschenko, V. V. & Knyazeva, E. E. (2005). Micro/mesoporous catalysts obtained by recrystallization of mordenite. *Studies in Surface Science and Catalysis*, *158*, 121-128.

[90] Kasyanov, Ivan. A. (2016). "Patterns of formation and catalytic action of micro-mesoporous materials based on zeolite MOR" PhD in Chemistry, Moscow State University.

[91] Horňáček, M., Hudec, P., Velebná, K. & Lovás, P. (2015). Positive effect of secondary structure creation in mordenites on alkylation of benzene with 1-tetradecene. *Catalysis Communications*, *64*, 1-5.

[92] Lee, Y. H., Kim, W. Y., Park, H., Choi, Y. H. & Lee, J. S. (2016). Highly active and coke-tolerant hierarchical mordenite catalysts synthesized by recrystallization for the isopropylation of naphthalene. *ChemCatChem*, *8* (18), 2996-3001.

In: An Introduction to Aluminosilicates
Editor: Nero Regina Blevins
ISBN: 978-1-53617-250-8

Chapter 11

PREPARATION AND CHARACTERIZATION OF CHITOSAN-ZEOLITE COMPOSITE@POLYACRYLAMIDE: ADSORPTIVE FEATURES FOR THORIUM

Onur Eninanç and Ulvi Ulusoy*
Cumhuriyet University, Department of Chemistry, Sivas, Turkey

ABSTRACT

The preparation of Chitosan (Ch)-Zeolite (Z) composite inter linked by ephycholorhydrine entrapped in polyacrylamide (PAAm) so called PAAmChZ in a ternary combination were achieved. The mineral (Z) was clinoptilolite with the SiO_2/Al_2O_3 ratio value of 4.7. For the structural characterization of the composite FT-IR, XRD, SEM, TGA, BET-Porosity, cation exchange capacity (CEC) and point of zero charge (PZC) analysis were used. Swelling characteristics in view of capacity and kinetics were tested in water and in solution of KNO_3 0.25 M.

FT-IR, XRD, SEM and PZC analysis confirmed the formation of composite in ternary combination. The results of swelling study showed that the adsorption kinetics obeyed Schott's second-order model the overall swelling mechanism was Fickian, i.e., the process was diffusion controlled.

Adsorptive features of PAAm-Ch-Z for Th^{4+} was investigated in view of its dependency on pH, concentration, time, temperature and ionic strength. The parameters derived from the compatibility of experimentally obtained data to Langmuir, Freundlich and Dubinin-Radushkevich (DR), van t'Hoff, pseudo frist-/second-order and Weber

* Corresponding Author's Email: ulusoy@cumhuriyet.edu.tr.

Morris (intra particle diffusion) models were utilized in the evaluation of adsorption and its thermodynamics and kinetics.

The results pointed out that the adsorption increased with increasing solution pH and that Th^{4+} had high affinity to the composite. The kinetics studies proved that the adsorption could be explained by the pseudo second-order model and the adsorption mechanism was via ion exchange and/or complex formation. The changes in values of adsorption enthalpy, entropy and Gibbs free energy were $\Delta H > 0$, $\Delta S > 0$ and $\Delta G < 0$ implying the overall adsorption process was endothermic, increasing entropy and spontaneous.

Keywords: composite, chitosan, zeolite, polyacrylamide, thorium

INTRODUCTION

Adsorption technique is the most preferable one amongst the conventional processes commonly used in the removal/recovery of species of interest from usually aquatic mediums. In principle, a good adsorbent should have a porous structure (high surface area), fast adsorption kinetics, strong affinity and high loading capacity beside which it should be cost effective. Fly ash, silica gel, zeolite, lignin, seaweed, wool wastes, agricultural wastes, chitin, chitosan, and clay materials are of natural adsorbents comparatively meet these requirements (Demirbaş, 2008; Gupta et al., 2009; Wu et al., 2010). Zeolite (Z) and Chitosan (Ch) are the considered materials of this work.

Natural Zeolite Clinoptilolite denoted as HEU by the Structure Commission of the International Zeolite Association is of aluminosilicate minerals. They are in three-dimensional framework structure containing AlO_4 and SiO_4 with tetrahedral geometry. They have rigid anionic frameworks formed well designated cavities and channels. The cavities host metal cations (calcium, sodium, potassium and magnesium) which are exchangeable, removable and/or replaceable ions. Zeolites have a general formula, $M2/n{:}Al_2O_3{:}xSiO_2{:}yH_2O$ where M stands for the extra-framework cation. As depending on the pore size and adsorption properties, zeolites are amongst the most important inorganic cation exchangers and used in a wide variety of industrial applications, e.g., water, waste water and nuclear waste treatments, catalysis, agriculture, animal feed additives, and in biochemical applications (Pavelich et al., 2018). Natural zeolites are the most abundant minerals on earth; the production of synthetic zeolites is also possible but it is economically not feasible in large scale applications.

Although, the usability of natural zeolite in adsorption practices have been demonstrated in a number of investigations (Wingenfelder et al., 2005; Oren and Kaya, 2006; Sprynskyy et al., 2006; Dimirkou, 2007; Oter and Akcay, 2007 and more therein the review by Wang and Peng, 2011), further researches require for enhancement of adsorptive characteristics and practical usage of zeolites, since the aggregation and coagulation of mineral particles under differing conditions of temperature and electrolytes may lead variations in flow properties of these minerals. These are undesired features for their practical implementations since it may result in e.g., hydrodynamic constraints in chromatographic applications (Baybaş and Ulusoy, 2011b).

Among the various natural polymers, as the interest of this study, chitosan (Ch) as a linear cationic polysaccharide (poly-β-(1, 4)-2-amino-2-deoxy-glucopyranose) is obtained by deacetylation of chitin that is the second most abundant natural biopolymer on earth after cellulose. Chitosan has exceptional features required for an ideal adsorbent such as its being low-cost and having amine and hydroxyl groups which acts as coordination sites to form complexes with various chemical species and ions. Beside this, its organic structure may also serve for adsorption of species requiring organic interactions. The use of Ch as adsorbent was well documented by Salehi et al. (2016) for membrane applications, by Kyzas and Bikiaris (2015) and Wang and Zhuang (2017) for modified chitosan adsorbents.

Ch is environmental friendly since it is biodegradable and non-toxic. However, its lonely use is not efficient because of its, e.g., solubility in acidic medium, small surface area, low porosity and poor selectivity. Such drawbacks could be overcome by incorporation of Ch as a component of a composite so that the obtained composition may have more desirable features (mechanical, thermal, adsorptive, dissolution and resistance to strong solution medium) than each component has its use alone (Humelnicu et al., 2011; Maity and Ray, 2018).

By considering the above mentioned limitations in adsorptive properties of both Z and Ch, their cooperative use as the composite components was proposed. The obtained composite with this approach might additionally have synergic feature that the components might not have individually. However, Dragan et al. (2010) materialized the incorporated use of Z and Ch to form a micro/nano composite with enhanced features particularly adsorption capacity; Z was entrapped in Ch matrix where epichlorohydrin and sodium tripolyphosphate were employed as cross linking agents.

After such successes gained by Dinu and Dragan (2010) and Ngah et al. (2011), this investigation was performed. We prepared the ChZ with reference to the previous prescriptions with some modifications and then it was embodied in poly-acrylamide (PAAm) to improve ChZ's adsorbent features further. Incorporation of ChZ into a hydrogel polymer provided suitable circumstances to avoid the aggregation and/or coagulation of composite, to strength mechanical and thermal stability of the structure, to increase resistance ability to electrolytic solution. Beside these, the obtained formation enhanced the swelling features and catalyzed the adsorption via micro-/nano-scale dispersion of mineral particles providing a wider sorption surface area (Baybaş and Ulusoy, 2011a and 2012). PAAm was used since useful polymer for such purpose should be a hydrogel capable of swelling in aquatic solutions to enable the diffusion and/or transfer of the ions towards the adsorbent components as its practical applicability was shown for its composite with bentonite and Z for Th^{4+} adsorption by Baybaş and Ulusoy (2011a).

The prepared PAAmChZ was characterized by FT-IR, XRD, SEM, TGA, BET-Porosity, cation exchange capacity (CEC) and point of zero charge (PZC) analysis.

Swelling characteristics in view of capacity and kinetics were tested in water and in solution of KNO_3. Adsorptive features were tested for Th^{4+}. Th was preferred since its importance in environmental and industrial concerns. Th is used as a fertile material for producing nuclear fuel. It may also be used directly as nuclear fuel instead of uranium, producing less transuranic waste. In the industry, Th is used as an alloying element in magnesium, used in manufacture of aircraft engines, imparting high strength and creep resistance at elevated temperatures and to coat tungsten wire utilized in electronic equipment, improving the electron emission of heated cathodes (Avedesian and Baker, 1999). Th is poisonous for living organisms due to its hazardousness associated with both radioactivity and toxicity. The principal concern from low to moderate level exposure to ionizing radiation is an increased risk of cancer (Url).

METHODS

2.1. Chemicals

Acrylamide monomer and ammonium persulfate (Merck), N,N'-methylbisacrylamide, $Th(NO)_3.4H_2O$ (Merck), N,N,N′,N′-Tetramethyl-ethylene-diamine, chitosan (≥75% deacylated) and ephycholorohydrin (Aldrich), sodium tripolyphosphate (ABCR), Aresenazo III (2,7-Bis(2-arsonophenylazo)-1,8-dihydroxynaphthalene-3,6-disulfonic acid disodium salt) (Fluka), and HCl, CH_3COOH, HNO_3, KBr and KNO_3 (Merck) were the chemicals used in the prescriptions. All were analytical grade and used without any further purification. The resistivity of distilled water was about 18 MΩ cm^{-1} at room temperature.

Clinoptilolite obtained from Central North Anatolian occurrences associated with Eocene submarine volcanism was in the composition of 71.89% SiO_2, 15.16% Al_2O_3, 6.51% CaO, 1.80% Fe_2O_3, 1.80% MgO, 1.06% Na_2O, and 0.59% SrO. The mineral was ≈90% clinoptilolite, as clinoptilolite $\{(Na,K)_6.(Al_6\ Si_{30}O_{72}).24H_2O\}$ and mordenite $\{Na_3KCa_2.(Al_8\ Si_{40}O_{96}).28H2O\}$, 5% quartz, 5% feldspar, and smectite in trace level. The Clinoptilolite content was consistent with its SiO_2/Al_2O_3 equal to 4.7 (Godelitsas and Armbruster, 2003).

2.2. Preparation of the Composite

The procedure provided by Dragan et al. (2010) was modified to prepare ChZ (Dragan et al., 2010). In obtaining 4 g of ChZ, 2 g of Z was suspended in 80 mL viscous solution containing 2% of acetic acid and 2 g of chitosan preconditioned by intense stirring for 24 h.

After addition of 80 mL of 0.01 M ephycholorohydrin solution with pH adjusted to 10, the mixture was additionally stirred for 2 h. The fractions of mixture withdrawn in a syringe were transferred in 250 mL of 0.05 M sodium tripolyphosphate under mild magnetic stirring as droplets. Each droplet instantly transformed into a spherical composite bead. The solution containing beads was stirred for another 3 h at 10 rpm. The ChZ beads were separated by decantation and washed several times with distilled water to remove unreacted chemicals, dried at 40^0C, ground and sieved to a mesh size 50.

The procedure for preparation of 3 g of PAAmChZ was as follows; after 1 g of ChZ was suspended in 10 mL of water under a vigorous magnetic stirring, 20 mL of a solution containing 2 g of AAm (monomer) and 0.2 g of N,N'-methylenebisacrylamide (cross-linking agent) was added. While the suspension was stirred at a rate to prevent ChZ precipitation and to provide homogeneous dispersion, the polymerization was started by addition of 1 mL ammonium peroxydisulfate (100 mg mL^{-1}) solution as initiator and 0.5 mL of N,N,N',N'-tetramethylenediamine as activator. The formed PAAmChZ was immediately dispersed in 100 mL of water by passing the composite throughout a syringe system (barrel and plunger).

The dispersion was waited for about 1 h with slight stirrings and the aqueous phase was decanted, the washing and decantation procedure was repeated to remove unreacted chemicals. The composites were spread out plastic trays and dried at ambient temperature, ground, and sieved to 50 mesh.

2.3. Characterization

FT-IR spectra were taken using a FT-IR spectrophotometer (Unicam, Mattson 1000) with 20 scans at a resolution of 4 cm^{-1}. Rigaku Ultima IV was employed for XRD patterns at 2θ/min scan rate using Cu K_α radiation. Nitrogen adsorption/de-sorption isotherms were performed at 77 K with AUTOSORB-6B (Quantachrome Instruments) apparatus. The samples were degassed about 24 h at 323 K prior to the measurements. BET and BJH methods were used to obtain the specific surface area, pore diameters and volumes. SEM views of gold-coated samples were obtained from TESCAN MIRA3 XMU Field Emission SEM with an accelerating voltage of 10 kV. TGA curves obtained for the mass losses at temperatures within the range of 20–600^0C at 10^0C min^{-1} heating rate under static air with Shimadzu DT-50.

Swelling dependence on time was investigated. 0.150 g of each dried materials was let to swell in water or 0.25 M KNO_3 and weighed with time intervals. Univalent cation exchange capacities (CEC) of the composite and its components were determined by ammonium acetate method (Chapman, 1965).

The point of zero charge (pH_{PZC}) of the composites and its associates was measured in presence of 0.1M KNO_3 as an inert electrolyte (Mullet et al., 1999). 0.1 g of samples was suspended in 10 mL of KNO_3 solutions with adjusted initial pHs (pHi) ranging from 1 to 12 by using 0.1 M of KOH or HNO_3. The equilibrium pHs (pH_e) was measured after 24 h solid solution interaction.

2.4. Adsorptive Features for Th^{4+}

The dependence of Th^{4+} adsorption on PAAmChZ to pH, concentration, time (kinetics), temperature (thermodynamic) and ionic intensity were examined. An Arsenazo III (Az)-Th^{4+} complex formation based spectrophotometric procedure was followed for Th^{4+} measurements (Khan et al., 2001); 50 µL of equilibrium solution transferred in 3 mL of complexation solution containing %0.04 Az in 3 M $HClO_4$. Absorbency of the formed complex with reference to the pure Az solution at 658 nm were recorded (Shimadzu-160A spectrophotometer).

Linear standard calibration curves obtained from absorbency of 25-200 ppm Th^{4+} solutions were used to determine equilibrium concentrations obtained from the experiments. With exception of the kinetics study, the interaction time for equilibrium was 24 h as a convention. Out of the thermodynamic study, all experiments were carried out at 298 K. Effect of the concentration change on Th^{4+} adsorption was studied for the range of 25-2000 ppm (1.1×10^{-4}-8.6×10^{-3} M). The studied concentration was constant at 2000 ppm (8.6×10^{-3} M) in all other adsorption tests. The initial pH range for its effect on adsorption was 0.5-4. The adsorption kinetics was examined by time interval sampling (50 µL) from the test solution during 24 h adsorbent-solution interaction. Ionic intensity impact on Th^{4+} adsorption was performed for Th^{4+} solution in 0.01-1.5 M KNO_3.

3. Results and Discussion

3.1. Characterization

3.1.1. Porosity, Surface Charge and Cation Exchange Capacity

The composite and its constituents were meso-porous structure according to the IUPAC classification, since the size of pore diameters were in the range of 2-50 nm (Table 1) (Yang, 2003). The magnitudes of BET and pore volume of the composite was considerably lower than that of mineral. This was attributed to the infiltration and filling of PAAm into the mineral pores after the polymerization as seen from the morphologic changes in SEM views of the composite (Figure. 1).

The PZC (point of zero charge) is the pH at which the surface of a solid adsorbent immersed in an electrolyte has zero charge referring the surface charge switches from positive at pH < PZC to negative at pH > PZC. The data obtained from PZC experiments provided PZC values and buffering features of the materials calculated from the linear equations derived from ΔpH (pH_e-pH_i) vs. pH_i (Figure 2) and the buffering pH ranges were obtained from the plateau region of pH_e vs. pH_i (inset). The PZC and CEC values of the composite and its associates were submitted in Table 2.

The CEC of ChZ was considerably higher than its components that attributable to the formation of additional active centers after ECH and tripolyphosphate insertion as the linkers. Comparison of the CEC values experimentally determined or that expected with reference to ChZ content of PAAmChZ with the value of ChZ signified that the entrapment of ChZ in PAAm lowered the CEC of PAAmChZ. This should be associated with the involvement of PAAm such that PAAm partially blocked the active centers by filling up the cages and holes present in the nature of mineralogical structure of Z.

As seen from Table 2 that PZC values of ChZ (PZC = 4.8 and CEC = 5.0) was noticeably different from its constituents (PZC = 7.1 and 7.7 for Z and Ch respectively) and from PAAm (PZC = 6.5) because of the ECH and tripolyphosphate involvement as discussed above. Contrarily, ChZ entrapped in PAAm (PAAmChZ) had lower CEC (PZC at 6.7) than its bare forms had. This might be attributed the partial blockage of attractive centers for the hydronium or ammonium ions by inter moleculer interactions, e.g., hydrogen bonding, between Ch and Z, and/or acrylamide molecules in the course of polymerization procedure.

The amphoteric pH range of the composite and its components were about 4-10. By considering the changes in the values of CEC and PZC, it could be foreseen that the use of PAAm as an inert hydrogel for the encapsulation caused significant changes in the physicochemical features of ChZ despite the fact that PAAm had a PZC at 6.5 with no buffering feature and its CEC was negligible.

Table 1. Physical characteristics of PAAmChZ and its components

Material	BET/m^2g^{-1}	Pore volume/cm^3g^{-1}	Pore diameter/nm
Z	10.4	0.07	3.8
PAAm	3.5	0.02	2.2
Ch	4.1	0.01	3.0
PAAmChZ	2.2	0.01	2.7

Figure 1. SEM appearance of Z with a view of agglomerations of mineral grains, Ch and PAAm are typically amorphous, and PAAmChZ is a combination of amorphous and mineral structures.

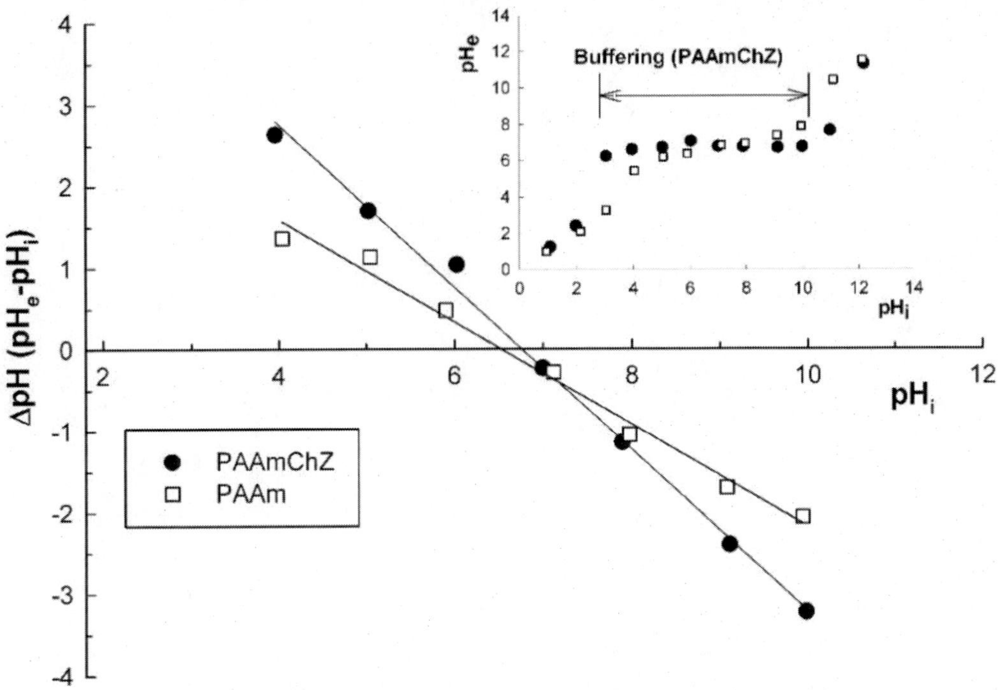

Figure 2. PZC and buffering range determination by pH_i vs ΔpH and pH_i vs pH_e depictions.

Table 2. CEC, PZC and buffering features of composite and its constituents

Material	CEC (mol NH_4^+/kg)	PZC	R^2	Buffering pH range
Z	2.8	7.1	0.986	4-9
Ch	3.4	7.7	0.998	4-10
ChZ	5.0	4.8	0.993	4-10
PAAm	-*	6.5	0.988	none
PAAmChZ	1.2	6.7	0.996	3-10

*PAAm had a negligible CEC.

3.1.2. FT-IR and XRD Analysis

FT-IR spectra of Ch, PAAm and Z were compared with those of PAAmChZ before and after Th^{4+} adsorption in Figure 3. The stretches were of N-acethylglucoseamine (1321 cm^{-1}), N-H and C = O in amides (1562 and 1661 cm^{-1}), -OH (3290 and 3430 cm^{-1}) and –NH (3500 cm^{-1}) in Ch spectrum (Wang et al., 2008; Humelnicu et al., 2011; Wan Ngah et al., 2011) PAAm was characterized by the peaks associated with C-N (1192 cm^{-1}), C-C (1323 cm^{-1}), $-CH_2$ (1452 cm^{-1}), $-CONH_2$ (1670 cm^{-1}), C-H in $-CONH_2$ (1570 and 1720 cm^{-1}), and N-H in $-CONH_2$ (3185 and 3450 cm^{-1}) (Cheim et al., 2005; Adhikary et al., 2007; Baybaş, 2009).

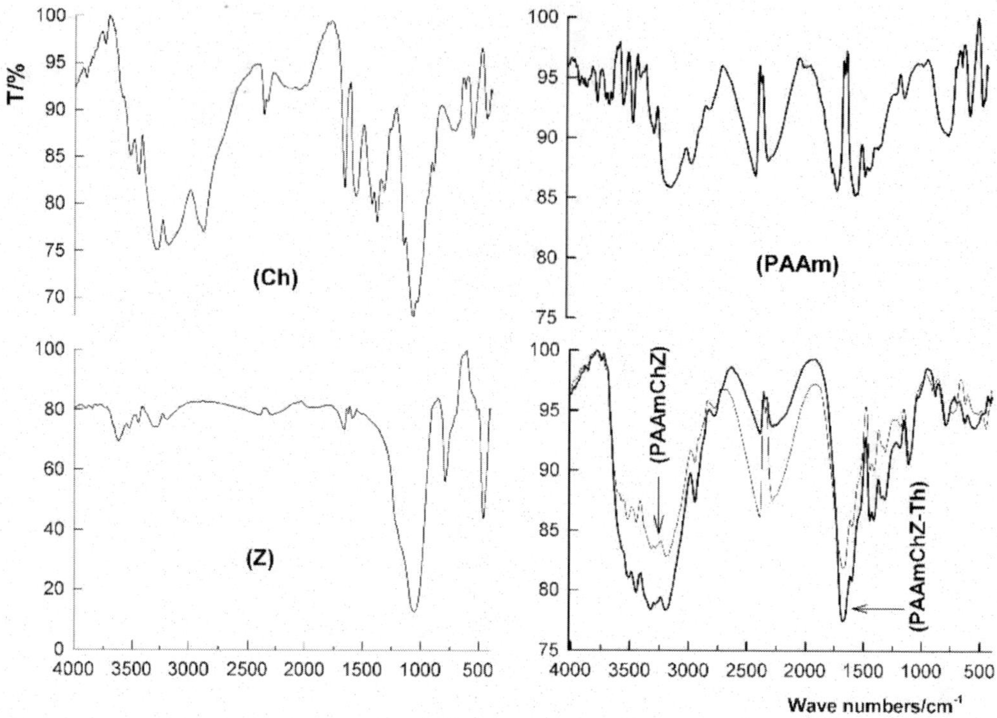

Figure 3. Comparison of FT-IR spectra of Ch, PAAm and Z with the composite before and after Th^{4+} adsorption.

The vibrations ascribed to Al-O (460 cm^{-1}), Si-O-Al (800 cm^{-1}) and Si-O-Si (1060 cm^{-1}) were of evidences defining clinoptililite (Z) as an aluminoslicate mineral (Doula and Ioannou, 2003; Doula, 2007 and Anirudhan et al., 2008).

Figure 4 shows the XRD patterns of the composite and its associates. The maximal reflections The reflections at 2θ = 9.1^0 and 22.2^0 in Ch were associated with two typical crystallite formations of Ch in semi-crystallite structure; the hydrated crystals of low crystallinity (Type I) and the relatively regular high crystallinity (Type II) in respective order (Huang et al., 2011). XRD pattern of pure PAAm defined a typical noncrystalline (amorphous) structure.

The reflections at 2θ = 9.7^0(d_{020}), 22.2^0 (d_{004}) and 25.6^0 were of clues characterizing clinoptilolite (Z) (Doula, 2007), whilst the presence of these peaks with no shift in PAAmChZ pattern was the proof of Z embodied in the composites without deformation in crystallite structure. (Baybaş and Ulusoy, 2011) This suggested that the constitution of PAAmChZ was a phase separated microcomposite (Alexandre and Dubois, 2000). The increased background in the patterns of composites was due to the irregular reflections arising from amorphous structure of PAAm and Ch.

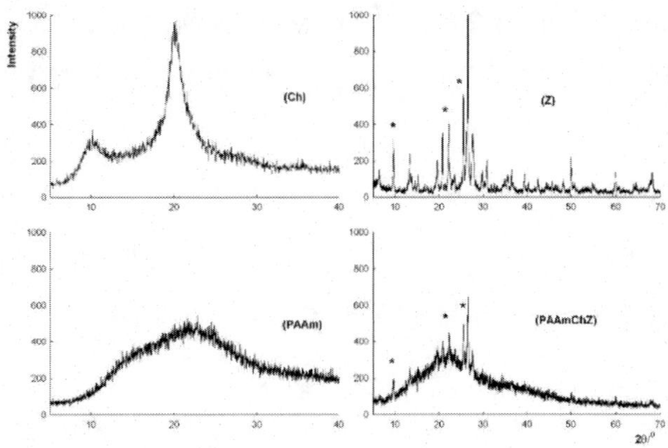

Figure 4. XRD patterns of the composite and its components (The reflections labeled with asterisks characterized Z as clinoptilolite).

3.1.3. Thermal Stability

The results obtained from the TGA and DTG analysis of the materials were submitted in Figure 5 and Table 3.

Zeolite had about 4-5% lost at 25-100^{0}C mainly resulting from the expulsion of humidity; water molecules weakly bound to surface of the mineral framework whereas the losses at higher temperatures were associated with the removal of water formed from the decomposition of indigenous hydroxyl groups (Doula, 2007).

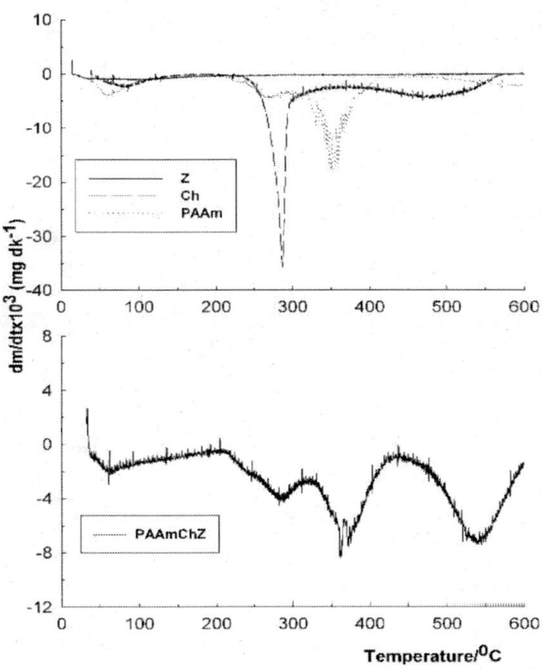

Figure 5. DTG patterns of the composite and its associates.

Table 3. Percentage weight lost of the composite and its components between various temperature ranges (The summed provided the total lost between 25-600^{0}C)

Material	25-200^{0}C	200-400^{0}C	400-600^{0}C	25-600^{0}C
Z	6.6 (67)[1]	2.3	0.8	9.7
Ch	8.3	63.7 (273)	24.7 (460)	96.1 (364)[2]
PAAm	11.9	52.4 (268, 353)	13.3 (596)	77.4 (300)
PAAm-Ch-Z	11.1	42.8 (285, 361)	33.4 (540)	87.3 (407)

[1] Characteristic DTG peaks at which the maximum lost was observed.

[2] Temperature by which half of the weight was lost.

The removal of moisture in PAAm and Ch also completed by 200^{0}C. As suggested by Wang et al., the highest mass lost was in the range of 200-400^{0}C due to the degradation and deacetylation whilst the other significant amounts observed at 400-600^{0}C was attributed to oxidative degradation of the carbonaceous residues (Wang et al., 2005). The characteristic peaks in PAAm were related to the emissions of H_2O, CO_2 and NH_3 after decomposition of amide groups at 200–350^{0}C, the losses of H_2O, CO_2, imides and nitriles after the main chain breakdown at 350–450^{0}C, and the releases during carbonization of the polymer at 450–600^{0}C (Baybaş and Ulusoy, 2011).

3.1.4. Swelling Features

The swelling features of the composite were tested to propose the kinetics and mechanism of swelling due to its importance in the practical implementation in biomedicine, pharmacy and environment. To accomplish this, the swelling characteristics and kinetics of the composite for distilled water and 0.25 M KNO_3 solution was investigated. Swelling characteristics of the composite and its associates were provided as water absorption capacity (S_e), Swelling% and water contents at equilibrium (S_{WCE}) in Table 4.

Table 4. Swelling characteristics of the composite and its associates (Water absorption capacity, $S_e = (m_{max}-m_0)/m_0$; Swelling% = (m_{max} x 100/m_0) and water contents at equilibrium $S_{WCE}\% = (m_{max}-m_0) \times 100/m_{max}$ with reference to weights of dry and swollen materials; m_0 and m_{max})

Material	S_e (g_{H2O}/g_{mat})	Swelling (%)	S_{WCE} (%)
Z	1.0	200	50
Ch	6.0	700	86
PAAm	8.1	910	89
PAAm-Ch-Z (17%Z,17%Ch, 66% PAAm)[1]	7.0 (6.5)[2]	800 (755)[2]	87 (82)[2]

[1] Constituent contents of the composite, [2]Expected values of the parameters calculated by taking into account of each constituent's contribution, e.g., (0.17 x 200 + 0.17 x 700 + 0.66 x 910 = 755%).

PAAmChZ had high water absorbency because of its constituents' contribution to the imbibing capacity. Moreover, the measured parameters for the composite were higher the calculated ones by considering the contribution of the feature of each constituent as much as its fraction in the composite. This should be ascribed to formation of a new synergetic texture with higher water affinity after the formation of PAAmChZ.

The parameters associated with the kinetics were derived from the compatibility of data to the linear forms of Schott's second-order and the exponential heuristic equations (Figure 6);

$$t/S_t = 1/(k_s S_m^2) + t/S_m \qquad (1)$$

$$S_t/S_m = F(t)^n \text{ or } \ln(S_t/S_m) = \ln(F) + n.\ln(t) \qquad (2)$$

where 'S_t 'and S_m are the amou nt of adsorbed water at time 't (min)' and the amount reached maximum (g g_{com}^{-1}), k_s is the Schott's rate constant (g_{comp} $g^{-1}dk^{-1}$), $1/(k_s S_m^2)$ is also defined as inverse initial adsorption rate ($1/r_0$). 'F' is a constant characterizing the hydrogel network and 'n' is diffusion constant peculiar to water transfer/swelling mechanism; the process controlled by solvent diffusion (Fickian) if $n < 0.5$, by solvent diffusion and the relaxation of hydrogel network (non Ficikian) if '$0.5 < n < 1$' or by relaxation of hydrogel if '$n > 1$' (Wang and Wang, 2009; Sun et al., 2013)

The results (Table 5) clearly showed that the experimentally obtained swelling data were well compatible to both Schott and Fick models with high significance ($R^2 > 0.99$). The compedance Schott's model was persistence with agreement between the values of S_e (7.0 g.g_{comp}^{-1}) in Table 4 and S_m (7.1 g.g_{comp}^{-1}) in Table 5.

When the values of parameters achieved for the composite in water and KNO_3 solution were compared, the higher value of S_m in KNO_3 solution might be associated with the osmotic pressure causing much water diffusion towards to the ChZ network embodied in PAAm. The initial rate of this process (r_0) realized in KNO_3 was slightly higher than that realized in water whereas the Schott's rate constant (k_s) for the composite in water was the higher.

These consequences might coherently indicate that the deriving force for the water diffusion was osmotic pressure and the rate limiting step was the diffusion of water into the composites. This should be consistent with the findings for exponents (n) to be smaller than 0.50 which confirmed that the overall swelling mechanism was Fickian, i.e., the process was diffusion controlled (Wang and Wang, 2009)

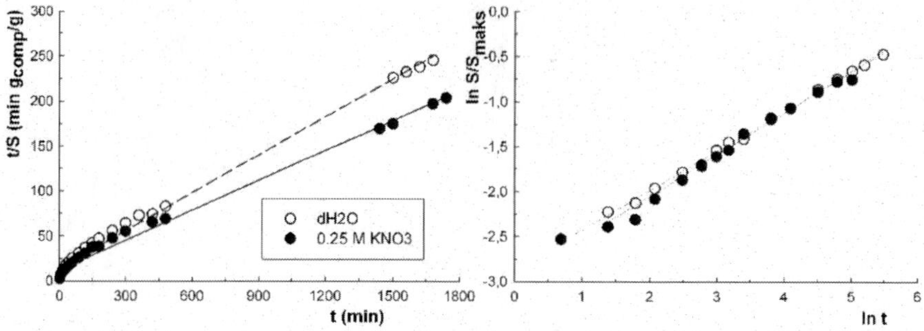

Figure 6. Compatibility of swelling kinetics of PAAmChZ in distilled H_2O and 0.25 M KNO_3 to Schott's second-order and exponential equations.

Table 5. Parameters obtained from swelling kinetics

Schott's Model	$PAAmChZ_{H2O}$	$PAAmChZ_{KNO3}$
k_s x 10^5/g_{comp} $g^{-1}s^{-1}$	2.5	1.7
S_m/g g_{comp}^{-1}	7.1 (6.9)*	9.0 (8.6)
r_0 x 10^3/g $g_{comp}^{-1}s^{-1}$	1.2	1.4
R^2	0.994	0.992
Fick model		
F x 10^3	9.4	7.6
N	0.44	0.46
R^2	0.998	0.987

* Experimentally obtained values of water absorption at equilibrium.

3.2. Adsorptive Features of PAAmChZ for Th^{4+} Adsorption

3.2.1. Data Assessment

The nomenclature and units used in the equations was as follows (Table 6).

Table 6. Nomenclature and units for the data processing of Th^{4+} adsorption

a_S: Sips isotherm model constant	**W**: Mass of adsorbent/kg
$C_{i,e}$: Th^{4+} concentrations at initial and equilibrium/mol L^{-1}	**X_{DR}**: D-R constant related to sorption capacity
E_{DR}: Free energy change/J mol^{-1}	**X_F**: Freundlich constant related to adsorption capacity
H: Initial adsorption rate/mol^{-1} kg min^{-1}	**X_L**: Langmuir monolayer sorption capacity/mol kg^{-1}
k, k_i: Rate constants/mol^{-1} kg min^{-1} and mol kg^{-1} $min^{-0.5}$	**β**: Freundlich constant (surface heterogeneity)
K_{DR}: Constant in DR model/mol^2 K J^{-2}	**$β_S$**: Sips isotherm model exponent
K_L: Langmuir adsorption equilibrium constant/L mol^{-1}	**ΔG**: Change in adsorption free energy/kJ mol^{-1}
	ΔH: Change in adsorption enthalpy/kJ mol^{-1}
K_s: Sips isotherm model constant	**ΔS**: Change in adsorption entropy (J mol^{-1} K^{-1})
	ε: Polanyi potential

Q: Adsorbed amounts/mol kg^{-1} at time t or equilibrium
R: Ideal gas constant (8.314 J mol^{-1} K^{-1})
T: Absolute temperature/K
$t_{1/2}$: Time required for adsorption of half of the concentration/min
V: Solution volume/L
Θe: Surface coverage ratio

The adsorbed amounts of Th^{4+} was calculated by using Eq. 3

$$Q = \frac{(C_i - C_e)V}{W} \tag{3}$$

Adsorption parameters were derived from the compatibility of experimentally obtained adsorption isotherms to Langmuir, Freundlich, Sips and Dubinin-Radushkevich (D-R) models expressed by the following equations (Foo and Hameed, 2010).

$$Q = \frac{K_L X_L C_e}{1 + KC_e} \text{ Langmuir} \tag{4}$$

$$Q = X_F C_e^{\beta} \text{ Freundlich} \tag{5}$$

$$Q = \frac{X_s a_s C_e^{\beta s}}{1 + a_s C_e^{\beta s}} \text{ Sips} \tag{6}$$

$$Q = X_{DR} e^{-K_{DR}\varepsilon^2}\ ;\ \varepsilon = RT \ln\left(1 + \frac{1}{C_e}\right) \text{ D-R} \tag{7}$$

$$E_{DR} = (-2K_{DR})^{\frac{1}{2}} \text{ D-R} \tag{8}$$

The following equation was employed to calculate filling ratio of adsorptive sites defined by the Langmuir model (Liu and Shen, 2008);

$$\Theta_e = \frac{K_L C_i}{1 + K_L C_i} \tag{9}$$

Thermodynamics parameters were ascertained from the values of distribution coefficients (K) at studied temperatures. By the depictions of 'ln K vs 1/T' to configure van t'Hoff equation provided magnitude of ΔH and ΔS followed by ΔG.

$$K = \frac{Q}{C_e} \tag{10}$$

$$\ln K = \frac{\Delta S}{R} - \frac{\Delta H}{R}\frac{1}{T} \tag{11}$$

$$\Delta G = \Delta H - T\Delta S \tag{12}$$

The kinetics of Th^{4+} adsorption was inspected by the compatibility of time dependent adsorption to pseudo first and second-order, and intra particle diffusion models (**Eq. 13-15**). The obtained parameters for first- and second-order models were forwarded to calculate initial adsorption rate (H) and the time elapsed to reach half of the Th^{4+} adsorption at equilibrium; half life ($t_{1/2}$) (Ho and McKay, 1998 and Basha and Murthy, 2007).

$$Q_t = Q_{e1}\left(1 - e^{-k_1 t}\right) \quad H_1 = k_1 Q_{e1} \quad \left(t_{1/2}\right)_1 = \frac{\ln 2}{k_1} \qquad \text{Eq. 13}$$

$$\frac{t}{Q_t} = \frac{1}{kQ_{e2}^2} + \frac{t}{Q_{e2}} \quad H_2 = k_2 Q_{e2}^2 \quad \left(t_{1/2}\right)_2 = \frac{1}{k_2 Q_{e2}} \qquad \text{Eq. 14}$$

$$Q_t = k_i t^{1/2} \qquad \text{Eq. 15}$$

3.2.2. Adsorption Dependence on pH and Ionic Intensity

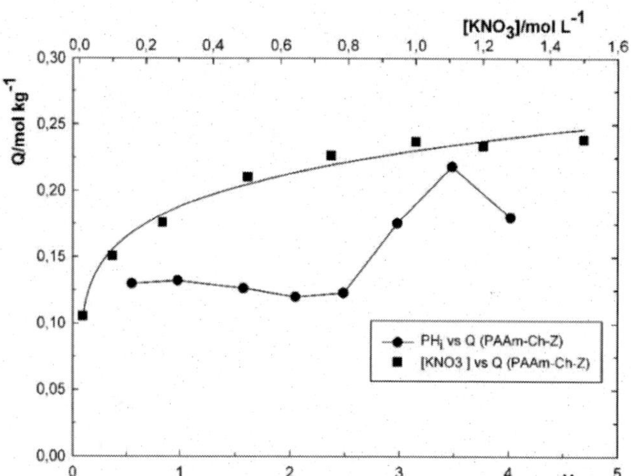

Figure 7. Effect of pH and ionic intensity on Th^{4+} adsorption.

As illustrated in Figure 7, Th^{4+} adsorption increased as the ionic strength of medium ascended such that the adsorption enhanced from 0.10 mol kg^{-1} (23 mg g^{-1}) from Th^{4+} solution containing no additional KNO_3 to 0.25 mol kg^{-1} (58 mg g^{-1}) from Th^{4+} solution in 1.0 M KNO_3. The initial pH of Th^{4+} solution was about 4 and it was not varied with the KNO_3 additions. The variation in adsorption by initial pH was negligible for pH at 0.5-2.5 (0.12 mol kg^{-1}) and increased to reach a maximal at pH = 4 (0.22 mol kg^{-1}). The studied initial pH range was 0.5-4.5 since the solubility of Th^{4+} salts after pH = 4.5 is lowered. The most dominant Th species in solutions at this pH range were Th^{4+} and $ThOH^{3+}$. Beside this, the active sites of the composite was mostly protonated form since solution pH < PZC (6.9); see Section 2.1.1. By considering the adsorbate species together with the surface charge, the adsorption should be governed by an ion exchange mechanism for which the following mechanisms might be foreseen (Moulin et al., 2001; Ekberg et al., 2000; Ulusoy and Baybaş 2011);

$$4(\equiv SOH) + Th^{4+} \leftrightarrow (\equiv SO)_4Th + 4H^+$$
$$3(\equiv SO^-) + (ThOH)^{3+} \leftrightarrow (\equiv SO^-)_3ThOH$$

3.2.3. The Adsorption Isotherms

The compatibility of experimentally obtained isotherm to Langmuir, Ferundlich, Sips and Dubinin-Radushkevich models by referring Eq. 3-6 and the parameters derived from the conformities were provided in Figure 8 and Table 6.

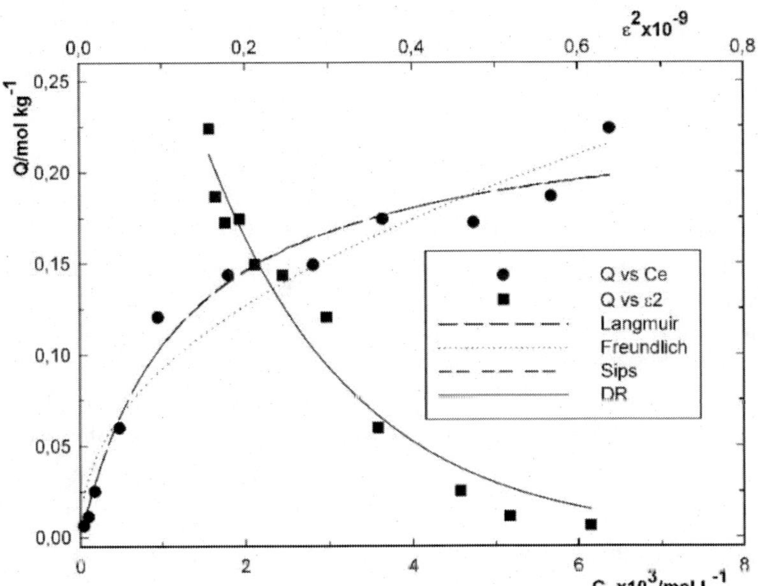

Figure 8. Fitness of Th^{4+} adsorption to Langmuir, Freundlich and Sips (C_e vs Q), and DR models (ε^2 vs Q).

Table 6. Adsorption parameters obtained from the isotherm models

Langmuir		**Sips**	
X_L/mol kg^{-1}	0.24 (0.72)*	X_s	0.23
K_L/L mol^{-1}	794	a_s	947
R^2	0.973	β_s	1.02
		R^2	0.973
Freundlich		**D-R**	
X_F	2.2	X_{DR}	0.5
B	0.46	$-K_{DR}$ x 10^9	5.7
R^2	0.947	E/kJ mol^{-1}	9.3
		R^2	0.963

* The extended value adsorption with reference to ChZ content in PAAmChZ (as a mass ratio of 1/3) because the inert feature of PAAm for Th^{4+} adsorption (Baybaş and Ulusoy, 2011a).

In accordance with the IUPAC physisorption classifications (Sing, 1982), the isotherm was Type-I indicating that PAAmChZ had high affinity to Th^{4+} ions. The values of coefficient of variation (R^2) implied that Langmuir and Sips models were best fitted to experimentally obtained adsorption for Th^{4+} to PAAmChZ.

This remarked that Langmuir model was the best description of the adsorption phenomenon such that the adsorptive centers available for the Th^{4+} ions were energetically equivalent and homogeneously distributed over the composite surface and monolayer adsorption capacity is defined by the amount of Th^{4+} required complete filling of these centers. For this study, the β_s (Sips model' exponent) was very close to one ($\beta_s = 1.02$). This stated the transformation of Sips equation back to Langmuir equation see Eq 4 and 6 (Foo and Hameed, 2010).

By using Eq. 9 (Liu and Shen, 2008) for hypothetic solutions containing 100, 1000 and 2000 mg L^{-1} Th^{4+}, the θ_e values were calculated to be 0.254, 0.773 and 0.872 respectively, predicating that the θ_e eventually reaches to 1 at $C_{i\infty}$ to provide K_L and X_L.

Adsorption free energy derived from the D-R model (E_{DR} = 9.3 kJ mol^{-1} as E_{DR} < 20 kJ mol^{-1}) verified that the physicochemical nature of the sorption was coulomb electrostatic interactions between negatively charged terminals on PAAmChZ and Th^{4+} ions. In brief, the adsorption occurred via physisorption process including ion exchange and/or complex formation as stated in Section 3.2.2 (Atkins and de Paula, 2006; Itodo and Itodo, 2010).

Langmuir monolayer Th^{4+} adsorption capacities obtained for PAAmChZ and for the adsorbents extracted from the literature was provided in Table 7. As seen, the adsorption capacity of PAAmChZ was amongst the ones having high values presented in the Table.

Table 7. Langmuir monolayer Th^{4+} sorption capacities of different adsorbents

Adsorbent	X_L (mol kg^{-1})	Reference
Activated carbon, natural and synthetic zeolites	0.24; 0.30; 0.71	(Metaxas et al., 2003)
Immobilized tannins onto collagen fiber	0.32; 0.08	(Liao et al., 2004)
Oxidized multi-wall carbon nanotubes	0.07	(Chen et al., 2007)
γ-Al_2O_3, γ-Al_2O_3 + HA, γ-Al_2O_3 + FA	0.06; 0.18; 0.25	(Chen and Wang, 2007)
TiO_2	0.22	(Tan et al., 2007)
Al-pillared rectorite	0.13	(Yu et al., 2008)
MX-80 bentonite	0.28	(Zhao et al., 2008)
Attapulgite	0.07	(Chen and Gao, 2009)
Moroccan oil shale-phosphoric acid	1.96	(Khouya et al., 2010)
Adsorbent	X_L (mol kg^{-1})	Reference
PAAm-Montmorillonite	0.33	(Baybaş and Ulusoy, 2011)
Z; PAAm-Z	0.07; 0.21	
ZY; PAAm-ZY	0.74; 0.89	
PAAm-Hydroxyapatite	0.55	(Baybaş and Ulusoy, 2012)
Phosphate-Enhanced Chitin	0.22	(Gado and Morsy, 2016)
Ginkgo leaf	0.45	(Huang et al., 2018).
Amorphous silica	0.07	(Kaynar and Şabikoğlu, 2018)
PAAmChZ	0.72	This study

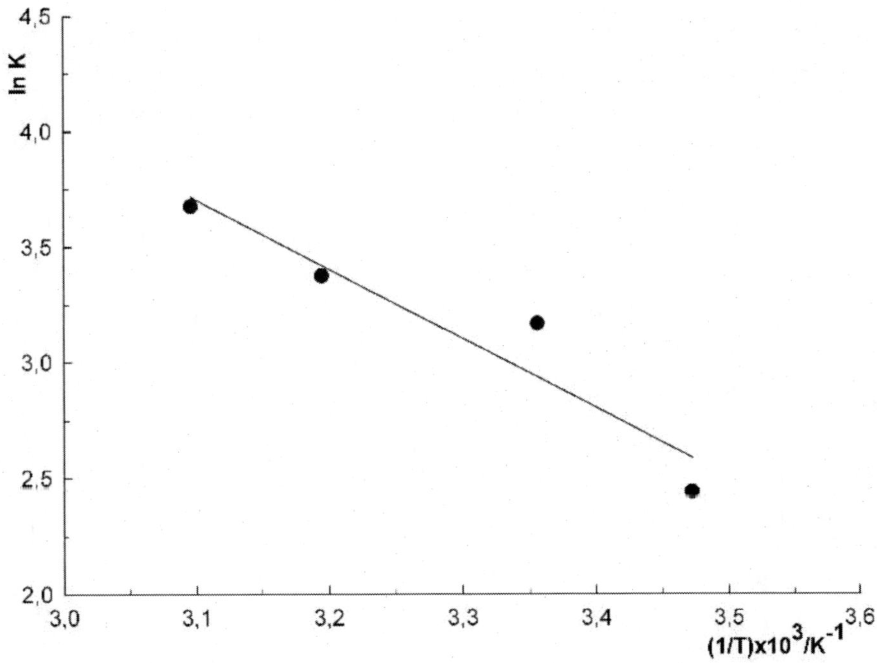

Figure 9. Temperature dependence of Th^{4+} adsorption onto PAAmChZ (T = 278, 288, 313 and 323 K).

Table 8. Thermodynamics of Th^{4+} adsorption onto PAAmChZ

ΔH/kJ mol^{-1}	ΔS/J molK^{-1}	ΔG/kJ mol^{-1} (T = 298K)
24.9	108	-9.1

The enthalpy and entropy changes were $\Delta H > 0$ and $\Delta S > 0$ for Th^{4+} sorption onto PAAmChZ. This signifies that the overall process was endothermic and the randomness throughout the adsorption process increased. Gibbs free enthalpy change was $\Delta G < 0$ confirming that the adsorption process was spontaneous under the entropy control ($T\Delta S > \Delta H$).

The changes appeared to be contrary to the adsorption phenomenon; the expected entropy and enthalpy changes should be negative because of the adsorbed molecules represents the transitions of regular and energetically stable states relative to their situations in solution. However, it should be stated that the determination of entropy and enthalpy changes special to the adsorbate-adsorbent system in a solution is almost impossible due to the involvement of multi coactions out of the system alone. The values discussed in here ultimately represent the changes in thermodynamic magnitudes throughout the whole process in the adsorption environment (Baybaş and Ulusoy, 2016).

3.2.4. The Adsorption Kinetics

Compatibility of experimentally obtained time dependent adsorption data to pseudo first/second-order rates and intra-particle diffusion models were delineated in Figure 10 and 11. The parameters calculated from the compatibility equations by referring Eq. 13-14 were tabulated in Table 9.

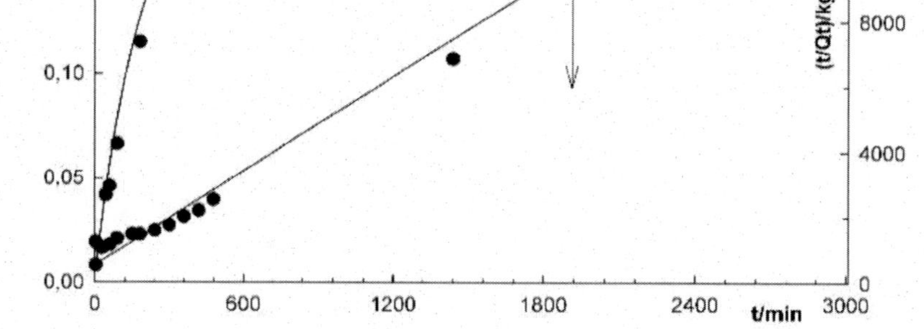

Figure 10. Fitness of Th^{4+} adsorption kinetics to pseudo first- and second-order (Q vs t and t/Q_t vs t) models.

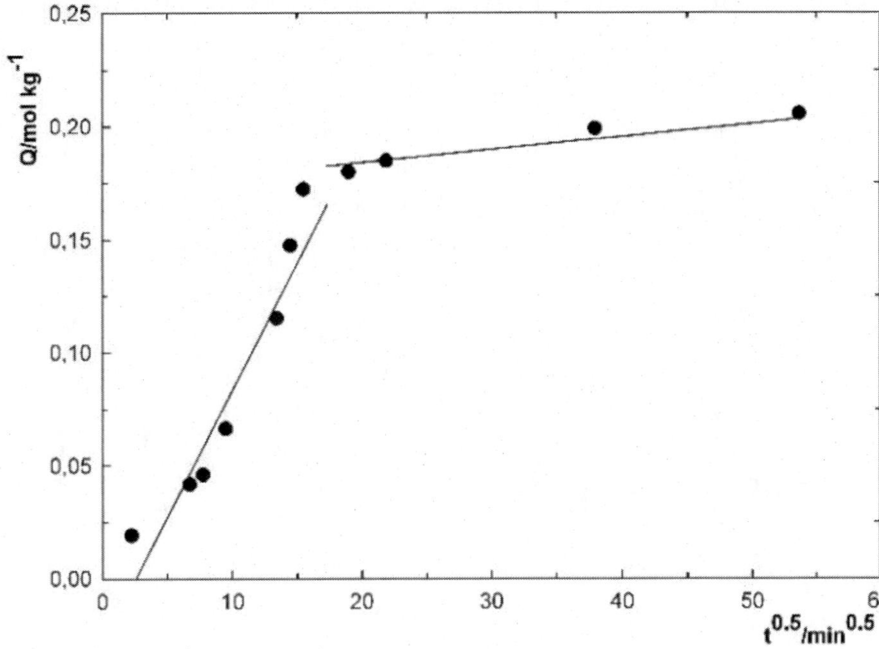

Figure 11. Fitness of Th^{4+} adsorption kinetics to intra particle diffusion model.

Table 9. Parameters of the adsorption kinetics

First order		Second order	
$k_1 \times 10^3$/min^{-1}	5.4	$k_2 \times 10^3$/mol^{-1} kg min^{-1}	45
Q_{M1}/mol kg^{-1}	0.20	Q_{M2}/mol kg^{-1}	0.21
Q_{exp}/mol kg^{-1}	0.21	R^2	0.992
R^2	0.973		
$H_1 \times 10^3$/mol kg^{-1}min^{-1}	0.24	$H_2 \times 10^3$/mol kg^{-1} min^{-1}	2.0
$t_{1/2}$/min	875	$t_{1/2}$/min	105
Weber-Morris			
$k_{ads} \times 10^3$/mol kg^{-1} min$^{0.5}$	12		
R^2	0.920		
$k_{idif} \times 10^3$/mol kg^{-1} min$^{0.5}$	0.84		
R^2	0.936		

As obviously seen from the table, the adsorption kinetics obeyed second-order model with a higher coefficient of variation (R^2 = 0.992) and the closeness of the adsorbed amounts at equilibrium obtained from the model (Q_{M2}) and from the experiment (Q_{exp}). Beside this, the $t_{1/2}$ (105 min) was in compliance with experimental observations (see figure 10, the value of $t_{1/2}$ derived from the first-order kinetics was unreasonably 875 min).

The evaluation of adsorption kinetics by Liu and Shen's approach (2008) also confirmed the obedience to the second order model since the ratio of kinetics constants found for the first and second order models (k_1/k_2 = 0.12) was significantly smaller than the filling ratio (θ_e = 0.872) calculated for the studied initial Th^{4+} concentration at 8.6 x 10^{-3} mol L^{-1} (2000 mg L^{-1}) by Eq 9. Well fitness to the second order model confirmed that the nature of adsorption was concentration dependent. Thus, the rate-controlling step was ion exchange and/or complex formation. This was also consistent with the results of intraparticle diffusion (Weber–Morris) model given by Eq. 15 and Figure 11. The plot did not provide an intercept at origin characterizing the adsorption process controlled by diffusion. But, the shape was curvature that could be evaluated in two linear parts. The initial part with a steep rise was associated with the initial rapid uptake by the boundary layer impact whereas a plateau like continuation characterized the intraparticle diffusion taking place after the initial uptake (Khambhaty et al., 2009). For the present study, the slope of linear line regarding the initial part (k_{ads}) was 10 times higher than that of second part (k_{idif}).

CONCLUSION

The composite of ChZ containing chitosan and zeolite was obtained by the aid of ephycholorohydrin and tripolyphosphate as cross-linking agents. PAAmChZ as a nanoparticle-hydrogel composite was then prepared by entrapment of ChZ in PAAm. Having had the PAAmChZ, the structural characteristics were investigated by FT-IR, XRD, SEM, TGA, BET-Porosity and point of zero charge (PZC) analysis. Swelling characteristics in view of capacity and kinetics were tested in water and in solution of KNO_3 0.25 M.

Adsorbent capability of the composite was tested for Th^{4+} in view of its dependency on pH, concentration, time (kinetics), temperature (thermodynamics) and ionic strength from which the associated parameters characterizing the adsorbent were obtained. Langmuir, Freundlich and Dubinin-Radushkevich were the models for the experimental isotherms, van't Hoff equation was for the thermodynamic, pseudo first order and pseudo second order, and Weber-Morris inrta-particle diffusion models for the kinetics evaluations.

The outcomes of FT-IR, XRD, SEM and PZC analysis confirmed the formation of the composite in ternary combination. Water imbibing kinetics obeyed Schott's second-order model the overall swelling mechanism was Fickian; the absorption process was diffusion controlled.

The investigation for adsorption dependency on pH showed that the adsorbed amount was increased with increasing pH. Isotherm acquired from the adsorption dependency on concentration was L type of Giles classification indicating that the adsorbent had high affinity to Th^{4+} ion. The parameters obtained from Langmuir, Freundlich and Dubinin-Radushkevich (D-R) models were proved the characteristics of this affinity. Adsorption free energy derived from the D-R model verified that the physicochemical nature of the sorption was Coulomb electrostatic interactions.

The values of enthalpy and entropy changes showed that the overall adsorption process had endothermic ($\Delta H > 0$) and increasing entropy ($\Delta S > 0$), and it was spontaneous ($\Delta G < 0$) as expected, and entropy controlled.

The adsorption kinetics indicated that the sorption process could be explained by the second order kinetics model. The fitness to the second order model and the provision of linear lines with two components from Weber-Morris model proved the facts that the adsorption process was concentration dependent and actualized via ion exchange/complex formation.

ACKNOWLEDGMENT

This investigation was supported by the Research Fund of Sivas Cumhuriyet University (CUBAP, Project no: F358) to which the authors are grateful.

REFERENCES

Adhikary, P., Tiwari, K. N. and Singh, R. P. (2007). Synthesis, characterization and flocculation characteristics of polyacrylamide-glycogen. *Journal of Applied Polymer Science,* 103: 773–778.

Alexandre M. and Dubois P. (2000). Polymer-layered silicate nanocomposites: preparations, properties and uses of a new class of materials. *Material Science and Engineering*, 28: 1-63.

Anirudhan, T. S., Suchithra, P. S. and Rijith, S. (2008). Amine–modified polyacrylamide–bentonite composite for the adsorption of humic acid in aqueous solutions. *Colloids and Surfaces A: Physicochemical Engineering Aspects*, 326: 147–156.

Atkins, P. and Paula, J. *Atkins' Physical Chemistry*, 8th ed. Oxford University Pres, GB, 2006, pp 916-917.

Avedesian, M. M., Baker, H. *ASM Specialty Handbook Mg and Mg Alloys.* ASM International, Materials Park, Ohio, 1999.

Basha, S. and Murthy, Z. V. P. (2007). Kinetic and equilibrium models for biosorption of Cr (VI) on chemically modified seaweed, Cystoseira indica. *Process Biochemistry*, 42: 1521–1529.

Baybaş Demet. "*The characterization of polyacrylamide-aluminosilicate composites and the investigations for their adsorptive features for thorium*" PhD thesis, Cumhuriyet University, 2009. https://tez.yok.gov.tr/UlusalTezMerkezi/tezSorguSonucYeni.jsp (245935)

Baybaş, D. and Ulusoy U. (2011a). Polyacrylamide-aluminosilicate composites for thorium adsorption, *Applied Clay Science,* 51: 138-146.

Baybaş, D. and Ulusoy, U. (2011b). Polyacrylamide–clinoptilolite/Y-zeolite composites: Characterization and adsorptive features for terbium. *Journal of Hazardous Materials*, 187: 241–249.

Baybaş, D. and Ulusoy, U. (2012) Polyacrylamide–hydroxyapatite composite: Preparation, characterization and adsorptive features for uranium and thorium. *Journal of Solid State Chemistry*, 194: 1–8.

Baybaş, D. and Ulusoy, U. (2016) Adsorptive features of polyacrylamide aluminosilicate composites for methylene Blue. *Turkish Journal of Chemistry.* 40: 147-162.

Chapman, H. D. "Cation-exchange capacity." In *Methods of Soil Analysis-Part 2. Chemical and Microbiological Properties. Agronomy Monograph* 9.2, edited by C. A. Black, 891-901. American Society of Agronomy, Soil Science Society of America. Inc., Madison, Wisconsin, 1965.

Chen, C. L. and Wang, X. K. (2007). Influence of pH, soil humic/fulvic acid, ionic strength and foreign ions on sorption of thorium (IV) onto γ-Al_2O_3. *Applied Geochemistry,* 22: 436-445.

Chen, C., Li, X., Zhao, D., Tan, X. and Wang, X. (2007). Adsorption kinetic, thermodynamic and desorption studies of Th(IV) on oxidized multi-wall carbon nanotubes. *Colloids Surfaces A-Physicochemical and Engineering Aspects*, 302: 449-454.

Chen, L. and Gao, X. (2009). Thermodynamic study of Th(IV) sorption on attapulgite. *Applied Radiation and Isotopes,* 67: 1–6.

Chiem, L. T., Huynh, L., Ralston, J. And Beattie, D. A. (2005). An in situ ATR-FTIR study of polyacrylamide adsorptionat the talc surface. *Journal of Colloid and Interface Science*, 297: 54-61.

Demirbas, A. (2008). Heavy metal adsorption onto agro-based waste materials: a review. *Journal of Hazardous Materials*, 157: 220–229.

Dimirkou, A. (2007). Uptake of Zn^{2+} ions by a fully iron-exchanged clinoptilolite. Case study of heavily contaminated drinking water samples. *Water Research*, 41: 2763–2773.

Dragan, E. S., Dinu, M. V. and Timpu, D. (2010). Preparation and characterization of novel composites based on chitosan and clinoptilolite with enhanced adsorption properties for Cu^{2+}. *Bioresource Technology*, 101: 812–817.

Doula, M. K. (2007). Synthesis of a clinoptilolite–Fe system with high Cu sorption capacity. *Chemosphere*, 67: 731–740.

Doula, M. K. and Ioannou, A. (2003). The effect of electrolyte anion on Cu adsorption–desorption by clinoptilolite. *Microporous and Mesoporous Materials*, 58: 115–130.

Dragan, E. S., Dinu, M. V. and Timpu D. (2010). Preparation and characterization of novel composites based on chitosan and clinoptilolite with enhanced adsorption properties for Cu^{2+}. *Bioresource Technology*, 101: 812–817.

Ekberg, C., Albinsson, Y., Comarmond, M. J. and Brown, P. L. (2000). Studies on the Complexation Behavior Thorium (IV). 1. Hydrolysis Equilibria. *Journal of Solution Chemistry*, 29(1): 63-86.

Foo, K. Y. and Hameed, B. H. (2010). Insights into the modelling of adsorption isotherm system. *Chemical Engineering Journal*, 156: 2-10.

Gado, M. A. and Morsy A. M. A. (2016). Thorium adsorption from waste effluents by phosphate-enhanced chitin radiochemistry. *Radiochemistry,* 59(5): 500–506.

Godelitsas, Ath. and Armbruster, Th. (2003). HEU-type zeolites modified by transition elements and lead. *Microporous and Mesoporous Materials,* 61: 3-24.

Gupta, V. K., Carrott, P. J. M., Ribeiro Carrott, M. M. L. and Suhas. (2009). Low-cost adsorbents: growing approach to wastewater treatment—a review. *Critical Review Environmental Science and Technology,* 39: 783–842.

Ho, Y. S. and McKay, G. A. (1998). Comparison of chemisorption kinetic models applied to pollutant removal on various sorbents. *Process Safety and Environmental Protect,* 76B: 332–40.

Huang, X.-Y., Bu, H.-T., Jiang, G.-B. and Zeng, M.-H. (2011). Cross-linked succinyl chitosan as an adsorbent for the removal of Methylene Blue from aqueous solution. *International Journal of Biological Macromolecules*. 49: 643– 651.

Huang, Y., Hu, Y., Chen, L., Yang, T., Huang, H., Shi, R., Lu, P. and Zhong, C. (2018) Selective biosorption of thorium (IV) from aqueous solutions by ginkgo leaf. *PLOS ONE,* 13(3): https://doi.org/10.1371/journal.pone.0193659.

Humelnicu, D., Dinu, M. V. and Dragan, E. C. (2011). Adsorption characteristics of UO_2^{2+} and Th^{4+} ions from simulated radioactive solutions onto chitosan/clinoptilolite sorbents. *Journal of Hazardous Materials,* 185(1): 447-55.

Itodo, A. U. and Itodo, H. U. (2010). Sorption energies estimation using Dubinin-Radushkevich and Temkin adsorption isotherms, *Life Science Journal*, 7(4): 31-39.

Kaynar, Ü. H. and Şabikoğlu I. (2018). Adsorption of thorium (IV) by amorphous silica; response surface modelling and optimization, *Journal of Radioanalytical and Nuclear Chemistry*, 318: 823–834.

Khambhaty, Y., Mody, K., Basha, S. and Jha, B. (2009). Kinetics, equilibrium and thermodynamic studies on biosorption of hexavalent chromium by dead fungal biomass of marine Aspergillus niger, *Chemical Engineering Journal,* 145: 489–495.

Khan, M., Ali, A. and Khan, N. (2001). Spectrophotometric determination of thorium with disodium salt of Arsenazo-III in perchloric acid. J*ournal of Radioanalytical and Nuclear Chemistry,* 250(2): 353–357.

Khouya, El H., Legrouri, K., Fakhi, S. and Hannache, H. (2010). Adsorption of uranium and thorium on new adsorbent prepared from Moroccan oil shale impregnated with phosphoric acid. *Nature Precedings*; http://hdl.handle.net/10101/npre.2010.5098.1. or <precedings.nature.com/documents/1526/version/1>

Kyzas G. Z. and Bikiaris D. N. (2015). Recent modifications of chitosan for adsorption applications: A critical and systematic review. *Marine Drugs,* 13: 312-337.

Liao, X., Li, L. and Shi, B. (2004). Adsorption recovery of thorium (IV) by Myrica rubra tanin and larch tannin immobilized onto collagen fibers. *Journal of Radionalytical and Nuclear Chemistry,* 260: 619-625.

Liu, Y. and Shen, L. (2008). From Langmuir kinetics to first- and second-order rate equations for adsorption. *Langmuir*, 24: 11625-11630.

Metaxas, M., Kasselouri-Rigopoulou, V., Galiatsatou, P., Konstanto-poulou, C. and Oikonomou, D. (2003). Thorium removal by different adsorbents. *Journal of Hazardous Materials,* 97(1-3): 71-82.

Maity, J. and Ray, S. K. (2018). Chitosan based nano composite adsorbent—Synthesis, characterization and application for adsorption of binary mixtures of Pb(II) and Cd(II) from water. *Carbohydrate Polymers*, 182: 159–171

Moulin, C., Amekraza, B., Hubert, S. and Moulin, V. (2001), Study of thorium hydrolsis species by electrospray-ionization mass spectrometry. *Analytica Chimica Acta*, 441: 269-279.

Mullet, M., Fievet, P., Szymczyk, A., Foissy, A., Reggiani, J-C. and Pagetti, J. (1999). A simple and accurate determination of the point of zero charge of seramic membranes, *Desalination*, 121: 41-48.

Oren, A. H. and Kaya, A. (2006). Factors affecting adsorption characteristics of Zn^{2+} on two natural zeolites. *Journal of Hazardous Materials,* 131: 59–65.

Oter, O. and Akcay, H. (2007). Use of natural clinoptilolite to improve, water quality: sorption and selectivity studies of lead(II), copper(II), zinc(II), and nickel(II). *Water Environment Research*, 79: 329–335.

Pavelic, S. K., Medica, J. S., Gumbarevic, D., Filoševic, A., Pržulj N. and Pavelic, K. (2018). Critical review on zeolite clinoptilolite safety and medical applications in vivo. *Frontiers in Pharmacology,* 9:1350. https://doi.org/10.3389/fphar.2018.01350

Salehi, E., Daraei, P. and Shamsabadi A. A. (2016). A review on chitosan-based adsorptive membranes, *Carbohydrate Polymers* 152: 419–432.

Sing, K. S. W. (1982) Reporting physisorption data for gas/solid systems with special reference to the determination of surface area and porosity, *Pure and Applied Chemistry*, 54(11): 22O1-22l8.

Sprynskyy, M., Buszewski, B., Terzyk, A. P. and Namiesnik, J. (2006). Study of the selection mechanism of heavy metal (Pb^{2+}, Cu^{2+}, Ni^{2+}, and Cd^{2+}) adsorption on clinoptilolite, *Journal of Colloid and Interface Science*, 304: 21–28.

Sun, X.-F., Jing, Z. and Wang, G. (2013). Preparation and swelling behaviors of porous hemicellulose-*g*-polyacrylamide hydrogels. *Journal of Applied Polymer Science,* 128(3): 1861-1870.

Tan, X., Wang, X., Fang, M. and Chen, C. (2007). Sorption and desorption of Th(IV) on nanoparticles of anatase studied by batch and spectroscopy methods. *Colloids and Surfaces A*: Physicochemical and Engineering Aspects, 296: 109-116.

Thoniyot, P., Tan, M. J., Karim, A. A., Young, D. J. and Loh, X. J. (2015). Nanoparticle–hydrogel composites: concept, design, and applications of these promising, multi-functional materials. *Advanced Science,* 2: 1400010, 1-13.

Url: http://www.epa.gov/rpdweb00/radionuclides/thorium.html#contact, Metaxas et al., 2003.

Wan Ngah, W. S., Teong, L. C., Wong, C. S. and Hanafiah, M. A. K. M. (2011). Preparation and characterization of chitosan-zeolite composites, *Journal of Applied Polymer Science,* 125: 2417-2425.

Wang J., Zheng, X., Wu, H., Zheng, Bin., Jiang, Z., Hao, X. and Wang, B. (2008). Effect of zeolites on chitosan/zeolite hybrid membranes for direct methanol fuel cell. *Journal of Power Sources*, 178: 9-19.

Wang, J. and Zhuang S. (2017). Removal of various pollutants from water and wastewater by modified chitosan adsorbents. *Critical Reviews In Environmental Science and Technology*, 47(23): 2331–2386.

Wang, S. and Peng, Y. (2010). Natural zeolites as effective adsorbents in water and wastewater treatment. *Chemical Engineering Journal*, 156: 11–24.

Wang, S. F., Shen, L., Tong, Y. J., Chen, L., Phang, I. Y., Lim, P. Q. and Liu, T. X. (2005). Biopolymer chitosan/montmorillonite nanocompo-sites: Preparation and characterization. *Polymer Degradation and Stability*, 90: 123-131.

Wang, W. and Wang, A. (2009). Synthesis and swelling properties of guar gum-g-poly(sodium acrylate)/Na-montmorillonite superabsorbent nanocomposite. *Journal of Composite Materials*, 43(23): 2805-2819.

Wingenfelder, U., Hansen, C., Furrer, G. and Schulin, R. (2005). Removal of heavy metals from mine waters by natural zeolites. *Environmental Science and Technology*, 39: 4606–4613.

Wu, F. C., Tseng, R. L. and Juang, R. S. (2010). A review and experimental verification of using chitosan and its derivatives as adsorbents for selected heavy metals. *Journal of Environmental Management,* 91 (2010) 798–806.

Yang, Ralph T., *Adsorbents: Fundamentals and Applications*, John Wiley & Sons, Inc., Canada, 2003.

Yu, S. M., Chen, C. L., Chang, P. P., Wang, T. T., Lu, S. S. and Wank, X. K. (2008). Adsorption of Th (IV) onto Al-pillared rectorite: Effect of pH, ionic strength, temperature, soil humic acid and fulvic acid. *Applied Clay Science* 38: 219-226.

Zhao, D. L., Feng, S. J., Chen, C. L., Chen, S. H., Xu, D. and Wang, X. K. (2008). Adsorption of thorium (IV) on MX-80 bentonite: Effect of pH, ionic strength and temperature. *Applied Clay Science,* 41: 17-23.

In: An Introduction to Aluminosilicates
Editor: Nero Regina Blevins
ISBN: 978-1-53617-250-8

Chapter 12

APPLICATION OF KAOLIN AS AN ALUMINOSILICATE SOURCE TO PRODUCE GEOPOLYMER AND ZEOLITIC MATERIALS

***Rozineide Aparecida Antunes Boca Santa*[1], *PhD*,**
***Humberto Gracher Riella*[1], *PhD*,**
***Domingos Lusitâneo Pier Macuvele*[1,2,*], *PhD*,**
Natan Padoin*[1], *PhD and Cíntia Soares*[1], *PhD
[1]Graduate Program in Chemical Engineering,
Federal University of Santa Catarina, Florianópolis, SC, Brazil
[2]Department of Natural Sciences and Mathematics,
University of Rovuma-Extension of Niassa, Lichinga, Mozambique

ABSTRACT

Kaolin is a source of aluminosilicate with a great technological potential of applications in several industrial sectors. This mineral is found in large quantities in the earth's crust. Its use has enabled the production of materials that involve different particle sizes, including nanotechnology. The aluminosilicate in the form of crystal present in kaolin can be transformed into amorphous materials through thermal treatments. Amorphous aluminosilicates, also called metakaolin, may react with other compounds and form new structures by means of the application of different physical and chemical processes, according to industrial interests. Thus, based on the importance and wide use of industrial kaolin, this chapter gives priority emphasis on two materials that may be

* Corresponding author email: lusitaneom24@gmail.com.

produced from metakaolin: geopolymers and synthetic zeolites. The application of kaolin to the production of geopolymers has been studied in order to obtain geopolymer matrices with different application potentials. The geopolymer technology may be applied in the production of cementitious matrices of high compressive strength and in the production of porous matrices for the application in the adsorption of substances. Zeolites are another promising technology that has been highlighted in the last decades, mainly for the application as molecular sieves. It is important to highlight that the aluminosilicate present in kaolin may also be used in the production of zeolitic materials. Some details of this application will also be explained in the course of this chapter.

1. INTRODUCTION

Aluminosilicates are present in large proportions on Earth. It is used for a wide range of industrial processes and has contributed to the economic development, quality of life, and well-being of the general population. SiO_2 and Al_2O_3 components may be easily found in various minerals. Although they are commonly found, these elements are also part of the gemstone composition depending on the structural arrangement in which the elements are linked. The most common elements that may be cited are feldspars, micas, quartz, and pyroxenes. The occurrence of Si (silicon) and Al (aluminum) elements coordinated with O (oxygen) and other elements present in the environment make different groups of aluminosilicate minerals be subject of study in research that aims to obtain products from structural modifications in the source material and/or in compositions with other products.

This chapter highlights some characteristics of kaolin mineral, which is used industrially in various processes as a source of aluminosilicate. Some of the main topics covered are some properties and processing of kaolin applied in the production of inorganic polymers and synthetic zeolites.

Kaolin is a mineral and its microstructure is composed of hydrated aluminum silicates. Generally, its composition occurs in oxides and may have more complex bonds in nature. It is usually white color and most of its structure is formed by hexagonal lamellae of kaolinite. The grain size of the kaolin is coarser and the resistance is low compared to other clays.

In 2030, the world production of kaolin is expected to reach 29.4 million tons. In Brazil, 2.9 million tons were produced in 2009 (MME, 2009). Among the applications of kaolin can be cited as one of the constituents of cement and ceramic tiles; as an adsorbent of digestive tract toxins, as a constituent of high quality rubbers, as a hardener in the textile industry; It is also applied in post-dentifrices; in the loading of wall plaster; in the composition of disinfectants, cosmetics and surgical gloves (MME, 2009). Kaolin is also used as a filler and coating in white paper production processes (Luz et al. 2008; R. A. A. Boca Santa et al. 2017).

The aluminosilicates present in kaolin have been studied by several researchers as one of the most attractive raw materials for the production of activated cements and some (synthetic) zeolites (Moreira et al. 2018; Bortolatto et al. 2017; Boca Santa, Soares, and Riella 2016; Boca Santa et al. 2018).

The activated cement is cited in several studies and dehydroxylated kaolin is one of the main raw materials used in the composition of this material. Davidovits 1994; Heah et al. 2012; Cheng et al. 2012; Boca Santa et al. 2013; Lancellotti et al. 2013; Sarkar, Dana, and Das 2015; and Boca Santa et al. 2019 are some of the researchers who used kaolin to produce geopolymer materials, but there are still many other authors that did the same. Activated cement is also known as geopolymer and their properties have been studied in order to insert it among the cement of the future. They are resistant materials (Davidovits 2002; Zhang et al. 2008; Pacheco-Torgal, Castro-Gomes, and Jalali 2008; Boca Santa et al. 2013; Antunes Boca Santa et al. 2017; Boca Santa et al. 2018) with adaptable formulations.

Geopolymers also contain industrial residues in their composition, especially those with amorphous aluminosilicates (Xu and Van Deventer 2000; Davidovits 2002b; Buchwald, Dombrowski, and Weil 2007; Duxson et al. 2007). In addition to the application of geopolymers in various processes as a binder, this new material has also attracked attention for its immobilizing and contaminant adsorbent properties (Yunsheng et al. 2007; Boca Santa, Soares, and Riella 2016).

Another material used for pollutant adsorption is the natural and synthetic zeolites (Luo et al. 2018; Doyle et al. 2017). There are several studies that highlight the use of kaolin as one of the raw materials in the production of synthetic zeolites (Liu et al. 2003; Kovo, Hernandez, and Holmes 2009; San Cristóbal et al. 2010; Kwapong et al. 2018; Li et al. 2010; Bortolatto et al. 2017; Moreira et al. 2018).

In the production of geopolymers, the mineral is heat-treated to undergo dihydroxylation and pass the metakaolin. This also happens in most of the zeolites synthesized that uses aluminosilicate from kaolin. The temperature used depends on the origin of the mineral (kaolin) and the equipment used. The dehydroxylation process makes the material amorphous and suitable for geopolymerization reactions (Davidovits 2002b) and for forming zeolitic crystals in hydrothermal processes (Kovo, Hernandez, and Holmes 2009).

Kaolin aluminosilicate sources, depending on the process, may be treated (Boca Santa et al. 2018) and served as a byproduct for the production of new materials. For example, kaolin present in the waste from the paper industry was reused for the processing of geopolymer (Santa et al. 2017) and synthetic zeolites (Moreira et al. 2018; Bortolatto et al. 2017).

In order to emphasize the importance of aluminosilicate for application in several products, this study presents information about kaolin mineral, its properties, and some of its applications. It also presents details on two different technologies, geopolymers and synthetic zeolites, and discusses the importance of using industrial by-products as a source of raw material for the production of new materials.

2. KAOLIN

Kaolin is composed of hydrated aluminum silicates such as kaolinite and halloysite (Luz and Damasceno 1993). The chemical formulas of kaolinite and halloysite are respectively following described according to the distribution and bonds of the constituents: ($Al_2O_3.2SiO_2.2H_2O$) and (Si_2O_5 (OH) $4Al_2nH_2O$). Among the properties of kaolin may be emphasized the whiteness, low electrical conductivity, non-abrasive, inert, among others. Some kaolin, depending on the region of occurrence of their mines, may have yellowish color due to proportions of iron oxide present.

Kaolin reserves are distributed in large quantities and various geographic regions. The largest reserves are located in the United States (53%), Brazil (28%), Ukraine (7%), and India (7%) accounting for 95% of approximately 15 billion tones (SIMERAL, 2015).

The ceramic industry, for example, classifies kaolin according to the degree of crystallinity and impurities. The classification includes the following categories: plastic (with illite impurity), modular flint (refractory and non-plastic, contains boehmite, and diasporus in its formulation), simi-flint, flint (pure and crystalline), and ball-clays (high plasticity, mechanically resistant, and viscous, being composed mainly by halloysite) (Luz, and Damasceno, 1993).

The lamellae, which forms the kaolin crystals, undergoes a structural water loss called kaolin dehydroxylation when subjected to temperatures of approximately 600°. This process results in metakaolin.

Figure 1 shows an image obtained by transmission electron microscopy (TEM) in a sample of kaolin supplied by the industry Colorminas Corificacao e Mineração S/A (COLORMINAS). The microscope used for the analysis was a JEOL JEM-2100 TEM (TEM 200kV) type from the Central Electron Microscopy Laboratory (LCME) of the Federal University of Santa Catarina (UFSC). The sample was evaluated using an FCF300-CU Formvar grid. / Carbon 300 mesh, Copper.

In X-ray diffraction (XRD) diffractogram (Figure 2) obtained from a kaolin sample (supplied and adapted from the fabricant COLORMINAS / Brazil), it is possible to observe the crystalline peaks referring to kaolinite and quartz present in the material.

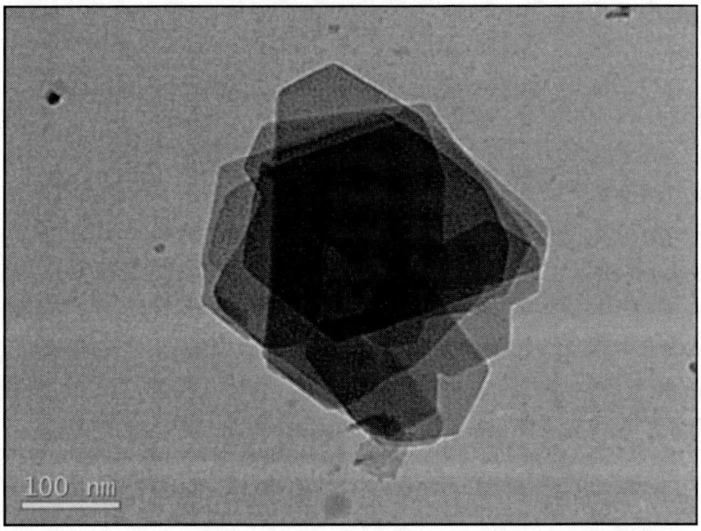

Figure 1. Transmission electron microscopy (TEM) obtained from a Brazilian kaolin sample.

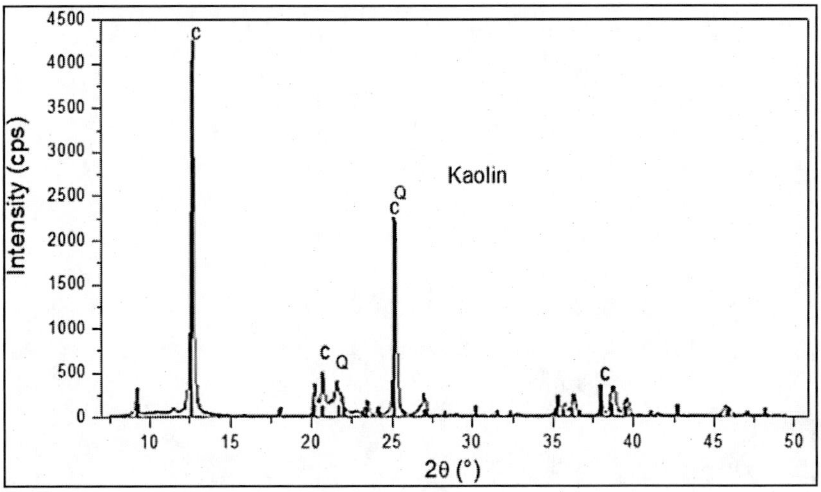

Figure 2. XRD of a kaolin sample. Adapted. Research and Technological Innovation Laboratory - COLORMINAS.

According to Boca Santa et al. (2016), Brazilian kaolin, used in their studies, has 47, 4% SiO_2 and 37.9% Al_2O_3. It also has other elements such as iron (0.45), K_2O (0.86%) among others in minority proportions.

2.1. Metakaolin

Metakaolin is obtained after heat treatment of kaolin at certain temperature ranges. The octahedral structure of kaolinite changes into a tetrahedral structure during the calcination process. The dehydroxylation process increases material reactivity (Panagiotopoulou et al. 2007). The transformations caused by the heat action on kaolinitic materials depend on the temperature range used in the process. The process of water loss through pores occurs at a temperature from 100 to 150°C. The organic impurities present in the material are oxidized at temperatures from 200 to 300°C.

At an increasing temperature that ranges from 400 to 600°C, the hydroxyls attached to the kaolinite structure undergo dihydroxylation. SiO_2 and Al_2O_3 components acquire high reactivity due to their amorphous state at temperatures from 600 to 900°C, and the crystalline structure is largely disintegrated at 900°C. Moreover, the crystallization of the mullite occurs when the material is subjected to temperatures above 950°C (Riella, 2010 apud Kirsch, 1972).

At a temperature range between 700 and 800°C, the material acquires high reactivity. In the metakaolinization process, the kaolin loses approximately 15% of mass (Nita, 2006 apud Shvarzman et al., 2003). By means of a thermogravimetric analysis performed on a kaolin sample, an exothermic peak was observed at a temperature of 969.9°C without mass change, showing that at temperatures close to 1000°C the material crystallization also may occur (R. A. A. Boca Santa et al. 2017). The melting points of these materials range from 1650 to 1775°C.

The Dehydroxylation of kaolin into metakaolin may be described by Equation 1.

$$\underset{\text{Kaolin}}{Al_2Si_2O_5(OH)_4} \longrightarrow \underset{\text{Metakaolin}}{Al_2O_3.2SiO_2} + 2H_2O \quad (1)$$

(a) (b)

Figure 3. Micrograph obtained by the scanning electron microscopy technique of a kaolin sample.

Figure 3 shows a sample of Brazilian kaolin obtained from the industry COLORMINAS. The equipment used for Scanning Electron Microscopy (SEM) analysis was the JEOL JSM-6390LV of the LCME-UFSC. The fresh kaolin sample (a) was subjected to 800°C temperature for 2h (b) for testing. Through the micrographs obtained by the SEM technique, it is possible to observe the structural changes of the material after the heat treatments.

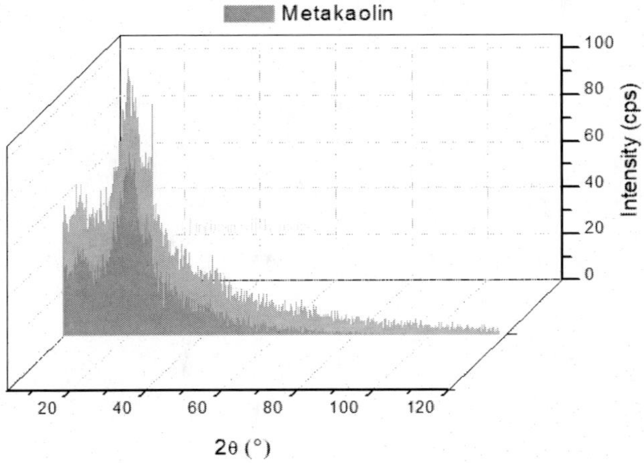

Figure 4. XRD of a heat-treated metakaolin sample at a temperature of 800°C for 2h.

After heat treatment at 800°C for 2h, the kaolin turns into metakaolin as may be observed in the Diffractogram, obtained by the X-ray diffraction technique shown in Figure 4. It is possible to note that the metakaolin has a large amorphous halo. It is also perceptible that the crystalline kaolinite peaks have disappeared. Besides that, quartz crystals remain present in the materialThe equipment used for XRD analysis was a Philips X'pert diffractometer.

2.2. Application of Kaolin and Metakaolin

Kaolin and its dehydroxylated form (metakaolin) are widely used as raw materials in industries. It was possible to verify a large number of publications involving kaolin and metakaolin through some studies published in the Science Direct data platform (Content for journals, articles, book chapters, and open-access content), Google Scholar, and Springer Science Books. In order to filter results, it wasused the Boolean expression "Kaolin OR metakaolin" and it was selected a range of publication year between 1996 and 2019.

Springer database presented 59.379 results, Science Direct presented 33.596.000, and Google Scholar presented 129,000,000. The Science Direct's and Google Scholar's results were organized by year .This is presented in the graph in Figure 5.

As shown in Figure 5, the number of works performed using kaolin or metakaolin has grown every year. The year 2019 is still in progress and already has a significant number of publications.

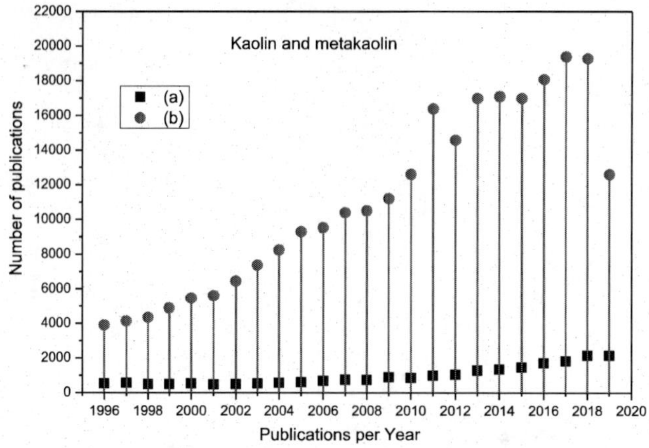

Figure 5. Number of studies published from 1996 to 2019 available in Science Direct (a) database and Google Scholar (b), using the Boolean expression "kaolin OR metakaolin."

Among the publications, we may mention the use of kaolin in studies involving geopolymer cement (Tchakouté et al. 2017; Kenne Diffo et al. 2015; Boca Santa et al. 2019); reinforcement of concretes (Song et al. 2019); mortar improvement; zeolite production (Ayele et al. 2016), (Luo et al. 2018), (Król and Mozgawa 2019); inorganic polymer synthesis (Balo et al. 2018); adsorption (Boca Santa et al. 2019); production high gloss clays (Jun, Everett Ben, and Andrews Windell 1998); among others.

Kaolin is also used as a filler and cover on paper in the manufacture of refractories, inks, rubbers, pesticides, and insecticides; as well as, in pharmaceuticals and cosmetics. It is also applied as a catalyst in clarifiers, chemicals, detergents, sealants, animal food; and in the manufacture of glass, sanitary parts, among others. (Silva, 2001, Bordonalli, 1995). Farias (2009) also mentions, in the Technical Report 39, the use of kaolin in the production of tiles, glass fibers, and polymers (nylon, unsaturated polyethylene, thermoplastic polyester, polyolefins, polyurethane, PVC).

In recent decades, kaolin has been widely used in the production of activated cement (geopolymers) and synthesis of synthetic zeolites. Therefore, in the following topics of this chapter will be highlighted the properties of these materials and the application of kaolin (metakaolin) to obtain these two products.

2.3. Geopolymeric Materials

Research involving inorganic polymers or geopolymers seeks to understand and identify the mechanisms that lead to the formation of the amorphous and resistant structure of these materials. Many findings related to geopolymerization are described in the literature, but there are still many doubts about the mechanisms of the reaction that occurs from the alkaline activation of one or more amorphous aluminosilicate sources.

Since a Davidovits' study published in the 70s, a large number of works have been presented using different sources of aluminosilicates. However, this chapter will only emphasize geopolymers made from metakaolin. It is important to highlight that the ratio between silica and alumina and among the activators used must be adequate for the microstructural formation and the obtaining of the expected properties of geopolymers.

P. J. Davidovits (2002) relied on old cement to find a heat-resistant binder in building fire situations. The researcher Davidovits uses in his studies calcined kaolinite (metakaolin) as one of the sources of aluminosilicates to perform the synthesis of geopolymeric materials. Metakaolin forms species that define the general structure of geopolymers (Komnitsas and Zaharaki 2007).

Geopolymers are considered inorganic polymers because they form oxygen-coordinated silicon (Si) and aluminum (Al) bonds. Geopolymeric materials are formed by the addition of an activator in a raw material containing amorphous aluminosilicates. The source of aluminosilicate may be natural, from industrial waste or various sources (Xu and Van Deventer 2003). Figure 6 shows a representation of the three-dimensional link structures between SI, Al, and O for the geopolymer structure formation.

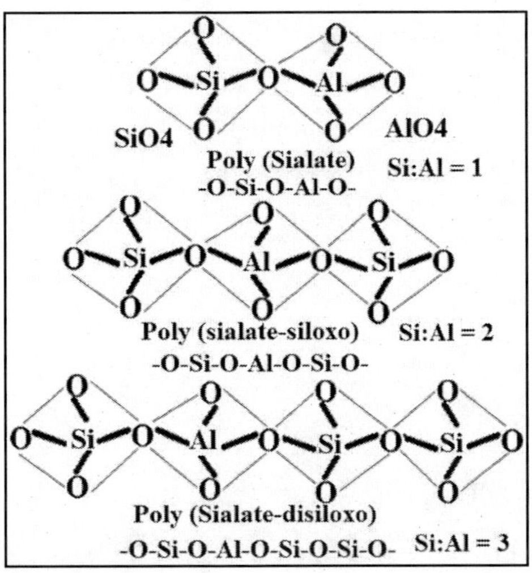

Figure 6. Links between Si, Al, and O elements for formation of different polysialates: PS, PSS, PSDS. Adapted. Geopolymer 2002 Conference, Melbourne, Australia (Davidovits, 2002).

The empirical formula of geopolymers (Equation 2) is similar to the formula of zeolites, but the structure comprises amorphous to semicrystalline phases. In the equation, Z represents 1, 2 or 3; M represents the alkali metal cation, and n represents the degree of polycondensation.

$$Mn[-(SiO_2)z\text{-}AlO_2]n.wH_2O \qquad (2)$$

The chemical reaction develops rapidly when amorphous aluminosilicate [Si-Al]-rich minerals are subjected to strong alkaline conditions (Xu and Van Deventer 2003). The elements involved generate a three-dimensional polymeric chain, in which silicon (Si) atoms alternate with aluminum (Al) at tetrahedral coordination, sharing the existing oxygen atoms (O), as shown in Figure 7. The main activators used are sodium (NaOH) and potassium (KOH) hydroxides, which are usually mixed with sodium silicate proportions. Na^+ or K^+, depending on the base used, binds to the structure by balancing loads of Si^{4+} and Al^{3+}. Other ions may be part of the process, depending on the activator used, for example, ions of Li^+, Ca^{++}, Ba^{++}, NH^{4+}, H_3O^+ (J. Davidovits 1994). After conformation of inorganic polymers, the reaction result is an amorphous alkaline aluminum silicate (Duxson et al. 2007; Fernández-Jiménez et al. 2006; Criado, Palomo, and Fernández-Jiménez 2005). The process releases water as noted in Equation 2.

Studies indicate great potential in the characteristics of geopolymeric materials, including fire resistance, acid resistance, compressive strength, weathering, and immobilization of hazardous waste (P. J. Davidovits 2002; J. Davidovits 2002; Davidovits, 2008; Rickard, Van Riessen, and Walls 2010).

The wide range of the potential application of geopolymer materials includes building and paving materials; fire-resistant materials; decoration artifacts; thermal insulators; low tech materials; ceramic tiles; heat shock resistant refractory items; foundry materials; composite materials for aircraft and automobile interiors; and high-tech resin systems. It also can used as a containment barrier for toxic and radioactive waste; as raw materials for making arts and decoration for cultural heritage, archeology, the history of science and others (Davidovits, 2008).

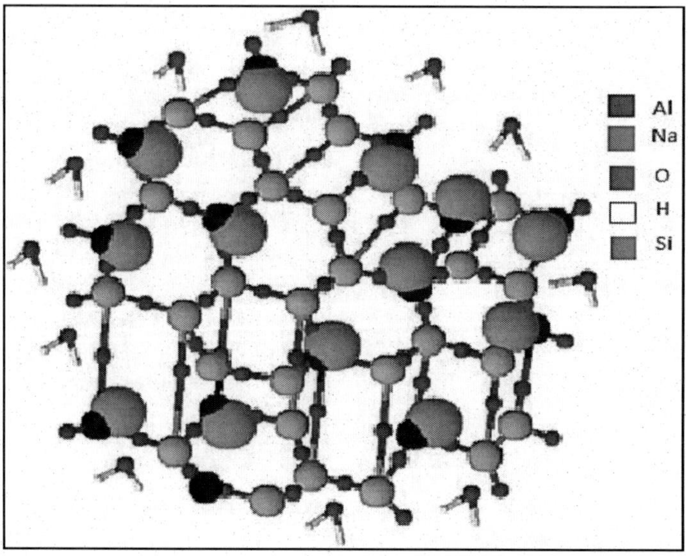

Figure 7. Model of the structural formation of the geopolymer matrix. Adapted Davidovits (2011).

Geopolymer materials may be considered extremely environment-friendly by allowing the use of waste as a raw material for its composition. For Duxson et al. (2007), despite the differences in the molecular structure and microstructure of materials synthesized from different sources, silicon and aluminum are present in all systems, just matching the materials with the conceptual model of geopolymer reaction. The production of cementitious materials, such as geopolymers from industrial waste, may assist in reducing industrial liabilities and the extraction of raw materials from natural sources for cement production.

Since the constant withdrawal of raw materials from nature may lead to shortages of important materials for the survival and development of future generations (Nuruddin et al., 2010). Another positive factor for investing in more technology for the production of geopolymer cement is the reduction of greenhouse gas emissions in the environment, such as CO_2 from burning limestone, clays, and other components for clinker production. Data shows that 5% of the total CO_2 emitted by the various man-made events comes from cement production, totaling approximately 1.5 billion tons per year (Part, Ramli, and Cheah, 2016).

An example of a material that may be reused is kaolin from the pulp residues of white paper. This material may be treated and the resulting kaolin may be used to obtain metakaolin by heat treatment process (Antunes Boca Santa et al. 2013; R.A.A. Boca Santa et al. 2017; Santa et al. 2017; Bortolatto et al. 2017; R.A.A. Boca Santa et al. 2018; Moreira et al. 2018). Figure 8 shows an image of a geopolymer sample produced from metakaolin.

As already mentioned, geopolymeric materials may be used in various applications since they are materials with technology and adaptable composition. They may be used as a cement substitute in constructions or the manufacture of different artifacts (compact structure) (Figure 10), and in the solidification/immobilization of toxic waste (R.A.A. Boca Santa, Soares, and Riella 2016); (R.A.A. Boca Santa, Soares, and Riella 2016). Geopolymers have shown potential application as an adsorbent material of different pollutants. However, for the use in adsorption processes, a microstructural adjustment to the pores of the material is required, i.e., directing the synthesis to obtain a porous structure geopolymer (Boca Santa et al. 2019).

Figure 9 presents a representation showing the characteristics of geopolymer with structure directed to increase porosity (porous geopolymers for the use in the adsorption of wastewater contaminated with toxic materials).

Figure 10 shows an effluent ($C_{16}H_{18}ClN_3S$) collected from an adsorption column prepared with a porous geopolymer bed. It is observed that the material has a great adsorption potential. It took more than 8 h of flow to saturate the column. Aliquots were collected at different time intervals. The results of this study and the formulation used to obtain porous geopolymers were published in the Journal Cleaner Production by Boca Santa et al. (2019).

Figure 8. Characteristics of geopolymers made from metakaolin treated at 800°C for 2h (compact structure for use as a binder in constructions and in the manufacture of artifacts, composites, among others).

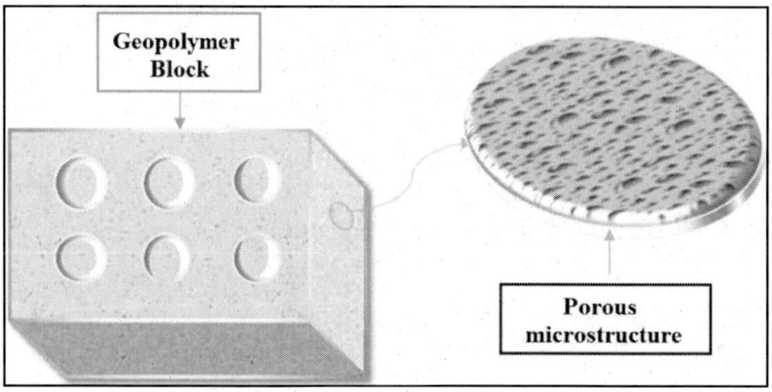

Figure 9. Representation of a block of porous geopolymers.

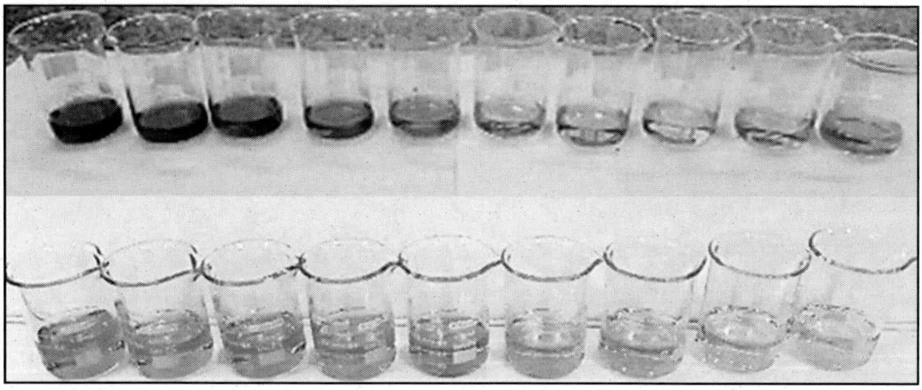

Figure 10. Samples of some aliquots of an effluent contaminated with methylene blue dye collected on fixed bed adsorption column of porous geopolymers.

2.3.1. Production of Geopolymer Materials from Kaolin/Metakaolin

In the composition of metakaolin, there are the main elements for the formation of geopolymer structure, where there is about 52% of SiO_2 and 40% of Al_2O_3. The other 8% is considered impurities such as quartz, iron, organic materials, among others (Nita, 2006).

A survey of 16 different aluminosilicate sources for geopolymer production revealed the need to add metakaolin in order to form a gel. It also found that there is good synergy between different aluminosilicate sources. The addition of kaolinite is necessary to increase the dissolution rate of alumina and to improve polymerization, but it does not cause softness in the structure in low quantities what makes its use for construction applications unfeasible (Komnitsas and Zaharaki 2007).

For the formation of geopolymers, the elements Si and Al, present in the amorphous microstructure of metakaolin after kaolinite dehydroxylation (Sperinck et al. 2011), undergo dissolution in the presence of an alkaline activator. Then, there is a microstructural reorganization and the formation of new structures. Metakaolin has fine particles that favor dissolution during the activation process (Panagiotopoulou et al. 2007). If the heat treatment performed on the kaolin is efficient, there will be large proportions of the amorphous material to react.

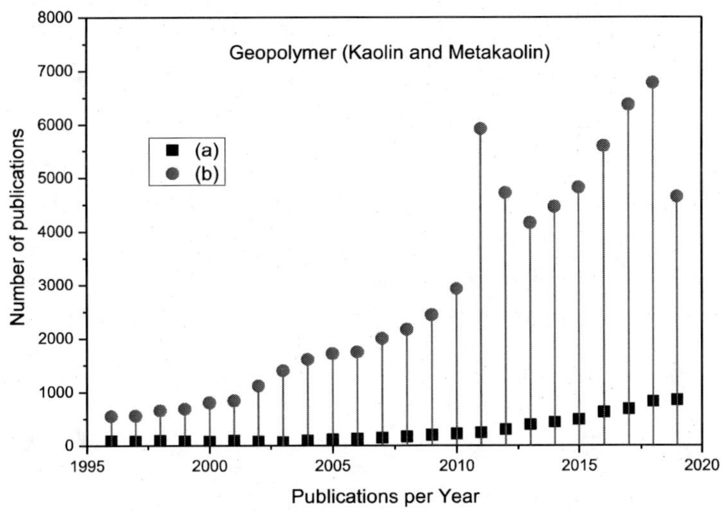

Figure 11. Number of works from 1996-2019 published in Science Direct (a) databases and Google Scholar (b) using the expression "kaolin OR metakaolin AND geopolymer OR inorganic polymer."

To obtain an amorphous material for geopolymer synthesis, a researcher (J. Davidovits 1999) used temperatures between 750 and 800°C for 6 hours. The researcher also reports that temperatures between 550 and 650°C or higher than 900°C may cause a deficit in dehydroxylation and that at temperatures above 1000°C may occurs the formation of mullite crystals ($3Al_2O_3.2SiO_2$). According to Vanderley (2003), geopolymers produced with metakaolin contain high proportions of amorphous material and minimal quantities of quartz crystals in the microstructure. Antunes Boca Santa et al. (2013) mixed coal bottom ash and metakaolin from the waste treatment of the white paper pulping industry and obtained a material with binder properties. The material was produced with aluminosilicates obtained from industrial waste. In this study, the heat treatment temperature of kaolin to obtain metakaolin was 850°C for 2h. R. A. A. Boca Santa, Soares, and Riella (2016) performed the immobilization of heavy metals in geopolymers using as a source heavy gray aluminosilicates and industrial kaolin. Kaolin was treated at 750°C for 2h to obtain metakaolin. In a study conducted by (R. A. A. Boca Santa et al. 2017), three heat treatment temperatures of kaolin, which was obtained from the treatment of white paper shavings to obtain metakaolin, were tested, among them: 800, 850, and 900°C. The results indicated that at the temperatures tested there were no significant differences to obtain amorphous metakaolin. Geopolymers with 100% residual kaolin were synthesized, and the prioritized calcination temperature was 850°C for 2h (Santa et al. 2017). It was observed in the mentioned studies that the heat treatment temperatures used to obtain an amorphous metakaolin varied between 650°C and 900°C, without compromising the material characteristics.

From a research using the Boolean expression "kaolin OR metakaolin AND geopolymer OR inorganic polymer" in the Science Direct databases (containing journals, articles, book chapters and, open access) and Google Scholar, it was obtained a significant number of publications from 1996 to 2019. As may be seen in the graph in Figure 11, there were about 6000 results in the Science Direct and18500 results in the Google Scholar platform.

Among the works published with the synthesis of geopolymers from metakaolin may be cited: geopolymers modified by polyacrylic emulsion and polipropylene fibers (Chen et al. 2018); chemical deformation in metakaolin-based geopolymers (Li et al. 2019); waterproof properties of geopolymers produced with metakaolin and rice husk ash (Liang et al. 2019); the role of kaolinite dehydroxylation and recrystallization in geopolymers (Wan, Rao, and Song 2017); mesoporous geopolymers (Barbosa et al. 2018); the effect of the addition of polyethylene glycol on geopolymers (Mikhailova and Rovnaník 2016); geopolymers with calcium carbonate incorporation (Aboulayt et al. 2017); control of geopolymeric microstructure (Kaddami and Pitois 2019); performance of geopolymers for Mn (II) and Co (II) removal (Kara et al. 2018); and many other studies.

2.4. Zeolitic Materials

Another material that may be made from metakaolin is the synthetic zeolites. Zeolites are of great importance for application in adsorption, catalysis, and ion exchange processes. They are also known as molecular sieves due to their microstructural properties (Mintova et al., 2016, Moreira et al. 2019). Its structure is formed by hydrated aluminosilicates organized in different conformations as well as different particle sizes. SiO4 and AlO4 form tetrahedral units in most of the zeolitic materials (Breck, 1974).

Zeolitic crystals may be found in nature or be synthetically produced. The IUPAC and the IZA Structure Committee have assigned a three-letter code for the structural topology of known zeolites, including Zeolite A (LTA) Zeolite P (GIS), Zeolite Faujasita (FAU), and others (Johnson and Arshad 2014). The known synthetic zeolites already reach more than 232 species (Thompson and Franklin, 2019; Moreira et al. 2019). The production of zeolites in the year 2018 was 1.800.000 tons (Johnson and Arshad 2014; Moreira et al. 2019).

Synthetic zeolites may be produced by analytical reagents and aluminosilicate sources, such as natural kaolins (Ayele et al. 2016; Luo et al. 2018; Król and Mozgawa 2019) and/or kaolins from industrial waste (Bortolatto et al. 2017; Moreira et al. 2015). For the production of zeolites from metakaolin, the heat treatments are similar to those used for the production of geopolymers. Moreira et al. (2018) synthesized zeolites A from kaolin obtained from white paper pulping residues. For the conversion to metakaolin, the kaolin was subjected to a temperature of 850°C for 2h. The same heat treatment conditions were used by Bortolatto et al. (2017) to obtain metakaolin from residual kaolin and produce zeolites Y. Ayele et al. (2016) obtained metakaolin from kaolin calcination at 600°C for 3h for the production of zeolite A. Gougazeh and Buhl (2014) synthesized zeolite A by performing kaolin metallinization at a temperature of 650°C for 2h. Loiola et al. (2012) synthesized NaA zeolite, using metakaolin as a source of alumina obtained from kaolin calcination at 900°C for 2h. As observed in the studies presented, the authors used different temperatures for kaolin dihydroxylation to obtain metakaolin, and to produce zeolitic materials, depending on the region and the kaolin used.

In order to evaluate the application of kaolin as a source of aluminosilicates in the production of synthetic zeolites, a search was made in the Science Direct and Google Scholar databases. . The results were filtered by the Boolean expression "Kaolin OR metakaolin AND zeolite," and a range of year of publication between 1996 and 2019. Therefore, the Science Direct database found 31,140 results and the Google Scholar database found 18,800 results. Data on numbers of papers published by each evaluated year is presented in the graph of Figure 12.

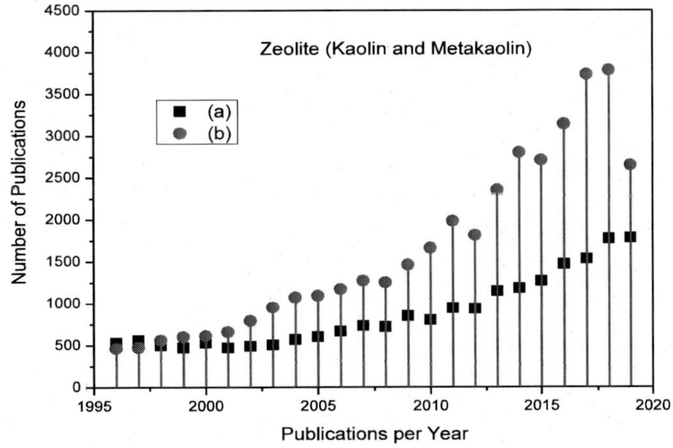

Figure 12. Results obtained from Science Direct and Google Scholar databases, filtering them through the year of publication (1996-2019) and the Boolean expression "Kaolin OR metakaolin AND zeolite."

3. Considerations

According to the data presented in this chapter, it was observed that kaolin has properties and proportions of aluminosilicates in its microstructure, which are attractive for the production of different materials. Kaolin, transformed into metakaolin by heat treatment, enables its use in processes that require amorphous material for synthesis reactions.

It was highlighted in this study the geopolymers and synthetic zeolites among several materials produced from kaolin and metakaolin. Some researchers have been presenting geopolymers as the cement of the future. Currently, this material is also being highlighted as a material with great potential in the application on toxic waste, immobilization/solidification processes, and pollutant adsorption processes. This study also emphasized the importance of different zeolites for application in different processes that require the presence of materials with molecular sieve properties.

Therefore, it was possible to present a review on the properties of kaolin and its main applications through some studies published in the Springer, Science Direct, and Google Scholar databases that may contribute to future research on aluminosilicates.

Acknowledgments

The authors wish to thank the National Council for Scientific and Technological Development (CNPq), Coordination for the Improvement of Higher Education Personnel (CAPES), the Central Laboratory of Electron Microscopy (LCME), the Laboratory of Materials and Scientific Computing, the Colorminas Colorifício e Mineração, and Federal University of Santa Catarina (UFSC).

References

Aboulayt, A., M. Riahi, M. Ouazzani Touhami, H. Hannache, M. Gomina, and R. Moussa. 2017. "Properties of Metakaolin Based Geopolymer Incorporating Calcium Carbonate." *Advanced Powder Technology*. https://doi.org/10.1016/j.apt.2017.06.022.

Antunes Boca Santa, R.A., A.M. Bernardin, H.G. Riella, and N.C. Kuhnen. 2013. "Geopolymer Synthetized from Bottom Coal Ash and Calcined Paper Sludge." *Journal of Cleaner Production* 57. https://doi.org/10.1016/j.jclepro.2013.05.017.

Antunes Boca Santa, R.A., C. Soares, and H.G. Riella. 2017. "Geopolymers Obtained from Bottom Ash as Source of Aluminosilicate Cured at Room Temperature." *Construction and Building Materials* 157. https://doi.org/10.1016/j.conbuildmat.2017.09.111.

Ayele, Lijalem, Joaquín Pérez-Pariente, Yonas Chebude, and Isabel Díaz. 2016. "Conventional versus Alkali Fusion Synthesis of Zeolite A from Low Grade Kaolin." *Applied Clay Science*. https://doi.org/10.1016/j.clay.2016.07.019.

Balo, A. Madi, H. Rahier, A. Mobili, A. Katsiki, N. Fagel, U. Melo Chinje, and D. Njopwouo. 2018. "Metakaolin-Based Inorganic Polymer Synthesis Using Cotton Shell Ash as Sole Alkaline Activator." *Construction and Building Materials*. https://doi.org/10.1016/j.conbuildmat.2018.10.047.

Barbosa, Tuany Ramos, Edson Luiz Foletto, Guilherme Luiz Dotto, and Sérgio Luiz Jahn. 2018. "Preparation of Mesoporous Geopolymer Using Metakaolin and Rice Husk Ash as Synthesis Precursors and Its Use as Potential Adsorbent to Remove Organic Dye from Aqueous Solutions." *Ceramics International*. https://doi.org/10.1016/j.ceramint.2017.09.193.

Boca Santa, R.A.A., J.C. Kessler, C. Soares, and H.G. Riella. 2018. "Microstructural Evaluation of Initial Dissolution of Aluminosilicate Particles and Formation of Geopolymer Material." *Particuology*. https://doi.org/10.1016/j.partic.2017.12.007.

Boca Santa, R.A.A., J.C. Moreira, L.L. Coelho, N.C. Kuhnen, and H.G. Riella. 2017. "Recovery Techniques of Industrial Waste Stemmed from White Paper Pulping Process to Obtain Metakaolin." *International Journal of Mineral Processing* 161. https://doi.org/10.1016/j.minpro.2017.02.010.

Boca Santa, R. A. A., C. Soares, and H. G. Riella. 2016. "Geopolymers with a High Percentage of Bottom Ash for Solidification/Immobilization of Different Toxic Metals." *Journal of Hazardous Materials* 318. https://doi.org/10.1016/j.jhazmat.2016.06.059.

Boca Santa, Rozineide Aparecida Antunes, Anderson Felipe Viana da Silva, Natan Padoin, Cíntia Soares, and Humberto Gracher Riella. 2019. "Novel Porous Geopolymeric Formulation as Green Material Applied to the Recovery of Contaminated Effluent Aiming Environmental Protection." *Journal of Cleaner Production*. https://doi.org/10.1016/j.jclepro.2019.05.213.

Bortolatto, L. B., R. A. A. Boca Santa, J. C. Moreira, D. B. Machado, M. A. P. M. Martins, M. A. Fiori, N. C. Kuhnen, and H. G. Riella. 2017. "Synthesis and Characterization of Y Zeolites from Alternative Silicon and Aluminium Sources." *Microporous and Mesoporous Materials* 248. https://doi.org/10.1016/j.micromeso.2017.04.030.

Buchwald, A., K. Dombrowski, and M. Weil. 2007. "Development of Geopolymer Concrete Supported by System Analytical Tools." In *12th International Congress on the Chemistry of Cement.*

Chen, Xiao, Mingkai Zhou, Weiguo Shen, Guorui Zhu, and Xuexiang Ge. 2018. "Mechanical Properties and Microstructure of Metakaolin-Based Geopolymer Compound-Modified by Polyacrylic Emulsion and Polypropylene Fibers." *Construction and Building Materials*. https://doi.org/10.1016/j.conbuildmat.2018.09.116.

Cheng, T. W., M. L. Lee, M. S. Ko, T. H. Ueng, and S. F. Yang. 2012. "The Heavy Metal Adsorption Characteristics on Metakaolin-Based Geopolymer." *Applied Clay Science*. https://doi.org/10.1016/j.clay.2011.11.027.

Criado, M., A. Palomo, and A. Fernández-Jiménez. 2005. "Alkali Activation of Fly Ashes. Part 1: Effect of Curing Conditions on the Carbonation of the Reaction Products." *Fuel*. https://doi.org/10.1016/j.fuel.2005.03.030.

Davidovits, Joseph. 1994. "Properties of Geopolymer Cements." *First International Conference on Alkaline Cements and Concretes*.

———. 1999. "Chemistry of Geopolymeric Systems, Terminology." In *Second International Conference on Geopolymer*.

———. 2002. "Environmentally Driven Geopolymer Cement Applications." *Geopolymer 2002 Conference*.

Davidovits, Prof Joseph. 2002. "30 Years of Successes and Failures in Geopolymer Applications. Market Trends and Potential Breakthroughs." *Geopolymer 2002 Conference*. https://doi.org/10.1017/CBO9781107415324.004.

Doyle, Aidan M., Ziad T. Alismaeel, Talib M. Albayati, and Ammar S. Abbas. 2017. "High Purity FAU-Type Zeolite Catalysts from Shale Rock for Biodiesel Production." *Fuel*. https://doi.org/10.1016/j.fuel.
2017.02.098.

Duxson, P., A. Fernández-Jiménez, J. L. Provis, G. C. Lukey, A. Palomo, and J. S. J. Van Deventer. 2007. "Geopolymer Technology: The Current State of the Art." *Journal of Materials Science*. https://doi.org/10.1007/s10853-006-0637-z.

Fernández-Jiménez, A., A. G. de la Torre, A. Palomo, G. López-Olmo, M. M. Alonso, and M. A. G. Aranda. 2006. "Quantitative Determination of Phases in the Alkaline Activation of Fly Ash. Part II: Degree of Reaction." *Fuel*. https://doi.org/10.1016/j.fuel.2006.04.006.

Gougazeh, Mousa, and J. Ch Buhl. 2014. "Synthesis and Characterization of Zeolite A by Hydrothermal Transformation of Natural Jordanian Kaolin." *Journal of the Association of Arab Universities for Basic and Applied Sciences*. https://doi.org/10.1016/j.jaubas.2013.03.007.

Heah, C. Y., H. Kamarudin, A. M. Mustafa Al Bakri, M. Bnhussain, M. Luqman, I. Khairul Nizar, C. M. Ruzaidi, and Y. M. Liew. 2012. "Study on Solids-to-Liquid and Alkaline Activator Ratios on Kaolin-Based Geopolymers." *Construction and Building Materials*. https://doi.org/10.1016/j.conbuildmat.2012.04.102.

Johnson, E. B.G., and Sazmal E. Arshad. 2014. "Hydrothermally Synthesized Zeolites Based on Kaolinite: A Review." *Applied Clay Science*. https://doi.org/10.1016/j.clay.2014.06.005.

Jun, Yuan, E. Everett Ben, and R. Andrews Windell. 1998. "5685900 Method for Beneficiating Discolored Kaolin to Produce High Brightness Coating Clay." *Minerals Engineering*. https://doi.org/10.1016/s0892-6875(98)90097-8.

Kaddami, Asmaa, and Olivier Pitois. 2019. "A Physical Approach towards Controlling the Microstructure of Metakaolin-Based Geopolymer Foams." *Cement and Concrete Research*. https://doi.org/10.1016/j.cemconres.2019.105807.

Kara, Ilknur, Dilek Tunc, Fatih Sayin, and Sibel Tunali Akar. 2018. "Study on the Performance of Metakaolin Based Geopolymer for Mn(II) and Co(II) Removal." *Applied Clay Science*. https://doi.org/10.1016/j.clay.2018.04.027.

Kenne Diffo, B. B., A. Elimbi, M. Cyr, J. Dika Manga, and H. Tchakoute Kouamo. 2015. "Effect of the Rate of Calcination of Kaolin on the Properties of Metakaolin-Based Geopolymers." *Journal of Asian Ceramic Societies*. https://doi.org/10.1016/j.jascer.2014.12.003.

Komnitsas, Kostas, and Dimitra Zaharaki. 2007. "Geopolymerisation: A Review and Prospects for the Minerals Industry." *Minerals Engineering*. https://doi.org/10.1016/j.mineng.2007.07.011.

Kovo, A. S., O. Hernandez, and S. M. Holmes. 2009. "Synthesis and Characterization of Zeolite y and ZSM-5 from Nigerian Ahoko Kaolin Using a Novel, Lower Temperature, Metakaolinization Technique." *Journal of Materials Chemistry*. https://doi.org/10.1039/b907554b.

Król, M., and W. Mozgawa. 2019. "Zeolite Layer on Metakaolin-Based Support." *Microporous and Mesoporous Materials*. https://doi.org/10.1016/j.micromeso.2019.03.028.

Kwapong, W.R., C. Peng, Z. He, X. Zhuang, M. Shen, and F. Lu. 2018. "Altered Macular Microvasculature in Neuromyelitis Optica Spectrum Disorders." *American Journal of Ophthalmology* 192. https://doi.org/10.1016/j.ajo.2018.04.026.

Lancellotti, Isabella, Michelina Catauro, Chiara Ponzoni, Flavia Bollino, and Cristina Leonelli. 2013. "Inorganic Polymers from Alkali Activation of Metakaolin: Effect of Setting and Curing on Structure." *Journal of Solid State Chemistry*. https://doi.org/10.1016/j.jssc.2013.02.003.

Li, Zhenming, Shizhe Zhang, Yibing Zuo, Wei Chen, and Guang Ye. 2019. "Chemical Deformation of Metakaolin Based Geopolymer." *Cement and Concrete Research*. https://doi.org/10.1016/j.cemconres.2019.03.017.

Liang, Guangwei, Huajun Zhu, Zuhua Zhang, Qisheng Wu, and Jianzhou Du. 2019. "Investigation of the Waterproof Property of Alkali-Activated Metakaolin Geopolymer Added with Rice Husk Ash." *Journal of Cleaner Production*. https://doi.org/10.1016/j.jclepro.2019.05.111.

Liu, Xinmei, Zifeng Yan, Huaiping Wang, and Yantuo Luo. 2003. "In Situ Synthesis of NaY Zeolite with Coal-Based Kaolin." *Journal of Natural Gas Chemistry*.

Loiola, A. R., J. C. R. A. Andrade, J. M. Sasaki, and L. R. D. da Silva. 2012. "Structural Analysis of Zeolite NaA Synthesized by a Cost-Effective Hydrothermal Method Using Kaolin and Its Use as Water Softener." *Journal of Colloid and Interface Science*. https://doi.org/10.1016/j.jcis.2010.11.026.

Luo, Hongwei, Wei Wee Law, Yichao Wu, Weiping Zhu, and En Hua Yang. 2018. "Hydrothermal Synthesis of Needle-like Nanocrystalline Zeolites from Metakaolin and Their Applications for Efficient Removal of Organic Pollutants and Heavy Metals." *Microporous and Mesoporous Materials*. https://doi.org/10.1016/j.micromeso.2018.06.015.

Mikhailova, Olesia, and Pavel Rovnaník. 2016. "Effect of Polyethylene Glycol Addition on Metakaolin-Based Geopolymer." In *Procedia Engineering*. https://doi.org/10.1016/j.proeng.2016.07.379.

Moreira, J. C., K. Vollmann, L. B. Bortolatto, L. Müller, H. G. Riella, and N. C. Kuhnen. 2015. "Síntese da zeólita 5a obtida por troca iônica a partir de resíduo da indústria de papel." In https://doi.org/10.5151/chemeng-cobeq2014-0310-26072-154022.

Moreira, Jarina Costa, Rozineide A. Antunes Boca Santa, Giorgio L. Miraglia, Cíntia Soares, and Humberto Gracher Riella. 2019. "Evaluation of Different Reaction Systems to Obtain Zeolite 4A via Reverse Microemulsion." *Microporous and Mesoporous Materials*. https://doi.org/10.1016/j.micromeso.2018.12.042.

Moreira, Jarina Costa, Rozineide Aparecida Antunes Boca Santa, Janaína Nones, and Humberto Gracher Riella. 2018. "Synthesis of Zeolite 4a for Obtaining Zeolite 5A by Ionic Exchange for Full Utilization of Waste from Paper Industry." *Brazilian Journal of Chemical Engineering*. https://doi.org/10.1590/0104-6632.20180352s20160395.

Pacheco-Torgal, Fernando, João Castro-Gomes, and Said Jalali. 2008. "Alkali-Activated Binders: A Review. Part 1. Historical Background, Terminology, Reaction Mechanisms and Hydration Products." *Construction and Building Materials*. https://doi.org/10.1016/j.conbuildmat.2007.10.015.

Panagiotopoulou, Ch, E. Kontori, Th Perraki, and G. Kakali. 2007. "Dissolution of Aluminosilicate Minerals and By-Products in Alkaline Media." *Journal of Materials Science*. https://doi.org/10.1007/s10853-006-0531-8.

Part, W. K., M. Ramli, and C. B. Cheah. 2016. "An Overview on the Influence of Various Factors on the Properties of Geopolymer Concrete Derived From Industrial Byproducts." In *Handbook of Low Carbon Concrete*. https://doi.org/10.1016/B978-0-12-804524-4.00011-7.

Qian, Xin, Jialai Wang, Liang Wang, and Yi Fang. 2019. "Enhancing the Performance of Metakaolin Blended Cement Mortar through In-Situ Production of Nano to Sub-Micro Calcium Carbonate Particles." *Construction and Building Materials*. https://doi.org/10.1016/j.conbuildmat.2018.11.134.

Rickard, W. D A, Arie Van Riessen, and Philip Walls. 2010. "Thermal Character of Geopolymers Synthesized from Class F Fly Ash Containing High Concentrations of Iron and α-Quartz." *International Journal of Applied Ceramic Technology*. https://doi.org/10.1111/j.1744-7402.2008.02328.x.

San Cristóbal, A. G., R. Castelló, M. A. Martín Luengo, and C. Vizcayno. 2010. "Zeolites Prepared from Calcined and Mechanically Modified Kaolins. A Comparative Study." *Applied Clay Science*. https://doi.org/10.1016/j.clay.2010.05.012.

Santa, Rozineide A. A. Boca, Giorgio L. Miraglia, Cíntia Soares, and Humberto G. Riella. 2017. "Development of Geopolymeric Structures to Prioritize the Use of Waste from Paper Industries." *Ciencia e Tecnologia Dos Materiais*. https://doi.org/10.1016/j.ctmat.2017.09.002.

Sarkar, Madhuchhanda, Kausik Dana, and Sukhen Das. 2015. "Microstructural and Phase Evolution in Metakaolin Geopolymers with Different Activators and Added Aluminosilicate Fillers." *Journal of Molecular Structure*. https://doi.org/10.1016/j.molstruc.2015.05.046.

Song, Qiulei, Rui Yu, Zhonghe Shui, Yunyao Wang, Suduan Rao, Shuo Wu, and Yongjia He. 2019. "Physical and Chemical Coupling Effect of Metakaolin Induced Chloride Trapping Capacity Variation for Ultra High Performance Fibre Reinforced Concrete (UHPFRC)." *Construction and Building Materials*. https://doi.org/10.1016/j.conbuildmat.2019.07.047.

Sperinck, Shani, Paolo Raiteri, Nigel Marks, and Kate Wright. 2011. "Dehydroxylation of Kaolinite to Metakaolin - A Molecular Dynamics Study." *Journal of Materials Chemistry*. https://doi.org/10.1039/c0jm01748e.

Tchakouté, Hervé K., Claus H. Rüscher, E. Kamseu, Jean N. Y. Djobo, and C. Leonelli. 2017. "The Influence of Gibbsite in Kaolin and the Formation of Berlinite on the Properties of Metakaolin-Phosphate-Based Geopolymer Cements." *Materials Chemistry and Physics*. https://doi.org/10.1016/j.matchemphys.2017.07.020.

Wan, Qian, Feng Rao, and Shaoxian Song. 2017. "Reexamining Calcination of Kaolinite for the Synthesis of Metakaolin Geopolymers - Roles of Dehydroxylation and Recrystallization." *Journal of Non-Crystalline Solids*. https://doi.org/10.1016/j.jnoncrysol.2017.01.024.

Xu, Hua, and J. S. J. Van Deventer. 2000. "The Geopolymerisation of Alumino-Silicate Minerals." *International Journal of Mineral Processing*. https://doi.org/10.1016/S0301-7516(99)00074-5.

Xu, Hua, and Jannie S. J. Van Deventer. 2003. "The Effect of Alkali Metals on the Formation of Geopolymeric Gels from Alkali-Feldspars." *Colloids and Surfaces A: Physicochemical and Engineering Aspects*. https://doi.org/10.1016/S0927-7757(02)00499-5.

Zhang, Jianguo, John L. Provis, Dingwu Feng, and Jannie S. J. van Deventer. 2008. "Geopolymers for Immobilization of Cr6+, Cd2+, and Pb2+." *Journal of Hazardous Materials*. https://doi.org/10.1016/j.jhazmat.2008.01.053.

In: An Introduction to Aluminosilicates
Editor: Nero Regina Blevins
ISBN: 978-1-53617-250-8

Chapter 13

OLIGOMERIZATION OF UNSATURATED COMPOUNDS IN THE PRESENCE OF HIERARCHICAL ZEOLITE Y

Sergey V. Bubennov[1,*], Nellya G. Grigorieva[1,2,†], Dmitry V. Serebrennikov[1], Marat R. Agliullin[1,2], Alfira N. Khazipova[1] and Boris I. Kutepov[1,2]

[1]Institute of Petrochemistry and Catalysis,
Russian Academy of Sciences, Ufa, Bashkortostan, Russian Federation
[2]Ufa State Petrochemical Oil Engineering University,
Ufa, Bashkortostan, Russia

ABSTRACT

The high activity and selectivity of high crystallinity zeolite Y with a hierarchical porous structure in the synthesis of practically important oligomers of various unsaturated compounds (n- and isopentenes, oct-1-ene, cyclohexene, cyclooctene, norbornene, indene, and α-methylstyrene) was established. The following advantages of the mentioned zeolite in comparison with the microporous zeolite H-Y are shown: high activity in the conversions of unsaturated compounds; oligomerization products obtained in his presence have a wider molecular weight distribution; provides highly selective dimers and trimers of cyclohexene and cyclooctene, in which the cycle of the initial monomer remains unchanged.

* Corresponding Author's Email: bubennov@list.ru.
† Corresponding Author's Email: ngg-ink@mail.ru.

Keywords: oligomerization, hierarchical Y zeolites, pent-1-ene, isopentenes, oct-1-ene, cyclohexene, cyclooctene, norbornene, indene, α-methylstyrene, dimers, oligomers

1. INTRODUCTION

Low molecular weight oligomers of unsaturated compounds are widely used in the petrochemical, chemical, and other industries. After hydrogenation, pentenes oligomers are used as components of gasoline (dimers) or diesel fuel (trimers and higher molecular weight) [1], additives [2], solvents [3], in the production of epoxides, amines, alkyl aromatic compounds and synthetic fatty acids [4]. Oligomers of higher linear alpha-α-olefins (PAO) are used as components of fuels and lubricants, plasticizers, surfactants, oil additives, extractants, resins, etc. [5-10]. Cycloolefin oligomers are used in the preparation of solvents, plasticizers, components of drugs and cosmetic materials [11-13]. Norbornene oligomerization products serve as the basis for high-density rocket fuels [14]. A variety of useful products are obtained from α-methylstyrene oligomers (mainly dimers): molecular weight regulators of polymers, plasticizers, dielectric liquids and solvents [15-21]. Hydrogenated linear dimers of a-methylstyrene are highly effective lubricants [8, 22-24], a cyclic dimer is used to obtain radiation-resistant heat carriers, monomers for the production of gas separation membranes [25].

Recent studies aimed at creating environmentally friendly, low-stage and low-waste heterogeneous-catalytic oligomerization technologies, including those based on the acidic forms of zeolites of various structural types [26-33].

The oligomerization of hex-1-ene [34-36] and α-olefins C_{12} – C_{18} [33, 37-39] in the presence of zeolites Y, MOR, ZSM-5 and Beta was studied.

High efficiency has been shown in the oligomerization of linear olefins of wide-porous zeolites with 12-membered rings, such as faujasite and mordenite. Authors [33], who studied the oligomerization of tetradec-1-ene in the presence of zeolite Y, note that the olefin conversion increases with increasing molar ratio Si/Al of the zeolite, and ultrastable zeolite Y (Si/Al = 60) provides the conversion of tetradec-1-ene, equal to 79.8%. It has been suggested that the cause of the high concentration of dimers is the formation of branched compounds, and the length of their alkyl chains is limited by the pore size of the zeolite.

Our previous study of oligomerization of oct-1-ene in the presence of zeolites Y, Beta, ZSM-12 and ZSM-5 [39] showed that in the temperature range 150-220°C, the activity of the catalysts is in the following order: HBeta > HY> HZSM-12 >> HZSM-5, and in selectivity formation of oligomers: HZSM-12 > HBeta ~ HY >> HZSM-5. Oligomers obtained in the presence of zeolite catalysts are characterized by a high content of dimers (83–100%) having an alkyl naphthenic structure. The content of unsaturated hydrocarbons with tri- and tetra-substituted double bonds is 2.2-3.2%.

When studying the catalytic transformations of dec-1-ene in the presence of zeolites Y (SiO_2/Al_2O_3 = 6.0), Beta, ZSM-12, and ZSM-5 in the H form [40], it was found that the main products formed on the above catalyst systems are dec-1-ene isomers and its dimers, and isomerization of dec-1-ene is predominantly occurs only on HZSM-5 zeolite.

The disadvantage of zeolites is the spatial limitations created by their microporous structure for the formation of large molecules of oligomers. Therefore, the processes of oligomerization of olefins on zeolite catalysts occur with the formation, as a rule, of dimers. The presence of "strong" acid sites leads to the fact that with increasing temperature the contribution of side reactions of cracking and coke formation increases. As a result, the catalyst rapidly loses activity due the blocking of its micropores [2, 39].

The spatial restrictions created by the microporous zeolite crystal lattice for bulk reacting molecules and reaction products can be reduced by forming a developed mesopore system in the porous structure of the zeolite.

To date the two alternatives of creating the transport meso- and macropores in zeolites were proposed. The first one was based on the application of templates during the process of crystallization, the former being surface-active substances [41-43], polymers [44, 45], starch [46], carbon nanotubes [47-49], coal aerogel [50-52], and nanosized $CaCO_3$ [53]. The application of templates enabled to form the zeolite mesopores of 0.15 thru 0.70 cm^3 g^{-1} volume and of 100 thru 500 m^2 g^{-1} surface area as well as to control the pore size within the range of 5-40 *n*m. The drawbacks of the technique were the low accessibility, high cost, and poor crystallinity degree of the materials obtained.

The second option of creating the transport mesopores in zeolites is based on a partial destruction if the zeolite crystalline framework by means of either steam, acid, or alkaline treatment resulting in the formation of mesopores. For example, the dealumination by the steam heating or acid treatment was proposed in refs [54-59], and the desilication with alkali in refs [60-65]. Depending on the zeolite structure and composition the above techniques enabled to create the volume of mesopores within the range of 0.12-0.50 cm^3 g^{-1}, the surface area of mesopores within the range of 100-400 m^2 g^{-1} and adjust the pore size within the range of 5-40 *n*m. Despite the accessibility and the ease of use the techniques in question were characterized by such drawbacks as the decreased acidity and crystallinity of the starting material as the result of the dealumination and desilication, as well as a large quantity of acidic and alkaline effluents.

We have developed a method for synthesizing granular zeolite Y with a high degree of crystallinity in the H-form with a hierarchical porous structure (HY-mmm) [67-69]. It was based on the selective crystallization into the integral cluster crystals of the preliminarily molded granules containing crystals of the zeolite in question and a porous binder matrix.

In this case, the porous structure of zeolite granules are formed not only from micropores but also from meso- and macropores (hierarchical porous structure). Subsequent regulation of the adsorption and catalytic properties of the obtained crystalline aluminosilicates is carried out by changing the modulus of the crystal lattice, the nature and concentration of the exchange metal cations, as well as the conditions of thermochemical treatments.

This paper presents the results of a study of the oligomerization of various unsaturated compounds - n- and isopentenes, oct-1-ene, cyclohexene, cyclooctene, norbornene, indene and α-methylstyrene in the presence of HY-mmm zeolite with the aim of synthesizing practically valuable oligomers. For comparison, the results of the oligomerization of the same compounds on microporous zeolite Y in the H-form (HY) are presented.

2. Synthesis and Physico-Chemical Properties of Hierarchical Y Zeolite

2.1. Synthesis of the Micro-Porous NaY Zeolites

NaY zeolite was crystallized from an amorphous alkaline aluminosilicate gel prepared by mixing sodium silicate, sodium aluminate, and sodium hydroxide solutions. The synthesis was performed from the reaction mix of the following composition: $2.5Na_2O \cdot Al_2O_3 \cdot 7SiO_2 \cdot 220H_2O$ at 98–100°C for 94–96 h. Upon completion of the synthesis the solid phase was separated from the mother liquor, washed off from the excess of alkali to the pH of ~8.0–9.0, and dried at 140–150°C.

2.2. Synthesis of the Hierarchical Y Zeolite

The synthesis of the hierarchical Y zeolite was performed by the crystallization of the preliminarily molded granules of d = 1.1 mm and l = 3.6 mm. The granules for the crystallization were prepared by mixing NaY zeolite with the kaolin binder in the VINCI Technologies MX 0.4 mixer followed by the granulation of the obtained mix using the VINCI Technologies VTE1 extruder. The NaY zeolite content in the starting granule amounted 60% by mass, and the binder material content was 40% by mass, respectively. The obtained granules were dried in the atmosphere of air at 30°C for 24 h followed by the calcination at 650°C for 4 h. During the calcination the amorphization of kaolin in the granule composition occurred and the granules gained the mechanical strength. The calcined samples were designated as NaY-binder, respectively.

The calcined granules of NaY-binder were subjected to the crystallization at 90–100°C for 46–52 h in a sodium silicate solution, the reaction mix composition being $2.5Na_2O \cdot Al_2O_3 \cdot 6.5SiO_2 \cdot 160H_2O$. After the crystallization the granules were separated form the liquid phase, washed off from the excess of alkali to the pH of ~8.0–9.0, and dried at 140–150°C. The Y zeolite samples in H-form were prepared by means of the ion exchange of Na^+ cations to NH_4^+ cations in the aqueous 70 g L^{-1} of ammonium nitrate solution (to provide the excess of the exchanging cation) under stirring at 70–90°C for 1 h. Upon completion of the ion exchange the sample was dried at 120–150°C and calcined at 540–550°C for 4 h in the atmosphere of dry air.

The powdered NaY zeolite sample was designated as micro-Y. The granulated samples of Y zeolite synthesized from the starting NaY-binder granules containing 60% of the crystalline phase respectively, were designated as NaY-mmm. After the conversion of the Na-form into the H-form the H index was added to the sample notation.

2.3. Physico-Chemical Studies of the Obtained Materials

Zeolite catalysts HY and HY-mmm were characterized using the following methods: X-ray fluorescence spectrometry, X-ray diffraction, and X-ray phase analysis, adsorption methods, low-temperature nitrogen adsorption, mercury porosimetry, ^{27}Al and ^{29}Si solid-state NMR spectroscopy (MAS NMR), scanning electron microscopy (SEM). The acidic properties of zeolites were studied by thermoprogrammed ammonia desorption (TPD NH_3) [70] and infrared spectroscopy (IR) using low- temperature adsorption of the CO probe molecule.

It was found that the Si-to-Al atomic ratio in all the synthesized samples of Y zeolite was close to 10 (Table 1).

In Figure 1 the XRD patterns of the synthesized samples were given. For all the samples the signals observed were characteristic just of Y zeolite. The obtained results witnessed a high phase purity of the obtained materials. The NaY and NaYmmm samples were characterized by the crystallinity degree close to 100% and 95%, respectively (Table 1).

Ion exchange of Na^+ cations for H^+ does not change the Si: Al ratio and the degree of crystallinity for all samples. It is noteworthy that in most works on the synthesis of hierarchical zeolites including those based on zeolite Y, it is rarely possible to produce a material with a high degree of crystallinity, as a rule, it does not exceed 70% [41-66].

It was known [67] that the presence of the extra-framework aluminum atoms in the zeolite resulted in the occurrence of the additional signal of 3 through 10 ppm in the MAS NMR ^{27}Al spectra. In Figure 2 the MAS NMR ^{27}Al spectra of the synthesized samples were shown. It was seen that for the NaY sample just one signal of 50 through 60 ppm was observed, characteristic of the aluminum atoms in the tetrahedral oxygen coordination. For the NaY mmm samples apart from the basic signal of 50 through 70 ppm, a weak signal of 0 through 20 ppm was observed; that was a reflection of some part of the extra-framework aluminum atoms present.

Table 1. Chemical compositions and degree crystallinity of zeolite Y

Sample	Chemical compositions, mol	Degree of crystallinity, %
NaY	$Si_{10.21}Al_{1.00}$	100
HY-mmm	$Si_{10.39}Al_{1.00}$	95

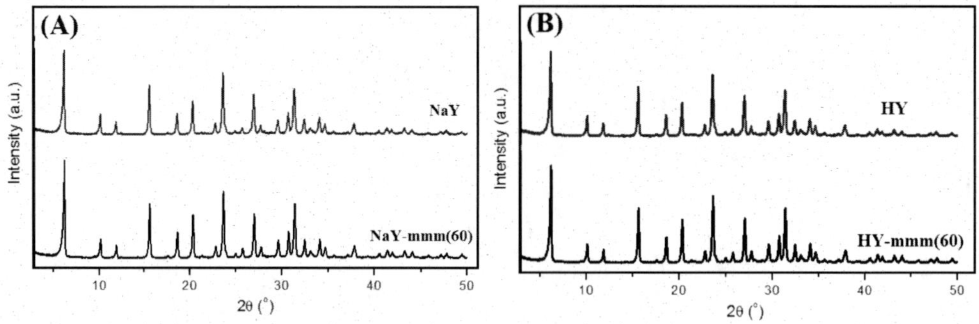

Figure 1. Powder X-ray diffraction of zeolites Y, (A) Na-form (B) H-form.

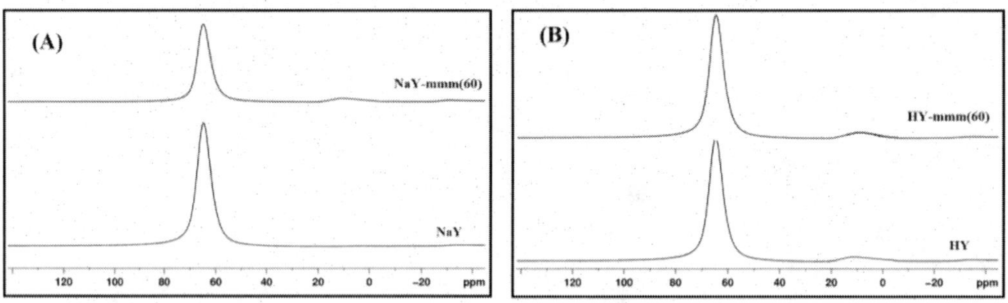

Figure 2. ^{27}Al MAS NMR spectra of zeolites Y, (A) Na-form (B) H-form.

An insignificant increase in the fraction of the signal from 0 to 20 ppm is observed in the transition from the Na-form to the H-form of the zeolite, which is apparently associated with a slight dealumination of the crystal lattice and the appearance of aluminum atoms in the octahedral coordination.

In Figure 3 the nitrogen adsorption–desorption isotherms and the pore size distribution for the NaY-mmm samples were shown. It is seen that for the NaY-type sample the isotherm of the type I, characteristic of the microporous materials is observed. For the NaY-mmm sample the type IV isotherm was observed yet, with the hysteresis loop of N1 type of the IUPAC classification, having an abrupt climb at the pressure of $P/P_0 \sim 1$, characteristic of the meso–macroporous materials. It should be noted that for the NaY-mmm sample there was also the characteristic abrupt climb at low temperatures being a reflection of the presence of macropores. The size of the mesopores determined by the BJH method was within the range of 20 through 30 nm for the NaY-mmm sample (Figure 3).

Similar isotherms of nitrogen adsorption and pore size distribution are also observed for the HY, HY-mmm samples.

In Table 2 the porous structure characteristics were given according to the data of the nitrogen adsorption–desorption and the mercury injection porosimetry. It is seen that according to the BET method the NaY sample is characterized by the micropore volume of 0.30 cm^3 g^{-1} and the surface area of 909 m^2 g^{-1}. The results obtained were in good agreement with the reference data for this zeolite reported in ref. [67, 71]. The porous structure of the NaYmmm sample was characterized by the specific surface of 741 m^2 g^{-1} and the volume of micro-, meso-, and macropores of 0.28, 0.15, and 0.15 cm^3 g^{-1}, respectively.

Table 2. Characteristics of the porous structure

Sample	S_{BET}[a] (m^2 g^{-1})	V_{micro}[b] (mL g^{-1})	V_{meso}[c] (mL g^{-1})	V_{macro}[d] (mL g^{-1})	V_{Σ}[e] (mL g^{-1})
NaY-Binder	550	0.18	0.18	0.50	0.68
NaY	909	0.30	0.03	-	0.33
NaY-mmm	741	0.28	0.15	0.15	0.58
HY-mmm	735	0.28	0.15	0.15	0.58

Conditions:

[a] surface area by method BET.

[b] micropore volume.

[c] mesopore volume.

[d] macropore volume.

[e] total volume.

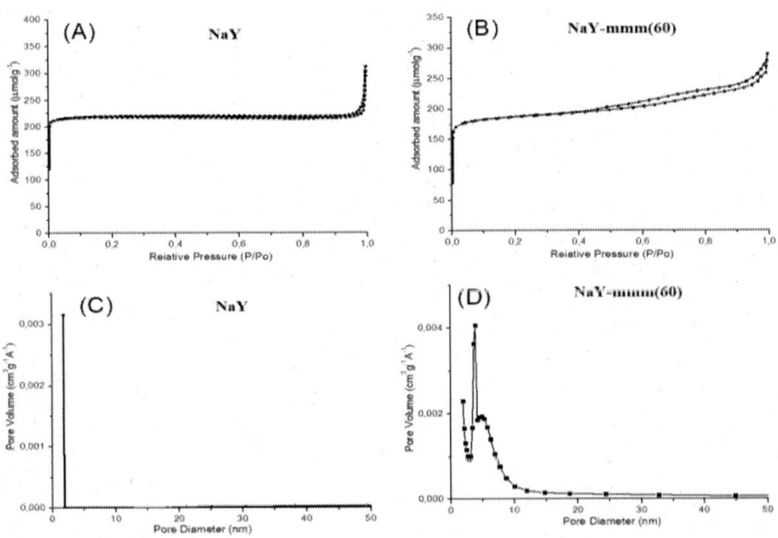

Figure 3. Nitrogen sorption isotherms and pore size distribution for zeolite Y: (A) micro-NaY; (B) meso-Y; (C) micro-NaY; (D) meso-Y.

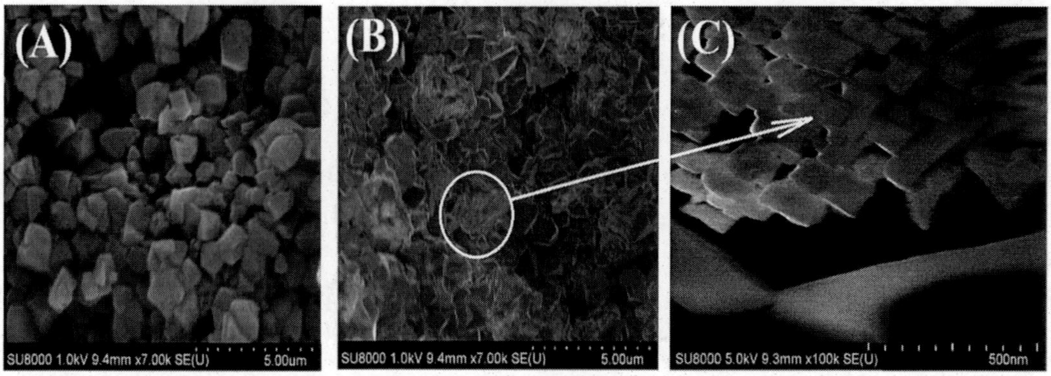

Figure 4. SEM images of zeolite Y: (A) micro-NaY; (B) meso-Y; (C) meso-Y. Reproduced by permission from ref. 67 Copyright 2017 The Royal Society of Chemistry.

It must be pointed out that the characteristics of the porous structure basically do not change during the transition from Na-form to H-form.

Thus, the porous structure of the NaYmmm sample was formed hierarchical porous structure contained not only micropores but meso- and macropores, too. It should be emphasized that in the late work on the synthesis of zeolites with the hierarchical porous structure [72–75] the majority of the obtained materials possessed the micro–mesoporous structure with little to no macro-pores.

In Figure 4 the NaY and NaY-mmm sample images obtained by means of the scanning electron microscopy at different magnification were given. It is seen that the powdered micro-NaY zeolite crystals have the regular cubic shape and the average size of 2 mm. Except for the starting Y zeolite crystals, in the NaY-mmm sample the nanocrystals within the range of 50–200 nm size were observed.

The porous structure of granules subjected to the crystallization in the process of the NaY-mmm sample preparation consisted of micro-pores in the separately taken zeolite crystals and macro-pores located between the separate crystals and the amorphous binder material particles (Table 2).

In the course of the amorphous component crystallization these pores disappeared. Meanwhile an additional quantity of micro-pores and the larger sized pores was formed, resulting from the cluster crystal growth. It followed from the results given above that the micro–meso–macroporous structure of the high crystallinity degree NaY zeolite was formed only when the crystallized granules contained a smaller fraction of the amorphous binder than the fraction of the high-dispersed NaY zeolite.

In Figure 5 the assumed mesopores formation mechanism in the NaY-mmm sample is shown. It was known that the size of crystals formed during the zeolite crystallization process depended on the seed oversaturation degree. The higher the latter, the smaller is the crystal size. Apparently the very high seed oversaturation degrees leading to the formation of the zeolite nanocrystals, mesopores being the cavities between them, occurred during the crystallization of the sample containing 60% by mass of NaY zeolite.

Therefore the primary reason of the mesopore formation in the NaY-mmm sample was the presence of the nanocrystals in the zeolite granule. And the macropores were represented by the cavities formed in the course of the larger crystals clustering.

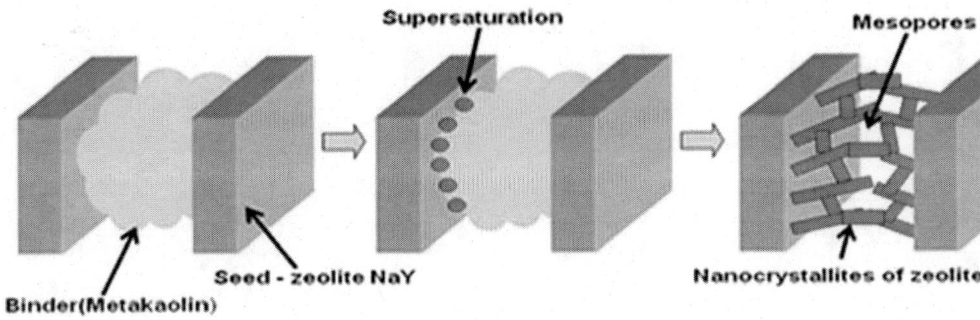

Figure 5. The mechanism of formation of a hierarchical porous structure in meso-Y. Reproduced by permission from ref.67 Copyright 2017 The Royal Society of Chemistry.

Table 3. Type and concentration of LAS and BAS according to FTIR spectroscopy of adsorbed CO and TPD NH_3

Type of acid sites	Lewis acid sites			Brønsted acid sites		TPD ΣNH_3
ν, cm^{-1}	ν_{CO} = 2231-223	ν_{CO} = 2212-2210	ν_{CO} = 2202-2190	ν_{OH} = 3615	ν_{OH} = 3725-3740	150-550°C
The strength of sites	-	-	-	$\Delta\nu_{OH}$ = 300	$\Delta\nu_{OH}$ = 270-320	-
Sample	Concentration (mmol g^{-1})					
micro-HY	7	20	62	78	53	940
HY-mmm	8	25	65	70	50	830

According to the TPD NH_3 data (Table 3) the gross acidity values for the HY and HY-mmm samples amounted 940 and 830 mmol g^{-1}, respectively. A 12% decrease in the acidity values of the HY-mmm samples as compared to the HY sample was explained by the fact that they consisted of the integral cluster crystals, which prevented the access of ammonia molecules to a part of the acid sites.

The nature of OH-groups over HY and HY-mmm zeolites was studied by FTIR spectroscopy. The IR spectrum of OH groups of HY sample revealed six clearly distinguished bands (Figure 6).

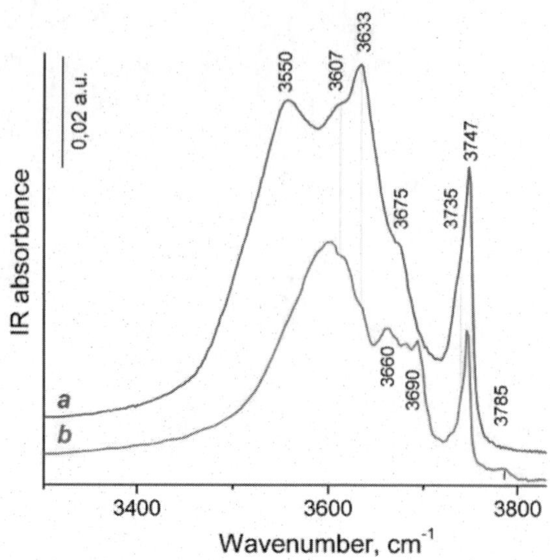

Figure 6. IR spectra of zeolite samples in the OH region: (a) HY; (b) HY-mmm. (Reproduced by permission from ref.67 Copyright 2017 The Royal Society of Chemistry).

The most intense signals at 3747, 3633 and 3550 cm^{-1} were typical for IR spectra of HY zeolites [76–78] and assigned to terminal Si–OH groups and two main kinds of "structural" hydroxyl groups – bridging Si–O(H)–Al groups in the supercages (HF, high frequency) and bridging OH in sodalite cages (LF, low frequency), respectively.

The less intense band at 3607 cm^{-1} was mostly attributed to HF groups polarized by Lewis acidic extraframework Al species (HF') [77–79]. The low intense band at 3675 cm^{-1} can possibly be assigned to the acid Al–O(H)–Al groups which are partially connected to the framework. The shoulder at 3735 cm^{-1} corresponded to internal or defect silanol OH groups typical for zeolites, which are located in the close vicinity to the lattice imperfection or Lewis acid site (e.g., tricoordinated Al atom, Si–OH/Al^{3+} groups) [80].

The HY-mmm zeolite sample synthesis led to the significant decrease in the signals of bridging OH-groups and to the small decrease in the peak intensity of Si–OH species in the IR spectrum (Figure 6b). The LF band was shifted to the shortwavelength region to 3600 cm^{-1}. The new bands at 3660, 3680, 3690 and 3785 cm^{-1} were related to the OH-groups bound to the extra-framework aluminum atoms (bridge Al–O(H)–Al and terminal Al–OH groups). This spectrum looked like for those of dealuminated HY zeolites [79, 81].

Progressive CO adsorption was carried out on both zeolites. During low temperature CO adsorption on HY sample, the HF bands fully disappeared, and a new band at 3330 cm^{-1} appeared, the HF' band was only partially perturbed, but LF band was not perturbed (the spectrum was not shown). This corresponded to a shift $\Delta\nu_{O\text{-}H\cdots CO}$ = 300 cm^{-1} for HF band, that attributed to strong Brønsted acid sites. This value was lower than for typical dealuminated HY zeolites ($\Delta\nu_{O\text{-}H\cdots CO}$ = 320– 400 cm^{-1}) and slightly higher than for typical HY zeolites ($\Delta\nu_{O\text{-}H\cdots CO}$ = 270–290 cm^{-1}) [78, 82]. The appearance of positive bands at 3330

cm^{-1} was accompanied by the broad intense band near at 3400–3450 cm^{-1} and small band at 3595 cm^{-1}.

However, the corresponded negative peak for the latter band was not identified. The appearance of the 3400–3450 broad band was caused by the perturbation of the acid OH groups attached to tricoordinated Al atom (bridge Al–O(H)–Al groups with band at 3675 cm^{-1}) and defect acid silanol OH groups (Si–OH/Al^{3+} groups with band at 3735 cm^{-1}). It demonstrates that these groups are less acidic than bridging Si–O(H)–Al groups in faujasite supercages according to the value of the low frequency shift of OH vibrations with adsorbed CO ($\Delta\nu_{O\text{-}H\cdots CO}$ = 275–285 cm^{-1}). Besides, the highest pressure of CO close to equilibrium resulted in the shift in the intensity of the silanol bands from 3747 cm^{-1} to *ca.* 3660 cm^{-1}: a shift of ~90 cm^{-1} is typical for weak acidic terminal Si–OH groups [81, 82]. The most intense signal in the carbonyl region of IR difference spectra was observed at 2175 cm^{-1}, this band shifted to 2170 cm^{-1} at additional CO dosage. These bands were attributed to the CO complex with strong and moderate acidic OH-groups [81, 82].

Additional bands at 2233–2223 cm^{-1} related to CO complexes with Al^{3+} ions in pentahedron environment being specific structure defects of zeolites (strong LAS), low intense bands at 2216–2190 cm^{-1} corresponded to CO complex with extraframework Al^{3+} species (weak LAS) and band at 2157 cm^{-1} was assigned to CO complex with Si–OH groups [80].

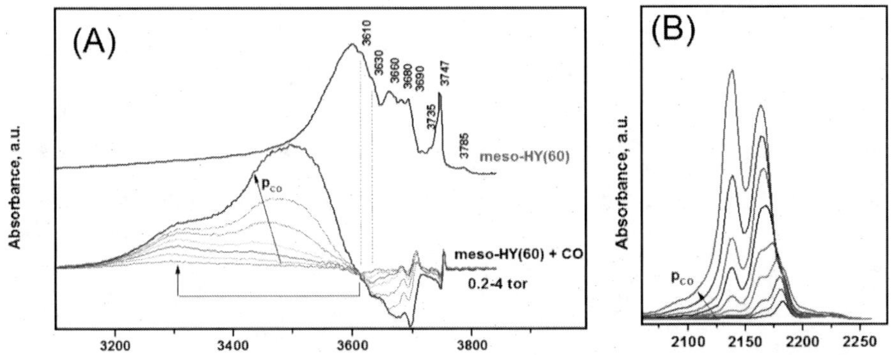

Figure 7. IR spectra of zeolite samples in the OH region: (a) HY; (b) HY-mmm. (Reproduced by permission from ref.67 Copyright 2017 The Royal Society of Chemistry).

Progressive low-temperature CO adsorption on HY-mmm sample revealed a more complex spectra than that of HY sample. During low temperature CO adsorption on HY-mmm sample, the HF and HF' bands (bridging Si–O(H)–Al groups in faujasite supercages) fully disappeared, and a new band at 3300 cm^{-1} appeared, but LF band at 3600 cm^{-1} was not perturbed (Figure 7A). It demonstrates that these HF groups are more acidic than those groups in HY-sample according to the value of the low frequency shift of OH vibrations

($\Delta\nu_{O\text{-}H\cdots CO}$ = 320 cm^{-1}). The concentration of strong BAS for the zeolites samples was given in Table 3. It could be seen that HY-mmm zeolite contained more than three times less strong BAS in faujasite supercages than HY zeolite. The appearance of positive bands at 3300 cm^{-1} 1 for HY-mmm zeolite was accompanied by the broad band near at 3430–3480 and very intense band at 3500–3520 cm^{-1}. The former signals were caused by the perturbation of the acid OH groups attached to tricoordinated Al atom (bridge Al–O(H)–Al groups with band at 3660 and 3680 cm^{-1}) and defect acid silanol OH groups (Si–OH/Al^{3+} groups with band at 3735 cm^{-1}).

The latter peaks were caused by the perturbation of the OH groups with band at 3690–3695 cm^{-1}; the low frequency shift of this OH groups ($\Delta\nu_{O\text{-}H\cdots CO}$ = 170–190 cm^{-1}) corresponded to moderate BAS. In the carbonyl region of IR difference spectra it could be observed the intense signal at 2178 cm^{-1} that shifted to 2167 cm^{-1} at additional CO dosage. These bands corresponded to the CO complex with strong and moderate acidic OH-groups (Figure 7B) [81]. According to IR spectra of adsorbed CO in carbonyl region, surface groups in HY-mmm sample contain three types of LAS with following absorbance bands: (1) bands at 2231–2223 cm^{-1} related to CO complexes with Al^{3+}ions in pentahedron environment being specific structure defects of zeolites (strong LAS), (2) low intense bands 2210–2208 cm^{-1} corresponding to LAS of medium strength, (3) bands at 2200–2192 cm^{-1} relating to CO complex with Al^{3+} ions of alumina clusters (weak LAS) [82]. The concentration of all types of LAS for the zeolite samples was given in Table 3. It could be seen that HY-mmm zeolite contained three times more moderate and weak LAS deal with extraframework Al species.

3. The Catalytic Properties of Zeolites HY-mmm and HY in the Oligomerization of Unsaturated Compounds

3.1. Synthesis of Oligomers

The catalytic transformations of pentenes, oct-1-ene [39] and cycloolefins [83, 84] were carried out in a temperature-controlled continuously rotating autoclave in a temperature range of 60–250ºC for 1-5 h. The weight of the initial olefin was from 1.0 to 2.0 g, the catalyst concentration was 10–30% of the weight of the feedstock.

The oligomerization of indene, α-methylstyrene [31, 85-87] was performed in a batch isothermal reactor equipped with a stirrer at 20°C for 1–3 h in chlorobenzene. The weight concentration of the catalyst was 10% per α-methylstyrene. Freshly distilled α-methylstyrene (99 wt %) preliminary dried over calcined barium oxide was used for the experiments.

After the experiments the reaction mixture was cooled down, separated by filtration from the catalyst and analyzed by gas-liquid chromatography (GLC) using an HRGS 5300 Mega Series Carlo Erba chromatograph equipped with a flame ionization detector (SE-30 capillary column with a length of 25 m, the analysis temperature of 50–280°C with programmed heating 8°C/min). To determine the composition of oligomers the products were analyzed by high- performance liquid chromatography (HPLC) on a Shimadzu LC-20 Prominence instrument. Analysis conditions were as follows: a Plgel 100Å polystyrene column, a toluene flow rate of 0.8 mL/min, and a RID 20A refractometric detector.

3.2. Oligomerization of Linear Olefins

The main scheme for the conversion of olefins on zeolite catalysts can be represented as follows:

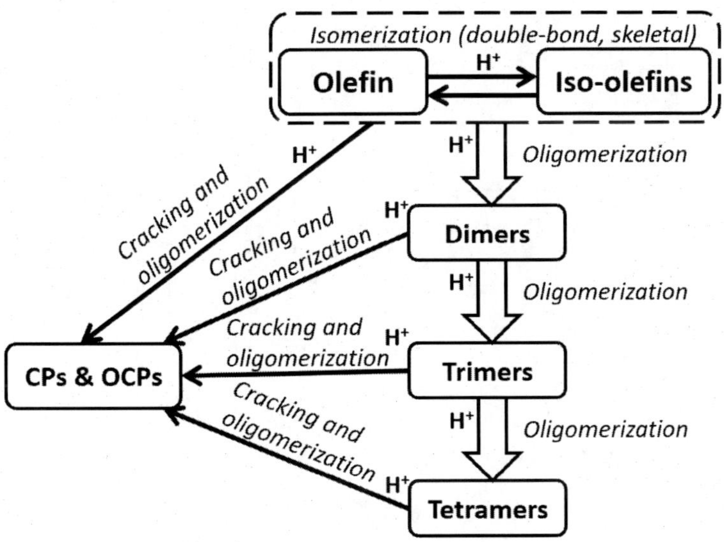

Figure 8. Basic transformation of linear olefins in the presence of zeolite catalysts.

Both literature [33, 34] and our data [39, 40] show that in the presence of zeolite catalysts olefins are converted to isomers and oligomers. Depending on the catalyst and reaction conditions oligomers contain dimers and higher molecular weight oligomers with oligomerization degree n ≥ 3. In addition, oligomers of olefins with molecular weights lower than that of the dimers (hereinafter they are designated as OCP – oligomers of cracking products) are present. These compounds are formed by the destruction of oligomers and the initial monomer or its isomers to produce lighter olefins (cracking products, CP), which are then oligomerized.

3.2.1. Oligomerization of Pentenes

In order to develop a heterogeneous-catalytic method for processing pentenes by oligomerization into environmentally friendly fuel components, the transformations of pent-1-ene and isopentenes fractions (2-methylbut-2-ene – 53%, 2-methylbut-1-ene – 28%, 3-methylbut-1-ene – 3%, n-pentenes – 5% and C5 alkanes – 11%) in the presence of HY-mmm and HY zeolites were studied.

The results of the oligomerization of pent-1-ene are given in Table 4. Under the studied conditions, pent-1-ene is easily isomerized on zeolites to cis- and trans-pent-2-enes, which are rapidly converted into oligomers. Oligomers are represented mainly by di- ($C_{10}H_{20}$), tri- ($C_{15}H_{30}$) and tetramers ($C_{20}H_{40}$) pentenes. The resulting oligomers, like the initial pentenes, at a temperature of ≥150°C can undergo destruction with the formation of "light" hydrocarbons, which are then oligomerized to form OCP.

Table 4. Oligomerization of the pent-1ene in the presence of HY-mmm and HY zeolites

Sample	Amount of the catalyst, % wt.	T, °C	Conversion, %	Selectivity, %				
				isomers	dimers	trimers	tetramers*	CP, OCP
HY-mmm	10	150	90	18	47	32	3	-
HY-mmm	30	150	100	-	46	39	15	-
HY-mmm	30	200	100	-	37	38	16	9
HY	10	150	95	34	40	24	2	-
HY	30	150	100	-	47	41	12	
HY	30	200	100	-	40	32	10	18

Reaction conditions 5 h.

Table 5. Oligomerization of the amylene fraction in the presence of zeolites HY-mmm and HY

Sample	% wt. of zeolite	T,°C	Conversion, %	Selectivity, %		
				dimers	trimers	CP, OCP
HY-mmm	10	80	83	84	15	1
HY-mmm	10	150	91	77	10	13
HY	10	80	80	90	9	1
HY	10	150	95	70	9	21

Reaction conditions 5 h.

Both catalysts are highly active in the reaction: pentene conversion at 150-200°C reaches 90–100%. The highest yield of oligomers (100%) was achieved at 150°C (30 wt % of the catalyst); at 200°C, the yield decreases due to the appearance of OCP in the reaction products. The amount of degradation products and their oligomers is significantly higher in the presence of microporous zeolite HY, which is associated with a higher acidity of this catalyst. Comparison of the composition of oligomers obtained over zeolites HY and HY-mmm shows that the products synthesized over a micro-meso-macroporous catalyst have a higher share of tri- and tetramers. With increasing temperature and the amount of catalyst in the oligomeric fraction the content of dimers decreases and the content of higher molecular weight compounds increases.

The oligomerization of the amylene fraction was carried out at a lower temperature – 80–150°C, because at a temperature above 100°C the number of cracking products and their OCP oligomers sharply increased (Table 5).

At 80°C, the conversion of isoamylenes was 80-83%, with an increase in temperature to 150°C it increased to 91-95%. The reaction products are di- and trimers of pentenes. Decenes are most selectively formed on microporous zeolite HY at 80°C, and the ratio of dimers: trimers is 10. Under the same conditions, over the hierarchical zeolite HY-mmm, the selectivity of trimer formation is higher, and the ratio of dimers: trimers is 5.6. with an increase in the reaction temperature (150°C), the content of low molecular weight oligomers of OCP in the reaction products increases; therefore, the selectivity of the formation of oligomers decreases on both catalysts. A higher concentration of acid sites in zeolite HY leads to the fact that the amount of OCP formed is almost two times higher than over zeolite HY-mmm. Therefore, if the selectivity for the formation of oligomers over HY-mmm zeolite at 150°C is 87%, then over the HY zeolite the selectivity decreases to 79%.

Thus, in the oligomerization of pentenes, the activity of micro- and micro-meso-macroporous zeolites HY and HY-mmm is close. Oligomers obtained over the hierarchical zeolite HY-mmm are characterized by a higher content of tri- and tetramers in comparison with microporous zeolite HY.

3.2.2. Oligomerization of Oct-1-ene

The initial oct-1-ene undergoes isomerization in the presence of zeolite catalysts; therefore, the oligomers formed from n- and i-octenes are a mixture of a large number of compounds. Along with oligomerization and isomerization, cracking of the initial monomers and oligomers can occur with the formation of lighter hydrocarbons and their subsequent oligomerization. These compounds are referred to as OCP.

Table 6 presents the results of the catalytic transformations of oct-1-ene in the presence of zeolites HY-mmm and HY.

It can be seen that under comparable conditions, the conversion of oct-1-ene over HY-mmm zeolite is higher than over HY zeolite. For example, at 180°C in the presence of 10-30%wt. of sample HY-mmm almost complete conversion of oct-1-ene is observed. At the same time, over microporous zeolite HY at 180°C and a catalyst content of 30%, the mass conversion is 95%. Thus, the hierarchical zeolite HY-mmm is somewhat more active than the microporous zeolite HY, although the latter is characterized by a higher concentration of acid sites. The results can be explained by the greater availability of acid sites in the zeolite sample HY-mmm.

The results presented in Table 6 shows that the qualitative composition of the products of the catalytic transformations of oct-1-ene formed in the presence of zeolites HY-mmm and HY is the same. At the same time, the content of individual compounds is significantly different. Over HY-mmm zeolite in the presence of 10% catalyst at 180°C, the total content of oligomers is 80%, while the selectivity for the formation of dimers, trimers and tetramers is 58, 17 and 5%, respectively. The ratio of oligomers: isomers is 4.3. Over zeolite HY, under the same conditions, mainly octene isomers and oligomer: isomer ratios are formed: 0.2.

With an increase in the amount of zeolite HY-mmm up to 30 wt % (180°C) or with an increase in temperature to 200°C (10% of the catalyst), the selectivity of the formation of oligomers increases slightly – up to 82–83%. It is important that the proportion of tri- and tetramers in the oligomeric fraction increases. So, at 180°C (10% of the catalyst), the content of dimers in the oligomer fraction is 73%, and that of tri- and tetramers is 27%; at 200°C, the content of dimers decreases to 66%, and the content of oligomers $n \geq 3$ increases to 34%. The composition of oligomers formed on zeolite HY changes in a similar way, however, the fraction of tri- and tetramers in the oligomeric fraction is lower than over zeolite HY-mmm.

Table 6. Oligomerization of oct-1-ene in the presence of zeolites HY-mmm and HY

Sample	% wt. of zeolite	T, °C	Conversion, %	Selectivity, %				
				isomers	dimers	trimers	tetramers	OCP
HY-mmm	20	130	71	40	44	12	4	-
HY-mmm	5	180	79	36	50	10	4	-
HY-mmm	10	180	99	19	58	17	5	1
HY-mmm	20	180	100	17	56	18	8*	1
HY-mmm	30	180	100	17	55	19	8*	1
HY-mmm	10	200	100	16	55	20	8*	1
HY	10	180	87	72	18	9	-	1
HY	20	180	90	67	21	10	-	2
HY	30	180	95	57	26	11	-	6
HY	10	200	90	46	37	10	2	5
HY	30	200	96	24	50	11	4	11
HY	30	250	100	23	38	12	4	23

* - the transformation products of oct-1-ene contain about 2-3% pentamers reaction time 5 h.

It should be noted that over the hierarchical zeolite HY-mmm, OCP compounds are formed in insignificant amounts. The selectivity of their formation does not exceed 1%. Over microporous zeolite HY, the selectivity of OCP formation reaches 6% at 180°C and increases with increasing temperature and the amount of catalyst. The propensity of this catalyst for destructive transformations is apparently due to a high concentration of acid sites, among which there are both strong and weak [88-90]. Cracking reactions are promoted, as is known [91], by stronger acid sites.

Thus, it follows from the obtained results that, under the studied conditions, the HY-mmm zeolite, the porous structure of which is characterized by the presence of micro-, meso-, and macropores, is much more effective in oct-1-ene oligomerization than the microporous HY zeolite.

3.3. Oligomerization of Cyclic Unsaturated Compounds

3.3.1. Oligomerization of Cyclohexene

Cyclohexene in the presence of zeolite catalysts is converted mainly into dimers, among which 1- (1´-cyclohexenyl) cyclohexane (1) predominates (Figure 9). Along with compound 1, other cyclohexene dimers are also contained: unsaturated 1- (2´-cyclohexenyl) (2) and 1- (3´-cyclohexenyl) cyclohexane (3) and saturated bicyclohexane (4). In addition to dimers, the products contain trimmers (5), cyclic cyclohexene isomers (methylcyclopentenes), as well as hexadiene, methylpentenes, dimethylbutenes and other hydrocarbons formed as a result of destructive transformations. The sum of the degradation products and their dimers with a molecular weight of 82–164+ g/mol, is designated as "others."

The results of studying the catalytic properties of zeolites HY-mmm and HY in the oligomerization of cyclohexene in the temperature range 120–200°C are presented in Table 7.

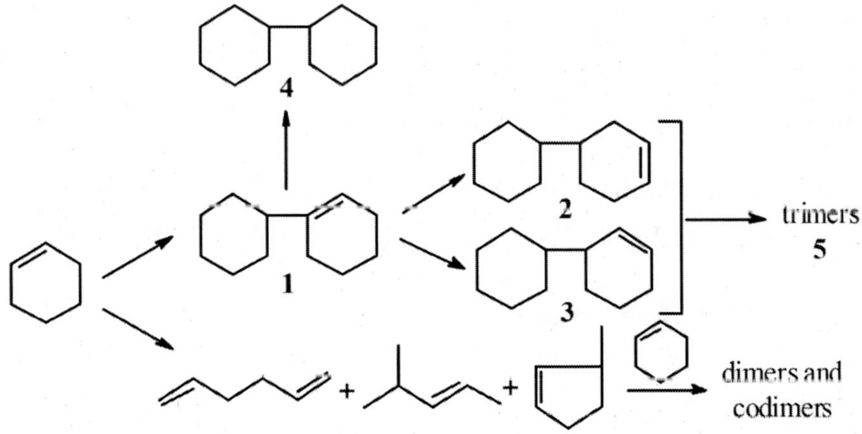

Figure 9. Transformations of cyclohexene in the presence of zeolites HY-mmm and HY.

Table 7. Oligomerization of cyclohexene over HY-mmm and HY zeolite samples

Catalyst	T,°C	Conversion, %	Selectivity, %			
			Dimers		Trimers	Others
			(1)	Σ (2-4)		
HY-mmm	120	23	76	traces	14	9
HY	120	19	77	7	traces	16
HY-mmm	150	57	44	1	15	40
HY	150	32	62	7	1	30
HY-mmm	200	74	34	1	16	49
HY	200	46	49	10	2	39

Reaction conditions: 20% wt. of catalyst, solvent- nonane, [M0] = 1.2 mol L-1, 5 h.

Comparison of cyclohexene conversion in the presence of HY-mmm and HY zeolites under the same conditions shows that the micro-meso-macroporous catalyst HY-mmm is more active than the microporous sample HY. With an increase in temperature from 120 to 200°C, the conversion of cyclohexene increases more significantly over the HY-mmm sample than over the HY zeolite.

At 120°C, the main reaction product in the presence of both catalysts is dimer 1. The selectivity of its formation over zeolites HY-mmm and HY is close and amounts to 76% and 77%. Dimers 2-4 over zeolite HY-mmm are formed in small amounts (<1%), and over zeolite HY, the selectivity of their formation is several times higher (7%). At the same time, trimers (5) with a selectivity of 14% are formed over HY-mmm zeolite, and their content over zeolite HY does not exceed 2%.

With an increase in temperature to 200°C, the rate of side reactions of skeletal isomerization and monomer destruction increases, as a result of which the selectivity for the formation of dimer 1 decreases both HY-mmm zeolite and HY zeolite, and the selectivity for the formation of destructive transformation products and their oligomers increases. Similar results were obtained by us earlier in the study of cyclohexene oligomerization in the presence of microporous zeolites of various structural types [84].

Thus, under the studied conditions, the maximum yield of cyclohexene oligomers (dimers and trimers) was achieved in the presence of HY-mmm zeolite with a hierarchical porous structure. The microporous structure of HY zeolite limits the possibility of the formation of cyclohexene trimers; therefore, dimers are the main reaction products on it.

3.3.2. Oligomerization of *Cis*-Cyclooctene

The scheme of catalytic transformations of cyclooctene is shown in Figure 10.

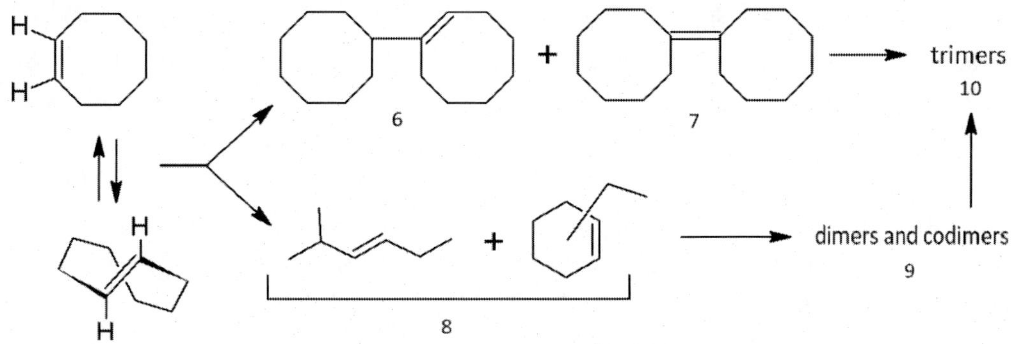

Figure 10. Transformation of cis-cyclooctene in the presence of zeolites HY-mmm and HY.

Table 8. Oligomerization of cis-cyclooctene over HY-mmm and HY zeolites

Catalyst	Amount of the catalyst, % wt.	T,°C	Conversion, %	Selectivity, %				
				trans-Isomer	Dimers		Trimers (10)	OCP (8+9)
					6	7		
HY-mmm	10	60	60	6	49	45	-	-
HY	10	60	6	54	24	-	-	22
HY-mmm	10	100	72	9	44	35	10	2
HY	10	100	32	28	48	-	2	22
HY-mmm	30	100	80	10	36	36	16	2
HY	30	100	71	10	39	-	8	43
HY-mmm	10	120	75	16	33	18	11	22
HY	10	120	68	9	16	-	6	69
HY-mmm	20	120	84	15	24	14	20	27
HY	20	120	71	7	13	-	12	68

Reaction condition: 5h.

OCP - the summary of structural isomers of cyclooctene and their homo- and codimers.

The products of *cis*-cyclooctene conversion mainly contain dimers (6, 7), trimers (10) and linear and cyclic cyclooctene isomers resulting from destructive transformations (8), as well as their homo- and codimers (9).

The results of studying the catalytic properties of HY-mmm and HY zeolites in the oligomerization of *cis*-cyclooctene in the temperature range of 60-120°C are presented in Table 8.

Over the entire temperature range studied (60–120°C), the conversion of cis-cyclooctene over the HY-mmm sample is higher than in the presence of the microporous zeolite. Note that at 60°C, HY zeolite is inactive (cyclene conversion 6%), while on a hierarchical sample, cyclooctene conversion is 60%.

The composition of the products varies significantly depending on the catalyst and reaction conditions.

HY-mmm allows the synthesis of cycloooctene dimers (6, 7) with high selectivity. The maximum value - 94% - was reached at 60ºC in the presence of 10% catalyst. With increasing temperature to 120ºC and the amount of catalyst up to 30%, the selectivity of the formation of dimers 6, 7 decreases due to the formation of trimers and products of the destructive transformations of the initial monomer, as well as their oligomers (OCP). At the same time, the composition of dimers 6, 7 also changes: the ratio of dimers 6: 7 with stricter reaction conditions changes in the direction of increasing isomer 6. For example, at 60ºC the ratio of dimers 6: 7 is 1: 1, at 100ºC -1.3: 1, at 120ºC -1.8: 1.

Over zeolite HY in the products of the conversion of cis-cyclooctene at 60ºC, more than half (54%) is accounted for by its trans-isomer. It should be noted that only dimer 6 is present in the reaction mass, and dimer 7 is not detected. Trimers at 60ºC are not formed. A characteristic feature of zeolite HY is the presence of a significant amount of OCP even at low temperature (22%), the selectivity of which increases with increasing temperature and the amount of catalyst. The high yield of OCP over HY zeolite can be explained, firstly, by a higher concentration of acid sites compared to HY-mmm; secondly, the fact that in the microporous structure of HY zeolite the destruction and isomerization reactions are taking place easier than dimerization with the formation of bulk molecules of compounds 6 and 7. The same factors (steric hindrance make by the microporous structure of HY zeolite) are apparently can cause lower the content in the reaction products of cyclooctene trimers, than over HY-mmm zeolite.

Thus, dimerization of cyclooctene while maintaining the ring structure over microporous HY zeolite is difficult due to steric hindrances arising from the movement of bulk molecules of monomers and dimers inside the cavity of the catalyst. The selective synthesis of dimers 6, 7 was carried out in the presence of HY-mmm zeolite, which has a micro-meso-macroporous structure. Previously, organometallic catalysts were used for the synthesis of 6, 7 dimers [92].

3.3.3. Oligomerization of Norbornene

The products of catalytic transformations of norbornene over HY and HY-mmm zeolites are the cyclic isomer of norbornene — nortricyclane 11, dimers 12 a-d, and trimers 13 (Figure 11).

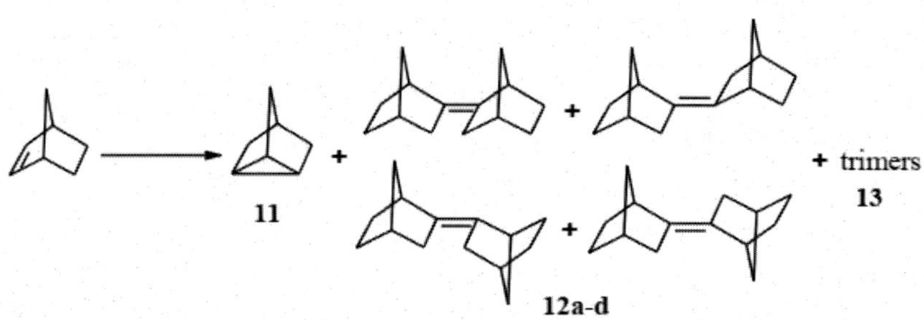

Figure 11. Transformations of norbornene over H-Ymmm and H-Y zeolites.

Norbornene 12 a-d dimers are represented by four compounds: (Z), syn — bis-2,2'-norbornylidene, (E), syn-bis-2,2'-norbornylidene, (Z), anti-bis-2,2 ' norbornylidene, (E), anti-bis-2,2'-norbornylidene.

The results of a study of the catalytic properties of HY-mmm and HY zeolites in the oligomerization of norbornene are presented in Table 9.

It can be seen that under comparable conditions, the conversion of norbornene over HY-mmm zeolite is higher than over HY zeolite, although with an increase in temperature from 60 to 80°C, the amount of catalyst from 10 to 20% wt. and the duration of the experiment is 5 to 300 minutes, the difference in conversion values decreases from 2.9 times to 1.4 times. The results are explained by the greater availability of acid sites of the HY-mmm zeolite, which has a combined micro-meso-macroporous structure.

Table 9. Oligomerization of norbornene over HY-mmm and HY zeolites

Sample	Amount of the catalyst, % wt.	T,ºC	Time, min	Conversion, %	Selectivity, %		
					Nortricyclane 11	D 12a-d	T 13
HY-mmm	10	60	5	46	88	12	-
HY	10	60	5	16	80	20	-
HY-mmm	10	60	60	80	50	35	15
HY	10	60	60	42	54	46	-
HY-mmm	20	80	300	93	40	46	14
HY	20	80	300	67	47	53	-

Reaction conditions: [M0] = 1.9 mol L-1 in CCl4. D- dimers, T- trimers.

At 60°C in the composition of the products obtained over both catalysts, the main product of the conversion of norbornene is its isomer - nortricyclane (11). With an increase in the reaction time from 5 to 60 minutes, the selectivities of the formation of oligomers and nortricyclane become close. Increasing the temperature, amount of catalyst and reaction time allows increasing the number of oligomers to 53% over HY and up to 60% over Ymmm. Moreover, the oligomers obtained over HY-mmm zeolite contain dimers and trimers, while for HY zeolite there were only dimers. Thus, in the presence of HY-mmm zeolite norbornene oligomerization proceeds with the formation of higher molecular weight oligomers than over HY zeolite.

3.4. Oligomerization of Indene

Indene oligomerization over zeolite catalysts leads to the formation of dimers and higher molecular weight compounds, the degree of oligomerization (n) of which depends on the reaction conditions. There are 5 compounds in the mixture of hydrocarbons with molecular weight = 232, called “indene dimers.” Of these, 4 compounds were identified as dimers of indene 14-17, the formation scheme of which is given below (Figure 12).

Another compound with a molecular weight of 232 - 6- (2,’ 3’-dihydro-1’-H-inden-1’-yl) -1H-indene (18) is the alkylation product of the aromatic indene ring with another molecule of olefin.

When carrying out solvent-free oligomerization, zeolite catalysts are rapidly deactivated by the indene oligomers formed. The indene conversion is 4-24% (10% wt. catalyst, 120°C).

Solvent-free indene oligomerization products are a mixture of compounds containing indene oligomers with n = 2-18, as well as hydrocarbons resulting from alkylation of indene on the aromatic ring with another olefin molecule (5-10%). In the reaction products obtained in the presence of 10% wt. HY zeolite, compounds with a molecular weight of 2090 prevail at 120°C, which corresponds to the degree of oligomerization of indene n ~ 18. The content of these hydrocarbons in the oligomers reaches 88%.

In the environment of hydrocarbon solvents (nonane), over HY zeolite the indene conversion remains low – 10% at 150°C, but the composition of the oligomers varies (Table 10). In this case, the products of oligomerization of indene are dimers; higher molecular weight compounds are absent.

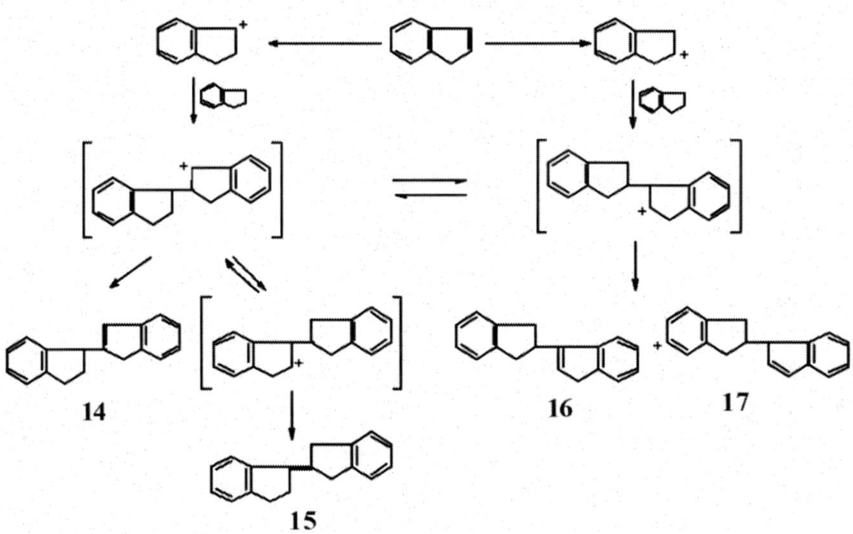

Figure 12. Dimerization of indene.

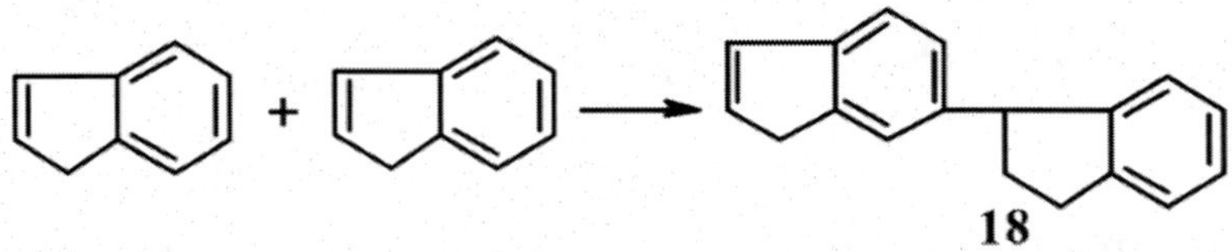

Figure 13. Alkylation of indene.

Table 10. Oligomerization of indene over zeolite HY-mmm and HY

Catalyst	T,°C	K, %	Selectivity, %		
			Dimers	Trimers	n≥4
HY-mmm	120	34	79	18	3
HY	120	5	100	-	-
HY-mmm	150	57	70	25	5
HY	150	10	100	-	-

Molar ratio of indene: nonane = 1:1, 10 wt.% catalyst, 5 h.

Unlike HY zeolite, hierarchical HY-mmm zeolite is more active in the oligomerization of indene and allows to obtain not only dimers, but also higher molecular weight oligomers - tri-, tetra-, pentamers.

3.5. Oligomerization of Vinylarene - α- Methylstyrene

The oligomerization products of α-methylstyrene over zeolite catalysts are linear (19, 20) and cyclic (21) dimers, as well as trimers (22) (Figure 14).

The results of the study of the oligomerization of α-methylstyrene over samples of HY-mmm and HY zeolites are shown in Table 11.

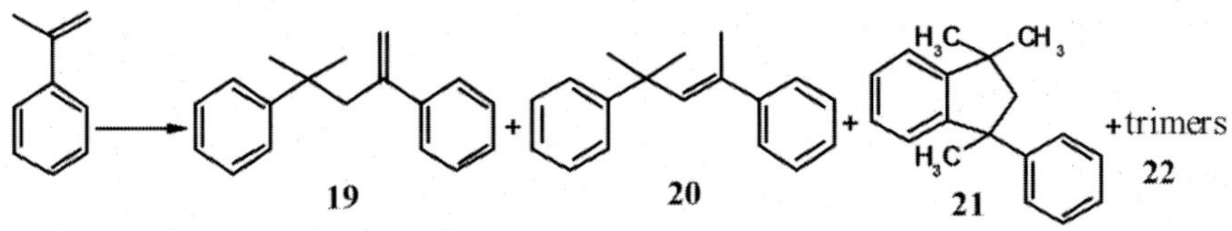

Figure 14. Oligomerization of α-methylstyrene in the presence of HYmmm and H-Y zeolites.

Table 11. Oligomerization of a-methylstyrene over HY-mmm and HY zeolites

Sample	T,°C	Reaction time, h	Conversion, %	Selectivity, %				
				21	19,20	Trimers	Tetramers	Pentamers
HY-mmm	20	5	100	63	-	24	9	4
HY	20	5	22	3	97	-	-	-
HY-mmm	120	1	100	68	-	31	1	-
HY	120	1	99	85	15			
HY-mmm	120	5	100	10	-	69	15	6

10% wt. of zeolite catalyst.

It was established that the hierarchical H-Ymmm allows the oligomerization of α-methylstyrene under very "mild" conditions, at 20°C. In this case, complete conversion of α-methylstyrene is achieved, oligomers with oligomerization degree n = 2–5 are formed.

The presence of only the cyclic isomer 21 in the composition of the dimers indicates that, under these conditions, the isomerization of linear dimers was complete.

The conversion of α-methylstyrene over microporous HY zeolites at 20°C even after 5 hours is only 22%. The reaction products are exclusively dimers, and the selectivity of the formation of linear dimers 19, 20 is very high - 97%.

The results are due to differences in the porous structure and acid properties of the HY and HY-mmm zeolite catalysts. The low activity of HY zeolite can be explained by the fact that the micropores of this catalyst at a low temperature (20°C) are quickly blocked by the formed oligomerization products, as a result of which the acid sites inside large cavities become inaccessible to the reacting molecules. Apparently, oligomerization over HY zeolite occurs at a few surface active sites, the strength of which is less than that of acid sites located in zeolite cavities. This assumption is confirmed by the composition of the products formed: linear dimers are formed at weak acid sites [32]. The high activity of HY-mmm in the oligomerization of α-methylstyrene is achieved due to the presence of meso- and macropores in its structure. Such a combined micro-meso-macroporous structure makes it possible to reduce diffusion restrictions for the movement of reagent molecules and reaction products, and to reduce the possibility of blocking pores. In addition, the concentration of reacting molecules in mesopores is higher, which leads to an increase in the acts of chemical interaction of the reactants and an increase in the degree of their conversion.

With an increase in the reaction temperature to 120°C, the conversion of α-methylstyrene over microporous HY zeolite increases to 99%, however, only dimers are still present in the oligomers. High temperature accelerates the isomerization reaction of linear dimers 19. 20; therefore, at 120°C in the dimeric fraction, the main compound is cyclic dimer 21.

Over HY-mmm zeolite, an increase in temperature to 120°C leads to an increase in the proportion of trimers in the composition of oligomers. An increase in the reaction time from 1 h to 5 h allows one to increase the yield of trimers to 69%, and tetra- and pentamers to 21%.

CONCLUSION

A simple and convenient method for the synthesis of granular Y zeolite with a high degree of crystallinity in the H-form with a hierarchical porous structure - zeolite HY-mmm was developed. The porous structure of the Ymmm sample was characterized by the specific surface of 741 m^2 g^{-1} and the volume of micro-, meso-, and macropores of 0.28, 0.15, and 0.15 cm^3 g^{-1}, respectively.

It was shown that the micro-meso-macroporous zeolite HY-mmm can be successfully used in the oligomerization of various unsaturated compounds (pentenes, oct-1-ene, cyclohexene, cyclooctene, norbornene, indene and α-methyl styrene).

In most of the reactions studied, HY-mmm zeolite is more active than traditional microporous HY zeolite and contribute to the formation of oligomers with a wider molecular weight distribution (degree of oligomerization n = 2–5) than the microporous catalyst, the main products of which are usually dimers.

REFERENCES

[1] Cruz, V. J., Izquierdo, J. F., Cunill, F., Tejero, J., Iborra, M. and Fité, C. (2005). Acid ion-exchange resins catalysts for the liquid-phase dimerization/etherification of isoamylenes in methanol or ethanol presence. *Reactive and Functional Polymers*, 65 (1-2): 149–160.

[2] Kulkarni, A., Kumar, A., Goldman, A. S. and Celik, F. E. (2016). Selectivity for dimers in pentene oligomerization over acid zeolites. *Catalysis Communications*, 75: 98–102.

[3] Pat. US 4335009 (1982). *Mixture of aliphatic C_{10} branched olefin epoxides and use thereof in augmenting or enhancing the aroma of perfumes.*

[4] Cejka, J., Zilkova, N. and Nachtigall, P. (2005). The applications of zeolite catalysts: production and uses of light olefins. *Industrial Studies in Surface Science and Catalysis,* 158:1201–1212.

[5] Pat. US7022784 (2006) Synthetic lubricant composition and process.

[6] Fink, J. (2015). *Petroleum Engineer's Guide to Oil Field Chemicals and Fluids*, Gulf Professional Publishing. 825 p.

[7] Toffen, G. E. (2003). *Fuels and Lubricants Handbook: Technology, Properties, Performance, and Testing*. USA, ASTM International. 1061p.

[8] Mang, Th. and Dresel, W. (2007). "*Lubricants and Lubrication.*" Second, Completely revised and extended edition. Wiley-VCH. Verlag GmbH & Co. KGaA, Weinheim.

[9] Kotov, S. V., Moiseev, I. K. and Shabanova, A. V. (2003). Olefin oligomers: manufacturing processes and use as fuel and lube oil components. *Petroleum Chemistry*, 43 (5): 289-298.

[10] Murazo, O. (2015) Maximizing diesel production through oligomerization: A landmark opportunity for zeolite research. *Industrial & Engineering Chemistry Research,* 54(3): 781-789.

[11] Jenkins, G. N., Roberts, S. M. and Turner, N. J. (1995) Recent Uses of Oxidoreductase Enzymes, Department of Chemistry, University of Exeter, Stocker

Road, Exeter, Devon, UK, EX4 4QD, http://www.ch.ic.ac.uk/ectoc/papers/04/4-Oxidoreductases.html.

[12] Davies, H. G., Dawson, M. J., Lawrence, G. C., Mayall, J., Noble, D., Stanley, M. R., Turner, M. K. and Wall, W. F. (1986). Microbial hydroxylation of cyclohexylcyclohexane: synthesis of an analogue of leukotriene-b3. *Tetrahedron Letters*, 27(9): 1089-1092.

[13] Pat. US 7572435 (2009). *Oil bodies for cosmetic compositions containing cyclohexylcyclohexane*.

[14] Eychenne, P., Perez, E., Rico, I., Bon, M., Lattes, A., Moisand, A. (1993) First example of latices synthesis via oligomerization of norbornene in aqueous emulsions, catalyzed by palladium chloride. *Colloid Polym Sci,* 271 (11):1049-1054.

[15] Kirk-Othmer (1978) Encyclopedia of Chemical Technology. 3rd ed., John Wiley and Sons, New York, NY: 819p.

[16] Qun Sun, Farneth W. E. and Harmer M. A. (1996) Dimerization of α-Methylstyrene (AMS) catalyzed by sulfonic acid resins: A Quantitative kinetic study. *Journal of Catalysis,* 164(1):62-69.

[17] Haiming Wang, Peng Cui, Gang Zou, Fan Yang and Jie Tang. (2006) Temperature-controlled highly selective dimerization of α-methylstyrene catalyzed by Brönsted acidic ionic liquid under solvent-free conditions. *Tetrahedron*, 62 (17):3985-3988.

[18] Han, Y., Cai, Q., Lu, B., Xiao, X. and Cui Sh. (2007) Selective dimerization of α-methylstyrene catalyzed by a solid acid. *Reaction Kinetics and Catalysis Letters* 91, 29-35.

[19] Zhen Lu Shen, Xi Jun He, Bao Xiang Hu, Nan Sun and Wei Min Mo. (2008). Study on Selective Dimerization of α-Methylstyrenes Promoted by Ionic Liquids. *Catalysis Letters, 122:*359-365.

[20] Tarlani, A., Zarabadi, M. P., Muzart, J., Lotfalipour, E., Darkhosh, F., Abedini, M. and Amini, M. M. (2011) Heteropolyacid-catalyzed dimerization of α-methylstyrene; on he efficiency and selectivity dependence. *Catal. Comm., 14*, 89-91.

[21] Yamada, B. and Inoue, K. (2004). Addition-fragmentation chain transfer reactivities of a-Methylstyrene unsaturated dimer. *Kobunshi Ronbunshu,* 61(4):275-281.

[22] Pat. Eur №1778607 *Fluids having partially hydrogenated susbstitued styrene linear dimers and method of making same*, (2007).

[23] Pat. JP №63192727 ***Production of unsaturated dimer of alpha-alkylstyrenes.*** (1988).

[24] Rudnik, L. R. (2006). *Synthetics, mineral oils, and bio-based lubricants. Chemistry and technology*. Second edition, CRC Press, Taylor&Francis Group, Boca Raton, 894p.

[25] Maier, G. (1998). Gas separation with polymer membranes. *Angew. Chem. Im. Ed.* 37:2960-2974.

[26] Flanigen, E. M.; Broach, R. W. and Wilson, S. T. (2010). *Zeolites in industrial separation and catalysis*; Kulprathipanja, S., Ed. Wiley-VCH: Weinheim, Germany, 618p.

[27] Martinez, C. and Corma, (2011) A. Inorganic Molecular Sieves: Preparation, Modification and industrial application in catalytic processes. *Coord. Chem. Rev.*, 255:1558−1580.

[28] Dai, P. S. E., Sanderson, J. R. and Knifton, J. F. (1994). Zeolites and Related Microporous Materials: State of the Art. *Studies in Surface Science and Catalysis*, 84:1701.

[29] Hartmann, M., Machoke, A. G. and Schwieger, W. (2016). Catalytic test reactions for the evaluation of hierarchical zeolites. *Chem. Soc. Rev.*, 45:3313-3330.

[30] Yoon, K. B., Lim, J. L. and Kochi, J. K. (1989). Zeolite catalysis in the selective oligomerization of styrene. *J. of Mol. Cat.* 52:375-386.

[31] Benito, A., Corma, A., Harcia, H. and Primo, J. (1994). Dimerization of styrene catalyzed by acid 12-membered ring zeolites. *Applied Catalysis A: General* 116:127-135.

[32] Grigor'eva, N. G., Paukshtis, E. A., Kutepov, B. I., Galyautdinova, R. R. and Dzhemilev, U. M. (2005). Type Y zeolites in the dimerization reaction of α-methylstyrene. *Petroleum chemistry*, 45(6):453-459.

[33] Knifton, J. F. and Sanderson, J. R. (1994). Olefin oligomerization via zeolite catalysis. *Catalysis Letters*, 28:223-230.

[34] Pater J. P. G., Jacobs P. A. and Martens J. A. (1999). Oligomerization of Hex-1-ene over Acidic Aluminosilicate Zeolites, MCM-41, and Silica-Alumina Co-gel Catalysts: A Comparative Study. *J. of Catalysis,* 184(1):262-267.

[35] Kofke, G. T. J. and Gorte, R. J. (1989). A temperature-programmed desorption study of olefin oligomerization in H-ZSM-5. *J. of Catalysis,* 115:233-243.

[36] Sanggil Moon, Ho-Jeong Chae and Min Bum Park (2018). Oligomerization of light olefins over ZSM-5 and Beta zeolite catalysts by modifying textural properties. *Applied Catalysis A.: General,* 553:15-23.

[37] Sanderson, J. R., Dai Pei S. E., Knifton J. F. *Pat. US* 5120891, (1992).

[38] Dileo, Th. J. *Pat. US* 5210346, (1992).

[39] Grigor'eva, N. G., Bubennov, S. V. and Kutepov, B. I. (2011). Oligomerization of α-Octene Catalyzed by Zeolites, *Catalysis in industry,* 3(2):144-150.

[40] Grigor'eva, N. G., Bubennov, S. V., Bikbaeva, V. R., Serebrennikov, D. V. and Kutepov, B. I. (2019). Development of an environmentally friendly way oligomerization of dec-1-ene on zeolite catalysts. *Chemistry for Sustainable development* (27):147–153.

[41] Choi, M., Cho, H., Srivastava S., R., Venkatesan, C., Choi, D. H. and Ryoo, R. (2006). Amphiphilic organosilane-directed synthesis of crystalline zeolite with tunable mesoporosity. *Nat. Mater.,* 5(9):718–723.

[42] Choi, M., Srivastava, R. and Ryoo R. (2006). Organosilane surfactant-directed synthesis of mesoporous aluminophosphates constructed with crystalline microporous frameworks. *Chem. Commun, 42:*4380–4382.

[43] Srivastava, R., Choi, M. and Ryoo, R. (2006). Mesoporous materials with zeolite framework: remarkable effect of the hierarchical structure for retardation of catalyst deactivation. *Chem. Commun.*, 43:4489–4491.

[44] Xiao, F. S. Wang, Yin, C. Y., Lin, K. F., Di, Y., Li, J. X., Xu, R. R. Su, D. S., Schlogl, R., Yokoi, T. and Tatsumi, T. (2006). Catalytic properties of hierarchical mesoporous zeolites templated with a mixture of small organic ammonium salts and mesoscale cationic polymers. *Angew. Chem., Int. Ed.,* 45(19): 3090–3093.

[45] Wang, H. and Pinnavaia, T. J. (2006) MFI zeolite with small and uniform intracrystal mesopores. *Angew. Chem., Int. Ed.,* 45(45): 7603–7606.

[46] Mei, C. S., Wen, P. Y., Liu, Z. C., Liu, H. X., Wang, Y. D., Yang, W. M., Xie, Z. K., Hua, W. M. and Gao, Z. (2008). Selective production of propylene from methanol: Mesoporosity development in high silica HZSM-5. *J. Catal.,* 258(1):243–249.

[47] Janssen, A. H., I. Schmidt, C. J. H. Jacobsen, A., Koster, J. and de Jong, K. P. (2003). Exploratory study of mesopore templating with carbon during zeolite synthesis. *Microporous Mesoporous Mater.,* 65:59–75.

[48] Jacobsen, C. J. H., Madsen, Houzvicka, C. J., Schmidt, I. and Carlsson, A. (2000). Mesoporous zeolite single crystals. *J. Am. Chem. Soc.,* 122:7116–7117.

[49] Schmidt, I., Boisen, A., Gustavsson, E., Stahl, K. Pehrson, S., Dahl, S., Carlsson, A. and Jacobsen, C. J. H. (2001). Carbon nanotube templated growth of mesoporous zeolite single crystals. *Chem. Mater.*, 13:4416–4418.

[50] Fang, Y. M., Hu, H. Q. and Chen, G. H. (2008). Zeolite with tunable intracrystal mesoporosity synthesized with carbon aerogel as a secondary template *Microporous Mesoporous Mater.,* 113:481–489.

[51] Tao, Y. S., Kanoh H. and Kaneko K. (2003). Uniform mesopore-donated zeolite Y using carbon aerogel templating. *J. Phys. Chem. B*, 107:10974–10976.

[52] Tao, Y. S., Tanaka, H., Ohkubo, T., Kanoh, H. and Kaneko, K. (2003). Pore structures of ZSM-5 synthesized in the mesopore spaces of a carbon aerogel. *Adsorpt. Sci. Technol.*, 21:199–203.

[53] Zhu, H., Liu, Z. Y., Wang, D. Kong, X. Yuan and Z. Xie, (2008). Nanosized $CaCO_3$ as hard template for creation of intracrystal pores within silicalite-1. *Chem. Mater.,* 20:1134–1139.

[54] S. van Donk, Janssen, A. H., Bitter, J. H. and de Jong K. P. (2003). Generation, characterization, and impact of mesopores in zeolite catalysts. *Catal. Rev.:* Sci. Eng., 45:297–319.

[55] Meyers, B. L., Fleisch, T. H., Ray, G. J., Miller, J. T. and Hall, J. B.. (1988). A multitechnique characterization of dealuminated mordenites *J. Catal.,* 110:82–95.

[56] Cartlidge, S., Nissen, H. U. and Wessicken, R. (1989). Ternary mesoporous structure of ultrastable zeolite CSZ-1 zeolites. *Zeolites*, 9(4):346–349.

[57] Horikoshi, H., Kasahara, S. T., Fukushima, K., Itabashi, T. Okada, O. Terasaki and Watanabe, D. (1989) Study of mesopores induced by dealumination in zeolite Y. *Nippon Kagaku Kaishi*, 398–404.

[58] Chauvin, B., Massiani, P., Dutartre, R., Figueras, F., Fajula, F. and Descourieres, T. (1990). Factors affecting the steam dealumination of zeolite omega. *Zeolites*, (10):174–182.

[59] Guisnet, M., Wang Q. L. and Giannetto, G. (1990). Dealumination of NH4NaY zeolites through hydrothermal treatment: Kinetic order with respect to aluminium. *Catal. Lett.*, 4:299–302.

[60] Groen, J. C., Moulijn J. A. and J. P´erez-Ram´ırez. (2006). Desilication: on the controlled generation of mesoporosity in MFI zeolites *J. Mater. Chem.*, 16, 2121–2131.

[61] Dessau, R. M., Valyocsik, E. W. and Goeke, N. H. (1992). Aluminum zoning in ZSM-5 as revealed by selective silica removal. *Zeolites*, (12):776–779.

[62] Lietz, G., Schnabel, K. H., Peuker, C., Gross, T. Storek, W and Volter, J. (1994). Modifications of H-ZSM-5 catalysts by NaOH treatment. *J. Catal.,* (148):562–568.

[63] Ogura, M., Shinomiya, S. Y., Tateno, J., Nara, Y., Kikuchi, E. and Matsukata, H. (2000). Formation of uniform mesopores in ZSM-5 zeolite through treatment in alkaline solution. *Chem. Lett.*, 8:882–883.

[64] Ogura, M., Shinomiya, S. Y., Tateno, J., Nara, Y., Nomura, M. Kikuchi E. and Matsukata, M. (2001). Alkali-treatment technique - new method for modification of structural and acid-catalytic properties of ZSM-5 zeolites. *Appl. Catal., A*, 219:33–43.

[65] Suzuki, T., Okuhara, T. (2001). Change in pore structure of MFI zeolite by treatment with NaOH aqueous solution. *Microporous Mesoporous Mater*., 43:83–89.

[66] Su, L. L., Liu, L., Zhuang, J. Q., Wang, H. X., Li, Y. G., Shen, W. J., Xu, Y. D. and Bao, X. H. (2003). Creating Mesopores in ZSM-5 zeolite by alkali treatment: A new way to enhance the catalytic performance of methane dehydroaromatization on Mo/HZSM-5 catalysts. *Catal. Lett.,* 91:155–167.

[67] Travkina, O. S., Agliullin, M. R., Filippova, N. A., Khazipova, A. N., Danilova, I. G., Grigor'eva, N. G., Nama Narender, Pavlov M. L. and Kutepov B. I. (2017). Template-free synthesis of high degree crystallinity zeolite Y with micro–meso–macroporous structure. *RSC Adv*. 7:32581–32590.

[68] Kutepov, B. I., Travkina, O. S., Pavlova, I. N., Khazipova, A. N., Grigor'eva, N. G. and Pavlov, M. L. (2015). New adsorbents and acid-base catalysts based on binder-free granulated zeolites of LTA and FAU types. *Russian Journal of Applied Chemistry,* 88(1):65-71.

[69] Kutepov, B. I., Travkina, O. S., Agliullin, M. R., Khazipova, A. N., Pavlova, I. N., Bubennov, S. V., Kostyleva, S. A. and Grigor'eva, N. G. (2019) High-Crystallinity Granular Zeolites of LTA, FAU, and MOR Structural Types with Hierarchical Porous Structure: Synthesis and Properties, *Petroleum Chemistry,* 59(3):297–309.

[70] Yushchenko, V. V. (1997). Calculation of the acidity spectra of catalysts from temperature-programmed ammonia desorption data. *Rus. J. Phys. Chem.*, 71(4):457-550.

[71] Cejka, J., Corma A. and Zones, S. Zeolites and catalysis: synthesis, reactions and applications, Wiley-VCH, Weinheim, (2010).

[72] Tian, F., Shen, Q., Fu, Z., Wu, Y. and Jia, C. (2014). Enhanced adsorption desulfurization performance over hierarchically structured zeolite Y., *Fuel Process. Technol.*, (128):176–182.

[73] Xinga, C., Yang, G., Wu, M., Yang, R., Tan, L., Zhu, P., Wei, Q., Li, J., Mao, J., Yoneyama, Y. and Tsubaki, N. (2015). Hierarchical zeolite Y supported cobalt bifunctional catalyst for facilely tuning the product distribution of Fischer–Tropsch synthesis. *Fuel*, 148:48–57.

[74] Fu, W., Feng, Y., Fang, Z., Chen, Q., Tang, T., Yu, Q. and Tang, T. (2016). Zeolite Y nanosheet assembled palladium catalysts with high catalytic activity and selectivity in the vinylation of thiophenes. *Chem. Commun.,* (52):3115–3118.

[75] Jacobs P. A. and Uytterhoeven, J. B. (1972). Assignment of the hydroxyl bands in the infrared spectra of zeolites X and Y. Part 1.-Na-H zeolites. *J. Chem. Soc., Faraday Trans.*, (69):359–371.

[76] Anderson M. W. and Klinowski J. (1986). Zeolites treated with silicon tetrachloride vapour. IV. Acidity *Zeolites*, 6:455–466.

[77] Kotrel, S., Lundsford, J. H. and Kn¨ozinger, H. (2001). Characterizing Zeolite Acidity by Spectroscopic and Catalytic Means: A Comparison *J. Phys. Chem. B*, 105:3917–3921.

[78] Daniell, W., Topsøe, N. Y. and Kn¨ozinger, H. (2001). An FTIR Study of the Surface acidity of USY zeolites: Comparison of CO, CD_3CN, and C_5H_5N Probe Molecules. *Langmuir,* 17: 6233–6239.

[79] Gabrienko, A. A., Danilova, I. G., Arzumanov, S. S., Toktarev, A. V., Freude D. and Stepanov, A. G. (2010). Strong acidity of silanol groups of zeolite Beta: Evidence from the studies by IR spectroscopy of adsorbed CO and ^{1}H MAS NMR. *Microporous Mesoporous Mater.,* 131:210–216.

[80] Cairon, O., Chevreau, T. and Lavalley, J.-C. (1998). Brnsted acidity of extraframework debris in steamed Y zeolites from the FTIR study of CO adsorption. *J. Chem. Soc., Faraday Trans.*, 94:3039–3047.

[81] Hadjiivanov, K. (2014). Identification and Characterization of Surface Hydroxyl Groups by Infrared spectroscopy. *Adv. Catal.,* 57:99–318.

[82] Paukshtis, E. A., Yurchenko, E. N. (1983) Study of the Acid – Base Properties of Heterogeneous Catalysts by Infrared Spectroscopy. *Russ. Chem. Rev.*, 52:242–258.
[83] Grigor'eva, N. G., Bubennov, S. V., Khalilov, L. M. and Kutepov, B. I. (2015). Dimerization of norbornene on zeolite catalysts. *Chinese Journal of Catalysis.* 36 268–273.
[84] Grigorreva, N. G, Bubennov, S. V., Khalilov, L. M. and Kutepov. B. I. (2013) Selective dimerization of higher cycloolefins in the presence of micro- and micromesoporous zeolite catalysts. *Russian Chemical Bulletin, Int. Ed.*, 62(2):444-449.
[85] Grigor'eva, N. G., Talipova, R. R., Bubennov, S. V., Khazipova, A.N., Kutepov, B. I. and Dzhemilev U. M. (2009). Dimerization of vinylarenes on zeolite catalysts. *Petroleum Chemistry*, 49(1):16–22.
[86] Balaev, A. V., Grigor'eva, N. G., Khazipova, A. N., Kutepov, B. I. and Dzhemilev U. M. (2012). A kinetic model of the dimerization of α-methylstyrene in the presence of high-silica zeolite Y. *Petroleun Chemistry*, 52(6):426–431.
[87] Grigor'eva, N. G., Galyautdinova, R. R., Vosmerikov, A. V., Korobitsina L. L., Kutepov, B. I. and Dzhemilev, U. M. (2008). Conversion of α-Methylstyrene over pentasil zeolites with various silica ratios. *Petroleum Chemistry*, 48(5):366-368.
[88] Suzuki, K., Katada, N. and Niwa, M. (2007). Detection and quantitative measurements of four kinds of OH in HY zeolite *J. Phys. Chem. C.*, 111(2):894–900.
[89] Suzuki, K., Noda, T., Katada, N. and Niwa, M. (2007). IRMS-TPD of ammonia: Direct and individual measurement of Bronsted acidity in zeolites and its relationship with the catalytic cracking activity. *J. Catal.* 250(1):151-160.
[90] Noda T., Suzuki K, Katada, N. and Niwa M. (2008). Combined study of IRMS-TPD measurement and DFT calculation on Brønsted acidity and catalytic cracking activity of cation-exchanged Y zeolites. *J. Catal.* 259(2): 203–210.
[91] Barthomeuf, D. (1987). Zeolite acidity dependence on structure and chemical environment. Correlations with catalysis. *Materials Chemistry and Physics*, 17(1-2):49-71.
[92] Bogdanovic, B., Henc, B. and Karmann, Hans-Georg (1970). Olefin Transformations Catalyzed by Organonickel Compounds. *Industrial and engineering chemistry*, 62(12):34-44.
[93] Cai, Q., Li, J., Bao, and F., Shan, Y. (2005). Tunable dimerization of α-methylstyrene catalyzed by acidic ionic liquids. *Appl. Catal. A: Gen.*, 279:139-143.
[94] T. R. Forbus, M. P. Hagemiester, Pat. EP 1778607 (2007).
[95] Y. Sakata, N. Soga, T. Kunugi, Pat. JP 9227888 (1997).
[96] Forbus, T. R., M. P. Hagemeister, Pat. US 069212, (2006).
[97] Chiu I-Ching, Pat. US 0242441 (2004).

[98] Serrano, D. P., Aguado, J. and Escola, J. M. (2011). Hierarchical zeolites: materials with improved accessibility and enhanced catalytic activity. *SPR Catalysis*, 23:253–283.

In: An Introduction to Aluminosilicates
Editor: Nero Regina Blevins
ISBN: 978-1-53617-250-8

Chapter 14

A DFT STUDY OF THE ACETYL TRANSFER REACTION CATALYZED BY H-ZSM-5 ZEOLITE USING A 3T-CLUSTER APPROACH

Pablo Corregidor[1]*, PhD and Emilce Ottavianelli[2], PhD
[1]Faculty of Engineering
[2]Faculty of Exact Sciences, Institute of Research for the Chemical Industry (INIQUI-UNSa-CONICET), National University of Salta, Salta, Argentina

ABSTRACT

Zeolites are well known microporous silicoaluminates with extensive uses as heterogeneous catalysts. Particularly, the acid form of ZSM-5 zeolite has left evidence of its high activity due to its participation in several chemical processes. Acyl-transfer reactions are among the most fundamental processes in biochemistry and organic chemistry in which zeolites play an efficient role due to its unique porous structure and acidity. It has been reported that an acylated surface has been generated as an active intermediate in several acylation reactions, but to the best of our knowledge, no rational mechanism has been suggested for its generation. Thus, in this chapter, we will consider a possible mechanism for the generation of an acetyl-zeolite intermediate due to transfer from different acetyl donors by theoretical studies using a cluster model of H-ZSM-5 zeolite designed by three TO_4 tetrahedral units (3T-cluster). Based on Intrinsic Reaction Coordinate (IRC) method, we discuss the atomic reorganization that evolves from acetyl donors adsorbed over 3T-cluster to the acetylated zeolite intermediate. Moreover, we confirm this arrangement and support bounded species and broken bonds based on the concept of Global Electron Density Transfer (GEDT) and Density of States (DOS),

* Corresponding Author's Email: pcorregidor@unsa.edu.ar.

Projected Density of States (PDOS) and Overlay Population Density of States (OPDOS) diagrams Thus, for the first time, a concerted mechanism has been proposed for the generation of the acetylated-zeolite intermediate, based on first-principle methods.

Keywords: zeolite, ZSM-5, acylation, acetyl-zeolite; DFT

1. INTRODUCTION

Zeolites are inorganic microporous crystalline materials constructed by tetrahedral TO_4 units, where T usually refers to Si and/or Al. These tetrahedra units can link together to form channels and cavities of molecular dimensions [1]. Related to the presence of Al, some zeolites exhibit great acidic properties, being ZSM-5 one of the most studied from the MFI family. The acidity strength of ZSM-5 has been compared with those of sulfuric acid solutions. Thus, it has been traditionally employed as a heterogeneous acidic catalyst in several reactions. It is well-known that the use of heterogeneous acid catalysts has reached significant importance, not only because they enable environmentally benign synthesis, but also due to the good yields, accompanied by excellent selectivities, that can frequently be achieved.

As most of the activity usually occurs inside the channels of ZSM-5, it is very difficult to obtain information at the molecular level by only planning simple experiments. In that sense, molecular simulations nicely complement the experimental efforts by offering a better understanding of a reaction inside the channels. In that sense, theoretical studies can simulate the adsorption of the reactants, diffusion of molecules to the acidic center, the chemical conversion, products diffusion away from the active site and desorption from the zeolite [2], thus contributing to a better knowledge on the influence of the reaction rates and products that are obtained.

On the other hand, acyl transfer reactions are very important transformations in organic synthesis as they provide an inexpensive and efficient way to obtain several organic compounds such as esthers, amides or ketones. In that sense, different acetyl donors were traditionally employed, being acetyl chloride and acetic anhydride two of the most useful agents in the presence of acidic catalysts [3].

Acyl transfer reactions including zeolite catalysts were the attention of several studies. Corma et al. [4] proposed that an acidic zeolite reacts with acylating agents to give an acylium ion or an acylium-like complex, which are important intermediates involved in acylation reactions of aromatic compounds. Moreover, Kresnawahjuesa et al. [5] reported that the gas adsorption of typical acylating agents reacted forming an acetyl-zeolite intermediate over Brønsted acidic sites of an H-ZSM-5 zeolite.

Furthermore, Gumidyala et al. [6] demonstrated the generation of stable acyl intermediates during ketonization of acetic acid over H-ZSM-5, under conditions that selectively yielded acetone. Other types of reactive intermediates have been proposed for reactions implicating acyl species and acylium ions, i.e., surface acyl species are in equilibrium with acylium ions, as well as gas–phase ketenes in low abundance. In that sense, ketenes were proposed as important reactive intermediates for reactions involving carboxylic acids and acyl groups [7].

Later, deuterium labeling studies performed by Bonati's group demonstrated that while ketenes were produced in parallel to acyl groups and acylium ions, they were not intermediates in acylation reactions [8]. While several reactive surface intermediates have been proposed over H-ZSM-5 zeolite, involving ketenes or ketene-like species and even protonated acyl donors, a consensus is slowly growing on the importance of an acyl-zeolite intermediate [5, 6, 9-11].

Quantum calculations via density functional theory (DFT) have been extensively employed in order to understand reactive and adsorptive processes in zeolites of different framework types, for that reason, finite cluster models of zeolite crystals have been proposed, as they are relatively simple to apply to a wide range of zeolites [12] with the advantage that avoids intractable periodic DFT calculations of zeolites having large unit cells, as well as numerical instabilities of QM/MM methods when computing transition states [13].

Consequently, in the present chapter we report the results of a set of calculations with the aim of contributing to the knowledge of possibles pathway for the generation of an acetylated-zeolite intermediate due to interaction of some of the most used acylating agents: acetic anhydride ($(CH_3CO)_2O$), acetyl chloride (CH_3COCl) and acetic acid (CH_3COOH), with a ZSM-5 cluster. We focus herein on computing the atomic rearrangement that occurs as a result of the mentioned interaction which naturally evolves to the acetyl-zeolite intermediate.

Moreover, we calculate energies associated to the mentioned interaction, thus leading to some thermodynamic properties and kinetic parameters. Finally, the concept of Global Electron Density Transfer (GEDT) and Density of States (DOS) has been utilized to corroborate the formation and breaking of chemical bonds observed during the proposed path of reaction.

The remainder of this chapter is organized as follows: in section 2 we provide a description of the system, methods and computational details for a rational study; in section 3 we perform a discussion based on results obtained from Intrinsic Reaction Coordinates (IRC) calculations, Global Electron Density Transfer (GEDT) and Density of States (DOS). Finally, section 4 deals with concluding remarks.

2. Computational Details

2.1. Zeolite Cluster Model

Within the cluster approach, the catalytic active site model is a small neutral zeolite fragment, with the dangling bonds satisfied using hydrogen atoms. Thus, a cluster model for the H-ZSM-5 having the following stoichiometry $H_3Si(OH)Al(H)_2OSiH_3$ has been used in the computation of interactions with acylating agents. This cluster consists of three tetrahedral units (TO_4) which have been extracted from the crystallographic information file containing the periodic crystal (downloaded from the International Zeolite Asociation website [14]). This cluster (3T-cluster model from now on), contains the T12 site, where the presence of a Brønsted acidic center has been contemplated (Figure 1) and thus allows to investigate its interaction with acetic acid, acetic anhydride, and acetyl chloride.

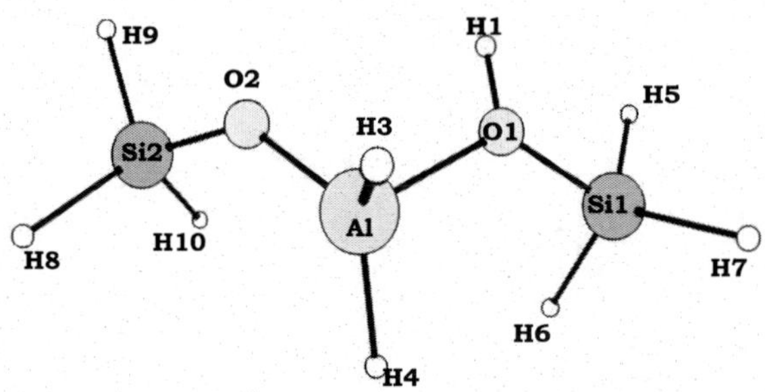

Figure 1. Structure of 3T-cluster model optimized at B3LYP/6-311+G(d) level.

2.2. Geometry Optimization of Reaction Complexes

All reactants, transition structures, and products were studied with cluster type calculations. During the geometry optimization, we look for a local minimum for reactants and products and a first–order saddle point for transition structures. On the stationary points, the eigenvalues of the Hessian matrix were computed to check whether it exhibits the proper number of imaginary frequencies: none for a minimum or one for a transition state (first–order saddle point). Zero–point energy (ZPE) corrections have been calculated for all optimized structures. In that sense, we studied the reaction path connecting the following species: a) isolated zeolite (cluster) and acetyl donor, b) adsorption complex formed by acetyl donor adsorbed over zeolite cluster, c) activated complex or transition state and d) Product complex which consist on acetylated zeolite and HG species, where G possesses different structures depending on the chemistry of acetyl donor (Figure 2). Thus, step 1 represents the adsorption of the acylating agent over the 3T-cluster model while step

2 deals with the generation of TS from adsorbed acetyl donor. Finally, the formation of acetylated zeolite has been expressed in step 3.

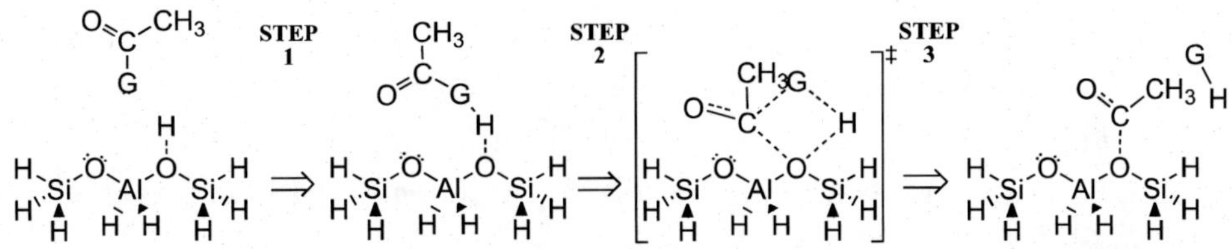

H-ZSM-5 + CH_3CO-G **H-ZSM-5----CH_3CO-G** **TS** **ZSM-5$^+$ CH_3CO^- + HG**

(a) **(b)** **(c)** **(d)**

Figure 2. Structure of systems under study: (a) Isolated zeolite and acetyl donor; (b) Adsorption complex: acetyl donor adsorbed over zeolite cluster; (c) Activated complex or transition state and (d) Product complex: acetylated zeolite and HG specie, where G = OH for acetic acid, Cl for acetyl chloride and $OCOCH_3$ for acetic anhydride.

2.3. Methods of Calculation

Cluster calculations have been performed with Gaussian16 package [15] using the 6-311+G(d) basis set and the Becke's three parameters hybrid density functional [16] with the Lee, Yang and Parr [17] correlation functional (B3LYP) within the density functional theory [18-20]. Geometry optimizations of the configuration of species interacting with the 3T-cluster were carried without constrains.

At first, the transition-state structures of the 3T-cluster model have been studied at two levels of theory: B3LYP/6-311+G(d) and semiempirical (AM1), using the reaction coordinate method referred to the synchronous transit-guided quasi-newton (STQN) calculation. After that, the complete reaction path (intrinsic reaction coordinate or IRC) has been calculated at the AM1 level [21]. First, the transition structure (TS) optimized at the AM1 level has been used as a starting point for the IRC calculations by following the steepest descent path toward reactants at one side and products at the other side. The AM1 level of calculation is the result of balancing the cost of IRC calculations and the quality of the obtained structures. As shown in the following section, the structures obtained by AM1 calculations resemble those obtained by B3LYP/6-311+G(d) close enough to give the same general picture of the course of the reactions studied here.

The reaction rate constants, k_r, have been calculated within the framework of Eyring's transition state theory as [22]:

$$k_r=\frac{k_BT}{hc^0}e^{\Delta G^{\#}/RT} \quad (1)$$

Here, c^0 is the standard-state concentration (often taken to be 1.00 mol·L^{-1}); k_B, h and R are Boltzman's, Planck's and the ideal gas constants, respectively. T is the absolute temperature and $\Delta G^{\#}$ the Gibbs free energy difference (the activation energy) between TS and adsorption complexes, for a given acyl donor. This energy contains the zero−point energy (ZPE) correction. Different rate constants were calculated for a temperature ranging from 273 to 500 K.

After that, macroscopic quantities linked to Arrhenius rate law were found by fitting to a set of k_r(T) values calculated from equation 1 in the mentioned range of temperatures, as reported previously [23-25]. Those macroscopic quantities are Arrhenius activation energy (E_{Arr}) and the pre exponential factor (A). Moreover, thermodynamic properties ΔG_R^0, ΔH_R^0 and ΔS_R^0 were obtained from frequency calculations at 273 K.

3. Results and Discussion

3.1. Intrinsic Reaction Coordinate (IRC) Calculations

3.1.1. Interaction with Acetic Acid

The IRC path obtained at AM1 level of theory has been calculated adopting 300 steps along the TS in the reverse direction and 295 steps following the forward one. Results obtained for the IRC calculation are presented in Figure 3. Moreover, Table 1 summarizes those bond lengths that are intrinsically related to the interaction, for the selected structures in the specified steps.

Thus, a possible structure for a TS has been obtained at B3LYP/6-311+G(d) level of theory, for a 3T-cluster model interacting with acetic acid, which evolves to an acetyl-zeolite intermediate. Moreover, IRC calculations allowed us to elucidate the atomic rearrangement that occurs during such interaction.

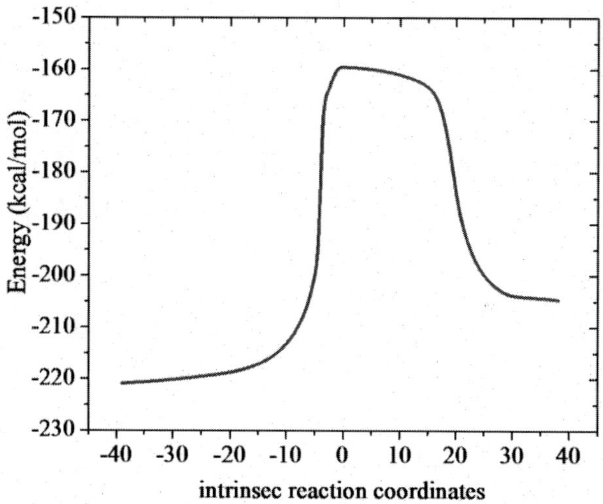

Figure 3. IRC diagram (AM1) for the interaction between the 3T-cluster and acetic acid.

Table 1. Bond lengths (angstroms) related with the interaction of acetic acid and zeolite 3T-cluster for selected geometries, starting from 0 (TS) in reverse (negative values) and forward (positive values) directions, between the following atoms: [1] H1 from the cluster and O from acetic acid's OH group. [2] HO−CO from acetic acid. [3] H1 and O1 from the cluster. [4] C from C = O and O1 of the cluster (see Figure 1 for atom labels)

Step	H–O[1]	C–O[2]	H–O[3]	Step	H–O[1]	C–O[4]	H–O[3]
-300	2.126	1.372	0.960	0	0.974	2.831	1.984
-250	2.132	1.371	0.959	5	0.971	2.828	2.007
-200	2.120	1.371	0.959	20	0.970	2.735	2.039
-180	2.106	1.371	0.959	50	0.969	2.546	2.075
-150	2.080	1.370	0.959	90	0.968	2.306	2.128
-100	2.042	1.372	0.958	150	0.963	1.606	2.453
-50	1.803	1.382	0.955	180	0.960	1.410	2.965
-40	1.646	1.391	0.961	200	0.962	1.397	3.123
-30	1.220	1.480	1.291	220	0.963	1.390	2.968
-20	1.014	1.573	1.788	250	0.963	1.388	3.179
-10	0.985	1.808	1.914	280	0.963	1.389	3.427
-5	0.977	1.928	1.956	295	0.963	1.390	3.533

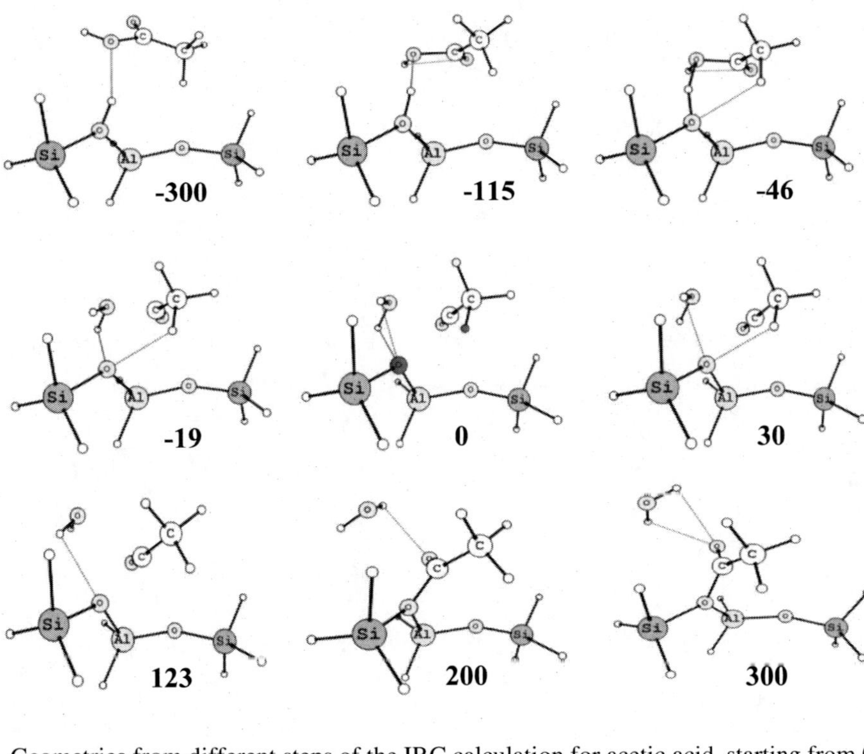

Figure 4. Geometries from different steps of the IRC calculation for acetic acid, starting from 0 (TS) in reverse (negative values) and forward (positive values) directions.

Bond lengths obtained for different steps in the IRC calculations (Table 1) evidenced that the distance between H1 of 3T-cluster and O from acetic acid's OH group decreases during the studied path. We name to this interaction as H---OH. Moreover, both the HO–CO bond from acetic acid and the H1–O1 bond from 3T-cluster increase, which probably indicate that both chemical bonds brake during such interaction, while C–O length between C=O and O1 of the cluster tends to decrease. Some geometries for selected IRC steps are shown in Figure 4.

Thus, starting from the adsorption complex with acetic acid, a decrease in the H---OH could probably justify the generation of water observed by some authors [5, 6] during gas–phase interaction of acetic acid and the acidic form of ZSM-5 zeolite. Moreover, related to the generation of water, the HO–CO bond must break, as evidenced by IRC calculations.

Table 2. Kinetic parameters and thermodynamic properties for the interaction of different acetyl donor with the 3T-cluster model.[a] rate constants at 273 K (up) and 500 K (down)

	Acetic acid	Acetyl chloride	Acetic anhydride
$\Delta G^{\#}$(kcal/mol)	50.87	36.92	39.40
[a]k_r (s^{-1})	$1.17 \cdot 10^{-28}$ $6.36 \cdot 10^{-10}$	$1.58 \cdot 10^{-17}$ $7.58 \cdot 10^{-4}$	$1.62 \cdot 10^{-19}$ $1.30 \cdot 10^{-5}$
E_{Arr} (kcal/mol)	51.54	37.64	39.27
A	$2.09 \cdot 10^{13}$	$2.09 \cdot 10^{13}$	$1.18 \cdot 10^{12}$
$\Delta H°_{step\ 1}$ (kcal/mol)	-11.08	-50.81	-14.90
$\Delta H°_{step\ 2}$ (kcal/mol)	18.16	7.05	13.03
$\Delta S°_{step\ 2}$ (cal/mol)	2.36	2.42	-4.87
$\Delta G°_{step\ 2}$ (kcal/mol)	18.81	7.05	11.70

Thus, from the frequency calculation, kinetic parameters, as well as thermodinamic properties, were obtained and are shown in Table 2. As was expected, an exothermic adsorption enthalpy ($\Delta H°_{step\ 1}$= -11.08 kcal/mol) has been calculated for the adsorption of acetic acid, followed by an endergonic step (18.81 kcal/mol) involving the generation of acetyl-zeolite intermediate. For the generation of such an intermediate, the system must climb 50.87 kcal/mol ($\Delta G^{\#}$) over the adsorption complex.

Thus, in agreement with results presented by Kresnawahjuesa et al. [5], the high value obtained for $\Delta G^{\#}$ may justify strong hydrogen-bond of acetic acid with the Brønsted-acid sites in H-ZSM-5.

3.1.2. Interaction with Acetyl Chloride

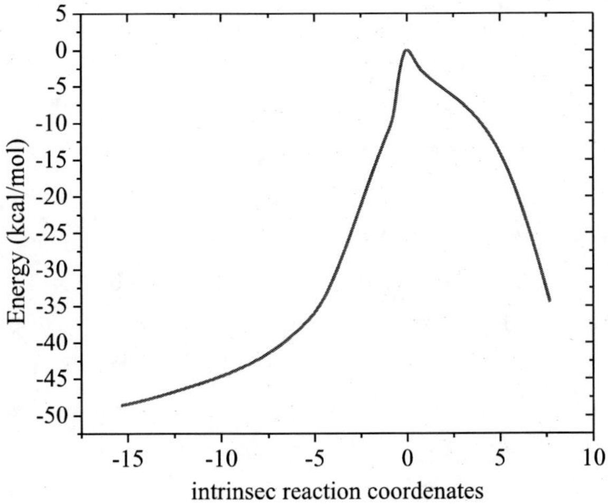

Figure 5. IRC diagram (AM1) for the interaction between the 3T-cluster and acetyl chloride.

The interaction with acetyl chloride and other acetyl donors has been studied with a similar methodology to the one applied for acetic acid. Thus, results in the actual and following sections will be organized and presented in the same order as described for acetic acid. In that way, Figure 5 is the result of IRC calculations (AM1) for the interaction of 3T-cluster and acetyl chloride. For this particular case, 300 steps have been obtained in each direction (forward and reverse).

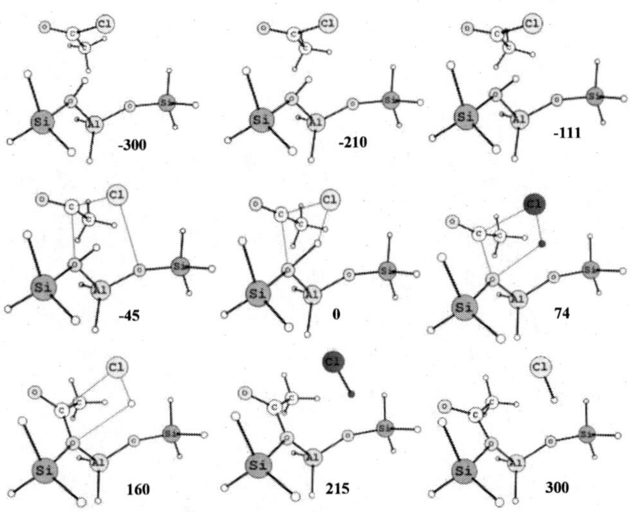

Figure 6. Geometries from different steps of the IRC calculation for acetyl chloride, starting from 0 (TS) in reverse (negative values) and forward (positive values) directions.

The optimized geometry (B3LYP/6-311+G(d)) for TS is similar to the one obtained for acetic acid. For this particular acetyl donor, Cl−CO length from acetyl chloride and H1−O1 from the cluster increase while H1---$ClCOCH_3$ (from zeolite cluster and acetyl chloride) and C---O1 (C from carbonyl carbon and O1 of the cluster) decrease. Thus, the mentioned changes in bond lengths are summarized in Table 3 and selected geometries for some IRC steps are shown in Figure 6.

Table 3. Bond lengths (angstroms) related with the interaction of acetyl chloride and zeolite 3T-cluster for selected geometries, starting from 0 (TS) in reverse (negative values) and forward (positive values) direction, between the following atoms: [1] H1 from the cluster and Cl from acetyl chloride. [2] Cl−CO from acetyl chloride. [3] H1 and O1 from the cluster. [4] C from C = O and O1 of the cluster (see Figure 1 for atom labels)

Step	H−Cl[1]	C−Cl[2]	H−O[3]	Step	H−Cl[1]	C−O[4]	H−O[3]
-300	2.243	1.765	0.969	0	1.458	2.360	1.487
-280	2.222	1.768	0.971	5	1.395	2.357	1.588
-250	2.191	1.769	0.972	10	1.354	2.352	1.701
-200	2.121	1.776	0.976	30	1.325	2.290	1.965
-150	2.034	1.789	0.982	50	1.318	2.200	2.046
-100	1.950	1.840	0.989	80	1.311	2.032	2.147
-90	1.931	1.892	0.991	100	1.307	1.904	2.216
-50	1.865	2.189	1.008	150	1.300	1.582	2.380
-30	1.840	2.351	1.025	200	1.298	1.401	2.616
-10	1.648	2.471	1.249	250	1.298	1.387	2.758
-5	1.537	2.478	1.382	300	1.298	1.386	2.738

The interaction between acetyl chloride and the 3T-cluster exhibits a 2.360 Å bond length for the C−O in the TS geometry (step 0 in the IRC calculation), compared with the value of 2.831 Å obtained for the interaction with acetic acid. This indicates a closer approach to the acyl donor which might be due to the presence of the electronegative chlorine atom.

According to this, the electrophilic carbonyl carbon must compensate its electronic deficiency getting closer to an oxygen of zeolite's cluster. Moreover, the O1−H1 bond is larger (1.984 Å) than for the acetic acid interaction (1.487 Å) as a result of the higher electronegativity value for the chlorine atom. Thermochemical properties, as well as kinetic parameters related to this interaction, are also resumed in Table 2. A more exothermic step for the interaction with acetyl chloride has been observed due to the strong interaction between both species, while the small barrier for $\Delta G^{\#}$ is in accordance with Kresnawahjuesa et al. who observed that the acetyl-zeolite intermediate generates easier with acetyl chloride than with acetic acid [5].

Rate constants calculated from equation 1 in the 273 K – 500 K range are much larger than the one calculated for acetic acid, corroborating that the acetyl-zeolite intermediate generates faster than the one with acetic acid. On the other hand, similar values for A and $\Delta S°_{step\ 2}$ reflect that those properties may be not dependent on the chemical structure of acyl donors.

3.1.3. Interaction with Acetic Anhydride

The IRC calculation obtained for the interaction with acetic anhydride is shown in Figure 7. We have calculated 32 steps in reverse direction and 80 in the forward one.

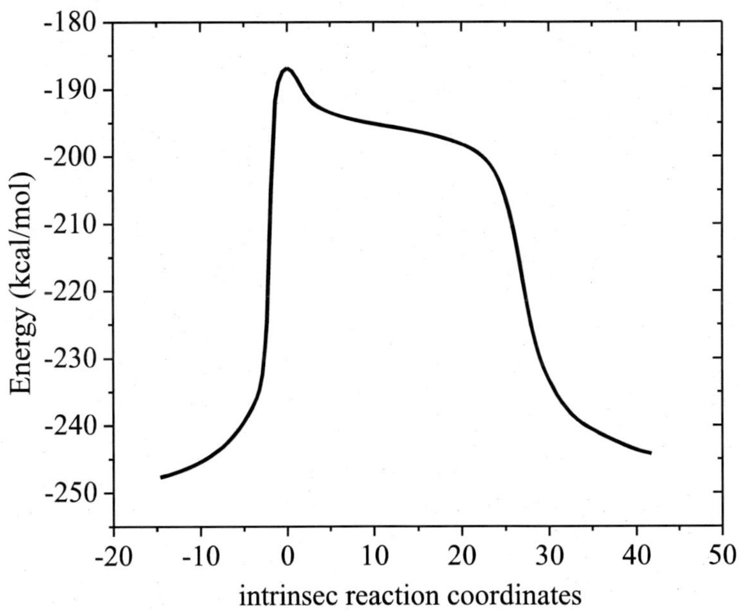

Figure 7. IRC diagram (AM1) for the interaction between the 3T-cluster and acetic anhydride.

From the shape observed in Figure 7, we evidenced a decrease of energy at the beginning (step 0 in Figure 7) due to the broken $CH_3OC{-}OCOCH_3$ bond followed by a plateau (steps 5 to 20 approximately) in which changes in the geometry of the anhydride are due to the rearrangement of some atomic groups to ease the interaction between the incipient acyl ion and the 3T-cluster.

This is evidenced by a decrease in the C---O length in which the C and the O atoms belong to the carbonyl group and to the 3T-cluster, respectively. As this occurred, a fragment of CH_3COOH is generated by protonation from the respective moiety in the acetyl donor. This acetic acid gradually moves away, increasing the O1–H1 length. Thus, a decrease in energy after the plateau is the result of the interaction of acyl ion with zeolite's cluster to generate the acetyl-zeolite intermediate. Relevant bond lengths/interactions during IRC calculation are resumed in Table 4.

Table 4. Bond lengths (angstroms) related with the interaction of acetic anhydride and 3T-cluster for selected geometries, starting from 0 (TS) in reverse (negative values) and forward (positive values) direction, between the following atoms: [1] H1 from the cluster and O(sp^3) from acetic anhydride. [2] HO−CO from acetic anhydride. [3] H1 and O1 from the cluster. [4] C from C = O and O1 of the cluster (see Figure 1 for atom labels)

Step	H–O[1]	C–O[2]	H–O[3]	Step	H–O[1]	C–O[4]	H–O[3]
-32	2.061	1.381	0.962	2	1.005	2.937	1.801
-30	2.050	1.382	0.962	5	0.990	3.011	1.904
-28	2.039	1.382	0.962	8	0.984	3.031	1.972
-25	2.020	1.382	0.961	10	0.983	2.996	2.000
-23	2.003	1.383	0.962	15	0.982	2.852	2.025
-20	1.968	1.384	0.961	20	0.982	2.726	2.047
-18	1.938	1.386	0.960	25	0.981	2.621	2.067
-15	1.874	1.386	0.960	30	0.980	2.518	2.090
-10	1.757	1.394	0.961	40	0.977	2.303	2.170
-8	1.712	1.443	0.962	50	0.973	1.920	2.344
-5	1.358	1.548	1.232	60	0.970	1.424	2.581
-2	1.071	1.619	1.603	70	0.970	1.397	3.002
0	1.016	1.783	1.756	80	0.970	1.391	3.590

On the other hand, the geometries of selected structures are presented in Figure 8. At this point, it is necessary to clarify that the structure of the reactant-complex converged in our IRC calculations is the one described in the previous paragraph, but when we optimized this structure at DFT level, we found that acetic anhydride adsorbs to 3T-cluster by an oxygen atom from carbonyl group, being the only from all acyl donors studied in this chapter that showed significant changes from its geometry optimized at AM1 and DFT level. Nonetheless, this observation doesn't affect kinetic parameters and thermodynamical properties calculated for step 2 but surely may produce different values for those properties evaluated in step 1, i.e., enthalpy of adsorption. As step 1 is not a direct object of our study, we will not take care of this behavior.

Figure 8 shows some geometries for the IRC calculation while Table 4 resumes some selected bonds/interactions lengths that are related to the rearrangements mentioned above. On the other hand, Table 1 presents thermodynamic properties and kinetics parameters related to the studied cooperation. The $\Delta G^{\#}$ obtained for this acetyl donor is lower than for acetic acid but slightly higher than for acetyl chloride, thus rate constants calculated in the 273 K – 500 K are higher than for acetic acid and smaller than those calculated for acetyl chloride. Moreover, pre−exponential factor are similar to those obtained for acetyl chloride and acetic acid.

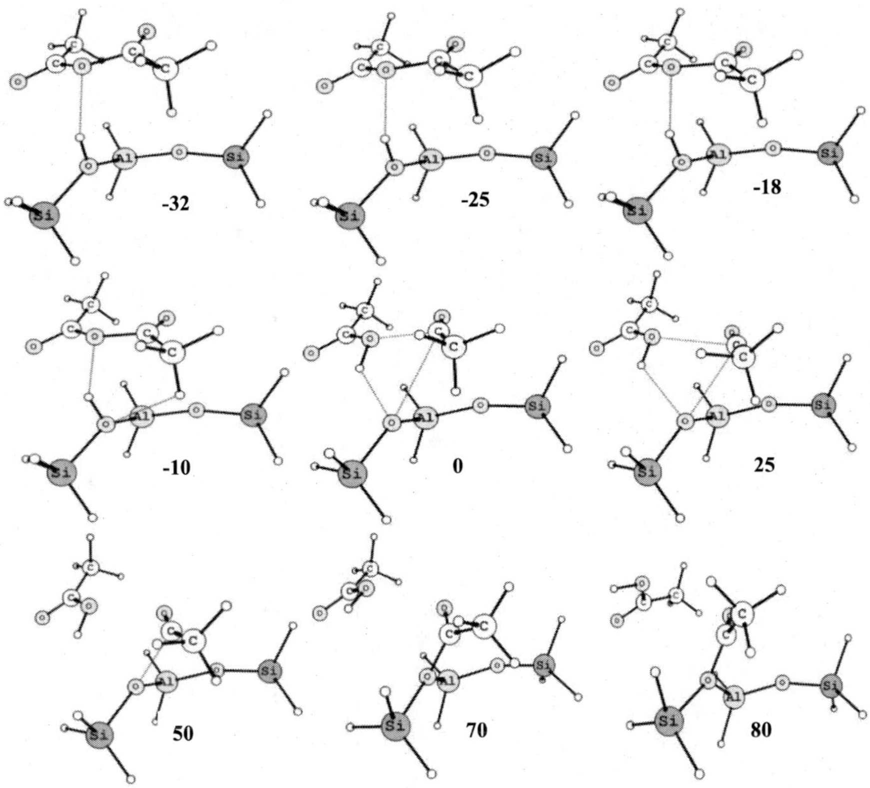

Figure 8. Geometries from different steps of the IRC calculation for acetic anhydride, starting from 0 (TS) in reverse (negative values) and forward (positive values) directions.

Thus, based on constant rates and $\Delta G^{\#}$, the following decreasing order for the generation of the acetyl-zeolite intermediate is proposed:

Acetyl chloride > acetic anhydride > acetic acid

This result is coincident with those found experimentally by Kresnawahjuesa et al. for the gas-solid interaction of the same acylating agents [5]. They proposed that interaction of different acyl donors over Brønsted acid sites of a ZSM-5 zeolite may be comparable with adsorption of several organic compounds like ketones, nitriles, and esters. Therefore, based on calorimetric determinations and ^{13}C-NMR, they affirm that molecules having a gas phase proton affinity (PA) higher or equal to ammonia (PA = 205 kcal/mol) must be protonated by an H-ZSM-5 zeolite. In that sense, molecules having PA smaller than ammonia stimulate hydrogen bonding with Brønsted acid centers and the strength of the interaction varies regularly with adsorbate's PA. Proton affinities for acetic acid (PA = 190.7 kcal/mol), acetyl chloride (PA = 176.7 kcal/mol) and acetic anhydride (PA = 197.7 kcal/mol) are smaller than ammonia and similar to simple alcohols, ketones, and esters. Hence, the methodology implemented in our theoretical studies is adequate for predicting a reactivity order for the generation of the acetyl-zeolite intermediate using acetic acid, acetyl chloride, and acetic anhydride.

3.2. Global Electron Density Transfer (GEDT)

To corroborate the interactions studied in the previous section, we apply the methodology proposed by Domingo [26, 27]. Thus, the Global Electron Density Transfer (GEDT) has been determined for the TS found for each acetyl donor. The GEDT at the TSs has been computed by sharing the natural charges by natural bond orbital (NBO) analysis [28, 29] between the electrophilic and the nucleophilic species at B3LYP/6-311+G(d) level of theory. The GEDT concept comes from the observation that the electron density transfer that takes place along polar and ionic reactions is not a local process, but a global flux of electron density taking place from the nucleophile to the electrophile, and not being dependent on the approach mode of both reagents [27]. Consequently, Domingo postulates that the whole system (including both interacting species) becomes one entity from the beginning of the reaction being, therefore, structurally indivisible [30]. For the calculation of GEDT, we applied the methodology suggested in Domingo's web page [31]. Moreover, GEDT values are hardly dependent on the computational method used to obtain the atomic charges as these values come from the integration of the electron density of two molecular frameworks that are usually not yet bound at the TS. By using this methodology, we expected a faster generation for transition states that are more stabilized by global electron density transfer as the stabilization reduces the activation energy and increases the kinetic constant value.

A non-symmetric electron density distribution takes place during the GEDT process in non-symmetric molecules. Therefore, while in the electrophilic species some atoms gather more electron density, in the nucleoophilic species some atoms lose less electron density. These relevant atoms correspond to the most electrophilic and most nucleophilic centers of the interacting epecies.

It can be seen in Table 5 that the GEDT values calculated for all the TS are in the range from 0.245e to 0.290e. This suggests a possible stabilization by the contribution of global electron density from acylating agents (donors) to the 3T-cluster (acceptor). Moreover, we observe that GEDT values follow the same sequence found for $\Delta G^{\#}$ and velocity of generation for the as-obtained TSs.

Table 5. GEDT calculated for TS in different acetyl donors

Acetic acid	Acetic anhydride	Acetyl chloride
0,249e	0,271e	0,281e

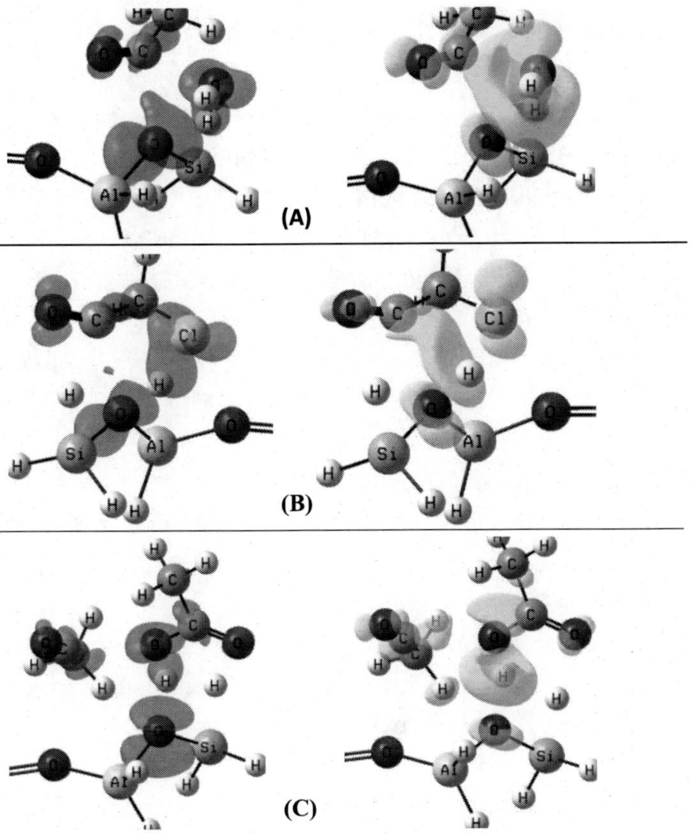

Figure 9. 3D representations for regions of electronic density transfer. Dark (left) and bright (right) isosurfaces representing increase and decrease, respectively, of electronic density. (A) acetic acid, (B) acetyl chloride and (c) acetic anhydride.

Therefore, GEDT values seems to be related to E_{Arr}. Consequently, the global electron density transfer could be a useful reactivity index to understand the relative stabilization of those TS calculated in this work. Figure 9 shows the 3D representations for the electronic density zones between the 3T-cluster and the different acetyl donors. We deduce that electronic density has increased in a region between the acidic hydrogen from the cluster and the respective atom from the acyl donor which we will denominate as Y. Atom Y is O from OH group in acetic acid, O sp^3 from acetic anhydride and Cl from acetyl chloride, in respective acyl donor. Moreover, an increase in electronic density has been evidenced in a region between the C atom from the carbonyl group and O1 from the cluster. On the other hand, pictures on the right side of Figure 9 (bright regions) represent a decrease in electron density in TSs. Such a decrease in the regions between 1) carbonilic C and atom Y and 2) H1 and O1 from the cluster, indicate a loss in the bond character.

In that way, results obtained from IRC calculations are confirmed with the global electron density transfer, thus corroborating the proposed path for the generation of acetyl-zeolite intermediate in a concerted mechanism where the generation of a TS can be described in terms of the electron density transfer from the nucleophile to the electrophile.

3.3. Density of States

DOS (Density Of States), PDOS (Projected Density Of States) and OPDOS (Overlap Population Density of States) give a pictorial representation for molecular orbitals composition and their contribution to chemical bonds [32, 33]. DOS, PDOS, and OPDOS for the acetic acid interaction in the HOMO-2 and LUMO+2 interval, are represented in Figure 10 where the vertical dotted line indicates the HOMO position. Contributions from C of carbonyl group, hybridized sp^3 O from acetic acid, O1 from the cluster and contributions from all other atoms (designed as other) were evaluated in the PDOS diagram. From the DOS curve, a 4.18 eV HOMO-LUMO gap is evidenced. In a region between -7.8 eV to -6.0 eV, the fragment designed as "others" is the one which mostly contributes to the DOS signal, while carbonyl carbon and O from the cluster contribute poorly. On the other hand, between -8.8 eV and -7.8 eV, a similar contribution from the mentioned atoms is observed. Moreover, LUMO is mainly formed by the C atom from carbonyl group and in less degree from O1 of the cluster.

From the OPDOS curve, a bonding interaction has been determined between C from C=O and O1 from the cluster in the -7.5 eV – -6.0 interval, while in the same region, an anti-bonding interaction is observed between the mentioned C and O sp^3 from acetic acid. On the other hand, in the -3.1 eV to -2.0 eV (the LUMO's region) a bonding interaction between H1 from the cluster and O sp^3 from acetic acid, as well as a similar interaction between H1 and O1 from the cluster has been evidenced. In the same region, a strong anti-bonding interaction (out of scale) between the carbonyl carbon and O1 from the zeolite is evidenced.

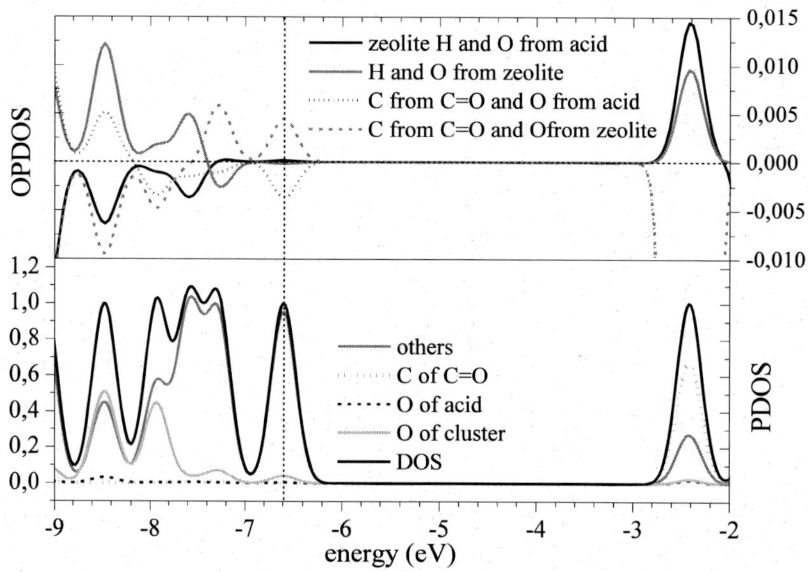

Figure 10. DOS, PDOS, and OPDOS for the interaction between the 3T-cluster and acetic acid.

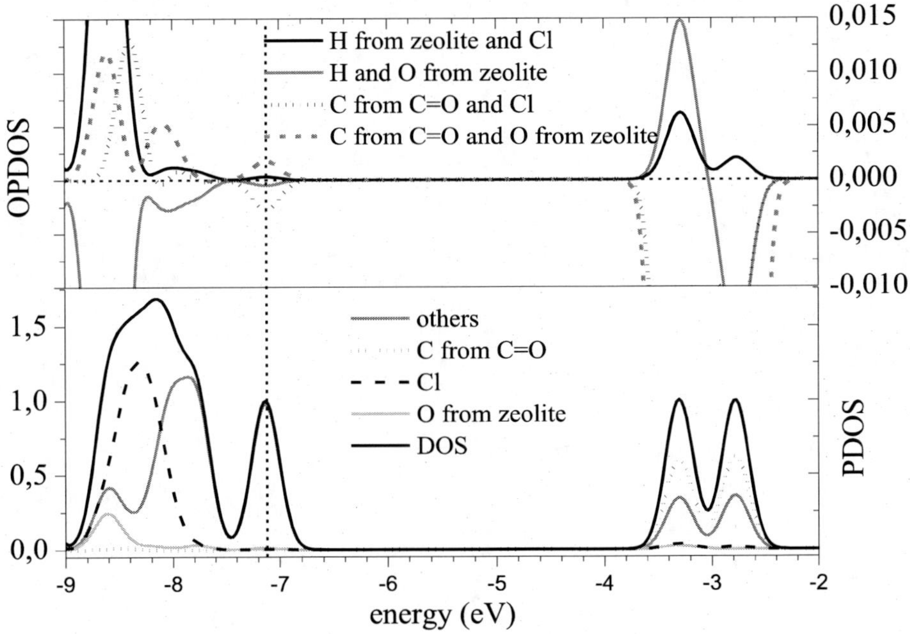

Figure 11. DOS, PDOS, and OPDOS for the interaction between the 3T-cluster and acetyl chloride.

Therefore, we affirm that in the HOMO region, bonding interaction between C from C=O and O1 prevails concerning H1 and O1 interaction from the cluster. In the LUMO region, the H---X interaction predominates over the C---O. Thus, these results agree with those exposed in the previous section for the interaction between acetic acid and the 3T-cluster.

On the other hand, Figure 11 presents DOS, PDOS, and OPDOS for the interaction between acetyl chloride and 3T-cluster. In this case, some differences are obvious concerning acetic acid interaction. At first, we observed that HOMO–LUMO gap is smaller (3.84 eV) than for acetyl chloride and acetic anhydride (as will be comented soon). Furthermore, PDOS between -9.0 eV and -7.8 eV shows a notorious contribution from chlorine atom in comparison with the little contribution from sp^3 oxygen from acetic acid in respective TS. Such differences between PDOS diagrams manifest a higher electronic population for chlorine in TS, with respect to the contribution of the sp^3 oxygen from acetic acid in the mentioned interval. Moreover, in the PDOS diagram between -9.0 eV and -6.5 eV we observe an intense signal due to a bonding interaction between H1 and chlorine, as well as a less intense signal from C (C=O) and O1. Thus, the as-evidenced interactions allowed us to confirm that rearrangement of the proposed TS geometry evolves to the acetyl-zeolite intermediate in accordance with results found in the IRC calculations and GEDT analysis.

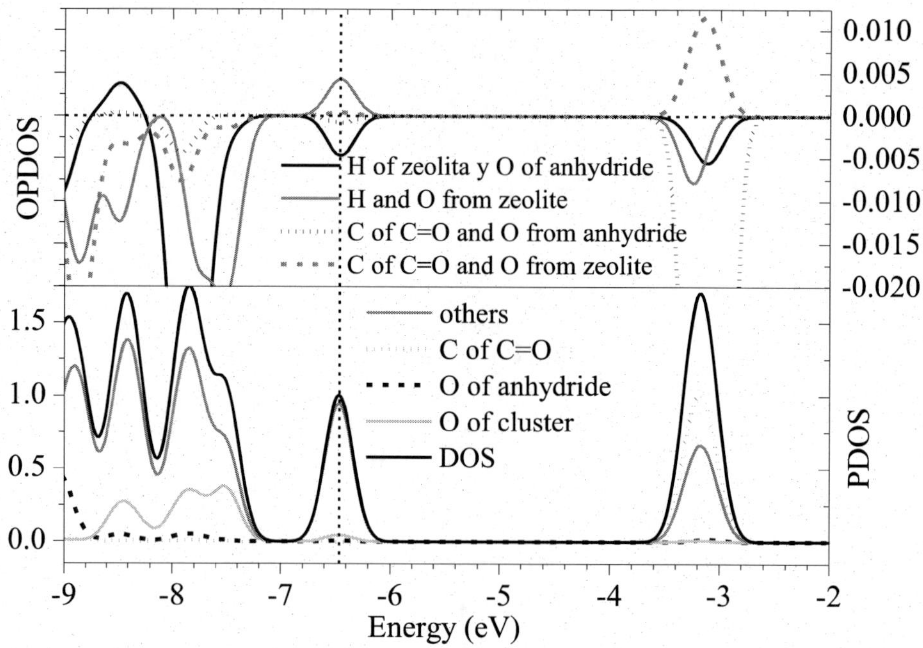

Figure 12. DOS, PDOS, and OPDOS for the interaction between the 3T-cluster and acetic anhydride.

Finally, Figure 12 exhibits the DOS, PDOS, and OPDOS curves for the interaction between the 3T-cluster and acetic anhydride. In this case, the HOMO−LUMO difference in energy results in a value of 3.22 eV (even lower than for acetyl chloride). The PDOS graph shows a lower contribution for selected atoms in the -9.0 eV to -6.0 eV interval, being the O1 from the cluster the only with significant importance. In the LUMO zone, similar behavior has been observed as explained previously for other acylating agents. In the OPDOS graph, we evidenced a big decrease in the overlap density of states, indicating a strong anti-bonding character between selected atoms. OPDOS curves representing the interactions between H1 (zeolite) and O sp^3 (anhydride) as well as H1 and O1 (from the cluster) are the only with bonding behavior in the -8.8 eV to -8.2 eV and -6.8 eV to -6.2 eV intervals. On the other hand, an increase in the overlap density of states in the LUMO zone is observed, therefore a strong bonding interaction between carbon from the carbonyl group and O1 from zeolite's cluster has been evidenced. Probably, the last one may make possible the interaction between acetic anhydride and 3T-cluster.

CONCLUSION

Based on IRC calculations, we proposed a possible mechanism for the generation of an acetyl-zeolite intermediate which was found in several acetyl transfer reactions catalyzed by H-ZSM-5 zeolite. We performed a theoretical study that allowed us to elucidate the atomic rearrangements that take place during the generation of the acetyl-zeolite intermediate. Based on that, we studied the interaction of acetic acid, acetyl chloride and acetic anhydride with a 3T-cluster model for a ZSM-5 zeolite. The reaction path modeled with every acylating agent allowed us to visualize both the bond braking and bond formation that takes place when passing from the adsorption complexes to the transition states and finally, to the acetyl-zeolite intermediate. We correctly found negative values for the adsorption enthalpies with all acyl donors and the following results by IRC calculations: 1) the HO−$COCH_3$, H_3CCOO−$COCH_3$ and Cl−$COCH_3$ bonds in the acetyl donor break, 2) O−H bond in zeolite's cluster breaks, 3) a C−O bond results from interaction of the carbonyl carbon and one of the oxygen atoms from the 3T-cluster, 4) an H−Y bound generates by interaction of the respective atom from the acetyl donor (sp^3 O from acetic acid or acetic anhydride or Cl from acetyl chloride) and the acidic hydrogen from zeolite cluster. In the same way, different thermodynamic properties and kinetic parameters were found which allowed us to suggest a reactivity order for the generation of the acetyl-zeolite intermediate. The reactivity order found by theoretical studies agrees with those reported by experimental methods. Moreover, the Global Electron Density Transfer could be an important factor that influences the generation of transition states in the proposed mechanism of reaction. Finally, from DOS, PDOS and OPDOS diagrams, we confirmed the interaction between atoms and/or group of atoms through bond breaking and bond formation processes.

ACKNOWLEDGMENTS

The authors acknowledge the SEAGrid platform (http://www.seagrid.org) for computational resources and services for the results used in this chapter. This work is supported by the Council of Research of the National University of Salta, project N° 2546.

REFERENCES

[1] Boronat, M., and Corma, A. (2019). What Is Measured When Measuring Acidity in Zeolites with Probe Molecules? *ACS Catalysis*, 9, 1539-1548.

[2] Smit, B., and Maesen, T. L. M. (2008). Molecular Simulations of Zeolites: Adsorption, Diffusion, and Shape Selectivity. *Chemical Reviews*, 108, 4125-4184.

[3] Krishna Mohan, K. V. V., Narender, N., and Kulkarni, S. J. (2006). Zeolite Catalyzed Acylation of Alcohols and Amines with Acetic Acid under Microwave Irradiation. 10.1039/B600031B. *Green Chemistry*, 8, 368-372.

[4] Corma, A., JoséCliment, M., García, H., and Primo, J. (1989). Design of Synthetic Zeolites as Catalysts in Organic Reactions: Acylation of Anisole by Acyl Chlorides or Carboxylic Acids over Acid Zeolites. *Applied Catalysis*, 49, 109-123.

[5] Kresnawahjuesa, O., Gorte, R. J., and White, D. (2004). Characterization of Acylating Intermediates Formed on H-ZSM-5. *Journal of Molecular Catalysis A: Chemical*, 208, 175-185.

[6] Gumidyala, A., Sooknoi, T., and Crossley, S. (2016). Selective Ketonization of Acetic Acid over HZSM-5: The Importance of Acyl Species and the Influence of Water. *Journal of Catalysis*, 340, 76-84.

[7] Rasmussen, D. B., Christensen, J. M., Temel, B., Studt, F., Moses, P. G., Rossmeisl, J., Riisager, A., and Jensen, A. D. (2015). Ketene as a Reaction Intermediate in the Carbonylation of Dimethyl Ether to Methyl Acetate over Mordenite. *Angewandte Chemie International Edition*, 54, 7261-7264.

[8] Bonati, M. L. M., Joyner, R. W., and Stockenhuber, M. (2007). On the Mechanism of Aromatic Acylation over Zeolites. *Microporous and Mesoporous Materials*, 104, 217-224.

[9] Resasco, D. E., Wang, B., and Crossley, S. (2016). Zeolite-Catalysed C–C Bond Forming Reactions for Biomass Conversion to Fuels and Chemicals. 10.1039/C5CY02271A. *Catalysis Science & Technology*, 6, 2543-2559.

[10] Corregidor, P. F., Cuesta, P. M., Acosta, D. E., and Destéfanis, H. A. (2019). Composite ZSM-5/MCM-41 Material Obtained from a Green Resource and Its Enhanced Catalytic Performance in the Reaction of Vinyl Acetate and Isoamyl Alcohol. *Applied Catalysis A: General*, 117262.

[11] Corregidor, P. F., Acosta, D. E., Gonzo, E. E., and Destéfanis, H. A. (2018). Isoamyl Acetate Preparation from Reaction of Vinyl Acetate and Isoamyl Alcohol Catalyzed by H-ZSM-5 Zeolite: A Kinetic Study. *Molecular Catalysis*.

[12] Migues, A. N., Vaitheeswaran, S., and Auerbach, S. M. (2014). Density Functional Theory Study of Mixed Aldol Condensation Catalyzed by Acidic Zeolites HZSM-5 and Hy. *The Journal of Physical Chemistry C*, 118, 20283-20290.

[13] Vaitheeswaran, S., Green, S. K., Dauenhauer, P., and Auerbach, S. M. (2013). On the Way to Biofuels from Furan: Discriminating Diels–Alder and Ring-Opening Mechanisms. *ACS Catalysis*, 3, 2012-2019.

[14] https://america.iza-structure.org/IZA-SC/framework.php?STC=MFI.

[15] Gaussian 16 Rev. C.01, Wallingford, CT.

[16] Becke, A. D. (1993). Density-Functional Thermochemistry. Iii. The Role of Exact Exchange. *The Journal of Chemical Physics*, 98, 5648-5652.

[17] Lee, C., Yang, W., and Parr, R. G. (1988). Development of the Colle-Salvetti Correlation-Energy Formula into a Functional of the Electron Density. *Physical Review B*, 37, 785-789.

[18] Parr, R. G., and Weitao, Y. *Density-Functional Theory of Atoms and Molecules.* Oxford University Press, 1994. https://books.google.com.ar/books?id=mGOpScSIwU4C.

[19] Hohenberg, P., and Kohn, W. (1964). Inhomogeneous Electron Gas. *Physical Review*, 136, B864-B871.

[20] Kohn, W., and Sham, L. J. (1965). Self-Consistent Equations Including Exchange and Correlation Effects. *Physical Review*, 140, A1133-A1138.

[21] Dewar, M. J. S., Zoebisch, E. G., Healy, E. F., and Stewart, J. J. P. (1985). Development and Use of Quantum Mechanical Molecular Models. 76. Am1: A New General Purpose Quantum Mechanical Molecular Model. *Journal of the American Chemical Society*, 107, 3902-3909.

[22] McQuarrie, D. A., Simon, J. D., Cox, H. A., and Choi, J. *Physical Chemistry: A Molecular Approach.* University Science Books, 1997. https://books.google.com.ar/books?id=f-bje0-DEYUC.

[23] Hemelsoet, K., Moran, D., Van Speybroeck, V., Waroquier, M., and Radom, L. (2006). An Assessment of Theoretical Procedures for Predicting the Thermochemistry and Kinetics of Hydrogen Abstraction by Methyl Radical from Benzene. *The Journal of Physical Chemistry A*, 110, 8942-8951.

[24] Vos, A. M., Rozanska, X., Schoonheydt, R. A., van Santen, R. A., Hutschka, F., and Hafner, J. (2001). A Theoretical Study of the Alkylation Reaction of Toluene with Methanol Catalyzed by Acidic Mordenite. *Journal of the American Chemical Society*, 123, 2799-2809.

[25] Vos, A. M., Schoonheydt, R. A., De Proft, F., and Geerlings, P. (2003). Dft Study on the Electrophilic Aromatic Substitution Catalyzed by lewis acids. *Journal of Catalysis*, 220, 333-346.

[26] Domingo, L., Ríos-Gutiérrez, M., and Pérez, P. (2016). Applications of the Conceptual Density Functional Theory Indices to Organic Chemistry Reactivity. *Molecules*, 21, 748.

[27] Domingo, L. R. (2014). A New C-C Bond Formation Model Based on the Quantum Chemical Topology of Electron Density. 10.1039/C4RA04280H. *RSC Advances*, 4, 32415-32428.

[28] Reed, A. E., Weinstock, R. B., and Weinhold, F. (1985). Natural Population Analysis. *The Journal of Chemical Physics*, 83, 735-746.

[29] Reed, A. E., Curtiss, L. A., and Weinhold, F. (1988). Intermolecular Interactions from a Natural Bond Orbital, Donor-Acceptor Viewpoint. *Chemical Reviews*, 88, 899-926.

[30] Domingo, L. R. (2016). Molecular Electron Density Theory: A Modern View of Reactivity in Organic Chemistry. *Molecules*, 21, 1319.

[31] http://www.luisrdomingo.com/p/tutot.html.

[32] Opalka, S. M., and Zhu, T. (2016). Influence of the Si/Al Ratio and Al Distribution on the H-ZSM-5 Lattice and Brønsted Acid Site Characteristics. *Microporous and Mesoporous Materials*, 222, 256-270.

[33] Sun, W. F., Sun, Y. X., Zhang, S. T., and Chi, M. H. (2019). Hydrogen Storage, Magnetism and Electrochromism of Silver Doped FAU Zeolite: First-Principles Calculations and Molecular Simulations. *Polymers*, 11, 279.

Reviewed by Dr. Reinaldo Pis Diez.

CEQUINOR, Centro de Química Inorgánica, Facultad de Ciencias Exactas, Universidad Nacional de La Plata, CONICET, La Plata, Argentina.

INDEX

A

B

D

G

H

I

N

O

P

Q

R

S

T

U